GALE BOOK OF AVERAGES

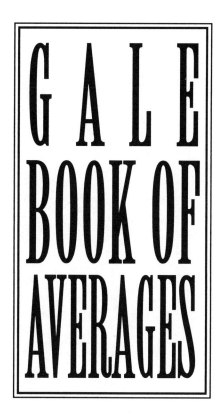

GALE BOOK OF AVERAGES

Kathleen Droste, Editor

Jennifer Dye, Associate Editor

Gale Research Inc. • DETROIT • WASHINGTON, D.C. • LONDON

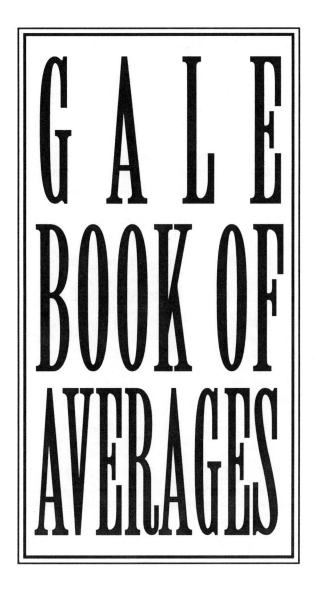

Kathleen Droste: *Editor*
Jennifer Dye: *Associate Editor*
Linda Metzger: *Senior Editor*

Mary Beth Trimper: *Production Director*
Catherine Kemp: *Production Assistant*
Cynthia Baldwin: *Art Director*
Doug Cannell: *Cover Designer*

Victoria Cariappa: *Research Manager*
Maureen Richards: *Research Supervisor*
Karen Farrelly: *Editorial Assistant*

With Special Thanks to James Bobick and Gregory Pomrenke,
Carnegie Library of Pittsburgh

Editorial Code and Data, Inc.
Nancy Ratliff: *Data Entry*
Gary Alampi: *Data Processing*

Library of Congress Cataloging-in-Publication Data
Gale book of averages / Kathleen Droste, editor; Jennifer Dye, associate editor.
p. cm.
Includes bibliographical references and index.
ISBN 0-8103-9138-4
1. Statistics. I. Droste, Kathleen D. II. Dye, Jennifer. III. Title: Gale book of averages.
HA155.G34 1993 519.5--dc20 93-29316
CIP

Library of Congress Catalog Number
ISBN 0-8103-9138-4

Printed in the United States of America
Published simultaneously in the United Kingdom
by Gale Research International Limited
(An affiliated company of Gale Research Inc.)

I(T)P™

The trademark **ITP** is used under license.

10 9 8 7 6 5 4 3 2 1

TABLE OF CONTENTS

CHAPTER 2 - BUSINESS AND LABOR continued:

CHAPTER 2 - BUSINESS AND LABOR continued:

CHAPTER 3 - CONSUMPTION continued:

CHAPTER 6 - EDUCATION continued:

CHAPTER 7 - EXPENDITURES continued:

CHAPTER 7 - EXPENDITURES continued:

CHAPTER 9 - HEALTH AND MEDICINE continued:

CHAPTER 9 - HEALTH AND MEDICINE continued:

CHAPTER 11 - INCOME continued:

CHAPTER 11 - INCOME continued:

CHAPTER 11 - INCOME continued:

CHAPTER 14 - VITAL STATISTICS AND LIFESPANS continued:

CHAPTER 15 - WEATHER AND NATURE continued:

CHAPTER 15 - WEATHER AND NATURE continued:

PREFACE

Gale Book of Averages (*GBOA*) is a compilation of national and international statistical, tabular, and textual data comprising thousands of specific average values arranged by broad subject categories.

The need for *GBOA* was prompted by the lack of a single reference source from which a broad spectrum of averages might be found. Anyone who has looked for averages will have experienced difficulties in identifying them and in gaining access to them. Averages appear randomly and in diverse publications, ranging from newspapers and popular magazines to books, journals, and government publications.

The material presented here is drawn from a wide range of sources—government and nongovernment, published and unpublished—in a single accessible compilation. *GBOA* provides the reader with:

- Over 1100 discrete tables, charts, graphs, and short narratives.
- Broad coverage of subject areas.
- Easy-to-access chapters, organized by topic and subtopic.
- A detailed subject index.
- A unique numerical values locator, which is a listing by numbers.

SCOPE

GBOA provides statistical and textual coverage on a wide range of subjects pertinent to everyday life. It will serve persons who need to know one particular average that can act as a benchmark, such as the average price of a new car or the average SAT score for 1992, as well as persons who might need to compile a range of information on a topic such as the average cost of raising a child at various stages, from birth through age 18.

Sources use the word "average" in at least three distinctive ways. The two most frequent are median, which is the point that divides a series of numbers so that half are on one side, half on the other, and mean, which is found by dividing the sum of a series by the number of its elements. For example, in the series, 1, 4, 8, 20, 24, 27, 42, the median is 20, the mean is 18. When a table in *GBOA* means median, this is indicated. Most of the tables in *GBOA* actually refer to mean. Mode is the third and less widely applied meaning of the word "average" and refers to the most frequent value of a set of data.

Because statistical data are often more meaningful when comparisons can be made, averages by sex, age, race/ethnicity, and date are presented in this book whenever feasible. Retrospective data, tracking averages across past decades and sometimes into the future, have been included as well. Conflicting averages, when found, are stated also.

ARRANGEMENT

Gale Book of Averages is divided into fifteen chapters, each of which focuses upon specific topics. Within each topic there appear individual tables and textual material organized by subtopic.

Each table title or headline is preceded by a table number. (All tables are numbered and may be accessed by headline or number from the Table of Contents. Data may also be accessed via the Subject Index or the Selective Numerical Values Locator.) Immediately beneath the headline, the statistical data is displayed as a table, a graph, or a prose statement. Where data requires explanation, and where such explanations are readily accessible, a brief head note is presented.

At the bottom of each table, the source of the data presented immediately above is cited in complete bibliographic detail. If the source has identified another source from which data was derived, that preceding source is also identified. Explanatory notes are placed after the source information. A Remarks section provides additional pertinent information found in the source where appropriate.

SOURCES

The data arranged herein have been collected from periodical literature, newspapers, books, government documents, reports and studies from companies, institutions, research centers, organizations, and other materials that provide averages regularly or incidentally. *GBOA* is a compilation of data from many sources; before appearing in this book, this data was not indexed in a manner useful to the reader interested in averages.

INDEXES

GBOA features both a comprehensive Subject Index and a Numerical Values Locator, an appendix containing a listing by number under special categories: area, energy, length, money, percentage, speed, temperature, time, volume, and weight. This type of arrangement enables the user to find equivalent or comparative numbers. For example:

Money
Dollars
$341 average social security benefit for retired workers, 1980.
$341 average loss per convenience store robbery, 1990.

Entries listed in these indexes are followed by the appropriate table number and page number.

COMMENTS ARE WELCOME

Comments and suggestions for future editions are always welcome and can be addressed: Editor, *Gale Book of Averages*, Gale Research Inc., 835 Penobscot Bldg., Detroit, Michigan 48226-4094.

GALE BOOK OF AVERAGES

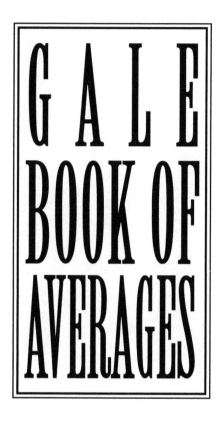

Chapter 1
ALLOCATION OF TIME

By Country

★1★

Time Allocation in Minutes for Selected Countries

Country	Housework and other household obligations		Child care	
	Employed women	Employed men	Employed women	Employed men
	Minutes per working day			
Belgium	163	34	14	6
Bulgaria	149	66	21	13
Czechoslovakia	255	78	30	18
France	173	58	24	8
Germany, Federal Republic[1]	216	48	26	7
German Democratic Republic	220	80	33	15
Hungary	217	78	26	17
Poland	200	60	27	20
United States[1]	162	46	17	8
USSR (former)	197	67	30	30
Yugoslavia[1]	231	76	25	13

Source: "Average daily time budget of employed men, employed women, and housewifes (sic) in 12 countries (in hours)," *The Economic Role of Women in the ECE Region: Developments 1975*, 1985, p. 110 (New York: United Nations, 1985). Primary source: A. Szalai (ed.) *The Use of Time; Daily Activities of Urban and Suburban Populations in Twelve Countries. Report on the Multinational Comparative Time-Budget Research Project*. The Hague, 1972; A. Szalai, "Women's Time. Women in the Light of Contemporary Time-Budget Research" in *Futures*, vol. 7, No. 5, pp. 385-399, October 1977. *Notes:* 1. Underweighted average of results of two investigations carried out under the multinational project on time use in these countries.

★2★

By Country

Time Allocation per Day in Selected Countries

Time allocation in twelve countries, in hours per day.

Activities	Employed Women	Employed Men	Housewives
	Hours per day		
On workdays (employed people) and weekends (housewives)			
Paid work and ancillary tasks (work brought home, journey to work, workplace chores, etc.)	7.9	9.4	0.2
Housework and household obligations (not including child care)	3.3	1.0	7.5
Child care	0.4	0.2	1.1
Sleep, meals, personal hygiene, and other personal needs	9.9	9.9	11.2
Free time (remaining disposable time)	2.5	3.5	4.0
Total	24.0	24.0	24.0
	Hours per day		
On days off (employed people) and Sundays (housewives)			
Paid work and ancillary tasks (work brought home, journey to work, workplace chores, etc.)	0.4	0.9	0.1
Housework and household obligations (not including child care)	5.1	2.3	5.2
Child care	0.6	0.3	0.7
Sleep, meals, personal hygiene, and other personal needs	11.9	12.2	11.7
Free time (remaining disposable time)	6.0	8.3	6.3
Total	24.0	24.0	24.0

Source: "Average daily time budget of employed men, employed women, and housewives in 12 countries (in hours)," *The Economic Role of Women in the ECE Region: Developments 1975*, 1985, p. 106 (New York: United Nations, 1985). Primary source: A. Szalai (ed.) *The Use of Time; Daily Activities of Urban and Suburban Populations in Twelve Countries. Report on the Multinational Comparative Time-Budget Research Project.* The Hague, 1972; A. Szalai, "Women's Time. Women in the Light of Contemporary Time-Budget Research" in *Futures*, vol. 7, No. 5, pp. 385-399, October 1977. Eleven countries were included in the project: Belgium, Bulgaria, Czechoslovakia, France, the German Democratic Republic, the Federal Republic of Germany, Hungary, Poland, the United States, the Soviet Union and Yugoslavia, and also Peru.

★3★

By Country

Time Allocation in Agricultural Societies

Time allocation in hours per week.

Activity	Women		Men	
	Rural Botswana 1975	Nepalese Villages 1981	Rural Botswana 1975	Nepalese Villages 1981
Total work	59.4	75.6	47.8	52.6
Marketable work	23.2	32.3	38.8	40.7
Housework	36.2	43.3	9.0	11.9
Leisure	42.0	27.8	55.5	46.3

Source: F. Thomas Juster and Frank P. Stafford, "Time Allocation in Agricultural Societies," *The Allocation of Time: Empirical Findings, Behavioral Models, and Problems of Measurement*, Table 2, p. 96 (contact Peter Seidman, Information Officer, News and Information Services, The University of Michigan, 412 Maynard, Ann Arbor, MI 48109-1399; 313-747-4416).

★4★

By Country

Time Spent on the Job in Selected Industrialized Countries, 1991

Country	Average Hours Worked per Year
Belgium	1,572
Canada	1,887
Denmark	1,595
France	1,610
Germany	1,603
Great Britain	1,856
Italy	1,858
Japan	2,155
Netherlands	1,592
Norway	1,614
Sweden	1,539
United States	1,951

Source: "Time Crunches American Workers," *Black Enterprise*, June 1993, p. 73.

★5★

By Country

Time Needed to Produce Steel in Selected Countries

Country	Average Worker Hours per Ton
South Korea	6.4
Britain	5.6
Germany	5.6
Japan	5.4
United States	5.3

Source: "U.S. Steel Industry Strong," *USA Today*, October 12, 1992, p. B1. Original source: WEFA Group.

★6★

By Country

Time Women Spend Working in Selected Areas of the World

According to a report by the U.N. International Labor Organization, women over most of the world work more hours than men.

Area	Average Number of Hours per Week	
	Women	Men
Africa	67	53
Asia (except Japan)	62	48
Japan	56	54
Western Europe	48	43
Latin America	60	54
North America and Australia	47.5	49

Source: Los Angeles Public Library/State of California, *Scan/Info*, 4, September 1992, p. 23. Primary source: *Los Angeles Times*, September 7, 1992, p. A2.

★7★

By Country

Vacation Time per Year in Selected Countries

Country	Average Number of Vacation Days
Mexico	6
United States	10
Japan	10
Germany	18

[Continued]

★7★

Vacation Time per Year in Selected Countries

[Continued]

Country	Average Number of Vacation Days
Britain	22
France	25
Sweden	30

Source: "The Leisure Lag," 120, *Newsweek*, February 17, 1992, p. 42. Original source: Hewitt Associates.

★8★

By Country

Weekly Time Allocation for Selected Activities in the Former Soviet Union

Activity	Men		Women	
	1965	1985	1965	1985
Active leisure	6.0	4.1	2.2	3.0
Adult ed.	6.2	1.0	5.0	2.6
Commuting	4.9	5.2	3.7	3.4
Housework	9.8	11.9	31.5	27.0
Sleep	56.0	56.9	55.5	58.2
Social interaction	8.6	7.8	8.9	9.6
TV	5.6	14.5	3.9	11.2
Work	49.7	48.6	40.1	35.9

Source: "The Allocation of Time: Empirical Findings, Behavioral Models, and Problems of Measurement," F. Thomas Juster and Frank P. Stafford, *Journal of Economic Literature*, 29, June 1991, pp. 471-522. **Remarks**: Also included in source: weekly allocation of time for selected activities in Japan, U.S., Norway, and Denmark.

★9★

By Country

Weekly Time Allocation for Selected Activities in Japan

Activity	Men		Women	
	1965	1985	1965	1985
Active leisure	3.4	5.3	1.8	3.6
Adult ed.	0.8	1.2	1.2	2.2
Commuting	3.6	4.5	1.0	1.2
Housework	2.8	3.59	31.5	31.0
Sleep	63.8	60.0	59.6	57.0
Social interaction	5.9	8.0	5.5	7.0
TV	16.8	17.3	19.4	21.4
Work	54.1	47.5	32.2	23.4

Source: "The Allocation of Time: Empirical Findings, Behavioral Models, and Problems of Measurement," F. Thomas Juster and Frank P. Stafford, *Journal of Economic Literature*, 29, June 1991, pp. 471-522. **Remarks**: Also included in source: weekly allocation of time for selected activities in the former Soviet Union, U.S., Norway, and Denmark.

★10★

By Country

Weekly Time Allocation for Selected Activities in the United States

Activity	Men		Women	
	1965	1981	1965	1981
Active leisure	2.8	5.6	3.2	4.2
Adult ed.	1.2	0.6	0.6	0.4
Commuting	4.8	3.5	1.6	2.0
Housework	11.5	13.8	41.8	30.5
Sleep	58.7	57.9	59.9	58.9
Social interaction	13.4	14.9	13.7	17.6
TV	11.8	12.7	9.3	11.5
Work	46.8	40.5	17.3	21.9

Source: "The Allocation of Time: Empirical Findings, Behavioral Models, and Problems of Measurement," F. Thomas Juster and Frank P. Stafford, *Journal of Economic Literature*, 29, June 1991, pp. 471-522. **Remarks**: Also included in source: weekly allocation of time for selected activities in the former Soviet Union, Japan, Norway, and Denmark.

★ 11 ★

By Country

Weekly Time Allocation for Selected Activities in Norway

Activity	Men		Women	
	1971	1980	1971	1980
Housework	15.4	16.8	41.3	33.0
Leisure	42.0	45.5	39.2	45.2
Sleep	72.8	71.4	74.2	72.1
Work	37.8	34.2	13.3	17.6

Source: "The Allocation of Time: Empirical Findings, Behavioral Models, and Problems of Measurement," F. Thomas Juster and Frank P. Stafford, *Journal of Economic Literature*, 29, June 1991, pp. 471-522. **Remarks**: Also included in source: weekly allocation of time for selected activities in the former Soviet Union, Japan, U.S., and Denmark.

★ 12 ★

By Country

Weekly Time Allocation for Selected Activities in Denmark

Activity	Men		Women	
	1964	1987	1964	1987
Housework	3.7	12.8	30.1	23.1
Leisure	53.8	54.3	51.7	53.7
Sleep	68.8	67.2	72.8	70.4
Work	41.7	33.4	13.3	20.8

Source: "The Allocation of Time: Empirical Findings, Behavioral Models, and Problems of Measurement," F. Thomas Juster and Frank P. Stafford, *Journal of Economic Literature*, 29, June 1991, pp. 471-522. **Remarks**: Also included in source: weekly allocation of time for selected activities in the former Soviet Union, Japan, U.S., and Norway.

Caregiving Time

★ 13 ★

Time Spent Caregiving During a Lifetime

According to the Older Women's League, women can expect to spend an average of 17 years caring for children and 18 years helping an elderly parent. Mainly due to caregiving responsibilities, women average 11.5 years out of the paid labor force; men average only 1.3 years.

Source: "Women Who Care Are Women Who Need Relief," Maryanne Sugarman Costa, *National Business Woman*, Summer 1989, p. 20. Primary source: "Failing America's Caregivers: A Status Report on Women Who Care," Older Women's League, 730 111th St., NW, Suite 300, Washington, DC 20001.

Children's Time

★14★

Allocation of Time for Children in the United States by Selected Activity, 1981-1982

Activity	Average Number of Hours per Week	
	Elementary school	High School
Eating	9.0	7.8
Household chores	2.7	4.8
Personal care	5.2	6.7
Playing	15.0	7.0
Reading	.9	1.6
Studying	1.8	3.8
Sleeping	68.2	60.3
School	25.2	26.2
TV	15.6	14.2
non-allocated	23.2	31.7

Source: "The Allocation of Time: Empirical Findings, Behavioral Models, and Problems of Measurement," F. Thomas Juster and Frank P. Stafford, *Journal of Economic Literature*, 29, June 1991, pp. 471-522. **Remarks**: Also included in source: Allocation of time for children in Japan, 1986.

★15★

Children's Time

Allocation of Time for Children in Japan by Selected Activity, 1986

Activity	Average Number of Hours per Week	
	Elementary school	High School
Eating	10.6	9.6
Household chores	3.0	4.1
Personal care	7.1	8.1
Playing	11.4	0.7
Reading	2.8	3.3
School	38.2	41.5
Sleeping	64.4	53.0
Studying	8.3	19.0

[Continued]

★15★

Allocation of Time for Children in Japan by Selected Activity, 1986
[Continued]

Activity	Average Number of Hours per Week	
	Elementary school	High School
TV	15.0	17.7
non-allocated	7.1	8.7

Source: "The Allocation of Time: Empirical Findings, Behavioral Models, and Problems of Measurement," F. Thomas Juster and Frank P. Stafford, *Journal of Economic Literature*, 29, June 1991, pp. 471-522. **Remarks**: Also included in source: Allocation of time for children in the United States, 1981-1982.

★16★

Children's Time

Days Spent in School in Selected Countries, 1991

Country	Average Number of Days per Year in School	Country	Average Number of Days per Year in School
Canada	188	Slovenia	190
France	174	South Korea	222
Hungary	177	Soviet Republics	198
Ireland	173	Spain	188
Israel	215	Switzerland	207
Italy	204	Taiwan	222
Jordan	191	United States	178
Scotland	191		

Source: U.S. Bureau of the Census. *Statistical Abstract of the United States 1992*. 112th edition. Washington, D.C.: U.S. Department of Commerce, 1992, p. 830. Primary source: National Center of Education Statistics, U.S. Department of Education. **Remarks**: Source also includes: average minutes per week spent in science and mathematics instruction.

Christmas Time

★17★

Time Spent on Select Christmas Activities

Results of a three-year study done by the Center for Lifestyle Management gathered from phone calls and mailed responses of 1,000 people in six cities: only nine minutes are spent by the average parent playing with his/her children Christmas morning; an average of 7 1/2 hours are used for baking and fixing holiday meals; an average of 17 minutes are spent planning holiday gifts, and six hours shopping for them; and an average of only 1/2 day is devoted to relaxing. Respondents also revealed that an average of 10 hours are spent the last week of December arguing and bickering with family members about holiday-related activities.

Source: "Christmas? Friends? Bah Humbug," *The Detroit Free Press*, December 20, 1992, p. 6A.

Household Labor and Child Care Time

★18★

Time Spent in Child Care and Household Labor per Week

Wife's Employment Status	Women			Men		
	Homemaker	Part-time	Full-time	Homemaker	Part-time	Full-time
	Preschool-age children in household					
Household labor	33.4	30.3	15.0	9.4	13.1	8.2
Child care	19.1	10.2	5.2	4.8	5.2	2.8
	School-age children in household					
Household labor	35.6	28.7	18.9	12.5	12.3	11.7
Child care	3.5	3.3	2.7	1.7	1.9	1.1
	No children in household					
Household labor	31.0	32.9	14.6	14.7	13.3	13.6
	All households					
Household labor	35.6	28.7	18.9	11.6	14.0	12.1
Child care	7.0	3.7	2.1	2.1	1.8	1.0
Total households	64	50	33	51	37	66

Source: Beth Anne Shelton, "The Distribution of Household Tasks," *Journal of Family Issues* 11:2 (June 1990, Table 2), p. 115-135. Primary source: data made available by the Inter-University Consortium for Political and Social Research, State University of New York at Buffalo. Data collected from 620 respondents and 376 of their spouses.

Housing

★ 19 ★

Time Spent Saving for a Home in Selected Cities

City	Average Number of Years to Save for a Home
Los Angeles	4.8
New York	4.2
Boston	3.7
Chicago	2.9
Detroit	2.6
Cleveland	2.4
Seattle	2.3

Source: "First Timers Fare Better Here," *Detroit Free Press*, February 28, 1993, p. 1J. Original source: Chicago Title and Trust Co.

Leisure Time

★ 20 ★

Time Spent on Crafts or Hobbies

According to a study by Hobby Industries of America, Americans spend an average of eight hours a week on a craft or hobby, needlecraft being number one. In 1992, in 82 percent of all U.S. households, there was at least one person involved in a craft or hobby. The growth areas were floral arranging and fashion-fabric painting and decorating.

Source: "Craft and Hobby Activity Grows," *Christian Science Monitor*, December 29, 1992, p. 7.

★ 21 ★

Leisure Time

Time Spent on a Honeymoon

In 1992 the average length of a honeymoon was eight days, remaining unchanged from 1987.

Source: "Business Reports," *American Demographics*, 14, August 1992, p. 17.

Men's Personal Care Time

★ 22 ★

Time Men Spend Grooming Themselves

According to a survey done by *GQ*, a men's magazine, there has been a significant increase in the amount of time men spend grooming themselves. In 1990, men spent an average of 44 minutes per day arranging their hair and clothes, an increase of 30 minutes over 1988. Men under 25 years spend the most time, an average of 53 minutes daily working on their outward appearance.

Source: "The Brave New World of Men," Diane Crispell, *American Demographics*, 14, January 1992, p. 42.

Parenting Time

★ 23 ★

Time Parents Spend with Their Children

According to priority Management Systems Inc., a management consulting firm in Belluvue, Washington, parents spend an average of 1.9 hours per day with their children, but ten hours working and commuting.

Source: "For Jacinta and Sam Mathis, Having It All Means Doing It All, with Barely Enough Time to Rest," Joan E. Rigdon, *The Wall Street Journal*, June 21, 1993, p. R13.

Shopping-Related Time

★ 24 ★

Time Spent by Grocery Store Shoppers in Selected Departments

Department	Average Time Spent in Seconds
Produce	181.0
Meat	154.0
Liquor	136.6
Baby Food	130.9
Bread	42.3
Seafood	40.5

Source: "Attention Shoppers," *Business Week*, 22, April 20, 1992, p. 44. Original source: Videocart Inc.

★ 25 ★

Shopping-Related Time

Time Spent Grocery Shopping per Trip

According to a study done by Market Growth Resources Inc., a retail consulting firm, Americans are spending less time in the grocery store. Shopping which used to take an average of 28 minutes per trip ten years ago, now takes an average of 22 minutes per trip.

Source: "Displays Pay Off for Grocery Marketers," *The Wall Street Journal*, October 15, 1992, p. B1. **Remarks**: Also included in source: number of items in the average grocery store, as well as the average number of new products added each year.

★ 26 ★

Shopping-Related Time

Time Spent in Checkout Lines

CNN and ActMedia's Checkout TV Channel views news, features, and ads and is targeted to those waiting in check-out lines. Its eight minute cycle coincides with the average time a grocery store shopper stands in line.

Source: "TV Takes on Tabloids at Checkout Line," *American Demographics*, 13, April 1991, p. 9.

★27★

Shopping-Related Time

Time Children Spend Shopping

According to a study conducted through personal interviews of 1,440 households in Texas, children accompany their parents to stores an average of 2.5 times per week; girls go with their parents 2.65 times a week, and boys 2.34 times. On the other hand, girls went to the store an average of 0.84 times a week by themselves, while boys went 1.11 times without their parents. Children aged 4 to 12 years old averaged 180 purchase-visits in 1991. This is down from an average of 280 in 1989.

Source: "Growing Up in the Market," James U. McNeal, *American Demographics*, 14, October 1992, p. 49. **Remarks**: Also included in source: amount of money children have to spend each year: the amount they do spend, including how much they spend during the Christmas season alone, and the amount they save.

★28★

Shopping-Related Time

Number of Houses Looked At Before Purchase, 1990-1992

Year	Average Number of Houses Looked at Before Purchase
1990	13.8
1991	15.0
1992	15.6

Source: Chicago Title and Trust Family of Title Insurers, Chicago, IL, *Who's Buying Homes in America*, 1993, p. 3. **Remarks**: Source also includes: median price of homes, average price of homes, average monthly payment, monthly payment as a percentage of income, average number of months buyers looked before deciding, all for 1990-1992.

★29★

Shopping-Related Time

Time Spent in Malls

Adults who visit large, regional shopping malls, do so an average of 3.9 times per month; those who visit small shopping centers, do so an average of 7.1 times per month.

Source: "Strip Malls: Plain but Powerful," Chip Walker, *American Demographics*, 13, October 1991, pp. 48-50.

★ 30 ★

Shopping-Related Time

Time Spent Looking at Homes Before Buying, 1990-1992

Year	Average Number of Months	
	First-time Buyer	Repeat Buyer
1990	5.2	4.7
1991	5.6	4.8
1992	5.4	4.7

Source: Chicago Title and Trust Family of Title Insurers, Chicago, IL, *Who's Buying Homes in America*, 1993, p. 3. **Remarks**: Source also includes: median price of homes, average price of homes, average monthly payment, monthly payment as a percentage of income, all for 1990-1992.

Sleep Time

★ 31 ★

Hours Spent Sleeping

Age	Average Number of Hours per Night
1-15 days	16-22
6-23 months	13
3-9 years	11
10-13 years	10
14-18 years	9
19-30 years	8
31-45 years	7.5
45-50 years	6
50+ years	5.5

Source: Kendig, Frank, and Richard Hutton. *Life-Spans*. New York: Holt, Rinehart and Winston, 1979, p. 25.

Tax Time

★32★

Time Spent on Taxes per Household

According to Prof. Joel Slemrod of the University of Michigan Business School, the average household spent 27.4 hours preparing income taxes in 1990, up 26% in seven years.

Source: "Tax Reports," *The Wall Street Journal*, April 17, 1991, p. A1. **Remarks**: Also included in source: the cost to filers of doing returns and the IRS of processing them.

★33★

Tax Time

Time Spent Filing Income Tax Forms per Person

According to Internal Revenue estimates, in 1991 it took taxpayers an average of nine hours to complete a 1040 tax return with Schedule A for itemized deductions, because of the new revisions.

Source: "1040 Form Has Many Running for Help," John Schmeltzer, *Chicago Tribune*, January 21, 1991, sec. 4, p. 3. **Remarks**: Also included in source: Figures for some tax preparation firm on numbers of tax returns filed by them and their average charge per return.

Teenagers' Time

★34★

Teenagers' and Young, Unmarried Adults' Weekly Allocation of Non-Free Time

Activity	Average Number of Hours per Week	
	Teenagers	Young Adults
Adult care	0.6	0.6
Child care	1.2	1.8
Classes	18.1	4.1
Cleaning, laundry	2.2	3.2
Cooking	0.9	2.2
Eat at home	5.6	4.2
Eat out	2.5	2.7
Garden, pets	0.5	0.2
Homework	3.1	3.4
Naps	1.3	1.7

[Continued]

★ 34 ★

Teenagers' and Young, Unmarried Adults' Weekly Allocation of Non-Free Time

[Continued]

Activity	Average Number of Hours per Week	
	Teenagers	Young Adults
Services	0.5	0.7
Shopping	2.4	2.4
Sleep	62.6	55.9
Travel, personal	0.7	1.6
Travel, school	2.7	0.8
Travel, shop	1.5	1.9
Wash, dress	6.3	6.8
Work-related	2.8	29.0
Yardwork	1.9	1.2

Source: "It's 10 O'Clock. Do You Know Where Your Teenager Is?" *American Demographics*, 13, October 1991, p. 8. Original source: Americans' Use of Time Project. **Remarks:** Also included in source: Weekly allocation of free time for teenagers and young unmarried adults, aged 18 to 29.

★ 35 ★

Teenagers' Time

Teenagers' and Young, Unmarried Adults' Weekly Allocation of Free Time

Activity	Average Number of Hours per Week	
	Teenagers	Young Adults
Church, synagogue	1.0	0.5
Conversation	2.6	3.1
Correspondence	0.4	0.4
Cultural events	0.3	0.4
Games	2.8	0.9
Hobbies, arts	1.1	0.7
Movies	0.6	0.7
Organizations, meetings	1.2	0.5
Radio, records	0.9	0.7
Reading	1.3	1.9
Sports	3.3	1.8
Sports events	0.8	0.3
Travel, organizations	0.4	0.3
Travel, social	2.1	2.5
TV	17.7	14.2

[Continued]

★35★

Teenagers' and Young, Unmarried Adults' Weekly Allocation of Free Time

[Continued]

Activity	Average Number of Hours per Week	
	Teenagers	Young Adults
Visiting, social	4.4	7.8
Walk, outdoors	2.0	1.9

Source: "It's 10 O'Clock. Do You Know Where Your Teenager Is?" *American Demographics*, 13, October 1991, p. 8. Original source: Americans' Use of Time Project. **Remarks**: Also included in source: Weekly allocation of non-free time for teenagers and young unmarried adults, aged 18 to 29.

★36★

Teenagers' Time

Minutes per Week Spent in Mathematics Instruction in Selected Countries, 1991

Country	Average Number of Minutes per Week in Mathematics Instruction	Country	Average Number of Minutes per Week in Mathematics Instruction
Canada	225	Slovenia	188
France	230	South Korea	179
Hungary	186	Soviet republics	258
Ireland	189	Spain	235
Israel	205	Switzerland	251
Italy	219	Taiwan	204
Jordan	180	United States	228
Scotland	210		

Source: U.S. Bureau of the Census. *Statistical Abstract of the United States 1992.* 112th edition. Washington, D.C.: U.S. Department of Commerce, 1992, p. 830. Primary source: National Center of Education Statistics, U.S. Department of Education, and the National Science Foundation, *Learning Mathematics and Learning Science*, February 1992. **Remarks**: Source also includes: average number of days spent in school, average minutes per week spent in science instruction.

★ 37 ★

Teenagers' Time

Minutes per Week Spent in Science Instruction in Selected Countries, 1991

Country	Average Number of Minutes per Week in Science Instruction	Country	Average Number of Minutes per Week in Science Instruction
Canada	156	Slovenia	283
France	174	South Korea	144
Hungary	207	Soviet republics	387
Ireland	159	Spain	189
Israel	181	Switzerland	152
Italy	138	Taiwan	245
Jordan	180	United States	233
Scotland	179		

Source: U.S. Bureau of the Census. *Statistical Abstract of the United States 1992*. 112th edition. Washington, D.C.: U.S. Department of Commerce, 1992, p. 830. Primary source: National Center of Education Statistics, U.S. Department of Education, and the National Science Foundation, *Learning Mathematics and Learning Science*, February 1992. **Remarks**: Source also includes: average number of days spent in school, average minutes per week spent in mathematics instruction.

Volunteer Time

★ 38 ★

Time Spent in Volunteer Work

According to a survey of Independent Sector, a nonprofit coalition in Washington, adults gave an average of 4.2 hours a week in volunteer work in 1991, a slight increase over 1989.

Source: "Recession Hurts Giving of Cash, Not Volunteering," Pamela Sebastian, *The Wall Street Journal*, October 16, 1992, p. B3A. **Remarks:** Also included in source: average amount of money Americans gave to charitable causes in 1989 and 1991.

Work-Related Time

★ 39 ★

Jobs Held by Young Adults

According to a 12-year study by the Bureau of Labor Statistics, the average worker, by the age of 29, has held eight jobs and worked an average of 36 weeks a year since the age of 18.

Source: "Odds and Ends," *Wall Street Journal*, Sept. 28, 1992, p. B1.

★ 40 ★

Work-Related Time

Length of Unemployment

Five to six months is the average length of unemployment, according to a survey done by the Conference Board. However, people were out of work 10 months or longer in 25 percent of all cases.

Source: "Unemployment Affects Families," *Christian Science Monitor*, July 12, 1991, p. 9.

★ 41 ★

Work-Related Time

Commuting Time

On a national average in metropolitan areas, commuting to work each day take 23.2 minutes; in New York it is 30.6, Washington 29.5, Chicago 28.1, Los Angeles 26.4, and Houston 26.1

Source: "Demographics," 113, *U.S. News & World Report*, August 10, 1992, p. 15.

★ 42 ★

Work-Related Time

Sick Days by Sex

According to a study by the National Center for Health Statistics for the years 1983 through 1985, men averaged 4.3 sick days per year, while women missed an average of 5.5 work days. Taking time off to give birth does not count in the statistics. Women lost the most days of work in the transportation and materials industry: an average of 9.3 days per year.

Source: Bluman, Allan. *Elementary Statistics*. Dubuque, IA: William C. Brown Publishers, 1992, p. 323.

★ 43 ★

Work-Related Time

Sick Days by Marital Status

According to a 1987 survey by the National Center for Health Statistics, couples are less likely to be sick than single people. Divorced or widowed men spend eight sick days in bed per year, while married men spend only 5.5 days. Widowed women spend 12 days sick in bed and divorced women 11, compared to seven days for married women.

Source: Bluman, Allan. *Elementary Statistics.* Dubuque, IA: William C. Brown Publishers, 1992, p. 435. Original source: *USA Today*, 1988.

★ 44 ★

Work-Related Time

Time in Minutes That Americans Worked to Purchase Selected Items

Item	Average Time in Minutes of Work			
	1962	1972	1982	1992
Barbie doll	81.1	60.8	33.2	31.3
Film	60.0	37.9	27.7	27.2
Ground chuck	15.3	13.9	13.0	11.1
Milk - 1/2 gal.	12.5	9.6	8.8	7.8
Chicken - 1 lb. cut	13.5	8.7	7.4	7.4
Gas - 1 gal.	7.5	5.2	9.1	6.4
Apples - 1 lb. Delicious	3.1	2.9	3.3	4.7
Phone call - long distance (N.Y. to L.A. for 3 min.)	60.8	23.5	13.4	4.3
Newspaper - NY Times daily	1.4	2.4	2.3	2.8
Postage - 1 oz. first class	1.1	1.6	1.7	1.6

Source: "Has Our Living Standard Stalled," *Consumer Reports*, 57, June 1992, pp. 392-393. **Remarks**: This table refers to before tax dollars and the actual prices for the years specified.

★ 45 ★

Work-Related Time

Time in Hours That Americans Worked to Purchase Selected Items

Item	Average Time in Hours of Work			
	1962	1972	1982	1992
Mattress	71.6	59.5	44.0	60.7
Refrigerator	168.0	99.2	83.7	59.8
Washing machine	92.3	52.7	58.3	37.8
Dishwasher	112.2	64.9	55.3	35.5
Television	85.1	121.6	42.6	21.7
Theater ticket	3.4	4.1	5.2	5.7
Electricity - 500 kwh	4.6	3.3	4.2	3.9

[Continued]

★ 45 ★

Time in Hours That Americans Worked to Purchase Selected Items

[Continued]

Item	Average Time in Hours of Work			
	1962	1972	1982	1992
Watch	3.1	2.1	2.2	2.4
Record album	1.8	2.2	1.2	1.6

Source: "Has Our Living Standard Stalled," *Consumer Reports*, 57, June 1992, pp. 392-393. **Remarks**: This table refers to before tax dollars and the actual prices for the years specified. Source gives brand names for items listed.

★ 46 ★

Work-Related Time

Time in Days That Americans Worked to Purchase Selected Items

Item	Average Time in Days of Work			
	1962	1972	1982	1992
House	1125.5	1330.7	1530.0	1777.3
College (private)	129.5	140.4	144.0	251.4
Car	203.1	131.0	161.0	197.8
College (public)	61.7	64.1	68.9	99.2
Child delivery	15.5	37.2	33.3	62.2
Income taxes - federal	50.0	48.3	63.8	49.0
Auto insurance	7.1	7.8	7.3	11.3

Source: "Has Our Living Standard Stalled," *Consumer Reports*, 57, June 1992, pp. 392-393. **Remarks**: This table refers to before tax dollars and the actual prices for the years specified.

★ 47 ★

Work-Related Time

Working at Home

According to a Bureau of Labor Statistics survey in May 1991, men and women who work partly at home average six hours per week of at-home work. For men and women who work entirely at home, men work an average of 39 hours a week, and women work an average of 33 hours. Self-employed men work an average of 13 hours of at-home work, and women an average of 21 hours.

Source: "Most Work at Home Goes Uncompensated," *The Wall Street Journal*, November, 27, 1991, p. B1. **Remarks**: Also included in source: other statistics related to at-home work.

★ 48 ★

Work-Related Time

Time Spent on the Job, 1969-1989

According to a study by Juliet Schor of Harvard University and Laura Leete-Guy of Case Western Reserve University in Cleveland, full-time American workers were on the job an average 138 hours more in 1989 than in 1969. It also found that paid time off has decreased while working time has increased. In 1981, American workers had an average of 19.8 days off, compared to 16.1 days in 1989.

Source: "Americans Are Working More Hours," *The Washington Post*, February 17, 1992, p. A10.

★ 49 ★

Work-Related Time

Time Spent on Business Trips

Companies are sending employees on fewer but longer business trips, in an effort to cut costs.

Year	Average Length of Trip (Days)
1989	3.8
1990	4.2
1991	4.4

Source: "Staying Longer on the Road," *USA Today*, September 8, 1992, p. E1. Original source: U.S. Travel Data Center.

★ 50 ★

Work-Related Time

Sick Days

In a 100-person office, the most frequent reasons given for taking sick days and the average number taken per year for each are: cold, 21 days; fractures, 23 days; sprains, 30 days; flu, 76 days.

Source: "Hello Boss...," *The Detroit Free Press*, March 14, 1993, p. 3A. Original source: National Center for Health Statistics.

★ 51 ★

Work-Related Time

Years Worked Before Retirement

According to a survey of 300 midsized and large companies by William M. Mercer, a benefits consulting firm in New York, the average worker retired in 1992 at age 62 or 63 years after 21 years with his/her employer. But by the year 2000, this average worker will not be able to afford retirement at this age, largely due to increasing health care costs.

Source: "Health Care Costs Becloud Plans for Retirement," *The Wall Street Journal*, June 9, 1993, p. B1.

★ 52 ★

Work-Related Time

Leave of Absence

Reason	Average Time Employees Would Take	
	Women	Men
Birth or adoption	8.5 weeks	2.7 weeks
Spouse/parent with critical illness	8.4	8.5

Source: "Women Likely to Take Longer Leave," *USA Today*, June 17, 1993, p. 1D. Primary source: BNA; Bruskin/Goldring.

★ 53 ★

Work-Related Time

Time Lost Due to Misplaced Items

According to a survey of 200 executives by Accountemps, office workers waste an average of six weeks per year, or 50 minutes per day, trying to locate mislabled, misfiled, or misplaced items.

Source: Working Women, January 1993, p. 34.

★ 54 ★

Work-Related Time

Time Executives Spend Waiting on Hold

According to a survey of 150 large companies done by OfficeTeam, a Menlo Park, California personnel agency, the average executive spends two weeks a year waiting on hold, or 15 minutes every day.

Source: "Bottom Line: Holding Patterns," *Detroit Free Press*, July 7, 1993, p. 1C.

★ 55 ★

Work-Related Time

Days of Disability by Family Income, 1985-1989

Income	Average Days of Disability per Person			
	1985	1987	1988	1989
Under $10,000	25.8	24.2	26.6	26.5
$10,000-$19,999	16.7	16.9	17.8	18.7
$20,000-$34,999	12.1	12.1	12.3	13.3
$35,000-34,999	9.9	9.9	9.7	9.9

Source: U.S. Bureau of the Census. *Statistical Abstract of the United States 1992.* 112th edition. Washington, D.C.: U.S. Department of Commerce, 1992, p. 123. Primary source: U.S. National Center for Health Statistics, *Vital and Health Statistics*, series 10, no. 176; and unpublished data.

★ 56 ★

Work-Related Time

Days of Disability by Sex and Race/Ethnicity, 1970-1989

	Average Days of Disability per Person					
	1970	1980	1985	1987	1988	1989
Male	13.2	17.1	12.8	12.7	12.7	13.2
Female	15.8	21.0	16.6	16.1	16.5	17.0
White	14.4	18.8	14.5	14.3	14.6	15.0
Black	16.2	22.7	17.4	16.4	16.6	17.1
Hispanic	NA	NA	13.2	12.0	13.0	13.2

Source: U.S. Bureau of the Census. *Statistical Abstract of the United States 1992.* 112th edition. Washington, D.C.: U.S. Department of Commerce, 1992, p. 123. Primary source: U.S. National Center for Health Statistics, *Vital and Health Statistics*, series 10, no. 176; and unpublished data.

Miscellaneous

★ 57 ★

Americans on the Move

The average American moves 11 times in a lifetime, or once every six years.

Source: "Future Work," Joseph F. Coates, et. al., *Futurist*, 25, May-June 1991, p. 13.

Chapter 2
BUSINESS AND LABOR

Advertising

★ 58 ★

Americans' Exposure to Advertising

Year	Average Number of Advertising Messages Americans Are Exposed To Each Day
1960s	1500
1990s	3000

Source: "Advertising Everywhere!" *Consumer Reports*, 57, December 1992, p. 752.

★ 59 ★

Advertising

Radio Advertising

In an average hour, radio listeners hear 12.7 commercials totaling 10 minutes and 6 seconds on AM radio and 11.9 commercials lasting 9 minutes and 17 seconds on FM.

Source: Rukeyser, Louis. *Louis Rukeyser's Business Almanac.* New York: Simon & Shuster, 1991, p. 509.

★ 60 ★

Advertising

Television Advertising and Teenagers

American teenagers are exposed to an average of three to four hours of television advertising per week. This amounts to around 100,000 ads between birth and high school graduation.

Source: "Leftfield: Enough is Enough," Alan During, *Dollars & Sense*, 167, June 1991, pp. 15-18.

★ 61 ★

Advertising

Advertising Expenditures

Year	Average Advertising Expenditures per Capita	
	United States	Worldwide
1950	$198	$15
1989	498	46

Source: "Leftfield: Enough is Enough," Alan During, *Dollars & Sense*, 167, June 1991, pp. 15-18.

★ 62 ★

Advertising

Direct-Mail Advertising

In the United States, 63.5 billion pieces of bulk mail, i.e. commercial and nonprofit solicitations, and 13.6 billion catalogs were sent out each day. This amounts to an average of 300 pieces of "junk" mail for each person living in the United States.

Source: "Ideas: How to Pare Your Share of 77 Billion Pieces of Junk Mail," Elizabeth Brown, *The Christian Science Monitor*, 83, January 15, 1991, p. 14.

★ 63 ★

Advertising

Help-Wanted Advertising

According to a survey by the Newspaper Advertising Bureau, American employers and agencies spend an average $2 billion on classified advertising each year.

Source: "The Relationship Between Advertisement Variables and Applicant Responses to Newspaper Recruitment Advertisements," *Journal of Business and Psychology*, 5, Spring 1991, pp. 383-395.

Airline Industry

★ 64 ★

Passengers Handled by O'Hare Airport, Chicago

One of the world's busiest airports, Chicago's O'Hare handles 55 million people every year, or an average of 6,700 passengers per hour. An average of 2,200 airplanes pass through O'Hare each day from 50 commercial airlines.

Source: How in the World. Pleasantville, NY: The Reader's Digest Association, 1990, p. 32.

★ 65 ★

Airline Industry

Cost of Airline Food by Selected Airlines

Airline	Average Spent on Food per Passenger	Airline	Average Spent on Food per Passenger
Southwest	$0.12	Continental	$5.37
America West	2.74	Northwest	5.71
USAir	3.55	United	6.16
Midway	3.57	TWA	6.70
Eastern	4.44	American	6.88
Delta	5.18	Pan Am	7.60

Source: Krantz, Les. *The Best and Worst of Everything.* New York: Prentice Hall General Reference, 1991, p. 201. Primary source: Yield Data Services.

★ 66 ★

Airline Industry

Cost of Airline Food per Passenger, 1989-1990

Year	Average Cost of Airline Food
1989	$4.39
1990	5.46

Source: "Airline News: 1990 Food Costs Rise 15 Percent but War Cuts into Passenger Loads," *Airline, Ship & Catering Onboard Services*, 23, July- August 1991, p. 8.

★ 67 ★

Airline Industry

Baggage Lost by Selected Airlines

Airline	Average Number of Bags Lost per 1,000 Passengers	Airline	Average Number of Bags Lost per 1,000 Passengers
Southwest	2.81	USAir	5.95
Continental	4.44	Delta	5.99
Pan Am	4.54	Northwest	6.47
American	4.98	TWA	7.52
Eastern	5.52	America West	8.61
United	5.64		

Source: Krantz, Les. *The Best and Worst of Everything.* New York: Prentice Hall General Reference, 1991, p. 199. Primary source: U.S. Department of Transportation.

Automobile Industry

★ 68 ★

Number of Workers Needed to Build One Car per Day

Year	Average Number of Workers Needed per Car per Day			
	General Motors	Ford	Chrysler	Japanese
1979	5.12	4.71	5.63	---
1989	4.88	3.25	4.58	---
1992	4.55	3.01	3.76	2.0 to 2.5

Source: "Hitting the Brakes Hard," *U.S. News & World Report*, November 9, 1992, pp. 78-79. Primary source: Economic Strategy Institute, Harbour & Associates Inc., the WEFA Group. **Remarks**: Source also includes: Labor costs per vehicle in the United States.

★ 69 ★

Automobile Industry

Labor Cost per Vehicle in the United States, 1992

Auto Maker	Average Labor Cost per Vehicle
General Motors	$2,388
Chrysler	1,872
Ford	1,629
Honda	920

Source: "Hitting the Brakes Hard," *U.S. News & World Report*, November 9, 1992, pp. 78-79. Primary source: Economic Strategy Institute, Harbour & Associates Inc., the WEFA Group. **Remarks**: Source also includes: average number of workers needed to build one car per day in the United States and Japan.

★ 70 ★

Automobile Industry

Automobile Braking Distance

Average stopping distance is directly related to vehicle speed. On a dry, level concrete surface, the minimum stopping distances are as follows (includes driver reaction time to apply brakes).

Miles per Hour	Reaction Time Distance	Braking Distance	Total Stopping Distance
10	11 Feet	9 Feet	20 Feet
20	22	23	45
30	33	45	78
40	44	81	125
50	55	133	188
60	66	206	272
70	77	304	381

Source: Automotive Encyclopedia. Rev. ed. South Holland, IL: Goodheart-Willcox Company, 1989, p. 637.

★ 71 ★

Automobile Industry

Monthly Vehicle Production in Selected Countries

Country	Average Monthly Vehicle Production
Canada	165,000
Spain	198,000
France	337,000
Germany	464,000

[Continued]

★ 71 ★

Monthly Vehicle Production in Selected Countries

[Continued]

Country	Average Monthly Vehicle Production
United States	791,000
Japan	1,000,000

Source: "Top Vehicle Producers," *USA Today*, October 28, 1992, p. B1. Primary source: Ward's Automotive Reports.

Communication

★ 72 ★

Telephone Usage per Business Day

There are about 9.5 billion minutes of telephone calling on an average business day.

Source: U.S. Department of Commerce. International Trade Administration. *U.S. Industrial Outlook '92*. Washington, D.C.: U.S. Department of Commerce, 1992, p. 28-2.

★ 73 ★

Communication

Telemarketing

On an average day, over 300,000 telemarketers call more than 18 million Americans.

Source: "Statements on Introduced Bills and Joint Resolutions," Larry Pressler, *Congressional Record*, July 10, 1991, p. S9498.

★ 74 ★

Communication

Fax Machine Users

The average *Fortune* 500 company received 428 pages per day by fax in 1993, compared with 300 a day in 1992; it sent an average of 260 pages (49 documents) per day in 1993, an increase of 41% over 1992.

Source: "Fax Users Paper Offices," *USA Today*, May 18, 1993, p. 2B.

★ 75 ★

Communication

Circulation of Newspaper Weeklies, 1990-1991

According to the National Newspaper Association, the number of newspaper weeklies fell from 7,550 in 1990 to 7,476 in 1991. While average circulation per weekly remained virtually unchanged from 7,309 in 1990 compared with 7,323 in 1991, total weekly circulation declined from 55.2 million to 54.7 million.

Source: U.S. Department of Commerce. International Trade Administration. *U.S. Industrial Outlook '92*. Washington, D.C.: U.S. Department of Commerce, 1992, p. 25-4.

★ 76 ★

Communication

United States Government Printing Office

The central office of the United States Government Printing Office in Washington, D.C. uses 98 million pounds of paper annually to process an average of 1,954 print orders daily. The Superintendent of Documents receives an average of 5,600 mail orders per day and sells around $75.7 million worth of U.S. government publications each year.

Source: Guinness Book of Records 1992. New York: Bantam Book, 1992, p. 465.

★ 77 ★

Communication

Mail Handled by the United States Post Office

According to the Postmaster General of the United States, in 1980 an average of 291,263,000 items passed through the country's 30,326 post offices each day.

Source: New York Public Library. *The Book of Answers*. New York: Prentice Hall Press, 1990, p. 20.

★ 78 ★

Communication

Greeting Cards Sold for Selected Occasions

Occasion	Average Number of Cards Sold
Halloween	28 million
Thanksgiving	40 million
Graduation	81 million
Father's Day	101 million
Mother's Day	150 million
Easter	165 million

[Continued]

★ 78 ★

Greeting Cards Sold for Selected Occasions
[Continued]

Occasion	Average Number of Cards Sold
Valentine's Day	1.0 billion
Christmas	2.3 billion

Source: "Or Better Yet, Give Him a Tie," *Business Week*, 22, June 22, 1992, p. 44. Primary source: Greeting Card Association.

Computers

★ 79 ★

Mainframes vs. Parallel Processing

Mainframe computing: can handle an average of 400-500 transactions per second. Over five years the capacity to process one transaction per second costs about $46,000.

Parallel Processing: can handle 1,000 transactions per second. Over five years the capacity to process one transaction per second costs about $10,000.

Source: "Improving on the Mainframe," William J. Cook, *U.S. News & World Report*, 113, June 8, 1992, p. 52-54.

Economic Growth

★ 80 ★

Economic Growth After a Recession

First Year After Recession	Growth	Length of Expansion (months)
1949-1950	14.2%	45
1954-1955	6.0	39
1958-1959	8.1	23
1961-1962	6.4	106
1970-1971	3.9	36
1975-1976	6.4	58
1980-1981	3.5	12
1982-1983	6.7	92
1992 (projected)	2.5	---
AVERAGE	6.9	52

Source: "Far From the Dundering Herd," John Liscio, *U.S. News & World Report*, 113, June 8, 1992, p. 58.

★ 81 ★

Economic Growth

Job Growth, Truman to Bush

President	Average Annual Job-Growth Rate
Truman	2.0%
Eisenhower	1.4
Kennedy	1.5
Johnson	3.7
Nixon/Ford	2.0
Carter	3.3
Reagan	2.0
Bush	0.7

Source: "Out of Work," *U.S. News & World Report*, 113, October 12, 1992, p. 59. Primary source: U.S. Department of Labor.

★ 82 ★

Economic Growth

Return on Stock, Bonds, and Real Estate, Truman to Bush

President	Average Return on Stocks, Bonds, and Real Estate
Truman	7.3%
Eisenhower's 1st	8.2
Kennedy/Johnson	6.6
Reagan's 1st	5.1
Regan's 2nd	11.1
Bush (estimate)	6.5

Source: "Riches in the Reagan Years," *USA Today*, November 2, 1992, p. B1. Primary source: PDI Strategies.

★ 83 ★

Economic Growth

Annual Return on Stocks, 1920s-1990s

Year	Standard & Poor's 500 Index Average Annual Return
1920s	19.2%
1930s	0
1940s	9.2
1950s	19.4
1960s	7.8
1970s	5.9
1980s	17.5
1990s	10.1

Source: "Stocks: A Strong Investment," *USA Today*, October 19, 1992, p. 1B. Primary source: Ibbitson Associates, National Association of Realtors, IBC/Donoghue's *Money Fund Report*, *USA Today* research. **Remarks**: Source also includes: average annual gain in the 1980s and 1990s for short-term money funds and real estate.

Energy and Gasoline

★ 84 ★

Manufacturing Energy Consumption by Type of Industry, 1988

Type of Industry	Energy Consumption Required to Produce Heat, Power, and Electricity per Employee (million BTUs)	Type of Industry	Energy Consumption Required to Produce Heat, Power, and Electricity per Employee (million BTUs)
Food	694	Leather	108
Tobacco	580	Stone, clay, & glass	2,037
Textile mill	451	Primary metal	3,782
Apparel	42	Fabricated metal	251
Lumber	551	Machinery	145
Furniture	131	Electric & electronic	144
Paper	3,498	Transportation	193
Printing & publishing	73	Instruments	104
Chemicals	3,601	Miscellaneous	110
Petroleum	28,414		
Rubber	329	Total	868

Source: U.S. Bureau of the Census. *Statistical Abstract of the United States 1992*. 112th edition. Washington, D.C.: U.S. Department of Commerce, 1992, p. 568. Primary source: U.S. Energy Information Administration, *Manufacturing Energy Consumption Survey: Consumption of Energy, 1988*. **Remarks**: Source also includes: total fuel consumption for selected industry groups by type of fuel, 1988.

★ 85 ★

Energy and Gasoline

Prices of Selected Fuels and Electricity, 1975-1990

In dollars per unit, except electricity, in cents per kWh. Represents price to end-users.

Item	Unit	1975	1980	1981	1983	1985	1986	1989	1990
Crude oil, composite	Barrel	10.38	28.07	35.24	28.99	26.75	14.55	17.97	22.23
Motor gasoline									
Leaded regular	Gallon	0.57	1.19	1.31	1.16	1.12	0.86	1.00	1.15
Unleaded regular	Gallon	(NA)	1.25	1.38	1.24	1.20	0.93	1.02	1.16
Unleaded premium	Gallon	(NA)	(NA)	1.47	1.38	1.34	1.09	1.20	1.35
No. 2 heating oil	Gallon	(NA)	0.79	0.91	0.92	0.85	0.56	0.59	0.73
No. 2 diesel fuel	Gallon	(NA)	0.82	1.00	0.83	0.79	0.48	0.59	0.73
Residual fuel oil	Gallon	(NA)	0.61	0.76	0.65	0.61	0.34	0.39	0.44

[Continued]

★ 85 ★

Prices of Selected Fuels and Electricity, 1975-1990

[Continued]

Item	Unit	1975	1980	1981	1983	1985	1986	1989	1990
Natural gas, residential	1,000 cu/ft.	1.71	3.68	4.29	6.06	6.12	5.83	5.64	5.77
Electricity, residential	kWh	3.5	5.4	6.2	7.2	7.8	7.4	7.6	7.8

Source: U.S. Bureau of the Census. Statistical Abstract of the United States 1992. 112th edition. Washington, D.C.: U.S. Department of Commerce, 1992, p. 483. Primary source: U.S. Energy Information Administration, *Monthly Energy Review*, July, 1991.

★ 86 ★

Energy and Gasoline

Price of Gasoline in the United States, 1960-1992

Prices for 1960-1975 are for leaded regular; 1980-1992 are for unleaded regular.

Year	Average Price of Gasoline per Gallon (adjusted for inflation)
1960	$1.45
1965	1.33
1970	1.23
1975	1.39
1980	2.10
1985	1.54
1990	1.25
1991	1.17
1992	1.12

Source: "Average Price per Gallon of Gasoline in the U.S.," *U.S. News & World Report*, 114, January 18, 1993, p. 61. Primary source: Energy Information Administration. **Remarks**: Source also includes: price of gasoline per gallon in selected countries, 1992.

★ 87 ★

Energy and Gasoline

Price of Gasoline in Selected Countries, 1992

Country	Average Price of Gasoline per Gallon
Italy	$4.71
Sweden	4.18
Japan	3.71
France	3.60
Germany	3.59
Great Britain	3.21
Canada	1.79
United States	1.05

Source: "Average Price per Gallon of Gasoline in the U.S.," *U.S. News & World Report*, 114, January 18, 1993, p. 61. Primary source: American Petroleum Institute, Energy Information Administration, OECD. **Remarks**: Source also includes: price of gasoline per gallon in the United States, 1960-1992.

Farming

★ 88 ★

States With the Lowest Farm-Land Values, 1989

State	Average Value per Acre of Farmland
Wyoming	$136
New Mexico	141
Montana	167
Nevada	199
South Dakota	204
U.S. average	597

Source: Bluman, Allan. *Elementary Statistics*. Dubuque, IA: Wiliam C. Brown Publishers, 1992, p. 449. Primary source: U.S. Department of Agriculture.

★89★

Farming

Size of farms, 1860-1986

An average of 5.4 million acres have been lost to farming in the United States each year between 1950 and 1982. But the cropland base has remained relatively stable since 1945: 475 million acres in 1945, compared with 469 million acres in 1982. Farm size has also been changing, increasing from an average of 199 acres in 1860 to 456 in 1986. In the last 20 years, crop failure on these farms has averaged 6 to 7 million acres per year, or about 2% of the cropland harvest.

Source: Council on Environmental Quality. *Environmental Trends.* Washington, DC: Council on Environmental Quality, 1989, pp. 76-80.

★90★

Farming

Size of Farms by State, 1987

State	Average size of farm (acres)	State	Average size of farm (acres)	State	Average size of farm (acres)
United States	462	Kentucky	152	North Dakota	1,143
Alabama	211	Louisiana	293	Ohio	189
Alaska	1,789	Maine	214	Oklahoma	449
Arizona	4,732	Maryland	162	Oregon	556
Arkansas	298	Massachusetts	99	Pennsylvania	153
California	368	Michigan	202	Rhode Island	84
Colorado	1,248	Minnesota	312	South Carolina	232
Connecticut	111	Mississippi	315	South Dakota	1,214
Delaware	205	Missouri	275	Tennessee	147
Florida	306	Montana	2,451	Texas	691
Georgia	247	Nebraska	749	Utah	710
Hawaii	353	Nevada	3,300	Vermont	240
Idaho	577	New Hampshire	169	Virginia	194
Illinois	321	New Jersey	99	Washington	480
Indiana	229	New Mexico	3,230	West Virginia	196
Iowa	301	New York	223	Wisconsin	221
Kansas	680	North Carolina	159	Wyoming	3,650

Source: U.S. Bureau of the Census. *Statistical Abstract of the United States 1992.* 112th edition. Washington, D.C.: U.S. Department of Commerce, 1992, p. 646. Primary source: U.S. Bureau of the Census, *Census of Agriculture: 1987.* **Remarks:** Source also includes: number of farms, amount of land in farming, and farms with sales of $10,000 or more, all by state, 1987.

★ 91 ★

Farming

Size of a Japanese Farm Vs. an American Farm

The average farm in the United States is 467 acres, compared with 3 acres for the average Japanese farm.

Source: "Sowing Seeds for a Global Trade War," William Cook, David Lawday, and Jim Impoco, *U.S. News & World Report*, 113, November 23, 1992, p. 72.

Government Employees

★ 92 ★

Cost of Reelection to the United States Senate

The cost of political campaigns is so high that the average member of Congress must raise $15,000 each week for six years in order to obtain the $4 million needed to win reelection to the United States Senate.

Source: "The Need for Comprehensive Reform of Congress," David L. Boren, *Congressional Record*, October 23, 1991, pS15043.

Hiring

★ 93 ★

Screening Costs of New Hires

Thirty percent of all resumes are falsified, according to Team Building Systems, an employee screener in Houston Texas. Businesses spend $100 for an average background check of a new hire.

Source: "Screening New Hires," *Inc.*, 14, August 1992, p. 82.

★ 94 ★

Hiring

Resumes Received by American Businesses

An average of 1,000 unsolicited resumes arrive daily at the Fortune 500 companies. After a quick look, four out of five are discarded.

Source: "Writing a Computer-Friendly Resume," *U.S. News & World Report*, 113, October 26, 1992, p. 90.

★95★

Hiring

Cost of Hiring per Person

According to the Department of Labor, businesses spend an average of $40,000 to hire a new worker. This figure includes hidden costs such as loss of productivity and overtime for remaining staff.

Source: "Are You Hiring the Right People?" Donald D. DeCamp, *Management Review*, 80, May 1992, p. 44.

Housing Industry

★96★

Timber in a Three-Bedroom Home

According to the United States Forest Service, builders use 10 to 14 average-size trees, or 10,000 board-feet of timber, to construct an average three-bedroom home.

Source: "The Recycled House," Marcella M. Soviero, *Popular Science*, 238, April 4, 1991, p. 69.

Industrial Accidents

★97★

Workers Killed or Injured on the Job, 1989

Workers killed on the job in 1989: a total of 10,400, which is an average of 40 workers per day, or one worker every 36 minutes.

Workers injured on the job in 1989: 1,700,000 workers suffered disabling injuries, an average of 6,538 each day, or one worker every 13 seconds.

Source: "Statements on Introduced Bills and Joint Resolutions," Howard Metzenbaum, *Congressional Record*, August 1, 1991, p. S11844.

★ 98 ★

Industrial Accidents

Mining Injuries

According to statistics of the federal Mine Safety and Health Administration, there were an average of 9,500 mining injuries each year in the mid-1980s. This figure increased to an average of 12,000 per year from 1987-1990. The higher figure is attributable to better reporting procedures rather than to an increase in injuries.

Source: "Business Labor: The Curse of Coal," Ted Gup, *Time*, 138, November 4, 1991, pp. 54-56, 61, 64.

Loss of Productivity

★ 99 ★

Weekly Absence Rate Among Full-Time Workers

Status of Worker	Average Percentage of Workers Absent per Week
Married men with children under 6 yrs.	3.3%
Married women with children under 6 yrs.	10.5
All workers	4.8

Source: "Absent, With Children," *USA Today*, March 19-21, 1993, p. 1A. Primary source: Bureau of Labor Statistics.

★ 100 ★

Loss of Productivity

PC Users and Lost Productivity

According to research done by Minnesota Mining & Manufacturing Co., 30% of PC users lose data and spend an average of one week to reconstruct it. This loss of productivity amounts to 24 million days per year nationally, at a cost of $4 billion. Another study by SBT Accounting Systems Inc., a software concern, estimates that workers spend an average of 5.1 hours per week "fiddling" (unproductive time) with their PCs.

Source: "Data Trap: How Using Your PC Can be a Waste of Time, Money," William M. Bulkeley, *The Wall Street Journal*, January 4, 1993, p. B5.

★ 101 ★

Loss of Productivity

Business Losses Due to Motor Vehicle Accidents

According to the Network of Employers for Traffic Safety (NETS), lost work time due to motor vehicle accidents costs businesses $16.4 billion annually. Workers compensation for each on-the-job motor vehicle-related fatality averages $110,500; each injury averages $2,400.

Source: "Safety First," *Business Week*, 22, November 2, 1992, p. 37.

★ 102 ★

Loss of Productivity

Workers' Compensation, 1980-1992

Year	Average Cost per Workers' Compensation Claim	Average Number of Days Lost to Workers' Compensation per 100 Employees
1980	$6,138	59
1992	24,000	87

Source: "Sticking It to Business," Warren Cohen, *U.S. News & World Report*, 114, March 8, 1993, pp. 59-61. **Remarks**: Source also includes: total cost of workers' compensation, 1980-1992.

★ 103 ★

Loss of Productivity

Business Losses Due to Poor Handwriting

According to The Writing Instrument Manufacturers Association of Marlton, New Jersey, businesses lose $200 million per year because of poor handwriting. At a cost of $4 million, the U.S. Postal Service sends 38 million pieces of illegible mail to the dead letter office each year. At the Los Angeles Post Office alone, 12 handwriting experts try to decipher 500 pieces of mail per day. Eastman Kodak was unable to return an average of 400,000 rolls of film per year because of illegible addresses.

Source: "Getting It Write the First Time," John Maines, *American Demographics*, 14, December 1992, p. 18.

★ 104 ★

Loss of Productivity

Cost of Traffic Congestion in Los Angeles

Traffic congestion is a great drain on the American economy. Lost productivity, wasted fuel, and air pollution due to stop-and-go conditions during rush hour traffic, costs $6 billion a year, or $3 per automobile, in Los Angeles alone.

Source: "Keep It Moving," Bruce Ingersoll, *The Wall Street Journal*, January 20, 1993, p. R12.

Malls

★ 105 ★

Mall of America, Bloomington, Minnesota

The Mall of America near Minneapolis, at 4.2 million square feet, was the largest enclosed shopping center in the United States in 1992. To be successful, every potential shopper must visit an average of four times per year and stay an average of four hours per visit.

According to CB Commercial Real Estate in Minneapolis, the total retail footage in Minneapolis-St. Paul is 45 million square feet, or an average of 18.3 square feet for every person in the metropolitan area. Compare this to other urban areas: Chicago with 20.3 square feet per person, Phoenix with 29.5, and Atlanta with 32.5.

Source: "The Biggest Mall of All," Judith Waldrop, *American Demographics*, 14, August 1992, p. 4.

Minorities in Business

★ 106 ★

Pregnancy Discrimination by American Businesses

The Equal Employment Opportunity Commission (EEOC), the federal agency that enforces the Pregnancy Discrimination Act (PDA) of 1978, handles 3,000 pregnancy-discrimination cases per year, but estimates that in the United States an average of 80,000 of the two million women who become pregnant each year lose their jobs, or at best are often denied raises and promotions.

Source: "The Mommy Backlash," Aaron Bernstein, *Business Week*, 22, August 10, 1992, p. 43.

★ 107 ★

Minorities in Business

Women in the Automobile Industry

Women who work for General Motors, Ford and Chrysler:

Average age: 39.
Average number of years with the company: 13.
Average number of hours worked per week: 49 (GM—56.5, Ford—49.5, Chrysler—51).

Source: "Auto Worker Women," *Detroit Free Press*, December 14, 1992, p. 13F. Primary source: *Free Press* and University of Michigan survey of professional women who work for GM, Ford, Chrysler.

★ 108 ★

Minorities in Business

On-Campus Recruitment of Minority Students

According to Hanigan Consulting Group, New York, one in eleven students, but only one in 20 minority students interviewed, are hired by the average employer.

Source: "Employers Go to School On Minority Recruiting," *The Wall Street Journal*, December 15, 1992, p. B1.

★ 109 ★

Minorities in Business

Minority-Owned Firms, 1987

Minority	Average Income	Average Number of Paid Employees
Asian	$93,200	3.79
Hispanic	58,600	3.19
Black	46,600	3.11
Native American	42,600	2.40

Source: "Reaching the Dream," William O'Hare, *American Demographics*, 14, January 1992, pp. 32-36. **Remarks**: Source also includes: figures for total number of minority-owned businesses, share of business by minority type, and percent of growth in businesses, 1982-1987.

★ 110 ★

Minorities in Business

Metropolitan Areas with the Highest Percentage of Women in the Labor Force

Metropolitan Area	Percentage
Washington, D.C.	68.8%
Minneapolis	67.3%
Atlanta	65.6%
Dallas	64.2%
Boston	62.3%
National metro average	58.2%

Source: "Demographics," *U.S. News & World Report*, 113, August 10, 1992, p. 15. Primary source: U.S. Census, 1990.

Movies Industry

★ 111 ★

Movie Production Cost, 1980-1990

Year	Average Movie Production Cost
1980	$10.0 million
1990	26.8 million

Source: "Movie Budgets Shake Hollywood," David Sterritt, *The Christian Science Monitor*, August 2, 1991, p. 12. Primary source: Motion Picture Association of America (MPAA).

★ 112 ★

Movies Industry

Advertising and Promotion Budget of a Film

In 1992, a film running for several months, had an average advertising and promotion cost of $12 million.

Source: "Warner's Sequel Weapon Cuts Down Promotion Costs," Geraldine Fabrikant, *The New York Times*, May 12, 1992, p. D23.

★ 113 ★

Movies Industry

Admission Price for Films, 1935-1990

Year	Average Admission Price	Year	Average Admission Price
1935	$0.24	1975	2.05
1940	0.24	1980	2.69
1945	0.35	1983	3.13
1950	0.53	1985	3.55
1955	0.50	1987	3.91
1960	0.69	1988	4.11
1965	1.01	1989	4.45
1970	1.55	1990	4.75

Source: The Universal Almanac 1992. Kansas City, MO: Andrews and McMeel, 1991. p. 253. **Remarks**: Source also includes: movie box office receipts in current and constant dollars and number of admissions, 1929-1990.

★ 114 ★

Movies Industry

Movie Videos Sold per Store

A movie that earns more than $100 million at the box office will do well in video stores, selling an average of eight copies at each video specialty outlet.

Source: "VCR Wars," *American Demographics*, 14, August 1992, p. 6. **Remarks:** Source also includes: number of movies seen by the average movie-goer, 1990 and 1991.

★ 115 ★

Movies Industry

Movie Stars Making the Most Money in the First Weekend After Their Films are Released

Movie Star	Average Amount Made (Millions) for Their Films in the First Weekend
Eddie Murphy	$16.9
Arnold Schwaarzenegger	15.4
Sigourney Weaver	15.2
Michael Keaton	14.2
Bill Murray	12.5
Mel Gibson	12.4
Harrison Ford	12.0
Tom Cruise	11.4

[Continued]

★115★

Movie Stars Making the Most Money in the First Weekend After Their Films are Released

[Continued]

Movie Star	Average Amount Made (Millions) for Their Films in the First Weekend
Julia Roberts	11.2
Michael J. Fox	10.8

Source: "Eddie Edges Out Arnold and Sigourney," *Parade* magazine, April 18, 1993, p. 13. Primary source: *Variety*, the show-business weekly.

Office Space

★116★

First-Class Office Rental per Square Foot

According to a survey of 13 major markets by Julien J. Studley Inc., commercial brokers in New York, midtown New York is the most expensive market for new office space, at $44.75 a square foot for rent in 1992, down from $47 in 1991; while the cheapest was Houston, at $19 a square foot.

But the results of another survey of 36 cities by Colliers International, a real-estate services company in Boston, points to Washington, D.C. as the costliest market for first-class office space: an average of $33.50 a square foot; compared to Memphis, TN, the cheapest in the nation: an average of $12.65 a square foot.

Source: "Odds and Ends," *The Wall Street Journal*, February 16, 1993, p. B1 and March 29, 1993, p. B1.

Productivity Improvements

★ 117 ★

Productivity of American Workers

According to a Department of Labor study, by the year 2000, the average worker between the ages of 21 and 25 will need to improve productivity skills by some 40% in order to satisfy the performance demands of the workplace. However, most corporate training is not designed for the front- line worker. Only one-third of the $30 million businesses spend annually on formal training is used for the noncollege-educated workforce. Thus, only 8% of these workers receive any training. In 1992, it took an average of three years for American workers to achieve the same level of productivity that was produced in one year prior to 1973.

Source: "Bill Takes Aim at Future Workers and Productivity," *Management Review*, 80, February 1992, pp. 7-8. Primary source: Commission on the Skills of the American Workforce report, *America's Choice: High Skills or Low Wages*. National Center on Education and the Economy.

★ 118 ★

Productivity Improvements

Productivity per Worker in Selected Countries

Country	Average Amount (dollars) in Goods & Services per Worker	Percentage Employed in Service Industries
United States	$49,600	74%
France	47,000	56
Germany	44,200	55
Japan	38,200	65
Great Britain	37,100	67

Source: "Still No. 1," *The New York Times*, October 13, 1992, p. D19. Primary source: McKinsey Global Institute.

★119★

Productivity Improvements

Money Businesses Save From Employee Suggestions, 1985-1991

Year	Average Amount Saved per Suggestion
1985	$4,397
1986	5,554
1987	8,075
1988	7,663
1989	6,114
1990	7,103
1991	6,405

Source: "Firms Look In-House for Ideas," S.C. Llewelyn Leach, *Christian Science Monitor*, November 17, 1992, p. 8.

★120★

Productivity Improvements

Time and Cost of Steel Production, 1982-1991

To produce and ship one metric ton of steel in 1982, Japanese required an average of 10.01 hours, and Americans, 10.59 hours. In 1991, the Japanese took 5.6 hours, and Americans, 5.4 hours to accomplish the same task. According to a security analyst at Paine Webber, the production cost of American steel decreased over the last decade by 6.2% to $490 a ton, while the cost of Japanese steel rose by 12.7% to $490.40 per ton.

Source: "The Big Threat to Big Steel's Future," Terence P. Pare, *Fortune*, 124, July 15, 1991, p. 107.

Railroad Industry

★ 121 ★

Cars per Freight Train, 1929-1987

Year	Average Number of Cars per Freight Train	Year	Average Number of Cars per Freight Train
1929	47.6	1980	68.3
1939	48.1	1983	70.4
1947	52.2	1984	71.4
1951	59.0	1985	71.8
1955	65.5	1986	70.3
1965	69.6	1987	71.1
1975	68.6		

Source: Association of American Railroads. *Railroad Facts*. Washington, D.C.: Information of Public Affairs Department, 1988, p. 35.

★ 122 ★

Railroad Industry

Weight of a Carload of Freight, 1929-1987

Year	Average Freight Carload (tons)	Year	Average Freight Carload (tons)
1929	35.4	1980	67.1
1939	36.8	1982	68.3
1947	41.0	1983	68.0
1951	42.0	1984	68.2
1955	42.4	1985	67.7
1965	48.9	1986	66.7
1970	54.9	1987	66.6
1975	60.8		

Source: Association of American Railroads. *Railroad Facts*. Washington, D.C.: Information of Public Affairs Department, 1988, p. 36.

★ 123 ★

Railroad Industry

Freight Train Load, 1929-1987

Year	Average Freight Train Load (net ton-miles per freight train-miles)	Year	Average Freight Train Load (net ton-miles per freight train-miles)
1929	804	1980	2,222
1939	806	1982	2,345
1947	1,131	1983	2,432
1951	1,283	1984	2,543
1955	1,359	1985	2,574
1965	1,685	1986	2,552
1970	1,820	1987	2,644
1975	1,938		

Source: Association of American Railroads. *Railroad Facts*. Washington, D.C.: Information of Public Affairs Department, 1988, p. 37.

★ 124 ★

Railroad Industry

Freight Car Capacity, 1929-1987

Year	Average Freight Car Capacity (tons)	Year	Average Freight Car Capacity (tons)
1929	46.3	1975	72.9
1939	49.7	1980	79.4
1947	51.5	1982	81.6
1951	52.9	1984	83.4
1955	53.7	1985	84.3
1965	59.7	1987	86.6
1970	67.1		

Source: Association of American Railroads. *Railroad Facts*. Washington, D.C.: Information of Public Affairs Department, 1988, p. 48.

★ 125 ★

Railroad Industry

Passenger Trains, 1987

Average miles of road operated	23,499
Average revenue per passenger	$27.39
Average revenue per passenger-mile	10.575 cents
Average trip per passenger	258.98 miles

Source: Association of American Railroads. *Railroad Facts.* Washington, D.C.: Information of Public Affairs Department, 1988, p. 61.

Research and Development

★ 126 ★

Research and Development per Employee by Type of Industry, 1992

Type of Company	R&D per Employee	Type of Company	R&D per Employee
Aerospace	$6,892	Housing	$2,764
Automotive	8,103	Leisure-time products	8,644
Chemicals	10,461	Manufacturing	4,143
Conglomerates	4,747	Metals & mining	2,006
Consumer products	2,297	Office equipment & services	14,728
Containers & packaging	1,247	Paper & forest products	1,971
Electrical & electronics	6,895	Service industries	916
Food	1,338	Telecommunications	5,678
Fuel	4,120	ALL-INDUSTRY, 1991	6,355
Health care	16,385	ALL-INDUSTRY, 1992	7,106

Source: "Who's Leading the Comeback," *Business Week*, 23, June 28, 1993, p. 102. Primary source: Standard & Poor's Compustat Services. **Remarks**: Source also includes: 1992 total spent on R&D, percentage change from 1991-1992, and R&D as a percent of sales for each type of company listed above.

Retail Business

★ 127 ★

Retail Sales per Household in Selected Metropolitan Areas, 1990

Metropolitan Areas	Sales per household (dol.)	Metropolitan Areas	Sales per household (dol.)
New York-Northern New Jersey-Long Island, NY-NJ-CT	20,028	St. Cloud, MN	36,640
Los Angeles-Anaheim-Riverside, CA	22,285	Portland, ME	32,650
Chicago-Gary-Lake County, IL-IN-WI	21,045	Honolulu, HI	31,340
San Francisco-Oakland-San Jose, CA	21,946	Anchorage, AK	29,963
Philadelphia-Wilmington-Trenton, PA-NJ-DE-MD	20,609	Terre Haute, IN	29,663
Detroit-Ann Arbor, MI	20,701	Portsmouth-Dover-Rochester, NH	29,032
Washington, DC-MD-VA	22,207	Manchester-Nashua, NH	28,502
Boston-Lawrence-Salem-Lowell-Brockton, MA	22,936	Rapid City, SD	26,777
Dallas-Fort Worth, TX	21,793	Bridgeport-Stamford-Norwalk-Danbury, CT	26,151
Miami-Fort Lauderdale, FL	21,918	Grand Forks, ND	25,139
Houston-Galveston-Brazoria, TX	19,940	Atlantic City, NJ	24,982
Atlanta, GA	22,233	Rochester, MN	24,934
Seattle-Tacoma, WA	21,340	Bangor, ME	24,808
Minneapolis-St. Paul, MN-WI	21,534	Orlando, FL	23,897
Cleveland-Akron-Lorain, OH	18,006	Burlington, VT	23,558

Source: U.S. Bureau of the Census. *Statistical Abstract of the United States 1992*. 112th edition. Washington, D.C.: U.S. Department of Commerce, 1992, p. 765. Primary source: Market Statistics, New York, NY, *The Survey of Buying Power Data Service*, annual. **Remarks**: Source also includes: total sales for the metropolitan areas listed above.

★ 128 ★

Retail Business

Retail Sales per Household by State, 1990

State	Sales per household Amount (dol.)	State	Sales per household Amount (dol.)	State	Sales per household Amount (dol.)
United States	19,488				
Alabama	17,435	Kentucky	17,283	North Dakota	18,630
Alaska	24,533	Louisiana	19,260	Ohio	17,880
Arizona	18,703	Maine	22,149	Oklahoma	16,780
Arkansas	17,205	Maryland	20,790	Oregon	20,159
California	21,245	Massachusetts	22,488	Pennsylvania	18,413
Colorado	18,864	Michigan	19,765	Rhode Island	19,276
Connecticut	22,435	Minnesota	20,050	South Carolina	18,685
Delaware	24,145	Mississippi	15,131	South Dakota	17,921
District of Columbia	15,360	Missouri	18,281	Tennessee	17,384
Florida	20,056	Montana	17,478	Texas	19,631
Georgia	19,417	Nebraska	17,113	Utah	19,501
Hawaii	31,121	Nevada	19,938	Vermont	21,223
Idaho	16,591	New Hampshire	28,373	Virginia	20,459
Illinois	19,840	New Jersey	22,579	Washington	19,370
Indiana	18,153	New Mexico	17,044	West Virginia	14,718
Iowa	17,724	New York	18,667	Wisconsin	19,691
Kansas	17,540	North Carolina	17,995	Wyoming	16,314

Source: U.S. Bureau of the Census. *Statistical Abstract of the United States 1992.* 112th edition. Washington, D.C.: U.S. Department of Commerce, 1992, p. 766. Primary source: Market Statistics, New York, NY, *The Survey of Buying Power Data Service*, annual. **Remarks**: Source also includes: total sales for all stores and sales by type of store, by state for 1990.

★ 129 ★

Retail Business

Supermarket Sales, 1990

In 1990, supermarkets averaged: $206,543 in sales per week; $8.43 of sales per week for every square foot in the store; and 1.19 cents of profit for every dollar of sales.

Source: Rukeyser, Louis. *Louis Rukeyser's Business Almanac.* New York: Simon & Shuster, 1991, p. 522.

★ 130 ★

Retail Business

Supermarkets, Rural vs. Urban, 1988

In 1988, the average of supermarkets per county in urban America was 29.2, compared with only 3.8 per county in rural America. Stated in other terms, urban America had an average of one supermarket every 27 square miles, compared with one every 265 square miles in rural America. Poor rural America has even fewer supermarkets: an average of 2.9 per county, compared with 3.9 per county in rural non-poor America.

Source: "Food Assistance in Rural Communities: Problems, Prospects, and Ideas from Urban Programs," U.S. House. Select Committee on Hunger. Hearing, April 5, 1991. Washington, DC, Government Printing Office, 1991.

★ 131 ★

Retail Business

Coupon Processing Fees

Coupon-issuing companies, in the last nine years, have paid an average of eight cents in handling fees to supermarkets and other retailers for each coupon given to them by consumers. But a recent study by Arthur Andersen & Co., the accounting firm, showed that seven cents would be more realistic. This study involved videotaping checkout clerks to determine that the average time spent per coupon was 6.8 seconds. In 1991, manufacturers paid distributors $596.8 million in coupon handling fees.

Source: "Study of Coupons Puts Manufacturers Against Grocers," Richard Gibson, *The Wall Street Journal*, May 15, 1992, p. 4A.

★ 132 ★

Retail Business

New Products Added to Grocery Store Shelves

There are about 20,000 items on the shelves of the average grocery store, and some 12,000 new products are added to these shelves each year.

Source: "Displays Pay Off for Grocery Marketers," *The Wall Street Journal*, October 15, 1992, p. B1.
Remarks: Source also includes: time spent on the average shopping trip.

★ 133 ★

Retail Business

Products in Supermarkets, 1960-1992

Year	Average Number of Products
1960	6,000
1992	30,000

Source: "What's in Store: Still More Advertising," *Consumer Reports*, December 1992, p. 755.

★ 134 ★

Retail Business

Retail Spending in Selected Metropolitan Areas

Metropolitan area	Average Amount Spent per $1,000 of Personal Spending	
	City	Suburbs
Boston	$468	$448
Chicago	365	433
Dallas	574	404
Detroit	348	459
Houston	643	285
Los Angeles	382	412
New York	298	395
San Francisco	422	366

Source: "New York's Lack of Retail Muscle," *New York Times*, March 8, 1993, p. B2. Primary source: City Planning Department, Census Bureau. **Remarks**, Source also includes: average number of residents per large retail store in selected New York counties.

Sales in Selected Businesses

★ 135 ★

Sales per Employee by Type of Industry, 1992

Type of Company	Sales per Employee (in thousands)	Type of Company	Sales per Employee (in thousands)
Aerospace	$155.1	Housing	$153.5
Automotive	203.7	Leisure-time products	151.9
Chemicals	245.1	Manufacturing	139.2
Conglomerates	181.0	Metals & mining	175.3
Consumer products	163.4	Office equipment & services	175.2
Containers & packaging	141.5	Paper & forest products	181.0
Electrical & electronics	115.8	Service industries	108.6
Food	181.3	Telecommunications	180.7
Fuel	549.4		
Health care	169.5	ALL INDUSTRY, 1992	192.2

Source: "R&D Scoreboard," *Business Week*, 23, June 28, 1993, pp. 105-123. Primary source: Standard & Poor's Compustat Services. **Remarks**: Source also includes: 1992 total spent on R&D, percentage change from 1991-1992, and R&D as a percent of sales for each type of company listed above.

★ 136 ★

Sales in Selected Businesses

Sales Per Capita by Type of Industry, 1989

Type of Business	Sales per Capita	
	1989	1985
Automotive	$1,556	$1,280
Food	1,400	1,203
General merchandise	829	---
Eating & drinking	705	---
Gasoline service stations	478	---
Apparel & accessories	371	---
ALL	7,064	5,804

Source: U.S. Department of Commerce. International Trade Administration. *U.S. Industrial Outlook '92*. Washington, D.C.: U.S. Department of Commerce, 1992, p. 39-2.

Temporary Employment

★ 137 ★

Temporary Workers on Assignment in Selected Countries

In Belgium in 1987, there was an average of 20,000 temporary workers on assignment per day, while in the Netherlands there was 100,000 per day. In France in 1989, 6,680,000 "assignment" contracts were signed for an average duration of 2.08 weeks. This is equivalent to 309,245 full-time jobs over the course of a year.

Source: "Temporary Work in Western Europe: Threat or Complement to Permanent Employment? *International Labour Review*, 130, n. 3, 1991, p. 295.

★ 138 ★

Temporary Employment

Temporary Workers in the Manufacturing Industry, 1992

According to a survey of companies that supply temporary workers conducted by the National Association of Temporary Services, an average of 348,000 temporary workers were supplied to manufacturers daily late in 1992, up from a daily average of 224,000 early in 1992. This figure was more than one-fifth of the total of temporary workers hired out in 1992 by temporary-help companies.

Source: "Temporary Workers Are on the Increase in Nation's Factories," Louis Uchitelle, *The New York Times*, July 6, 1993, p. D2.

Travel

★ 139 ★

Business Trips in One Household, 1985-1990

Based on monthly telephone survey of 1,500 U.S. adults.

Household characteristics	1980	1985	1989	1990
Average household members on trip	1.5	1.4	1.4	1.4
Average nights per trip	---	3.6	3.2	3.7
Average miles per trip in U.S.	---	1,180	1,090	1,020

Source: U.S. Bureau of the Census. *Statistical Abstract of the United States 1992*. 112th edition. Washington, D.C.: U.S. Department of Commerce, 1992, p. 245. Primary source: U.S. Travel Center, Washington D.C., *National Travel Survey*, annual. **Remarks**: Source also includes: pleasure trips in one household, 1985-1990.

★ 140 ★

Travel

Commuting to Work, 1985

Type of Transportation	Average One-Way Distance Traveled (miles)	Average One-Way Commute Time (minutes)
Car, Truck, or Van	11.36	19.01
Bus or Streetcar	8.68	38.27
Subway or Elevated rail	6.70	27.62
Railroad	24.82	64.33
Taxicab	2.92	12.66
Motorcycle	9.62	16.45
Bicycle	2.06	13.76
Other Vehicle	6.25	23.23
Walked Only	0.19	8.33
Total	10.73	19.66

Source: MVMA Motor Vehicle Facts and Figures '91. Detroit, MI: Motor Vehicle Manufacturing Association, 1991, p. 52. Primary source: U.S. Department of Energy, *Transportation Energy Data Book: Edition 11, 1991*.

Unemployment

★ 141 ★

Length of Unemployment, 1980-1991

Year	Average Duration of Unemployment (weeks)
1980	11.9
1984	18.2
1985	15.6
1987	14.5
1989	11.9
1990	12.1
1991	13.8

Source: U.S. Bureau of the Census. *Statistical Abstract of the United States 1992*. 112th edition. Washington, D.C.: U.S. Department of Commerce, 1992, p. 399. Primary source: U.S. Bureau of Labor Statistics, *Employment and Earnings*, monthly, January issues. **Remarks**: Source also includes: number of workers unemployed, 1980-1991, by age, sex, and race/ethnicity.

★ 142 ★

Unemployment

Number of Unemployed by Length of Unemployment, 1990-1992

Numbers in thousands.

Weeks of Unemployment	Average Number of Unemployed		
	1990	1991	October 1992
Less than 5 weeks	3,269	3,380	3,176
5 to 14 weeks	2,201	2,724	2,642
15 weeks and over	1,504	2,323	3,522
15 to 26 weeks	809	1,225	1,436
27 weeks and over	695	1,098	2,086
Mean duration in weeks	12.1	13.8	19.4
Median duration in weeks	5.4	6.9	9.3

Source: Monthly Labor Review, December 1992, p. 73.

★ 143 ★

Unemployment

Unemployment Rate By the Political Party of the President-in-Office

Refers to presidents since World War II.

Party of the President	Average Unemployment Rate
Democratic	4.6%
Republican	6.2%

Source: "A New View of Who is Better for the Economy," *U.S. News & World Report*, 113, September 14, 1992, p. 21. **Remarks**: Source also includes: the average inflation rate and average growth in corporate profits during the Republican and Democratic years.

★ 144 ★

Unemployment

Corporate Jobs Lost Daily

There are an average of 1,500 corporate jobs lost every day in the United States.

Source: "The Best Places to Live in America," Marguerite Smith and Debra Englander, *Money*, 21, September 1992, p. 117.

★ 145 ★

Unemployment

Unemployment and Unemployment Insurance Benefits

Year	Average Number Unemployed per Month	Average Number Receiving Unemployment Benefits
1989	6,520,000	2,220,000
1990	6,900,000	2,600,000

Source: "Unemployment Insurance and the Recession," William J. Cunningham, U.S. House Committee on Ways and Means, Hearing, February 6, 1991. Washington, DC, Government Printing Office, 1991.

★ 146 ★

Unemployment

Annual Unemployment in Selected Countries, 1982-1992

Country	Average Unemployment, 1982-1992	Average Inflation 1982-1992	Country	Average Unemployment, 1982-1992	Average Inflation, 1982-1992
Australia	7.8%	6.4%	Italy	10.1%	7.4%
Austria	3.5	3.0	Japan	2.5	1.8
Belgium	10.9	3.5	Netherlands	8.7	1.9
Canada	9.8	4.3	New Zealand	6.1	7.9
Denmark	9.4	4.2	Norway	3.7	5.7
Finland	5.8	5.3	Portugal	6.6	14.9
France	9.6	4.4	Spain	18.7	7.6
Germany	6.0	2.2	Sweden	2.4	6.7
Great Britain	9.6	5.5	Switzerland	0.9	3.1
Greece	7.6	18.7	United States	7.1	3.8

Source: "How Instability Hurts Economic Performance," *Business Week*, 22, June 7, 1993, p. 86. Primary source: DRI/McGraw-Hill, OECD, Alberto Alesina, Harvard University. **Remarks**: Source also includes: average misery index and index of political stability for the countries listed above.

Miscellaneous

★ 147 ★

Price of Hotels in the United States, 1987-1991

Year	Average Selling Price of Hotels per Room
1987	$20.7
1988	23.6
1989	20.7
1990	21.5
1991	18.4

Source: "Value of US Hotels Drops," *The Christian Science Monitor*, October 13, 1992, p. 8. Primary source: Hotel & Motel Brokers of America.

★ 148 ★

Miscellaneous

Crayola Crayons Produced Daily

Each year more than two billion Crayola crayons are produced, or an average of five million daily, enough to circle the globe four and a half times.

Source: Binney & Smith, Division of Hallmark Cards, Inc., 1100 Church Lane, P.O. Box 431, Easton, PA 18044-0431, Phone: 1-800-Crayola.

★ 149 ★

Miscellaneous

Cost of Packaging in the United States

Four percent of consumer expenditures on goods in the United States goes to packaging, or an average of $225 per person.

Source: Durning, Alan. *How Much is Enough?* New York: W.W. Norton & Co., 1992.

Chapter 3
CONSUMPTION

Automobile and Fuel Usage

★150★

States With the Most Automobile Miles Driven per Capita

State	Average Miles per Capita
Delaware	10,165
Georgia	10,083
New Mexico	9,950
Vermont	9,657
Virginia	9,638
South Dakota	9,491
Alabama	9,417
Maine	9,277
Oklahoma	9,246
Montana	9,164

Source: The 1992 Information Please Environmental Almanac. Boston, MA: Houghton Mifflin Company, 1992, p. 191. Primary source: U.S. Department of Transportation, *Highway Statistics 1988.*

★151★

Automobile and Fuel Usage

Automobiles With the Best Mileage (MPG)

Make/Model	Average Mileage	
	City	Highway
Geo Metro XFI	53	58
Honda CRX HF	49	52
Geo Prizm	45	50
Suzuki Swift	45	50
GEO Metro LSI	45	50
Honda CRX HF	43	49

[Continued]

★ 151 ★

Automobiles With the Best Mileage (MPG)

[Continued]

Make/Model	Average Mileage	
	City	Highway
Volkswagen Jetta	37	43
Daihatsu Charade	38	42
Ford Festiva	35	42
Honda Civic	33	37
Toyota Tercel	33	37
Ford Escort	31	41
Suburu Justy	33	37
Pontiac Lemans	31	40

Source: Krantz, Les. *The Best and Worst of Everything.* New York: Prentice Hall General Reference, 1991, p. 209. Primary source: Environmental Protection Agency. **Remarks**: Source also includes: automobiles with the worst mileage.

★ 152 ★

Automobile and Fuel Usage

Automobiles With the Worst Mileage (MPG)

Make/Model	Average Mileage	
	City	Highway
Rolls-Royce	10	13
Lamb. DB 132/Diablo	9	14
Ferrari, Testarossa	10	15
Ferrari, F40	12	17
BMW M5	11	20
Jaguar	13	17
Mercedes	14	17
Porsche	13	19

Source: Krantz, Les. *The Best and Worst of Everything.* New York: Prentice Hall General Reference, 1991, p. 209. Primary source: Environmental Protection Agency. **Remarks**: Source also includes: automobiles with the best mileage.

★ 153 ★

Automobile and Fuel Usage

Automobile Fuel Consumption per Year by Miles per Gallon (MPG)

Gas Mileage (MPG) per Automobile Based on 15,000 miles	Average Number of Gallons
20	750
27.5	545
35	429

Source: "Fuel Facts," *Consumer Reports*, 56, April 1991, p. 210.

★ 154 ★

Automobile and Fuel Usage

Fuel Consumption per Vehicle by Type of Vehicle, 1970-1990

Year	Avg. Fuel Consumption per Vehicle (gal.)		
	Cars	Buses	Trucks
1970	760	2,172	1,257
1975	716	2,279	1,217
1980	591	1,926	1,243
1981	576	1,938	1,219
1982	566	1,756	1,191
1983	553	1,507	1,229
1984	536	1,359	1,308
1985	525	1,407	1,302
1986	526	1,463	1,320
1987	514	1,500	1,357
1988	509	1,496	1,345
1989	509	1,518	1,328
1990	505	1,436	1,305

Source: U.S. Bureau of the Census. *Statistical Abstract of the United States 1992*. 112th edition. Washington, D.C.: U.S. Department of Commerce, 1992, p. 614. Primary source: U.S. Federal Highway Administration, *Highway Statistics Summary to 1985*, and *Highway Statistics*, annual. **Remarks**: Source also includes: miles per gallon by type of vehicle, 1970-1990, and fuel consumption for all vehicles, cars only, buses only, and trucks only.

★ 155 ★

Automobile and Fuel Usage

Miles per Gallon by Type of Vehicle, 1970-1990

Year	Avg. Miles Per Gallon		
	Cars	Buses	Trucks
1970	13.52	5.54	7.85
1975	13.52	5.75	8.99
1980	15.46	5.95	9.54
1981	15.94	5.92	9.59
1982	16.65	5.93	9.80
1983	17.14	5.92	9.77
1984	17.83	5.85	9.83
1985	18.20	5.84	9.79
1986	18.27	5.84	9.81
1987	19.20	5.89	9.87
1988	19.95	5.93	10.16
1989	20.40	5.96	10.41
1990	21.00	5.36	10.62

Source: U.S. Bureau of the Census. *Statistical Abstract of the United States 1992*. 112th edition. Washington, D.C.: U.S. Department of Commerce, 1992, p. 614. Primary source: U.S. Federal Highway Administration, *Highway Statistics Summary to 1985*, and *Highway Statistics*, annual. **Remarks**: Source also includes: fuel consumption per vehicle by type of vehicle, 1970-1990, and fuel consumption for all vehicles, cars only, buses only, and trucks only.

★ 156 ★

Automobile and Fuel Usage

Automobile Mileage, 1978-1990

Year	Average MPG
1978	19.9
1979	20.3
1980	23.5
1981	25.1
1982	26.0
1983	25.9
1984	26.3
1985	27.0
1986	27.9
1987	28.1
1988	28.6
1989[1]	28.1
1990[1]	27.8

Source: The Christian Science Monitor, March 18, 1991, p. 6, from the Environmental Protection Agency. *Note:* 1. Preliminary figure.

★ 157 ★

Automobile and Fuel Usage

Motor Fuel Consumption per Day, 1987-1990

Year	Motor fuel consumption (thousand barrels per day)	
	Gasoline	Other fuels
1978	7,555	837
1979	7,291	913
1980	6,820	896
1981	6,726	969
1982	6,679	972
1983	6,731	1,043
1984	6,850	1,127
1985	7,020	1,158
1986	7,229	1,202
1987	7,359	1,242
1988	7,405	1,306
1989	7,437	1,385
1990	7,474	1,408

Source: The 1992 Information Please Environmental Almanac. Boston, MA: Houghton Mifflin Company, 1992, p. 67. Primary source: Energy Information Administration, *Annual Energy Review 1990.* **Remarks**: Source also includes: total number of motor vehicle registrations, and broken down by type: passenger cars, motorcycles, and buses and trucks.

★ 158 ★

Automobile and Fuel Usage

Petroleum Used per Day, 1979-1989

Year	Petroleum Use by Sector in the United States (million barrels per day)			
	Residential and commercial	Industrial	Transportation	Electric utilities
1979	1.73	5.34	10.01	1.44
1981	1.33	4.27	9.49	0.96
1983	1.29	3.85	9.41	0.68
1985	1.35	4.03	9.87	0.48
1987	1.37	4.26	10.49	0.55
1989	1.40	4.26	10.85	0.74

Source: The 1992 Information Please Environmental Almanac. Boston, MA: Houghton Mifflin Company, 1992, p. 68. Primary source: Energy Information Administration, *Annual Energy Review 1989.*

★ 159 ★

Automobile and Fuel Usage

Automobile Tire Usage per Year

Americans wear out an average of 240 million automobile tires each year. That is almost one tire for every man, woman, and child in the United States.

Source: "Tired Out," *Consumer Reports*, 56, April 1991, p. 211.

By Type of Food

★ 160 ★

Per Capita Consumption of Breakfast Cereal in Selected Countries, 1988-1993

Country	Average per Capita Consumption (pounds)			
	1988	1990	1992	1993
Australia	6.94	7.01	6.95	7.00
Canada	6.15	6.08	5.95	6.02
France	1.10	1.75	1.70	1.78
Germany	1.75	1.73	1.76	1.79
Great Britain	6.70	7.42	7.41	7.34
Netherlands	8.06	7.58	7.29	7.32
New Zealand	8.29	8.48	8.44	8.51
South Africa	0.73	0.62	0.63	0.67
South Korea	0.06	0.05	0.06	0.07
Taiwan	0.26	0.42	0.55	0.59
United States	12.63	12.84	11.88	11.90

Source: U.S. Department of Commerce. International Trade Administration. *U.S. Industrial Outlook '92*. Washington, D.C.: U.S. Department of Commerce, 1992, p. 32-17. Primary source: *United Nations Yearbooks of Demographics, Industrial Statistics, and International Trade*. **Remarks**: Source also includes: figures for cereal consumption in kilograms.

★ 161 ★

By Type of Food

Per Capita Consumption of Breakfast Cereal in the United States by Type, 1991

Type	Average per Capita Consumption (ounces)
Corn	54.0
Oat (cold)	27.0
Oatmeal	16.0
Rice (cold)	14.0
Wheat	59.0
Other (cold)	27.0
Other (hot)	7.6

Source: "Cereal for Breakfast," *Detroit Free Press*, July 15, 1992. Primary source: International Trade Administration, Commerce Dept.

★ 162 ★

By Type of Food

Per Capita Consumption of Bread and Related Products, 1988-1993

Product	Average per Capita Consumption (pounds)			
	1988	1990	1992	1993
Breads	48.68	49.87	50.50	51.19
Rolls (all)	21.70	22.81	23.47	23.91
Hamburg & hotdog rolls	13.05	13.30	13.40	13.46
Bagels	2.47	2.99	3.33	3.56
Doughnuts	1.29	1.50	1.63	1.71
Soft cakes	7.21	7.73	8.13	8.34
Pies	1.74	1.69	1.66	1.64

Source: U.S. Department of Commerce. International Trade Administration. *U.S. Industrial Outlook '92*. Washington, D.C.: U.S. Department of Commerce, 1992, p. 32-25. Primary source: *Annual Survey of Manufacturers*, 1988 and 1989. **Remarks**: Source also includes: figures for consumption of bread and related products in kilograms.

★ 163 ★

By Type of Food

Per Capita Consumption of Frozen Bakery Products (except bread), 1988-1993

Product	Average per Capita Consumption (pounds)			
	1988	1990	1992	1993
Frozen pies	1.81	1.56	1.27	1.19
Sweet yeast goods	0.51	0.36	0.24	0.20
Soft cakes	1.55	1.66	1.73	1.77
Doughnuts	0.35	0.35	0.35	0.35
Other	1.65	1.46	1.25	.85
Total	5.87	5.39	4.84	4.36

Source: U.S. Department of Commerce. International Trade Administration. *U.S. Industrial Outlook '92*. Washington, D.C.: U.S. Department of Commerce, 1992, p. 32-27. Primary source: *Annual Survey of Manufacturers*, 1988 and 1989. **Remarks**: Source also includes: figures for consumption of frozen bakery products in kilograms.

★ 164 ★

By Type of Food

Per Capita Consumption of Cookies and Crackers, 1988-1993

Product	Average per Capita Consumption (pounds)			
	1988	1990	1992	1993
Cookies	12.23	12.58	12.19	12.29
Crackers	7.97	8.08	8.15	8.23
Pretzels	1.11	1.21	1.21	1.22

Source: U.S. Department of Commerce. International Trade Administration. *U.S. Industrial Outlook '92*. Washington, D.C.: U.S. Department of Commerce, 1992, p. 32-26. Primary source: *Annual Survey of Manufacturers*, 1988 and 1989. **Remarks**: Source also includes: figures for consumption of cookies and crackers in kilograms.

★ 165 ★

By Type of Food

Per Capita Consumption of Cookies and Crackers in Selected Countries, 1988-1993

Country	Average per Capita Consumption (pounds)			
	1988	1990	1992	1993
Netherlands	51.63	54.76	56.83	58.28
Belgium/Luxembourg	38.34	41.63	43.28	45.23
Great Britain	27.93	28.47	29.06	29.34
New Zealand	21.67	22.50	23.38	23.95
Australia	18.42	18.33	18.28	18.25
Austria	15.62	16.47	17.41	17.90
Canada	16.34	16.39	16.47	16.51

[Continued]

★ 165 ★

Per Capita Consumption of Cookies and Crackers in Selected Countries, 1988-1993

[Continued]

Country	Average per Capita Consumption (pounds)			
	1988	1990	1992	1993
Germany	15.09	15.25	15.53	15.68
United States	12.23	12.58	12.19	12.29

Source: U.S. Department of Commerce. International Trade Administration. *U.S. Industrial Outlook '92*. Washington, D.C.: U.S. Department of Commerce, 1992, p. 32-17. Primary source: *United Nations Yearbooks of Demographics, Industrial Statistics, and International Trade*. 1988 and 1989. **Remarks**: Source also includes: figures for cookies and crackers in kilograms.

★ 166 ★

By Type of Food

Per Capita Consumption of Red Meat and Poultry, 1955, 1992

In 1955, the average per capita consumption of red meat and poultry in the United States was 137 pounds; in 1992, 178 pounds.

For a family of four, this would equate to a half a steer, a whole pig, and 100 chickens per year.

Source: "Fat of the Land: We Can't Keep Eating the Way We Do," Alan Thein Durning, *USA Today* (magazine), November 1992, p. 25.

★ 167 ★

By Type of Food

Per Capita Consumption of Red Meat and Poultry, 1970-1990

Type	Average per Capita Consumption (pounds)					
	1970	1975	1980	1985	1987	1990
Red meat, total	132.0	125.3	126.4	124.9	117.4	112.3
Beef	79.6	83.0	72.1	74.6	69.5	64.0
Veal	2.0	2.8	1.3	1.5	1.3	0.9
Lamb	2.1	1.3	1.0	1.1	1.0	1.1
Pork	48.2	38.2	52.1	47.7	45.6	46.3
Chicken	27.7	27.5	34.3	39.9	43.4	49.3
Turkey	6.4	6.7	8.3	9.6	12.6	14.4

Source: U.S. Bureau of the Census. *Statistical Abstract of the United States 1992*. 112th edition. Washington, D.C.: U.S. Department of Commerce, 1992, p. 132. Primary source: U.S. Department of Agriculture, Economic Research Service, *Food Consumption, Prices, and Expenditures*, annual. **Remarks**: Source also includes: per capita consumption of eggs, dairy products, fats and oils, flour and cereal products, and coloric sweeteners, 1970-1990.

★ 168 ★

By Type of Food

Per Capita Consumption of Eggs, 1970-1990

Year	Average per Capita Consumption					
	1970	1975	1980	1985	1988	1990
Number	309	276	271	255	246	233

Source: U.S. Bureau of the Census. *Statistical Abstract of the United States 1992*. 112th edition. Washington, D.C.: U.S. Department of Commerce, 1992, p. 132. Primary source: U.S. Department of Agriculture, Economic Research Service, *Food Consumption, Prices, and Expenditures*, annual. **Remarks**: Source also includes: per capita consumption of red meat, poultry, dairy products, fats and oils, flour and cereal products, and caloric sweeteners, 1970-1990.

★ 169 ★

By Type of Food

Per Capita Consumption of Dairy Products, 1970-1990

Total includes others not shown separately.

Type	Average per Capita Consumption (pounds)					
	1970	1975	1980	1985	1987	1990
Milk (fluid)	269.1	254.0	237.4	229.7	226.5	221.5
Yogurt	0.8	2.1	2.6	4.1	4.4	4.1
Cream	3.8	3.3	3.4	4.4	4.7	4.6
Sour cream & dip	1.1	1.6	1.8	2.3	2.4	2.5
Cheese (not cottage)	11.4	14.3	17.5	22.5	24.1	24.7
Cottage cheese	5.2	4.7	4.5	4.1	3.9	3.4
Ice cream	17.8	18.6	17.5	18.1	18.4	15.7
Total (milk equivalent)	563.8	539.1	543.3	593.7	601.3	570.6

Source: U.S. Bureau of the Census. *Statistical Abstract of the United States 1992*. 112th edition. Washington, D.C.: U.S. Department of Commerce, 1992, p. 132. Primary source: U.S. Department of Agriculture, Economic Research Service, *Food Consumption, Prices, and Expenditures*, annual. **Remarks**: Source also includes: per capita consumption of red meat, poultry, eggs, fats and oils, flour and cereal products, and coloric sweeteners, 1970-1990.

★ 170 ★

By Type of Food

Per Capita Consumption of Fats and Oils, 1970-1990

Total includes others not shown separately.

Type	Average per Capita Consumption (pounds)					
	1970	1975	1980	1985	1987	1990
Butter	5.4	4.7	4.5	4.9	4.7	4.4
Margarine	10.8	11.0	11.3	10.8	10.5	10.9
Salad & cooking oils	15.4	17.9	21.2	23.5	25.4	24.2
Total (fat content)	52.6	52.6	57.2	64.3	62.9	62.7

Source: U.S. Bureau of the Census. *Statistical Abstract of the United States 1992.* 112th edition. Washington. Washington, D.C.: U.S. Department of Commerce, 1992, p. 132. Primary source: U.S. Department of Agriculture, Economic Research Service, *Food Consumption, Prices, and Expenditures,* annual. **Remarks**: Source also includes: per capita consumption of red meat, poultry, eggs, dairy products, flour and cereal products, and coloric sweetners, 1970-1990.

★ 171 ★

By Type of Food

Per Capita Consumption of Pork in Selected Countries, 1991

Country	Average per Capita Consumption (pounds)
Hungary	147.5
Denmark	144.2
Germany	121.5
Austria	116.4
Poland	116.4
Belgium-Luxembourg	104.7
Spain	103.2
Bulgaria	97.7
Netherlands	92.4
United States	65.5

Source: U.S. Bureau of the Census. *Statistical Abstract of the United States 1992.* 112th edition. Washington, D.C.: U.S. Department of Commerce, 1992, p. 883. Primary source: U.S. Department of Agriculture, Foreign Agriculture Service, *World Poultry Situation,* August 1991, and *World Livestock Situation,* November 1991. **Remarks**: Source also includes: per capita consumption of poultry, beef and veal in selected countries, 1991.

★ 172 ★

By Type of Food

Per Capita Consumption of Poultry in Selected Countries, 1991

Country	Average per Capita Consumption (pounds)
United States	94.8
Israel	81.8
Hong Kong	76.5
Singapore	75.2
Canada	63.5
Saudi Arabia	56.7
Australia	56.2
Taiwan	51.6
Spain	50.7
Hungary	49.2

Source: U.S. Bureau of the Census. *Statistical Abstract of the United States 1992*. 112th edition. Washington, D.C.: U.S. Department of Commerce, 1992, p. 883. Primary source: U.S. Department of Agriculture, Foreign Agriculture Service, *World Poultry Situation*, August 1991, and *World Livestock Situation*, November 1991. **Remarks**: Source also includes: per capita consumption of pork, beef and veal in selected countries, 1991.

★ 173 ★

By Type of Food

Per Capita Consumption of Beef and Veal in Selected Countries, 1991

Country	Average per Capita Consumption (pounds)
Argentina	153.9
Uruguay	122.8
United States	97.0
Australia	84.0
Canada	80.3
New Zealand	76.7
France	66.2
Soviet Union (former)	65.9
Italy	58.9
Switzerland	56.2

Source: U.S. Bureau of the Census. *Statistical Abstract of the United States 1992*. 112th edition. Washington, D.C.: U.S. Department of Commerce, 1992, p. 883. Primary source: U.S. Department of Agriculture, Foreign Agriculture Service, *World Poultry Situation*, August 1991, and *World Livestock Situation*, November 1991. **Remarks**: Source also includes: per capita consumption of pork and poultry in selected countries, 1991.

★ 174 ★

By Type of Food

Per Capita Consumption of Fish, 1981-1990

Year	Average per Capita Consumption (pounds)			
	Fresh & frozen	Canned	Cured	Total
1981	7.8	4.6	0.3	12.7
1983	8.4	4.7	0.3	13.4
1985	9.8	5.0	0.3	15.1
1987	10.7	5.2	0.3	16.2
1988	10.0	4.9	0.3	15.2
1990	10.1	5.1	0.3	15.5

Source: U.S. Department of Commerce. International Trade Administration. *U.S. Industrial Outlook '92*. Washington, D.C.: U.S. Department of Commerce, 1992, p. 32-8. Primary source: *Fisheries of the United States*, 1990, National Marine Fisheries Services, p. 71.

★ 175 ★

By Type of Food

Per Capita Consumption of Cheese in France, 1983

One of the world's biggest consumers of cheese is France, where the average person ate 43.6 pounds of cheese in 1983.

Source: Guinness Book of World Records 1992. New York: Bantam Book, 1992, p. 431.

★ 176 ★

By Type of Food

Per Capita Consumption of Potatoes, 1960, 1989

Year	Average Per Capita Consumption (pounds)		
	Fresh	Frozen	Total
1960	81	7.6	106
1989	50	46.0	126

Source: "Potato Producers Finding Big Market in Frozen Fries," Doug Martinez, *Famine News Service*, October 1991, pp. 2-6. Total consumption includes potatoes chips and products made from dehydrated potatoes.

★ 177 ★

By Type of Food

Per Capita Consumption of Broccoli, 1980, 1990

In 1980, Americans ate only 1.4 pounds of broccoli per person, compared with 3.4 pounds in 1990.

Source: U.S. News & World Report, 113, November 16, 1992, p. 22.

★ 178 ★

By Type of Food

Per Capita Consumption of Fresh Fruits, 1970, 1989

Due to the rise in popularity of fresh fruits, such as bananas, apples, grapes, pears and strawberries, the per capita consumption of fresh fruit rose to an average of 94 pounds in 1989, from 18 pounds in 1970-1974.

Source: "Food and Nutrient Consumption: Food Consumption, 1970-90," Judith Jones Putnam, *Food Review*, 14, July-September 1991, pp. 2, 4-12.

★ 179 ★

By Type of Food

Per Capita Consumption of Canned Food

Each year, Americans eat an average of 150 pounds of canned food, or 11% of food consumed. Because of its popularity, some 1,400 different canned food items are produced each year; 37 billion metal and glass containers are packed in more than one billion cases.

Source: Rukeyser, Louis. *Louis Rukeyser's Business Almanac*. New York: Simon & Shuster, 1991, p. 441.

★ 180 ★

By Type of Food

Per Capita Consumption of Pasta, United States and Italy

Americans consume an average of 19 pounds of pasta per person per year. By some estimates this is twice as much as was consumed 20 years ago. By contrast, Italians eat an average of 60 pounds per person each year. According to a survey by *Consumer Reports*, readers of this magazine eat pasta an average of once a week, usually topped with a red sauce.

Source: "Spaghetti and Spaghetti Sauce," *Consumer Reports*, 57, May 1992, p. 322.

★ 181 ★

By Type of Food

Per Capita Consumption of Peanut Butter

The average American eats 3.36 pounds of peanut butter each year, of which 47% is creamy, 33% is crunchy and 18% is natural.

Source: "Peanut Butter Choice is Creamy," *USA Today*, April 19, 1993, p. 1D. Primary source: The Adult Peanut Butter Lovers Fan Club, Peanut Advisory Board.

★ 182 ★

By Type of Food

Per Capita Consumption of Candy, 1990

Average per capita consumption of candy was 20.5 pounds in 1990 and 20.7 pounds in 1991. Total candy consumption was 20.5 billion pounds in 1990, and 20.7 billion pounds in 1991.

Source: U.S. Department of Commerce. International Trade Administration. *U.S. Industrial Outlook '92.* Washington, D.C.: U.S. Department of Commerce, 1992, p. 32-29.

★ 183 ★

By Type of Food

Female Dormitory Students' Weekly Consumption of Pizza

According to a study done by Domino's Pizza Inc., students' choice of college is based not only on their love of knowledge, but on their love of pizza as well. The study revealed that colleges with less than 10,000 student order 12% more pizza than larger institutions; freshman dorm student order 15% more pizza than upper classmen; and female dorm students order the most pizza. Two or more female roommates order pizza an average of three times per week, while singles order it twice a week.

Source: "You Can Call This Study Cheesy, But it Gives a Slice of Campus Life," *The Wall Street Journal*, September 14, 1992, p. B1.

★ 184 ★

By Type of Food

Americans' Consumption of Pizza

According to the National Association of Pizza Operators, Americans as a whole consume an average of 90 acres of pizza each day. While figures from Domino's Pizza revealed that Americans eat an average of 4,248,489 pounds of green peppers and seven times that much pepperoni as pizza toppings.

Source: "Nutrition: Pizza Has Pizzazz," *Your Health and Safety*, 13, October- November 1991, pp. 18-19.

★ 185 ★

By Type of Food

Jars of Baby Food Consumed per Baby

Year	Average Number of Jars (dozens) per Baby
1972	66
mid-1980's	47-49
1992	53-54

Source: "Baby Food is Growing Up," *American Demographics*, 15, May 1993, p. 20.

★ 186 ★

By Type of Food

Annual Consumption of Junk Food per Person

Type	Average amount
Cookies & cakes	50 pounds
Doughnuts	63 dozen
Refined sugar	100 pounds
Fat & Oil	55 pounds
Soda pop	300 containers
Chewing gum	200 sticks
Ice cream	20 gallons
Potato chips	5 pounds
Candy	10 pounds

Source: Elkort, Martin. *The Secret Life of Food.* Los Angeles, CA: Jeremy P. Tarcher, 1991, p. 125.

By Type of Liquid

★ 187 ★

Per Capita Consumption of Selected Beverages, 1970-1990

Type	Average per Capita Consumption (gallons)					
	1970	1975	1980	1985	1988	1990
Milk	31.2	29.5	27.6	26.7	25.8	25.7
Tea	6.8	7.5	7.3	7.1	7.1	6.9
Coffee	33.4	31.4	26.7	27.4	25.7	26.7
Soft drinks	20.8	22.2	35.0	35.8	41.1	42.5
Citrus juice	3.6	5.2	5.1	5.2	5.4	4.0
Beer	30.6	33.9	36.3	34.7	34.1	34.4
Wine	2.2	2.7	3.2	3.5	3.2	2.9
Distilled spirits	3.0	3.1	3.0	2.5	2.2	2.2

Source: U.S. Bureau of the Census. *Statistical Abstract of the United States 1992.* 112th edition. Washington, D.C.: U.S. Department of Commerce, 1992, p. 133. Primary source: U.S. Department of Agriculture, Economic Research Service, *Food Consumption, Prices, and Expenditures*, annual.

★ 188 ★

By Type of Liquid

Per Capita Consumption of Caffeine

Each day Americans consume an average of 227 milligrams of caffeine, which is the equivalent of two to three cups of coffee. Tea and cola drinks are other sources of caffeine. There are an average of 40 milligrams in a 6 ounce cup of tea and 45 milligrams in a 12 ounce can of cola.

Source: "Headache? You Skipped Your Coffee," Elisabeth Rosenthal, *The New York Times*, October 15, 1992, p. A18.

★ 189 ★

By Type of Liquid

Per Capita Consumption of Sports Drink

According to the *New Age & Sports Beverages in the US: 1992* report from Beverage Marketing Corporation, per capita consumption of sports drink was 1.2 gallons in 1991. Gatorade accounted for more than 80% of this market.

Source: "Is Gatorade a Sleeping Giant," Eric Sfiligoj, *Beverage World*, August 1992, p. 32.

★ 190 ★

By Type of Liquid

Per Capita Consumption of Apple Cider

Americans consumed an average of 13.2 pounds of apples in the form of cider or juice annually during the last two decades. this is more than three times the amount consumed in 1974.

Source: "Cider Stages Modern-Day Comeback," *The Christian Science Monitor*, October 29, 1992, p. 15.

★ 191 ★

By Type of Liquid

Teenage "Binge" Drinkers' Weekly Consumption

According to a report of the Surgeon General, Antonia Coello Novello, 10.6 of the nation's 20.7 million teenagers drink alcoholic beverages, and of these, 8 million drink at least once a week. A half-million of them are "binge drinkers," consuming an average of 15 drinks each week, or five times more than the average for all teenagers. Some drink as many as 33 beers per week.

Source: "10.6 Million Teen-Agers Found to Drink," Marlene Cimons, *Los Angeles Times*, June 7, 1991, p. A29.

★ 192 ★

By Type of Liquid

Weekly Alcohol Consumption of College Students

The findings of a 1990 survey of 58,000 students at 78 college and universities:

1. By number of students—the average number of drinks consumed at institutions with 2,500 or fewer students was 6.6 drinks per week, or twice the number taken by students at institutions of 20,000 or more.
2. By geographic area—the average number of drinks per week per student in the Northeast—7.1; North Central—5.3; South—3.9; West—2.9.

Source: "Alcohol Abuse by Students Is Found Most Severe on Campuses in Northeast," *The Chronicle of Higher Education*, May 26, 1993, A28. Primary source: "Alcohol and Drugs on American College Campuses: Use, Consequences, and Perceptions of the Campus Environment."

★ 193 ★

By Type of Liquid

Per Capita Consumption of Wine in Selected Countries, 1989 (gallons)

Country	Per Capita Consumption	Country	Per Capita Consumption
Ireland	1.1	Hungary	5.3
Finland	1.6	Yugoslavia (former)	5.6
USSR (former)	1.7	Bulgaria	5.8
Norway	1.7	Belgium	6.9
Iceland	1.8	West Germany (former)	6.9
Poland	2.0	Uruguay	7.4
United States	2.1	Romania	7.4
Canada	2.3	Greece	7.9
South Africa	2.4	Chile	9.2
East Germany (former)	3.2	Austria	9.3
United Kingdom	3.4	Spain	10.0
Sweden	3.4	Argentina	13.1
Cyprus	3.6	Switzerland	13.2
Czechoslovakia (former)	3.6	Luxembourg	16.2
New Zealand	3.8	Italy	17.4
Netherlands	4.2	Portugal	20.8
Australia	5.1	France	21.2
Denmark	5.1		

Source: Jobson's Wine Marketing Handbook, 1991. New York: Jobson Publishing, 1991, p. 131.

★ 194 ★

By Type of Liquid

Consumption of Cognac in Hong Kong

According to Jardine Riche Monde, agent for Hennessy, in 1991, the residents of Hong Kong consumed 3.6 million bottles of Cognac, an average of more than half a bottle for every man, woman, and child living there. The French Trade Commission, reports that since the mid-1980s Hong Kong has been the world leader in the per-capita consumption of these distilled spirits.

Source: "Where the V.S.O.P. Goes Down A.S.A.P.," Barbara Basler, *The New York Times*, October 28, 1992, p. C10.

★ 195 ★

By Type of Liquid

Per Capita Consumption of Soft Drinks in China

The Chinese drink an average of just eight soft drinks per year, but only one is a Coca-Cola product.

Source: "Company Faces Hard Task in China, Nick Driver, *Atlanta Constitution*, June 14, 1992, H6.

★ 196 ★

By Type of Liquid

Daily Consumption of Milk in South Korea, 1969, 1989

In South Korea, each person bought an average of one quarter cup of milk each day in 1989, compared with one-quarter teaspoon in 1969.

Source: "Asians Following in Westerners' Fatty Footsteps," Denis D. Gray, *Detroit Free Press*, January 5, 1993, p. 3D.

Disposable Diapers

★ 197 ★

Disposable Diaper Usage per Baby

Every baby who wears disposable diapers uses an average of 4,500 in his or her infancy. In 1991 alone, 17 million disposable diapers were sold in the United States.

Source: "Among the Earth Baby Set, Disposable Diapers are Back," Michael Specter, *The New York Times*, October 23, 1992, p. A1.

Eating Out

★ 198 ★

Weekly Meal Consumption Away from Home by Sex

Men eat out an average of 4.3 times per week; women, 3.6 times per week.

Source: "Business Reports," *American Demographics*, 14, October 1992.

★ 199 ★

Eating Out

Americans' Meal Consumption Away From Home

According to a Survey by the National Restaurant Association, in 1991, the average American ate 3.8 meals per week away from home, down from 3.9 in 1985. By age: 13-24 year-olds consumed an average of 5.1 meals a week away from home, and those 65 and older, 2.4 meals per week. By income: those with an income of $75,000 or more consumed an average of 4.9 meals per week away from home, and those with incomes of less than $15,000, 3.1 meals per week.

Source: "Eating Out, Going Up," Judith Waldrop, *American Demographics*, 14, January 1992, p. 55.

★ 200 ★

Eating Out

Meals Carried Away From Home

According to Harry Balzer of NPD Group, a consumer-research firm in Chicago, 11% of all Americans carry their lunch away from home each day. In 1984 the average American carried 42 meals from home, of which 71% contained a sandwich. In 1992, Americans carried an average of 53 meals away from home, of which only 58% contained a sandwich.

Source: "Firms See a Fat Opportunity in Catering to Americans' Quest for 'Easy' Lunches", Kathleen Deveny, *The Wall Street Journal*, November 3, 1992, p. B1.

Energy Usage

★ 201 ★

Annual Household Energy Usage, 1971-1991

Year	Average Used per Household (kilowatt hrs.)
1971	7,380
1981	8,825
1991	9,738

Source: "Energy Use Surges," *USA Today*, March 15, 1993, p. B1.

★ 202 ★

Energy Usage

Per Capita Energy Consumption by Selected Countries, 1989

Country	Energy Consumed (coal equiv.) Per capita (kilograms) 1989	Country	Energy Consumed (coal equiv.) Per capita (kilograms) 1989	Country	Energy Consumed (coal equiv.) Per capita (kilograms) 1989
World, total	1,975	France	3,915	Poland	4,531
United States	10,124	Greece	3,113	Portugal	1,811
Algeria	936	Hong Kong	1,996	Romania	4,486
Argentina	1,899	Hungary	3,697	Saudi Arabia	6,362
Australia	7,208	India	307	South Africa	2,644
Austria	4,014	Indonesia	311	South Korea	2,195
Bahrain	15,601	Iran	1,650	Soviet Union (former)	6,553
Bangladesh	69	Iraq	1,059	Spain	2,485
Belgium	5,787	Ireland	3,632	Sudan	64
Brazil	791	Israel	3,040	Sweden	5,070
Bulgaria	5,052	Italy	3,813	Switzerland	3,662
Burma	62	Japan	3,995	Syria	978
Canada	10,927	Kuwait	8,825	Tanzania	36
Chile	1,206	Libya	4,286	Thailand	637
China:Mainland	819	Malaysia	1,278	Trinidad & Tobago	6,160
Taiwan	3,074	Mexico	1,736	Tunisia	761
Colombia	773	Morocco	366	Turkey	1,027
Cuba	1,532	Netherlands	6,639	United Arab Emirates	20,361
Czechoslovakia (former)	6,003	New Zealand	5,053	United Kingdom	5,043
Denmark	4,413	Nigeria	207	Venezuela	3,012
East Germany (former)	7,631	North Korea	2,814	Vietnam	132
Ecuador	662	Norway	7,181	West Germany (former)	5,391
Egypt	739	Pakistan	265	Yugoslavia (former)	2,656
Ethiopia	24	Peru	505	Zaire	67
Finland	5,839	Philippines	295	Zambia	198

Source: U.S. Bureau of the Census. *Statistical Abstract of the United States 1992.* 112th edition. Washington, D.C.: U.S. Department of Commerce,1992, pp. 844-845. Primary source: Statistical Office of the United Nations, New York, NY, *Energy Statistics Yearbook*, annual. **Remarks**: Source also includes: total energy consumed, electric energy production, crude petroleum production, and coal production, all by country, 1980 and 1989.

★ 203 ★

Energy Usage

Annual Energy Consumption for Selected Household Electrical Products

Household product	Typical energy consumption (kilowatt hrs/yr.)	Household product	Typical energy consumption (kilowatt hrs/yr.)
Aquarium/terrarium	200-1,000	Freezer (frost free)	1,820
Bottled water dispenser	200-400	Furnace fan	300-1,500
Ceiling fan	10-150	Garbage disposer	20-50
Clock	17-50	Humidifier	20-1,500
Clothes washer	103	Iron	20-150
Coffee maker	20-300	Refrigerator (frost-free)	1,591
Color television	75-1,000	Toaster/toaster oven	25-120
Computer	25-400	VCR	10-70
Dehumidifier	200-1,000	Ventilation fan	2-70
Dishwasher	165	Waterbed heater	500-2,000
Electric mower	5-50	Well pump	200-800
Electric clothes dryer	993	Whole house fan	20-500
Electric range/oven	650	Window fan	5-100

Source: The 1992 Information Please Environmental Almanac. Boston, MA: Houghton Mifflin Company, 1992, p. 64. Primary source: Leo Rainer, Steve Greenberg, and Alan Meier, Lawrence Berkley Laboratory.

★ 204 ★

Energy Usage

Per Capita World Energy Consumption by Selected Regions, 1960-1988

Region	Per Capita (kilograms)			
	1960	1970	1980	1988
World, total	1,302	1,748	1,919	1,959
North America	5,951	7,825	7,480	7,011
United States	8,047	10,811	10,386	10,015
South America	535	749	1,046	1,042
Europe	2,443	3,745	4,433	4,418
Asia	276	386	607	752
Japan	1,025	3,246	3,726	3,921
Soviet Union (former)	2,777	4,132	5,549	6,888
Oceania	2,597	3,675	4,572	5,189

Source: U.S. Bureau of the Census. *Statistical Abstract of the United States 1992.* 112th edition. Washington, D.C.: U.S. Department of Commerce, 1992, p. 570. Primary source: Statistical Office of the United Nations, New York, NY, *Energy Statistics Yearbook*, annual. **Remarks**: Source also includes: total energy consumption and percent distribution.

★ 205 ★

Energy Usage

Per Capita World Energy Consumption by Source of Energy, 1960-1988

Energy source	Per Capita (kilograms)			
	1960	1970	1980	1988
Solid fuels	644	586	591	645
Liquid fuels	433	770	848	794
Natural gas	197	351	412	449
Electricity	28	42	68	95

Source: U.S. Bureau of the Census. *Statistical Abstract of the United States 1992.* 112th edition. Washington, D.C.: U.S. Department of Commerce, 1992, p. 570. Primary source: Statistical Office of the United Nations, New York, NY, *Energy Statistics Yearbook*, annual. **Remarks:** Source also includes: total energy consumption and percent distribution.

★ 206 ★

Energy Usage

Per Capita Energy Consumption by State, 1989

State	Per capita (mil. Btu)	State	Per capita (mil. Btu)	State	Per capita (mil. Btu)
United States	327.6				
Alabama	399.0	Kentucky	396.0	North Dakota	483.0
Alaska	1,075.0	Louisiana	804.0	Ohio	354.0
Arizona	258.0	Maine	279.0	Oklahoma	401.0
Arkansas	343.0	Maryland	269.0	Oregon	323.0
California	245.0	Massachusetts	232.0	Pennsylvania	298.0
Colorado	274.0	Michigan	298.0	Rhode Island	206.0
Connecticut	237.0	Minnesota	307.0	South Carolina	331.0
Delaware	346.0	Mississippi	377.0	South Dakota	297.0
District of Columbia	290.0	Missouri	294.0	Tennessee	357.0
Florida	236.0	Montana	434.0	Texas	570.0
Georgia	315.0	Nebraska	327.0	Utah	318.0
Hawaii	269.0	Nevada	345.0	Vermont	231.0
Idaho	367.0	New Hampshire	226.0	Virginia	302.0
Illinois	303.0	New Jersey	302.0	Washington	396.0
Indiana	446.0	New Mexico	366.0	West Virginia	430.0
Iowa	326.0	New York	198.0	Wisconsin	290.0
Kansas	409.0	North Carolina	295.0	Wyoming	803.0

Source: U.S. Bureau of the Census. *Statistical Abstract of the United States 1992.* 112th edition. Washington, D.C.: U.S. Department of Commerce, 1992, p. 564. Primary source: U.S. Energy Information Administration, *State Data Report, 1960-1989.* **Remarks**: Source also includes: total energy consumed; energy used for residential, commercial, industrial, and transportation; energy used by source—petroleum, gas, coal, hydroelectric power, and nuclear electric power.

★ 207 ★

Energy Usage

Annual Household Energy Usage, 1980-1985

Year	Average Used per Household (mil Btu.)
1980	126
1981	114
1982	114
1983	103
1985	105

Source: U.S. Bureau of the Census. *Statistical Abstract of the United States 1992.* 112th edition. Washington, D.C.: U.S. Department of Commerce, 1992, p. 567. Primary source: U.S. Energy Information Administration, *Residential Energy Consumption Survey*, 1987. **Remarks**: Source also includes: household energy used by region, expenditures for energy per household, and average price for energy by source of energy.

★ 208 ★

Energy Usage

Annual Household Energy Usage by Region, 1987

Region	Average Used per Household (mil Btu.)
Northeast	124
Midwest	123
South	84
West	78

Source: U.S. Bureau of the Census. *Statistical Abstract of the United States 1992.* 112th edition. Washington, D.C.: U.S. Department of Commerce, 1992, p. 567. Primary source: U.S. Energy Information Administration, *Residential Energy Consumption Survey*, 1987. **Remarks**: Source also includes: energy used per household, 1980-1985; expenditures for energy per household; and average price for energy by source of energy.

★ 209 ★

Energy Usage

Household Energy Consumption by Year House Built, 1987

Year house Built	Average Used per Household (mil Btu.)
1939 or earlier	120
1940-1949	104
1950-1959	110
1960-1969	100

[Continued]

★ 209 ★

Household Energy Consumption by Year House Built, 1987

[Continued]

Year house Built	Average Used per Household (mil Btu.)
1970-1974	95
1975-1979	86
1980 or later	71

Source: U.S. Bureau of the Census. *Statistical Abstract of the United States 1992*. 112th edition. Washington, D.C.: U.S. Department of Commerce, 1992, p. 567. Primary source: U.S. Energy Information Administration, *Residential Energy Consumption Survey*, 1987. **Remarks**: Source also includes: energy consumption per residence by family income; total energy consumption; expenditures per household; and expenditures by type of energy.

★ 210 ★

Energy Usage

Household Energy Consumption by Type of House, 1987

Type	Average Used per Household (mil Btu.)
Single-family detached	115
Single-family attached	99
Two to four-unit bldg.	93
Five or more unit bldg.	64
Mobile home	76

Source: U.S. Bureau of the Census. *Statistical Abstract of the United States 1992*. 112th edition. Washington, D.C.: U.S. Department of Commerce, 1992, p. 567. Primary source: U.S. Energy Information Administration, *Residential Energy Consumption Survey*, 1987. **Remarks**: Source also includes: energy consumption per residence by family income; total energy consumption; expenditures per household; and expenditures by type of energy.

★ 211 ★

Energy Usage

Household Energy Consumption by Family Income, 1987

Income	Average Used per Household (mil Btu.)
Less than $5,000	83
$5,000-$9,999	90
$10,000-$14,999	91
$15,000-$19,999	92

[Continued]

★211★

Household Energy Consumption by Family Income, 1987

[Continued]

Income	Average Used per Household (mil Btu.)
$20,000-$34,999	96
$25,000-$34,999	99
$35,000-$49,000	112
$50,000-	129

Source: U.S. Bureau of the Census. *Statistical Abstract of the United States 1992*. 112th edition. Washington, D.C.: U.S. Department of Commerce, 1992, p. 567. Primary source: U.S. Energy Information Administration, *Residential Energy Consumption Survey*, 1987. **Remarks**: Source also includes: energy consumption per residence by year house built; total energy consumption; expenditures per household; and expenditures by type of energy.

★212★

Energy Usage

Energy Consumption of an Electric Iron

The average household clothes iron consumes 1 kilowatt of power for two hours of usage. 146 million irons in household use (from estimating populations of U.S., Canada, Western Europe, and Australia divided by five—the average number of people per household) for two hours per week would equal 292 million kilowatt hours per week, or more than 15 billion per year. This is equivalent to one-third of the total annual net electricity consumption of Switzerland, or one-half that of Greece (1990 estimates).

Source: "How Crumpled Clothes Could Save the World," Vitali Matsarski, *New Scientist*, 136, October 3, 1992, p. 46.

★213★

Energy Usage

Energy Needed to Produce Meat and Poultry

Energy equal to 50 gallons of gasoline is needed to produce the red meat and poultry eaten by the average American every year. Livestock production takes almost one-half of the energy used in U.S. agriculture.

Source: "Fat of the Land" We Can't Keep Eating the Way We Do," Alan Thein Durning, *USA Today* (magazine), November 1992, p. 26.

★214★

Energy Usage

Energy Usage of Developing Countries

In developing countries, the average person uses two barrels of oil per year for fuel; the average European and Japanese, 10 to 30 barrels per year; and the average Americans, 40 barrels.

Source: "Survey: Energy and the Environment: A Power for Good, a Power for Ill," *Economist*, 320, August 31, 1991, p. 44.

Water Usage

★215★

Per Capita Water Usage for Selected Activities

Activity	Average Used (gallons)
Shower	25-30
Brush teeth	2
Shave	10-15
Wash Dishes	20
Dishwasher	10
Flush toilet	5-7
Individual (daily)	168
Residence (year)	107,000

Source: The World Almanac and Book of Facts 1992. New York: World Almanac, 1991, p. 662. Primary source: American Water Works Association.

★216★

Water Usage

Daily per Capita Water Usage, 1960-1985

Year	Average Used Daily (gallons)
1960	339
1965	403
1970	427
1975	451

[Continued]

★216★

Daily per Capita Water Usage, 1960-1985

[Continued]

Year	Average Used Daily (gallons)
1980	440
1985	380

Source: U.S. Bureau of the Census. *Statistical Abstract of the United States 1992.* 112th edition. Washington, D.C.: U.S. Department of Commerce, 1992, p. 212. Primary source: 1940-1960, U.S. Bureau of Domestic Business Development, based principally on committee prints, *Water Resources Activities in the United States,* for the Senate Committee on National Water Resources, U.S. Senate, thereafter U.S. Geological Survey, *Estimated Use of Water in the United States in 1985,* circular 1004, and previous quinquennial issues. **Remarks**: Source also includes: total water used, water used for irrigation, rural, industrial, and steam electric utilities.

★217★

Water Usage

Daily per Capita Water Usage by State, 1985

State	Water Withdrawn Per capita (gal. per day) fresh	State	Water Withdrawn Per capita (gal. per day) fresh	State	Water Withdrawn Per capita (gal. per day) fresh
United States	1,400	Kentucky	1,130	North Dakota	1,690
Alabama	2,140	Louisiana	2,210	Ohio	1,180
Alaska	727	Maine	733	Oklahoma	386
Arizona	1,960	Maryland	321	Oregon	2,450
Arkansas	2,500	Massachusetts	1,070	Pennsylvania	1,210
California	1,420	Michigan	1,270	Rhode Island	152
Colorado	4,190	Minnesota	676	South Carolina	2,040
Connecticut	375	Mississippi	885	South Dakota	956
Delaware	222	Missouri	1,210	Tennessee	1,770
District of Columbia	556	Montana	10,500	Texas	1,230
Florida	554	Nebraska	6,250	Utah	2,540
Georgia	899	Nevada	3,860	Vermont	235
Hawaii	1,100	New Hampshire	688	Virginia	853
Idaho	22,200	New Jersey	307	Washington	1,600
Illinois	1,250	New Mexico	2,320	West Virginia	2,810
Indiana	1,470	New York	508	Wisconsin	1,400
Iowa	960	North Carolina	1,260	Wyoming	12,200
Kansas	2,310				

Source: U.S. Bureau of the Census. *Statistical Abstract of the United States 1992.* 112th edition. Washington, D.C.: U.S. Department of Commerce, 1992, p. 211. Primary source: 1940-1960, U.S. Bureau of Domestic Business Development, based principally on committee prints, *Water Resources Activities in the United States,* for the Senate Committee on National Water Resources, U.S. Senate, thereafter U.S. Geological Survey, *Estimated Use of Water in the United States in 1985,* circular 1004, and previous quinquennial issues. **Remarks**: Source also includes: total water used, water used for irrigation, public supply, industrial, and thermoelectric all by state.

★ 218 ★

Water Usage

Daily Family Use of Water for Selected Activities

Activity	Average Used Daily for Family of Four (gallons)
Utility sink	5
Bathroom sink	8
Dishwashing	15
Laundry	35
Bathing	80
Flush toilet	100

Source: The 1992 Information Please Environmental Almanac. Boston, MA: Houghton Mifflin Company, 1992, p. 102.

★ 219 ★

Water Usage

Daily Water Usage for Flushing Toilets

	Average (billions)
Gallons Flushed in America Daily	4.800
Gallons that would be flushed if toilets were replaced by ultra-low flush models	1.536

Source: Garbage, January/February 1990, p. 12. Primary source: U.S. Department of Housing and Urban Development and *Garbage* staff research.

★ 220 ★

Water Usage

Hot Water Usage

Activity	Averaged Use Hot water (gal/min)
Bathtub	3.6
Laundry	3.3
Shower	2.5
Kitchen sink	1.6
Dishwasher	1.5
Bathroom sink	0.3

Source: Consumer's Research, August 1989, p. 19.

★ 221 ★

Water Usage

Water Usage for Selected Activities

Activity	Average Used (gallons)
Bath	30-40
Shower	5 per minute
Wash clothes	20-30
Cooking	8
Flush toilet	3-4
Water lawn	10 per minute

Source: Cunningham, William P., and Barbara Woodhouse Saigo. *Environmental Science: A Global Concern*. Dubuque, IA: Wm. C. Brown Publishers, 1990, p. 298. **Remarks**: Source also includes: water usage given in liters.

★ 222 ★

Water Usage

Water Usage for Agriculture and Food Processing

Activity	Average Used (gallons)
1 egg	40
1 ear of corn	80
1 loaf of bread	160
1 pound of beef	2,500

Source: Cunningham, William P., and Barbara Woodhouse Saigo. *Environmental Science: A Global Concern*. Dubuque, IA: Wm. C. Brown Publishers, 1990, p. 298. **Remarks**: Source also includes: water usage given in liters.

★ 223 ★

Water Usage

Water Usage for Industrial and Commercial Products

Activity	Average Used (gallons)
1 Sunday paper	280
1 pound steel	32
1 pound synthetic rubber	300
1 pound aluminum	1,000
1 automobile	100,000

Source: Cunningham, William P., and Barbara Woodhouse Saigo. *Environmental Science: A Global Concern*. Dubuque, IA: Wm. C. Brown Publishers, 1990, p. 298. **Remarks**: Source also includes: water usage given in liters.

★ 224 ★

Water Usage

Monthly Water Usage in Selected Cities

City	Average Monthly Household Use (gallons)	Cost per Gallons (cents)
Santa Barbara (dry)	4,488	14.6
Tucson (dusty)	8,600	13.7
New York City (Catskill-dependent)	7,650	12.6
Miami (aquifer-fed)	8,750	10.3-11.9
Chicago (lake-fed)	7,480	8.93

Source: "The Cost of Water," *Garbage*, 3, Mar/Apr 1991, p. 66.

★ 225 ★

Water Usage

Water Needs per Person

The average person needs 2 1/2 quarts of water each day, 1 1/2 quarts are obtained from liquids, and 1 quart from the water content of food, in particular fruits and vegetables. The percentage of water in selected types of food: Bread, 35%; meat, 50-70%; pineapple, 87%, and a ripe tomato, 95%.

Source: Tufty, Barbara. *1001 Questions About Earthquakes, Avalanches, Floods and Other Natural Disasters*. New York: Dover Publications, Inc., 1978, p. 245.

Miscellaneous

★ 226 ★

Protein Needs of a Chicken, Pig, and Cow

According to a 1967 study by the President's Science Advisory Committee, a cow needs an average of 21.4 pounds of protein to make one pound of edible protein; a chicken needs 5.5 pounds; and a pig needs 8.3 pounds.

Source: Garbage 4:34 (January/March 1992).

★ 227 ★

Miscellaneous

Per Capita Timber Consumption, 1970-1988

Product	Unit	1970	1975	1980	1982	1983	1985	1987	1988
All products	Cu. ft.	61.1	54.0	70.8	65.7	72.0	76.2	81.0	79.5
Industrial roundwood	Cu. ft.	58.5	51.4	57.2	51.3	58.2	61.8	68.0	65.9
Lumber	Cu. ft.	29.1	25.8	28.3	24.9	29.5	32.2	36.3	34.2
Plywood and veneer	Cu. ft.	5.7	5.9	5.5	5.2	6.2	6.5	7.2	6.9
Pulp products	Cu. ft.	21.4	17.9	21.4	19.3	20.5	21.1	22.4	22.7
Other products	Cu. ft.	2.1	1.8	1.8	1.9	1.9	1.9	2.0	2.1
Fuelwood	Cu. ft.	2.6	2.6	13.6	14.4	13.8	14.4	12.9	13.6
Lumber	Bd. ft.	193	171	188	161	190	207	233	221
Plywood and veneer	Bd. ft.	34	32	30	28	34	35	39	37
Pulp products	Cords (128 cu. ft)	0.3	0.2	0.3	0.2	0.3	0.3	0.3	0.3

Source: U.S. Bureau of the Census. *Statistical Abstract of the United States 1992*. 112th edition. Washington, D.C.:U.S. Department of Commerce, 1992, p. 673. Primary source: U.S. Forest Service, U.S. *Timber Production, Trade, Consumption, and Price Statistics, 1960-1988*, annual.

★ 228 ★

Miscellaneous

Animal Feed Needed per Pound of Food Product Generated

Animal	Average Amount of Feed (pounds)	Average Amount Generated (pounds)
Pigs	7 of corn and soy	1 (meat)
Cattle	4.8 of grain and soy	1 (meat)
Chickens	2.8 of feed	1 (meat)
Dairy cows	3 of grain and soy	1 (cheese)

Source: "Fat of the Land" We Can't Keep Eating the Way We Do," Alan Thein Durning, *USA Today* (magazine), November 1992, p. 26.

★ 229 ★

Miscellaneous

Cropland Area Worldwide per Person

On a steady decline since mid-century, the average cropland area available worldwide in 1990 was 0.69 acre per person; by 2025 it is expected to drop to 0.42 acre per person.

Source: The 1992 Information Please Environmental Almanac. Boston, MA: Houghton Mifflin Company, 1992, p. 13.

★ 230 ★

Miscellaneous

Hong Kong Residents' Annual Consumption of American Goods

Residents of Hong Kong consume an average of $1,100 worth of American goods each year.

Source: "It's Time to Clarify US-Hong Kong Relations," Perry Bechky, *The Christian Science Monitor*, April 17, 1992, p. 19.

★ 231 ★

Miscellaneous

Woman's Daily Nutrients

The average woman's daily nutrient intake recommended by government guidelines is: 2200 calories, 20 grams of fiber, 250 grams of complex carbohydrates, 55 grams of sugar, and 2400 miligrams of sodium. The average man's recommended intakes are higher; children's are lower.

Source: "How the Nutrients Stack Up," *Consumer Reports*, 57, November 1992. p. 692.

★ 232 ★

Miscellaneous

Americans' Calorie Consumption from Fat

	Calorie Consumption	
	Total Fat	Saturated Fat
Average intake	36%	12%
Government guidelines	30%	10%
"Ideal"	25%	7%

Source: "Are You Eating Right," *Consumer Reports*, 57, October 1992, pp. 644-651.

★ 233 ★

Miscellaneous

Calories Consumed in a Kiss

According to Italian nutritionists in an article in the *Environmental Nutrition* newsletter, the energy used in a kiss averages between six and 12 calories. At an average of nine calories per kiss, three kisses per day for 365 days would burn up 9,855 calories.

Source: Bluman, Allan. *Elementary Statistics*. Dubuque, IA: William C. Brown Publishers, 1992, p. 81.

Chapter 4
CONTEMPORARY LIFE AND SOCIETY

America's Affluent

★ 234 ★

Net Worth of America's Wealthiest Families, 1983, 1989

In 1983, the average net worth of America's wealthy elite (one-half of one percent or half a million families), was $5.86 million per family, or 24.1% of total household wealth. In 1989, this average net worth increased to $10.3 million and 29.1% of household wealth.

Source: "The Supperrich Have Been on a Roll," *Business Week*, 22, May 18, 1992, p. 22. Primary source: Arthur B. Kennickell, Federal Reserve Board; and R. Louise Woodbaum, Internal Revenue Service.

★ 235 ★

America's Affluent

America's Affluent

Affluent households: those with 1989 incomes of $50,000 or more. The growing affluent: the income of this group increased between 1980 and 1990. The truly affluent: have incomes of $100,000 or more.

Households	Average Percent of Population Who Are Affluent
Affluent Blacks	13.2%
Affluent Asians	35.0
Affluent Hispanics	16.1
Young Affluence	17.0
Middle-Aged Affluence	36.0
Seniors' Affluence	17.8

[Continued]

★ 235 ★

America's Affluent
[Continued]

Households	Average Percent of Population Who Are Affluent
Growing Affluent	4.8
Truly Affluent	4.4

Source: "American Affluence," Judith Waldrop and Linda Jacobsen, *American Demographics*, 14, December 1992, pp. 29-42. **Remarks**: Source also includes: Counties in the United States with the highest percentages of affluence in the above-named categories, giving the number of households and the percent who are affluent in each country.

Conceptions and Abortions

★ 236 ★

Conceptions and Abortions per Day

According to the World Health Organization, there are more than 100 million acts of sexual intercourse every day, resulting in 350,000 cases of sexually transmitted diseases. An average of 910,000 conceptions occur each day, of which 150,000 end in abortion. Due to complications during pregnancy or childbirth, one woman worldwide dies each minute.

Source: "U.N. Agency on Sex: Pitfalls and Promise," *The New York Times*, June 25, 1992, p. A12.

★ 237 ★

Conceptions and Abortions

Abortions per Year

During the 1980s, an average of 1.6 million abortions took place each year. Even before the broad right to obtain an abortion was made legal in 1973 with *Roe v. Wade*, the number of legal abortions per year was near 600,000. Even more than a century ago, several hundred thousand women were aborting fetuses each year. Twenty-five percent of all abortions occur among teenagers.

Source: "Sound and Fury Signifying Little," Ted Gest et al., *U.S. News & World Report*, 113, July 13, 1992, pp. 3238.

Housing Size

★ 238 ★

Size of Single Family Homes Vs. Multifamily Apartments

The average square footage of single-family homes in 1986 was about 1800; the average square footage for apartments in multifamily units was 900.

Source: Council on Environmental Quality. *Environmental Trends*. Washington, DC; Council on Environmental Quality, 1989, p. 126.

★ 239 ★

Housing Size

Size of Japanese Homes

The average size of a Japanese home is 650 square feet, or less than one-half the size of the average American home.

Source: "Small-Scale Success Story," Hal Foster, *Los Angeles Times*, July 15, 1991, p. D1.

★ 240 ★

Housing Size

Newly-Built American House

According to a federal government survey, the average newly-built American home has about 1,900 square feet of floor space (up six percent in five years), three bedrooms, central air conditioning, and a two-car garage. To heat the homes: 65% used natural gas, 29% electricity; 4%, fuel oil; 2%, other types or none. The exteriors were: 33%, wood; 23%, vinyl; 21%, brick; 14%, stucco; 5%, aluminum; 4%, cinder block, stone and other material.

Source: "In Typical New Home, More Room to Roam," *International Herald Tribune*, February, 27-28, 1993, p. 5.

Mail

★ 241 ★

Correspondence Received by the White House

In 1993, the White House received a daily average of 75,850 pieces of correspondence for President Clinton, of which there were 250 telegrams, 600 computer messages, 25,000 letters, and 50,000 phone calls.

Source: "Public Clamors for Clinton's Ear," *USA Today*, April 2-4, 1993, p. 1A. Primary source: White House, Western Union.

★ 242 ★

Mail

Mail-Order Catalogs Received per Household, 1981-1991

In 1981, an average of 59 mail-order catalogs were received per household; in 1991, the average number increased to 142. Fifty-five percent of adults shop by mail or phone.

Source: "Database," *U.S. News & World Report*, 113, December 21, 1992, p. 26.

★ 243 ★

Mail

Christmas Cards Sent per Family, 1991

In 1991, an average of 44 Christmas cards were sent per family, of which 41% were sent to non-relatives.

Source: "Database," *U.S. News & World Report*, 113, December 21, 1992, p. 26.

★ 244 ★

Mail

Holiday Cards Sent per Day in the United States

According to the United States Postal Service, from Thanksgiving Day to December 25, Americans send out more than 2.3 billion holiday cards, or an average of 78 million cards per day.

Source: "Rituals to Match the Ways People Live Now," Catherine S. Manegold, *The New York Times*, December 24, 1992, p. C1.

★ 245 ★

Mail

Mail Sent in the United States and Switzerland

In 1988, each person in Switzerland sent an average of 655 letters and packages; in the United States in that same year, each person mailed 645 letters and packages.

Source: Guinness Book of Records 1992. New York: Bantam Book, 1992, p. 421.

Phones and Phone Calls

★ 246 ★

Phone Calls to the Capitol Switchboard

The Capitol switchboard, which handles phone calls for 535 senators and representatives and five congressional delegates, receives 84,000 phone calls on an average day.

Source: "Americans Burn up Capital's Phone Lines," Christopher Scanlan, *Detroit Free Press*, January 28, 1993, p. 9A.

★ 247 ★

Phones and Phone Calls

Waiting Period for a Telephone in Russia and Albania

According to a report by the Geneva-based International Telecommunications Union, the average waiting period to start telephone service in Russia is 32 years. Conditions are far worse in Albania, where it would take 100 years to install telephones just for the people who have already requested service.

Source: "Thirty-Two Years Is the Average Wait for Phone in Russia," Richard L. Hudson, *The Wall Street Journal*, October 2, 1992, p. A5B. **Remarks**: Source also includes: number of phone lines in selected countries.

★ 248 ★

Phones and Phone Calls

Phones in Selected Countries

According to a report by the Geneva-based International Telecommunications Union, for each 100 inhabitants, Russia has 14.3 telephones; Turkey has 12.11; and Latvia has close to the level of Ireland and Portugal, or 24.04. The average for Western Europe is 43.

Source: "Thirty-Two Years Is the Average Wait for Phone in Russia," Richard L. Hudson, *The Wall Street Journal*, October 2, 1992, p. A5B. **Remarks**: Source also includes: average wait for phone service in Russia and Albania.

★ 249 ★

Phones and Phone Calls

Phones in China

In China there is an average of one phone for every 133 people, but in the countryside this density falls to only one phone for every 500 residents. Compare this to the developing world's average density of three telephones for every 100 citizens.

Source: "China's Economic Dilemmas in the 1990s," Ken Zita, U.S. Congress. Joint Economic Committee. Study Papers, April 1991. Washington, D.C.: Government Printing Office, 1991.

Profiles

★ 250 ★

Profile of a Peace Corps Volunteer

Thirty years ago when the Peace Corps was founded, the average volunteer was a 22-year-old college graduate with a liberal arts degree sent to teach or help improve agriculture in Third World countries. By 1990, this profile had changed to a 32-year-old man or woman with a master's degree in business or in technical skills for solving environmental problems, and likely to be found in an urban setting in a developing country or a country once behind the Iron Curtain.

Source: "Peace Corps Volunteers Getting Down to Business," Carol Jouzaitis, *Chicago Tribune*, May 27, 1991, sec. 2, p. 7.

★ 251 ★

Profiles

Profile of a Gardener

Characteristics	Dabblers	Decorators	Cultivators	Masters	Total
Average age	45	49	50	53	47
Average number of years gardening	13.2	14.6	15.0	16.0	13.9
Average annual gardening expenditures	$128	$237	$140	$271	$156
Percent female	59.7%	59.0%	47.0%	54.0%	56.1%
Percent in large urban areas	50.6%	51.3%	37.1%	45.0%	48.1%
1991 median household income	$36,896	$37,159	$27,939	$28,287	$35,081

Source: "Gardening Tips," *American Demographics*, 14, April 1993, p. 46. Primary source: Organic Gardening/1992 Gardening in America Survey conducted by NFO Research.

★ 252 ★

Profiles

Profile of a Colorado Tourist, 1991

Characteristics	Summer Vacationer	Winter Vacationer
Head of household age	37	37
Married	68%	55%
Never married	17%	30%
Have children	45%	36%
White collar worker	44.9%	81.6%
College graduate	50.2%	73.5%
Median income	$41,500	$57,700
Type of travel	66% by car	66% by air
Median distance traveled	797 miles	891 miles
Average stay	5.3 days	5.8 days
Daily amount spent per person	$44	$83

Source: "When the Season Changes, So Do the Tourists," *The New York Times*, April 13, 1993, p. A12. Primary source: Colorado Tourism Board.

★ 253 ★

Profiles

Profile of a Sewing Enthusiast

According to a *Sew News* readership survey, the average sewing enthusiast is 25-45 years-old, is a college-educated female, has an annual household income of $35,000, and often has a child at home.

Source: "New-Generation Sewing Machines," Anita M. Samuels, *Consumer's Digest*, September/October 1992, p. 57.

★ 254 ★

Profiles

Profile of a Motorcyclist

According to the Motorcycle Industry Council, the average motorcyclist is a male, aged 32.5 years (up from the 1980s' average of 27.5 years), is married, and has some college education. His household income is $33,200, 12% higher than the national average.

Source: "Bikers Ride into Middle Age," *American Demographics*, 13, December 1991, p. 15.

★ 255 ★

Profiles

Profile of a Franchisee

According to a survey by Francorp, a consulting firm in Olympia Fields, Illinois, the average franchisee is a white male, 40 years-old, has some college education, and enjoys an income of $50,000-$70,000.

Source: "Small Stores with Big Names," Dan Frost and Susan Mitchell, *American Demographics*, 14, November 1992, pp. 52-57.

★ 256 ★

Profiles

Profile of a CEO

The average CEO is a white male protestant who has been with his company since 1970, and at age 56, has been a CEO for eight years. He is a college graduate, but not usually a varsity type. In college he mainly studied economics or other business-related subjects, such as business administration and engineering. Over one-half of them received graduate degrees. The highest percentage, or 13%, came from New York state; then Illinois, Pennsylvania, and Ohio. Most graduated from Yale, Princeton, Harvard, University of Michigan, University of Pennsylvania, Northwestern, Stanford, and Cornell, in that order. In 1991, the average CEO made $878,000 in salary and bonuses. Their most popular hobbies are reading, collecting, and gardening.

Source: "Portrait of the Boss," Laurel Touby and Fred F. Jespersen, *Business Week*, 22, October 12, 1992, pp. 108-109.

★ 257 ★

Profiles

Profile of a Condominium Timesharer

According to D.K. Shifflet & Associates Ltd. in McLean, Virginia, the average timesharer is 47 years-old. Twenty-three percent are retired, forty-five percent are college graduates, forty-four percent have children living at home, and their median household income is $51,000.

[Continued]

★ 257 ★

Profile of a Condominium Timesharer
[Continued]

According to Plog Research of Los Angeles, timeshare owners' spend an average of 26 nights in leisure travel per year over 5.6 trips. They spend an average of $1,029 on their timeshare vacations (which tend to be longer and involve more family members), and $719 on other trips.

Source: "Timesharing Sheds a Shady Image," W. Lynn Seldon, Jr., *American Demographics*, 14, May 1993, pp. 14-15.

★ 258 ★

Profiles

Profile of a Casino Visitor

According to a survey of the general public as well as casino gamblers done for Harrah's Casino Hotels by Home Testing Institute, Yankelovich Clancy Shulman, and Communication Development Co., the average visitor to a casino (who is just as likely to be a man as a woman), is between the ages of 40 and 64, lives in a one- or two-member household, has a median household income of $35,000 a year (higher than the average American), and is slightly better educated than the average American.

Source: "Survey Reveals a Recent Jump in Casino Goers," *The Wall Street Journal*, January 8, 1993, p. B1.

★ 259 ★

Profiles

Profile of a Business Traveler

The average business traveler is a 39-year-old male, married, has a median household income of $40,000, and does managerial or professional work.

Source: "How Business Travelers Are Changing,"Deborah Schoeder and Judith Waldrop, *American Demographics*, 14, November 1992, p. 10.

★ 260 ★

Profiles

Profile of a Senior Citizen

Characteristics of the average senior citizen, i.e. those over the age of 65:

Two-thirds of them are women who have a median age of 72 years.

One in 25 Americans were over age 65 when she was born; by 1990, this number had changed to one in eight.

[Continued]

★ 260 ★

Profile of a Senior Citizen

[Continued]

She lives with her 76-year-old husband who will die of heart disease in three years.

She and her husband are high school graduates who receive $566 a month in Social Security, and have a total household income of $20,000.

They own their own home, have no debts, and more than half of their income is spent on medical care, food, and housing.

She will die of heart disease at age 84.

Source: "Meet the Typical Senior," *American Demographics*, 13, April 1990, p. 17. Primary source: *Fact Book on Aging*.

★ 261 ★

Profiles

Profile of a Gay Person

The median age of the average gay man is 37 years old; the average gay woman is 35. They have 15.7 years of education, compared with 12.7 for the overall adult population in the United States. Thirty-two percent of gays have attended some college, compared with only twenty-one percent overall. The median household income of gay men is $42,689, and $36,072 for gay women, an amount considerably higher than the overall median income in the United States.

Source: "Gay Consumers Come Out Spending," Joe Schwartz, *American Demographics*, 14, April 1992, p. 11.

★ 262 ★

Profiles

Characteristics of Readers of Women's Magazines by Selected Magazines

Title	Median Age	Median Household Income
Allure	35.4	$57,100
Better Homes & Gardens	42.5	34,968
Cosmopolitan	31.2	34,853
Elle	29.2	45,522
Family Circle	43.3	33,188
Glamour	29.6	36,461
Good Housekeeping	42.3	33,256
Harper's Bazaar	35.6	37,775

[Continued]

★ 262 ★

Characteristics of Readers of Women's Magazines by Selected Magazines

[Continued]

Title	Median Age	Median Household Income
Ladies Home Journal	43.6	33,120
Lear's	43.1	67,200
Mademoiselle	29.1	36,682
McCalls	43.1	31,584
Redbook	39.9	34,026
Vogue	31.6	38,420
Woman's Day	43.4	32,524
ALL	43.0	33,000

Source: "The Magazines and the Readers," *American Demographics*, 13, December 1991, pp. 44-46. Primary source: Mediamark Research, Inc. **Remarks**: Source also includes: circulation figures of the magazines.

Miscellaneous

★ 263 ★

Birthdays in the United States

On an average day in the United States, some three million birthday gifts are purchased for the 673,693 people who are celebrating their birthday. An average of one in 68 individuals turn 18 every day; one in 61 turn 40.

Source: Krantz, Les. *What the Odds Are: A to Z Odds on Everything You Hoped or Feared Could Happen.* New York: Prentice Hall General Reference, 1992, p. 41.

★ 264 ★

Miscellaneous

Sunday Church Attendance at Selected Protestant Churches

Church/city	Average Attendance
First Baptist Church Hammond, IN	20,000
Willow Creek Church S. Barrington, IL	14,605
Calvary Chapel Costa Mesa, CA	12,000
Thomas Road Baptist Church Lynchburg, VA	11,000
First Assembly of God Phoenix, AZ	10,000
North Phoenix Baptist Church Phoenix, AZ	9,500
Chapel in University Park Akron, OH	8,700
Second Baptist Church Sun Valley, CA	8,000
Mt. Paran Church of God Atlanta, GA	7,850

Source: Krantz, Les. *The Best and Worst of Everything.* New York: Prentice Hall General Reference, 1991, p. 313.

★ 265 ★

Miscellaneous

Sexually Active Teens

Some statistics on teen sexual behavior in the United States

Each year an average of:
-**1 out of 10** teenage women in the U.S. become pregnant
- About **30,000** pregnancies occur to girls under the age of 15

Each day an average of:
- More than **3,000** girls become pregnant
-**1,300** babies are born to teen-age girls
-**500** girls have induced abortions
-**26** 13-and 14-year-old girls have their first child
-**13** 16-year-olds have their second child

Sexual activity:
-**1 of 5** 15-year-old girls are having intercourse
-**5 million** teenage females are sexually active
-**one third** of sexually active girls report using contraceptives consistently
-**one half** of unintended teen pregnancies occur during the first six months after the initiation of sexual activity

Source: Hutzel Hospital, "Statistics," Detroit, Michigan.

★ 266 ★

Miscellaneous

Fire Losses, 1960-1990

1960 excludes Alaska and Hawaii. Includes allowance for uninsured and unreported losses but excludes losses to government property and forests. Represents incurred losses.

Year	Per capita	Year	Per capita	Year	Per capita
1960	$6.19	1973	$12.49	1982	$25.41
1965	7.51	1974	14.95	1983	26.98
1966	7.64	1975	14.81	1984	32.15
1967	8.66	1976	16.35	1985	32.47
1968	9.20	1977	17.13	1986	35.21
1969	9.71	1978	18.05	1987	35.44
1970	11.41	1979	21.60	1988	39.11
1971	11.20	1980	24.55	1989	38.33
1972	11.01	1981	24.50	1990	34.61

Source: U.S. Bureau of the Census. *Statistical Abstract of the United States 1992*. 112th edition. Washington, D.C.: U.S. Department of Commerce, 1992, p. 201. Primary source: 1960, National Board of Fire Underwriters, New York, NY, *Report of the Committee on Statistics and Origin of Losses*; thereafter, Insurance Information Institute, New York, NY, *Insurance Facts*, annual. **Remarks**: Source also includes: total amount of fire losses in millions of dollars.

★ 267 ★

Miscellaneous

Square Feet Painted per Gallon of Paint

An ordinary roller covers an average of 650 square feet with one gallon of paint, although manufacturers recommend 400 to 450 square feet for every gallon to obtain a "one coat" coverage.

Source: "New Paint Roller Lays It on Thick," *Consumer Reports*, 57, October 1992, p. 626.

★ 268 ★

Miscellaneous

Radios per Household

The average American household uses six radios, which include portable and car models.

Source: "Small, Portable and Digital Are Dazzling, and a Bit Dizzying at Electronics Show," *The Wall Street Journal*, January 8, 1993, p. B1.

★ 269 ★

Miscellaneous

Typical Woman's Handbag

According to a survey of women's handbags conducted by *Glamour* magazine and gathered from 143 women in 13 cities around the country: half of those surveyed carried a shoulder bag which weighed an average of 3 to 5 pounds, 29% weighed under three pounds, 16% over five pounds. The average handbag contained: keys, wallet, checkbook, calendar, glasses, lipstick, pen, gum or mints. Three quarters of the commuters carried a second bag, usually a tote, attache or briefcase containing work-related materials and personal items. The second bag was lighter for those taking public transportation, or an average of eight pounds, but as high as 20 pounds for those driving.

Source: "How Heavy Is Your Bag and What's in It?" *Glamour* December 1992, p. 126.

★ 270 ★

Miscellaneous

Children in Trouble

In the United States, on average:

A child drops out of school	every 8 minutes
A child runs away from home	every 26 seconds
A teenager has a baby	every 67 seconds

[Continued]

★ 270 ★

Children in Trouble

[Continued]

A child is arrested for drug offenses	every 7 minutes
100,000 children are homeless	every day
135,000 children bring guns to school	every school day

Source: "Kids First!" Marian Wright Edelman, *Mother Jones*, 16, May-June 1991, p. 31.

Chapter 5
CRIME AND LAW ENFORCEMENT

Type of Crime

★ 271 ★

Crime Clock

Type of Crime	Average frequency
Property crime	every 2 seconds
Larceny-theft	every 4 seconds
Burglary	every 10 seconds
Violent crime	every 17 seconds
Motor vehicle theft	every 19 seconds
Aggravated assault	every 29 seconds
Robbery	every 46 seconds
Forcible Rape	every 5 minutes
Murder	every 21 minutes

Source: U.S. Department of Justice. Federal Bureau of Investigation. *Crime in the United States 1991*. Washington D.C.: U.S. Department of Justice, 1992, p. 4.

★ 272 ★

Type of Crime

Violence Against Women

The following statistics were prepared for a congressional hearing on legislation to reduce the growing problem of violent crime against women: an average of 16 women confront rapists every hour, one is raped every six minutes; an average of three to four million women are battered each year, one every 18 seconds; each year 1,000,000 women seek medical assistance for injuries caused by battering. Other statistics reported in this study: since 1974, the rate of assault against women between the ages of 20 and 24 has risen almost 50%; three out of four women will be victims of at least one violent crime during their lifetimes; a woman is 10 times more likely to be raped than to die in a car crash; the crime rate against women in the United States is significantly higher than in other countries.

Source: U.S. Senate, Committee on the Judiciary, *One Hundred First Congress Second Session on Legislation to Reduce the Growing Problem of Violent Crime Against Women*, June 20, 1990, Part 1, serial No. J-101-80, p. 12.

★ 273 ★

Type of Crime

Robbery Rates and Characteristics of the Victims

Personal or household characteristics	Average annual rate of robbery per 1,000 persons	
	White	Black
Sex		
Male	7.2	18.5
Female	3.8	8.5
Age		
12-15	9.0	16.5
16-19	9.5	18.3
20-24	10.5	19.9
25-34	6.4	14.8
35-49	3.8	9.1
50-64	2.8	8.3
65 or older	2.1	6.3
Marital status[1]		
Never married	9.9	18.5
Divorced or separated	12.3	16.9
Widowed	3.1	6.2
Married	2.7	6.9
Location of residence		
Metropolitan area		
Central city	10.1	18.7
Suburb	5.0	7.9
Nonmetropolitan area	2.6	3.8

[Continued]

★ 273 ★

Robbery Rates and Characteristics of the Victims

[Continued]

Personal or household characteristics	Average annual rate of robbery per 1,000 persons	
	White	Black
Family income[2]		
Less than $7,500	9.5	17.1
$7,500-14,999	6.3	13.0
$15,000-24,999	4.6	10.4
$25,000-49,999	4.1	9.1
$50,000 or more	4.1	6.5

Source: "Robbery Rates, by Selected Personal or Household Characteristics and Race of Victim, 1979-86," *Black Victims*, p. 4. Primary source: U.S. Department of Justice, Bureau of Justice Statistics Special Report, April, 1990. *Notes:* Victimization rates are average annual rates per 1,000 persons. 1. The category "marital status not ascertained" is not displayed. 2. The category "family income not ascertained" is not displayed.

★ 274 ★

Type of Crime

Missing Children

In the United States from 1984 to 1991, an average of ten children have disappeared every day. Of the 26,000 children that have been abducted, 17,481 have been found alive, 225 dead, and 9,039 are still missing.

Source: "Petitions and Memorial," *Congressional Record*, November 15, 1991, pS16845.

★ 275 ★

Type of Crime

Domestic Violence

Every 15 seconds in the United States, on average, a woman is physically assaulted in her own home. Every day, four women are beaten to death. One out of every five women in the emergency rooms of hospitals have been beaten by their husbands.

Source: "Viewpoint: No More Rules of Thumb," *Glamour*, 89, October 1991, p. 108.

★ 276 ★

Type of Crime

Illegal Aliens

In 1991, Border Patrol agents arrested an average of 1,800 illegal aliens each day in the San Diego area, the busiest illegal border-crossing zone in the United States. The Immigration and Naturalization Service (INS) estimates that for every illegal alien arrested, 2-3 escape across the border.

Source: "Growth in Illegal Immigration Causes Stir Over Sanctions," Scott Armstrong, *The Christian Science Monitor*, April 23, 1991, p. 1.

★ 277 ★

Type of Crime

Shopping Carts Stolen From Supermarkets

In 1988, an average of 11 shopping carts were stolen per supermarket.

Source: The Food Market Industry Speaks. The Food Marketing Institute, Washington, DC, 1989.

★ 278 ★

Type of Crime

Handguns and Crime

According to Department of Justice figures, there were 639,000 violent crimes committed with handguns each year from 1979-1987. This includes 9,200 murders, 15,000 maimings, and 12,100 rapes. An average of ten American children under the age of 18 are killed with handguns every day; a teenager commits suicide with a handgun every three hours. In 1988, a below-average year for deaths due to handguns, 8,915 were murdered in the United States, compared with seven in Great Britain.

Source: "Tom Foley, Gun Nut," *New Republic*, 204, January 28, 1991, p. 7.

Cost of Crime

★ 279 ★

Murder

In 1990, 20,930 murders known to police were committed. The average loss in dollars per offense: $108.

Source: U.S. Department of Justice. Bureau of Justice Statistics. *Sourcebook of Criminal Justice Statistics 1991*. Washington, D.C.: U.S. Department of Justice, 1992, p. 396. Primary source: U.S. Department of Justice. Federal Bureau of Investigation. *Crime in the United States 1991*. Washington, D.C.: U.S. Department of Justice, 1992, p. 162. **Remarks**: Also included in source: number of offenses, 1990, percentage change over 1989, and average loss in dollars for: forcible rape, robbery, burglary, larceny-theft, and motor vehicle theft.

★ 280 ★

Cost of Crime

Forcible Rape

In 1990, 94,644 cases of forcible rape known to police were committed. The average loss in dollars per offense: $29.

Source: U.S. Department of Justice. Bureau of Justice Statistics. *Sourcebook of Criminal Justice Statistics 1991*. Washington, D.C.: U.S. Department of Justice, 1992, p. 396. Primary source: U.S. Department of Justice. Federal Bureau of Investigation. *Crime in the United States 1991*. Washington, D.C.: U.S. Department of Justice, 1992, p. 162. **Remarks**: Also included in source: number of offenses, 1990, percentage change over 1989, and average loss in dollars for: murder, robbery, burglary, larceny-theft, and motor vehicle theft.

★ 281 ★

Cost of Crime

Burglary

Offense	Number of Offenses 1990	Average Loss (dollars)
Offenses known to police		
Burglary	2,793,477	$1,133
Residence	1,847,493	1,143
Night	586,057	943
Day	807,126	1,224
Unknown	454,310	1,263
Non-residence	945,954	1,110
Night	445,491	982
Day	238,822	1,072
Unknown	261,641	1,362

Source: U.S. Department of Justice. Bureau of Justice Statistics. *Sourcebook of Criminal Justice Statistics 1991*. Washington, D.C.: U.S. Department of Justice, 1992, p. 369. Primary source: U.S. Department of Justice. Federal Bureau of Investigation. *Crime in the United States 1991*. Washington, D.C.: U.S. Department of Justice, 1992, p. 162. **Remarks**: Also included in source: number of offenses, 1990, percentage change over 1989, and average loss in dollars for: murder, forcible rape, robbery, larceny-theft, and motor vehicle theft.

★ 282 ★

Cost of Crime

Robbery

Offense	Number of Offenses 1990	Average Loss (dollars)
Offenses known to police		
Robbery	599,100	$783
Street/highway	335,991	633
Commercial house	68,686	1,341
Residence	58,502	1,049
Convenience store	36,331	341
Gas or service station	16,436	442
Bank	8,855	3,244
Miscellaneous	74,299	735

Source: U.S. Department of Justice. Bureau of Justice Statistics. *Sourcebook of Criminal Justice Statistics 1991*. Washington, D.C.: U.S. Department of Justice, 1992, p. 369. Primary source: U.S. Department of Justice. Federal Bureau of Investigation. *Crime in the United States 1991*. Washington, D.C.: U.S. Department of Justice, 1992, p. 162. **Remarks**: Also included in source: number of offenses, 1990, percentage change over 1989, and average loss in dollars for: murder, forcible rape, burglary, larceny-theft, and motor vehicle theft.

★ 283 ★

Cost of Crime

Larceny-theft (except motor vehicle theft)

Offense	Number of Offenses, 1990	Average Loss (dollars)
Offenses known to police		
Larceny-theft, by type	7,286,075	$480
Pocket-picking	73,931	355
Purse-snatching	74,812	278
Shoplifting	1,183,826	115
From motor vehicles	1,599,349	541
Motor vehicle accessories	1,086,289	319
Bicycles	406,322	215
From buildings	1,024,807	791
From coin-operated machines	57,940	147
All others	1,778,799	671
By value		
Over $200	2,606,031	1,243
$50 to $200	1,738,324	114
Under $50	2,941,720	20

Source: U.S. Department of Justice. Bureau of Justice Statistics. *Sourcebook of Criminal Justice Statistics 1991*. Washington, D.C.: U.S. Department of Justice, 1992, p. 369. Primary source: U.S. Department of Justice. Federal Bureau of Investigation. *Crime in the United States 1991*. Washington, D.C.: U.S. Department of Justice, 1992, p. 162. **Remarks**: Also included in source: number of offenses, 1990, percentage change over 1989, and average loss in dollars for: murder, forcible rape, robbery, burglary, and motor vehicle theft.

★ 284 ★

Cost of Crime

Motor Vehicle Theft

In 1990, 1,515,364 cases of motor vehicle theft known to police were committed. The average loss in dollars per offense: $5,032.

Source: U.S. Department of Justice. Bureau of Justice Statistics. *Sourcebook of Criminal Justice Statistics 1991*. Washington, D.C.: U.S. Department of Justice, 1992, p. 396. Primary source: U.S. Department of Justice. Federal Bureau of Investigation. *Crime in the United States 1991*. Washington, D.C.: U.S. Department of Justice, 1992, p. 162. **Remarks**: Also included in source: number of offenses, 1990, percentage change over 1989, and average loss in dollars for: murder, robbery, burglary, and larceny-theft.

★ 285 ★

Cost of Crime

Arson

Average cost of property damage to structures, 1990.

Target	Number of Offenses	Average Damage
Total structure	46,478	$19,763
Single occupancy	20,233	13,602
Other residential	8,062	12,964
Storage	4,384	12,388
Industrial/manufacturing	750	84,942
Other commercial	5,226	43,730
Community/public	4,701	36,436
Other	3,122	6,712
Total mobile	23,595	4,202
Motor vehicle	21,917	3,920
Other mobile	1,678	7,884
Other	16,074	895
Total	86,147	11,980

Source: U.S. Department of Justice. Federal Bureau of Investigation. *Crime in the United States 1991*. Washington D.C.: U.S. Department of Justice, 1992, p. 55.

Cost of Incarceration

★ 286 ★

Cost per Prisoner

The cost to maintain each prisoner averages $20,000 per year, 80% of whom are high school dropouts.

Source: "Economy: Business Leaders Say Dropout Problem Calls for Action," David R. Francis, *The Christian Science Monitor*, March 8, 1991, p. 8.

★ 287 ★

Cost of Incarceration

Health Care Cost per Prison Inmate, by State, 1989

State	Annual health cost per inmate	State	Annual health cost per inmate	State	Annual health cost per inmate
Alabama	$792	Maine	1,870	Oklahoma	909
Alaska	3,381	Maryland	1,226	Oregon	1,868
Arizona	1,913	Massachusetts	2,379	Pennsylvania	1,429
Arkansas	1,595	Michigan	2,636	Rhode Island	1,711
California	1,953	Minnesota	2,157	South Carolina	1,387
Colorado	1,154	Missouri	907	South Dakota	787
Connecticut	2,108	Montana	1,665	Tennessee	1,962
Delaware	1,524	Nebraska	1,795	Texas	2,262
Florida	2,706	Nevada	1,764	Utah	1,174
Georgia	1,648	New Hampshire	1,941	Vermont	1,558
Idaho	1,560	New Jersey	2,016	Virginia	1,500
Illinois	1,570	New Mexico	2,900	Washington	2,664
Iowa	1,618	New York	2,249	West Virginia	1,035
Kansas	1,640	North Carolina	1,973	Wisconsin	1,695
Kentucky	1,210	Ohio	1,366	Wyoming	1,264
Louisiana	831				

Source: U.S. Department of Justice. Bureau of Justice Statistics. *Sourcebook of Criminal Justice Statistics 1991.* Washington, D.C.: U.S. Department of Justice, 1992, p. 14. **Remarks**: Also included in source: Total department of corrections expenditure, total health expenditure, percent of departmental expenditures devoted to health, and total inmate population, by state, 1989.

★ 288 ★

Cost of Incarceration

Health Care of Elderly Prison Inmates

The average health care cost of an elderly inmate is $69,000, not including special diets, necessary ramps and expansion of doors for wheelchairs, and expensive alterations of facilities built for incarceration, not convalescence. This is three times as much as that spent on younger inmates. Elderly prisoners develop an average of three chronic illnesses during their time behind bars.

Source: "A Solution to Prison Overcrowding," Jonathan Turley, *USA Today*, (magazine) November 1992, p. 80.

★ 289 ★

Cost of Incarceration

Operating Costs of Public Juvenile Facilities, by State, 1988

State	Average cost per resident for 1 year	State	Average cost per resident for 1 year	State	Average cost per resident for 1 year
United States	$29,600	Kentucky	32,100	North Dakota	30,300
Alabama	19,400	Louisiana	24,600	Ohio	25,300
Alaska	54,500	Maine	31,400	Oklahoma	40,800
Arizona	20,700	Maryland	33,200	Oregon	34,300
Arkansas	34,100	Massachusetts	55,500	Pennsylvania	45,700
California	23,300	Michigan	42,900	Rhode Island	78,800
Colorado	26,500	Minnesota	37,100	South Carolina	23,600
Connecticut	57,500	Mississippi	18,500	South Dakota	17,600
Delaware	33,300	Missouri	26,200	Tennessee	29,400
District of Columbia	42,600	Montana	27,700	Texas	31,400
Florida	22,900	Nebraska	22,000	Utah	40,400
Georgia	25,800	Nevada	28,800	Vermont	50,000
Hawaii	34,400	New Hampshire	45,900	Virginia	32,700
Idaho	39,500	New Jersey	35,700	Washington	41,100
Illinois	33,400	New Mexico	24,600	West Virginia	25,700
Indiana	21,500	New York	55,300	Wisconsin	34,800
Iowa	29,300	North Carolina	29,000	Wyoming	36,600
Kansas	33,100				

Source: U.S. Department of Justice. Bureau of Justice Statistics. *Sourcebook of Criminal Justice Statistics 1991*. Washington, D.C.: U.S. Department of Justice, 1992, p. 608. Primary source: U.S. Department of Justice, Office of Juvenile Justice and Delinquency Prevention, *Children in Custody, 1989*, NCJ-127189 (Washington, DC: U.S. Department of Justice, January 1991), p. 9. Table adapted by *Sourcebook* staff. **Remarks**: Also included in source: number of facilities and total operating cost of public juvenile facilities for each state.

★ 290 ★

Cost of Incarceration

Per Resident Cost of Juveniles Held in Public and Private Custody, 1985-1989

	Public custody			Private custody		
	1985	1987	1989	1985	1987	1989
Per Resident Operating Cost	25,200	27,000	29,600	24,300	27,800	31,300

Source: U.S. Bureau of the Census. *Statistical Abstract of the United States 1992*. 112th edition. Washington, D.C.: U.S. Department of Commerce, 1992, p. 196. Primary source: U.S. Bureau of Justice Statistics, *Census of Public and Private Juvenile Detention, Correctional, and Shelter Facilities, 1975-85*; thereafter, U.S. Office of Juvenile Justice and Delinquency Prevention, *1987* and *1989 Census of Public and Private Juvenile Custody Facilities*, and *1987 Children in Custody* and *Juvenile Taken into Custody: Fiscal Year 1990 Report*. **Remarks**: Also included in source: average age of juveniles, number of residents by sex, and number of personnel in public and private facilities, 1985-1989.

Prison Population

★ 291 ★

Number of Square Feet per Jail Inmate, 1983, 1988

	Average square feet per inmate	
	1983	1988
All jails	54.3	50.9
Region		
Northeast	55.6	54.2
Midwest	57.9	58.4
South	53.4	48.3
West	52.3	48.8
Persons per unit		
1 person	66.8	68.2
2 persons	48.4	39.2
3 to 5 persons	45.9	40.7
6 to 49 persons	52.0	48.9
50 or more persons	49.1	47.5
Housing unit use		
General	52.8	49.8
Special	57.8	55.0
Other	71.9	61.9
Size of facility		
Fewer than 50	64.4	60.6
50 to 249	54.8	53.5
250 to 499	57.2	52.3
500 to 999	50.2	47.8
1,000 or more	45.1	45.7

Source: U.S. Department of Justice. Bureau of Justice Statistics. *Sourcebook of Criminal Justice Statistics 1991*. Washington, D.C.: U.S. Department of Justice, 1992, p. 99. Primary source: U.S. Department of Justice, Bureau of Justice Statistics, *Population Density in Local Jails, 1988*, Special Report NCJ-122299 (Washington, DC: U.S. Department of Justice, 1990), p. 4, Table 4. **Remarks**: Also included in source: average number of inmates per unit and average number of hours inmates are confined in unit, 1983 and 1988.

★ 292 ★

Prison Population

Number of Jail Inmates by Unit, 1983, 1988

	Average number of inmates per unit	
	1983	1988
All jails	2.4	2.5
Region		
Northeast	1.6	1.8
Midwest	2.0	2.0
South	2.6	2.7
West	3.7	3.3
Persons per unit		
1 person	1.0	1.0
2 persons	2.0	2.0
3 to 5 persons	3.6	3.6
6 to 49 persons	12.7	13.4
50 or more persons	71.6	78.5
Housing unit use		
General	2.5	2.6
Special	1.4	1.6
Other	2.6	2.6
Size of facility		
Fewer than 50	1.8	1.9
50 to 249	2.4	2.4
250 to 499	2.5	2.5
500 to 999	2.5	2.3
1,000 or more	3.2	3.4

Source: U.S. Department of Justice. Bureau of Justice Statistics. *Sourcebook of Criminal Justice Statistics 1991*. Washington, D.C.: U.S. Department of Justice, 1992, p. 99. Primary source: U.S. Department of Justice, Bureau of Justice Statistics, *Population Density in Local Jails, 1988*, Special Report NCJ-122299 (Washington, DC: U.S. Department of Justice, 1990), p. 4, Table 4. **Remarks**: Also included in source: average square feet per inmate and average number of hours inmates are confined in unit, 1983 and 1988.

★ 293 ★

Prison Population

Number of Hours Inmates Confined in Unit, 1983, 1988

	Average number of hours inmates confined in unit	
	1983	1988
All jails	13.5	13.5
Region		
Northeast	11.8	11.8
Midwest	12.5	13.0
South	14.4	14.3
West	13.9	13.8
Persons per unit		
1 person	12.4	12.6
2 persons	12.2	12.2
3 to 5 persons	13.6	13.6
6 to 49 persons	15.4	15.1
50 or more persons	12.0	13.3
Housing unit use		
General	13.4	13.4
Special	18.3	17.7
Other	12.8	12.5
Size of facility		
Fewer than 50	13.2	14.3
50 to 249	13.1	13.6
250 to 499	14.1	13.6
500 to 999	13.8	13.3
1,000 or more	14.0	13.3

Source: U.S. Department of Justice. Bureau of Justice Statistics. *Sourcebook of Criminal Justice Statistics 1991*. Washington, D.C.: U.S. Department of Justice, 1992, p. 99. Primary source: U.S. Department of Justice, Bureau of Justice Statistics, *Population Density in Local Jails, 1988*, Special Report NCJ-122299 (Washington, DC: U.S. Department of Justice, 1990), p. 4, Table 4. **Remarks**: Also included in source: average square feet per inmate and average number of inmates per unit, 1983 and 1988.

★ 294 ★

Prison Population

Daily Population in Jails by State, 1978 and 1988

State	Average daily jail population		State	Average daily jail population	
	1978	1983		1978	1983
United States	157,930	336,017	Montana	398	596
Alabama	5,052	4,900	Nebraska	545	1,110
Alaska	37	28	Nevada	988	2,184
Arizona	2,408	5,808	New Hampshire	389	785
Arkansas	1,217	1,959	New Jersey	3,930	10,978
California	26,199	63,359	New Mexico	738	2,155
Colorado	1,632	4,492	New York	10,926	25,484
District of Columbia	1,360	1,551	North Carolina	2,623	5,553
Florida	10,317	27,029	North Dakota	146	280
Georgia	8,070	16,172	Ohio	5,516	9,062
Idaho	532	820	Oklahoma	1,735	2,717
Illinois	5,993	9,453	Oregon	1,876	2,807
Indiana	2,552	5,061	Pennsylvania	6,326	13,563
Iowa	719	1,062	South Carolina	2,260	3,521
Kansas	942	1,864	South Dakota	306	514
Kentucky	2,310	4,711	Tennessee	4,623	10,082
Louisiana	5,084	11,092	Texas	10,859	29,124
Maine	310	651	Utah	647	1,354
Maryland	3,619	7,300	Virginia	4,396	9,111
Massachusetts	2,269	5,412	Washington	2,457	5,806
Michigan	5,815	9,444	West Virginia	1,057	1,406
Minnesota	1,396	3,001	Wisconsin	1,967	4,626
Mississippi	2,273	3,247	Wyoming	246	487
Missouri	2,870	4,296			

Source: U.S. Department of Justice. Bureau of Justice Statistics. *Sourcebook of Criminal Justice Statistics 1991*. Washington, D.C.: U.S. Department of Justice, 1992, p. 613. Primary source: U.S. Department of Justice, Bureau of Justice Statistics, *Census of Local Jails, 1988*, Special Report NCJ-121101 (Washington, DC: U.S. Department of Justice, 1990), p. 4, Table 6. table 1.

★ 295 ★

Prison Population

Daily Population of Juveniles in Local Jails by State, 1978 and 1988

By region and State, 1978 and 1988.

Region and State	Average daily jail population		State	Average daily jail population	
	1978	1988		1978	1988
United States	1,740	1,451	Montana	28	5
Alabama	17	7	Nebraska	33	16
Alaska	3	0	Nevada	14	4
Arizona	9	33	New Hampshire	1	2
Arkansas	48	66	New Jersey	1	10
California	131	35	New Mexico	31	58
Colorado	26	12	New York	82	28
District of Columbia	15	10	North Carolina	30	0
Florida	49	563	North Dakota	6	18
Georgia	16	5	Ohio	68	6
Idaho	37	7	Oklahoma	27	22
Illinois	27	19	Oregon	37	3
Indiana	154	132	Pennsylvania	3	0
Iowa	19	1	South Carolina	30	33
Kansas	53	15	South Dakota	14	14
Kentucky	73	32	Tennessee	86	2
Louisiana	16	0	Texas	80	6
Maine	8	10	Utah	4	0
Maryland	0	62	Virginia	186	46
Massachusetts	0	0	Washington	28	1
Michigan	27	8	West Virginia	15	0
Minnesota	24	7	Wisconsin	55	59
Mississippi	0	56	Wyoming	23	32
Missouri	42	6			

Source: U.S. Department of Justice. Bureau of Justice Statistics. *Sourcebook of Criminal Justice Statistics 1991*. Washington, D.C.: U.S. Department of Justice, 1992, p. 613. Primary source: U.S. Department of Justice, Bureau of Justice Statistics, *Census of Local Jails, 1988*, Special Report NCJ-121101 (Washington, DC: U.S. Department of Justice, 1990), p. 10, appendix table 1. *Notes:* Data are for the annual periods ending Feb. 15, 1978 and June 30, and 1988.

★ 296 ★

Prison Population

Age of Juveniles Held in Public and Private Custody, 1985-1989

	Public custody			Private custody		
	1985	1987	1989	1985	1987	1989
Average age	15.4	15.8	16.0	14.9	14.8	14.9

Source: U.S. Bureau of the Census. *Statistical Abstract of the United States 1992*. 112th edition. Washington, D.C.: U.S. Department of Commerce, 1992, p. 196. Primary source: U.S. Bureau of Justice Statistics, *Census of Public and Private Juvenile Detention, Correctional, and Shelter Facilities, 1975-85*; thereafter, U.S. Office of Juvenile Justice and Delinquency Prevention, *1987* and *1989 Census of Public and Private Juvenile Custody Facilities*, and *Juvenile Taken into Custody: Fiscal Year 1990 Report*. **Remarks**: Also included in source: number of juvenile residents by sex, average daily residents, number of personnel, and per resident operating cost of public and private facilities, 1985-1989.

★ 297 ★

Prison Population

Daily Residents Held in Public and Private Custody, 1985-1989

	Public custody			Private custody		
	1985	1987	1989	1985	1987	1989
Average Daily Residents	49,480	54,036	56,595	33,121	38,172	37,944

Source: U.S. Bureau of the Census. *Statistical Abstract of the United States 1992*. 112th edition. Washington, D.C.: U.S. Department of Commerce, 1992, p. 196. Primary source: U.S. Bureau of Justice Statistics, *Census of Public and Private Juvenile Detention, Correctional, and Shelter Facilities, 1975-85*; thereafter, U.S. Office of Juvenile Justice and Delinquency Prevention, *1987* and *1989 Census of Public and Private Juvenile Custody Facilities*, and *1987 Children in Custody* and *Juvenile Taken into Custody: Fiscal Year 1990 Report*. **Remarks**: Also included in source: average age of juveniles, number of residents by sex, number of personnel, and per resident operating cost of public and private facilities, 1985-1989.

Time in Prison

★ 298 ★

Length of Imposed Prison Sentence by Type of Offense, 1980-1990

Most serious conviction offense	Average sentence length imposed for convicted offenders sentenced to prison (in months)						
	1980	1985	1986	1987	1988	1989	Preliminary 1990
All offenses	44.3	50.7	52.7	55.2	55.1	54.5	57.4
Violent offenses	125.4	135.4	132.0	126.2	110.7	90.6	89.8
Murder	123.4	212.0	196.3	154.6	162.7	180.1	134.8

[Continued]

★ 298 ★

Length of Imposed Prison Sentence by Type of Offense, 1980-1990

[Continued]

Most serious conviction offense	Average sentence length imposed for convicted offenders sentenced to prison (in months)						
	1980	1985	1986	1987	1988	1989	Preliminary 1990
Negligent manslaughter	36.6	33.5	40.2	25.9	29.2	23.3	19.2
Assault	40.7	45.6	44.6	48.4	39.7	34.4	33.3
Robbery	141.5	151.1	153.2	148.1	131.4	100.4	101.3
Rape	85.7	123.3	143.8	114.4	95.8	90.1	85.8
Other sex offenses	39.7	81.0	68.2	59.5	75.1	44.7	33.0
Kidnaping	262.6	254.5	242.9	293.9	190.5	147.9	170.3
Threats against the President	39.4	48.2	26.2	45.0	44.1	35.9	25.0
Property offenses	29.4	33.0	34.3	32.5	31.5	26.0	22.3
Fraudulent offenses	27.1	31.1	32.6	31.1	31.0	26.1	22.3
Embezzlement	20.1	22.9	21.9	22.1	19.6	16.5	17.7
Fraud	24.9	30.0	33.8	32.1	32.9	29.8	23.9
Forgery	32.5	36.5	32.8	30.6	32.1	18.3	17.1
Counterfeiting	35.0	38.2	43.3	37.2	29.1	20.1	19.5
Other offenses	32.9	36.6	37.9	36.5	32.7	25.7	22.5
Burglary	46.0	67.0	41.9	59.0	55.6	41.7	35.5
Larceny	29.2	31.7	33.6	33.8	27.5	22.7	18.7
Motor vehicle theft	38.1	42.8	42.4	44.3	38.0	28.6	28.2
Arson	34.3	34.8	24.8	51.7	19.0	45.3	25.5
Transportation of stolen property	47.7	48.0	56.0	37.3	51.1	33.3	31.2
Other	9.3	15.5	24.1	11.7	17.7	12.2	12.1
Drug offenses	47.1	58.2	62.2	67.8	71.3	74.9	81.2
Trafficking	48.1	60.8	63.9	69.1	73.6	77.3	83.5
Possession and other	14.7	34.5	41.0	48.0	13.6	8.1	13.1

Source: U.S. Department of Justice. Bureau of Justice Statistics. *Sourcebook of Criminal Justice Statistics 1991*. Washington, D.C.: U.S. Department of Justice, 1992, p. 506. Primary source: U.S. Department of Justice, Bureau of Justice Statistics, *Federal Criminal Case Processing, 1980-89, With Preliminary Data for 1990*, NCJ- 130526 (Washington, DC: U.S. Department of Justice, 1991), p. 17. Table adapted by *Sourcebook* staff.

★ 299 ★

Time in Prison

Length of Probation by Type of Offense, 1988

Most serious offenses of conviction	Probation (in months)
All offenses	37.3
Violent offenses	43.0
Assault	30.2

[Continued]

★ 299 ★

Length of Probation by Type of Offense, 1988

[Continued]

Most serious offenses of conviction	Probation (in months)
Robbery	51.8
Rape	45.8
Other sex offenses	43.9
Property offenses	40.4
Fraudulent offenses	42.7
Embezzlement	41.5
Fraud	43.6
Forgery	40.1
Counterfeiting	41.2
Other offenses	33.8
Burglary	41.5
Larceny	32.1
Motor vehicle theft	46.8
Transportation of stolen property	47.1
Other property offenses	24.0
Drug offenses	40.6
Trafficking	49.5
Possession and other	20.5

Source: U.S. Department of Justice. Bureau of Justice Statistics. *Sourcebook of Criminal Justice Statistics 1991*. Washington, D.C.: U.S. Department of Justice, 1992, p. 503. Primary source: U.S. Department of Justice, Bureau of Justice Statistics, *Compendium of Federal Justice Statistics, 1988*, NCJ-130474 (Washington, DC: U.S. Department of Justice, 1991), pp. 39, 40. Table adapted by *Sourcebook* staff. **Remarks**: Also included in source: total number of offenders sentenced; percent of offenders convicted and sentenced to incarceration, probation, mixed sentence, and fine; and average sentence length, 1988.

★ 300 ★

Time in Prison

Length of Sentence by Offense and Sex

Offender characteristics	Average sentence length (in months) for offenders convicted of:						
	All offenses	Violent offenses	Property		Drug offenses	Public-order offenses	
			Fraudulent	Other		Regulatory	Other
Sex							
Male	58.6	112.3	33.0	35.0	74.8	32.8	34.1
Female	36.6	82.6	22.7	20.9	50.5	17.9	19.8

Source: U.S. Department of Justice. Bureau of Justice Statistics. *Sourcebook of Criminal Justice Statistics 1991*. Washington, D.C.: U.S. Department of Justice, 1992, p. 507. Primary source: U.S. Department of Justice, Bureau of Justice Statistics, *Compendium of Federal Justice Statistics, 1988*, NCJ-130474 (Washington, DC: U.S. Department of Justice, 1991), p. 43. **Remarks**: Also included in source: length of sentence by race/ethnicity, age, education, marital status, employment status, criminal record, and drug abuse.

★ 301 ★

Time in Prison

Length of Sentence by Offense and Race/Ethnicity

Offender characteristics	Average sentence length (in months) for offenders convicted of:						
	All offenses	Violent offenses	Property		Drug offenses	Public-order offenses	
			Fraudulent	Other		Regulatory	Other
Race							
White	54.2	115.1	33.4	34.8	68.8	34.3	31.3
Black	63.2	123.7	26.0	31.6	82.8	20.8	42.3
Other	46.8	59.4	18.5	18.3	68.2	*	24.1
Ethnicity							
Hispanic	56.8	114.5	24.8	35.3	74.6	23.7	21.2
Non-Hispanic	56.1	110.5	31.7	32.9	71.3	33.0	37.3

Source: U.S. Department of Justice. Bureau of Justice Statistics. *Sourcebook of Criminal Justice Statistics 1991*. Washington, D.C.: U.S. Department of Justice, 1992, p. 507. Primary source: U.S. Department of Justice, Bureau of Justice Statistics, *Compendium of Federal Justice Statistics, 1988*, NCJ-130474 (Washington, DC: U.S. Department of Justice, 1991), p. 43. **Remarks**: Also included in source: length of sentence by sex, age, education, marital status, employment status, criminal record, and drug abuse. *Note:* * Too few cases to obtain statistically reliable data.

★ 302 ★

Time in Prison

Length of Sentence by Offense and Age

Offender characteristics	Average sentence length (in months) for offenders convicted of:						
	All offenses	Violent offenses	Property		Drug offenses	Public-order offenses	
			Fraudulent	Other		Regulatory	Other
Age							
16 to 18 years	38.0	65.7	*	*	49.9	*	9.9
19 to 20 years	44.2	70.4	12.4	27.2	59.3	*	17.9
21 to 30 years	52.6	107.8	23.2	29.6	64.1	25.8	27.1
31 to 40 years	60.6	124.9	30.2	37.6	74.8	39.5	39.5
Over 40 years	57.6	109.0	38.8	33.5	86.6	32.4	35.5

Source: U.S. Department of Justice. Bureau of Justice Statistics. *Sourcebook of Criminal Justice Statistics 1991*. Washington, D.C.: U.S. Department of Justice, 1992, p. 507. Primary source: U.S. Department of Justice, Bureau of Justice Statistics, *Compendium of Federal Justice Statistics, 1988*, NCJ-130474 (Washington, DC: U.S. Department of Justice, 1991), p. 43. **Remarks**: Also included in source: length of sentence by sex, race/ethnicity, education, marital status, employment status, criminal record, and drug abuse. *Note:* * Too few cases to obtain statistically reliable data.

★ 303 ★

Time in Prison

Length of Sentence by Offense and Education

Offender characteristics	Average sentence length (in months) for offenders convicted of:						
	All offenses	Violent offenses	Property		Drug offenses	Public-order offenses	
			Fraudulent	Other		Regulatory	Other
Education							
Less than high school graduate	57.7	108.5	28.8	31.3	73.9	32.1	31.6
High school graduate	65.9	132.1	41.0	36.7	72.8	30.6	48.0
Some college	55.2	111.1	33.0	33.9	70.8	37.0	35.6
College graduate	43.5	79.7	33.7	42.8	68.0	25.9	28.6

Source: U.S. Department of Justice. Bureau of Justice Statistics. *Sourcebook of Criminal Justice Statistics 1991*. Washington, D.C.: U.S. Department of Justice, 1992, p. 507. Primary source: U.S. Department of Justice, Bureau of Justice Statistics, *Compendium of Federal Justice Statistics, 1988*, NCJ-130474 (Washington, DC: U.S. Department of Justice, 1991), p. 43. **Remarks**: Also included in source: length of sentence by sex, race/ethnicity, age, marital status, employment status, criminal record, and drug abuse.

★ 304 ★

Time in Prison

Length of Sentence by Offense and Marital Status

Offender characteristics	Average sentence length (in months) for offenders convicted of:						
	All offenses	Violent offenses	Property		Drug offenses	Public-order offenses	
			Fraudulent	Other		Regulatory	Other
Marital status							
Never married	55.2	100.6	28.4	31.0	67.3	19.2	30.9
Divorced/separated	59.8	128.1	35.4	31.7	72.7	38.3	40.1
Married	54.8	109.8	31.0	38.0	75.1	33.2	32.5
Common law	56.4	99.6	26.5	26.2	75.1	22.6	26.0
Other	69.8	*	27.2	*	79.8	*	46.9

Source: U.S. Department of Justice. Bureau of Justice Statistics. *Sourcebook of Criminal Justice Statistics 1991*. Washington, D.C.: U.S. Department of Justice, 1992, p. 507. Primary source: U.S. Department of Justice, Bureau of Justice Statistics, *Compendium of Federal Justice Statistics, 1988*, NCJ-130474 (Washington, DC: U.S. Department of Justice, 1991), p. 43. **Remarks**: Also included in source: length of sentence by sex, race/ethnicity, age, education, employment status, criminal record, and drug abuse. *Note:* * Too few cases to obtain statistically reliable data.

★ 305 ★

Time in Prison

Length of Sentence by Offense and Employment Status

Offender characteristics	Average sentence length (in months) for offenders convicted of:						
	All offenses	Violent offenses	Property		Drug offenses	Public-order offenses	
			Fraudulent	Other		Regulatory	Other
Employment status							
Unemployed	59.3	105.3	36.9	30.7	71.3	42.6	41.7
Employed	50.0	122.1	32.2	52.8	89.5	*	45.7

Source: U.S. Department of Justice. Bureau of Justice Statistics. *Sourcebook of Criminal Justice Statistics 1991*. Washington, D.C.: U.S. Department of Justice, 1992, p. 507. Primary source: U.S. Department of Justice, Bureau of Justice Statistics, *Compendium of Federal Justice Statistics, 1988*, NCJ-130474 (Washington, DC: U.S. Department of Justice, 1991), p. 43. **Remarks**: Also included in source: length of sentence by sex, race/ethnicity, age, education, marital status, criminal record, and drug abuse. *Note:* * Too few cases to obtain statistically reliable data.

★ 306 ★

Time in Prison

Length of Sentence by Offense and Criminal Record

Offender characteristics	Average sentence length (in months) for offenders convicted of:						
	All offenses	Violent offenses	Property		Drug offenses	Public-order offenses	
			Fraudulent	Other		Regulatory	Other
Criminal record							
No convictions	47.8	71.9	28.6	25.5	64.9	32.3	28.6
Prior conviction							
Misdemeanor only	73.7	147.7	50.3	38.5	90.1	44.7	50.7
Felony	58.7	100.4	37.9	31.5	73.2	31.9	40.2

Source: U.S. Department of Justice. Bureau of Justice Statistics. *Sourcebook of Criminal Justice Statistics 1991*. Washington, D.C.: U.S. Department of Justice, 1992, p. 507. Primary source: U.S. Department of Justice, Bureau of Justice Statistics, *Compendium of Federal Justice Statistics, 1988*, NCJ-130474 (Washington, DC: U.S. Department of Justice, 1991), p. 43. **Remarks**: Also included in source: length of sentence by sex, race/ethnicity, age, education, marital status, employment status, and drug abuse.

★ 307 ★

Time in Prison

Length of Sentence by Offense and History of Drug Abuse

Offender characteristics	Average sentence length (in months) for offenders convicted of:						
	All offenses	Violent offenses	Property		Drug offenses	Public-order offenses	
			Fraudulent	Other		Regulatory	Other
Drug abuse							
No known abuse	50.8	87.6	32.3	29.6	70.6	33.2	32.5
Drug history	58.8	105.5	36.1	28.4	62.7	25.3	48.4

Source: U.S. Department of Justice. Bureau of Justice Statistics. *Sourcebook of Criminal Justice Statistics 1991*. Washington, D.C.: U.S. Department of Justice, 1992, p. 507. Primary source: U.S. Department of Justice, Bureau of Justice Statistics, *Compendium of Federal Justice Statistics, 1988*, NCJ-130474 (Washington, DC: U.S. Department of Justice, 1991), p. 43. **Remarks**: Also included in source: length of sentence by sex, race/ethnicity, age, education, marital status, employment status, and criminal record.

★ 308 ★

Time in Prison

Sentence Length and Time Served by Female Prison Inmates, 1986

Most serious offense	Average maximum sentence of women admitted in 1986 (in months)	Average time served by female inmates released in 1986 (in months)
All offenses	66	16
Violent offenses	108	27
Murder	200	42
Negligent manslaughter	107	27
Kidnaping	106	38
Sexual assault	146	24
Robbery	91	26
Assault	67	19
Other violent	71	19
Property offenses	53	13
Burglary	57	14
Larceny/theft	52	13
Motor vehicle theft	34	10
Arson	86	16
Fraud	52	13
Stolen property	43	14
Other property	39	13
Drug offenses	54	14
possession	57	11
Trafficking	57	17
Other or unspecified drug	47	12

Source: U.S. Department of Justice. Bureau of Justice Statistics. *Sourcebook of Criminal Justice Statistics 1991*. Washington, D.C.: U.S. Department of Justice, 1992, p. 649. Primary source: U.S. Department of Justice, Bureau of Justice Statistics, *Women in Prison*, Special Report NCJ-127991 (Washington, DC: U.S. Department of Justice, 1991), p. 4, Table 6.

★ 309 ★

Time in Prison

Maximum Sentence Length by Type of Offense, 1989

Most serious offense	Maximum sentence length (in months)	
	Median	Average
All offenses	6	17
Violent offenses	12	35
Homicide	23	73
Sexual assault	24	35
Robbery	24	56
Assault	9	23
Other violent	12	18
Property offenses	10	21
Burglary	12	36
Larceny/theft	6	14
Motor vehicle theft	6	15
Fraud	12	18
Stolen property	8	13
Other property	3	11
Drug offenses	9	19
Possession	7	15
Trafficking	12	22
Public-order offenses	5	8
Weapons	6	13
Obstruction of justice	3	5
Traffic	3	4
Driving while intoxicated	6	7
Drunkenness or morals	4	6
Violation of parole or probation	12	18
Other public-order	6	6
Other offenses	6	15

Source: U.S. Department of Justice. Bureau of Justice Statistics. *Sourcebook of Criminal Justice Statistics 1991*. Washington, D.C.: U.S. Department of Justice, 1992, p. 627. Primary source: U.S. Department of Justice, Bureau of Justice Statistics, *Profile of Jail Inmates, 1989*, Special Report NCJ-129097 (Washington, DC: U.S. Department of Justice, 1991), p. 7, Table 12. **Remarks**: Also included in source: total number of jail inmates, and time served since admission, 1989.

Law Enforcement

★ 310 ★

Expenditures per Capita for Police Protection and Correction, by State, 1988

Employment as of October. Expenditures for fiscal years ending between July 1987 and June 1988. Local government data are estimates subject to sampling variation.

State	Per capita (dol.)	State	Per capita (dol.)	State	Per capita (dol.)
United States	230	Kentucky	140	North Dakota	123
Alabama	145	Louisiana	193	Ohio	196
Alaska	559	Maine	142	Oklahoma	149
Arizona	300	Maryland	267	Oregon	212
Arkansas	103	Massachusetts	245	Pennsylvania	171
California	343	Michigan	248	Rhode Island	203
Colorado	233	Minnesota	186	South Carolina	161
Connecticut	230	Mississippi	101	South Dakota	127
Delaware	260	Missouri	163	Tennessee	157
District of Columbia	958	Montana	153	Texas	177
Florida	262	Nebraska	141	Utah	170
Georgia	187	Nevada	367	Vermont	145
Hawaii	248	New Hampshire	177	Virginia	204
Idaho	147	New Jersey	275	Washington	202
Illinois	210	New Mexico	211	West Virginia	90
Indiana	131	New York	383	Wisconsin	194
Iowa	144	North Carolina	177	Wyoming	227
Kansas	159				

Source: U.S. Bureau of the Census. *Statistical Abstract of the United States 1992*. 112th edition. Washington, D.C.: U.S. Department of Commerce, 1992, p. 191. Primary source: U.S. Bureau of the Census, *Public Employment*, series GE No. 1, annual, and *Government Finances*, series GF, No. 5, annual. **Remarks**: Also included in source: number employed in police protection and correction, total and separate counts; total expenditures for police protection and correction, shown separately.

★ 311 ★

Law Enforcement

Operating Expenditures of Local Police Departments per Employee, 1990

Size of Population	Average Operating Expenditure per Employee
All sizes	$42,300
1,000,000 or more	48,300
500,000 to 999,999	48,400
250,000 to 499,999	46,800
100,000 to 249,000	44,000

[Continued]

★311★

Operating Expenditures of Local Police Departments per Employee, 1990

[Continued]

Size of Population	Average Operating Expenditure per Employee
50,000 to 99,999	40,600
25,000 to 49,999	42,100
10,000 to 24,999	40,400
2,500 to 9,999	31,800
Under 2,500	27,400

Source: U.S. Department of Justice. Bureau of Justice Statistics. *Sourcebook of Criminal Justice Statistics 1991*. Washington, D.C.: U.S. Department of Justice, 1992, p. 44. Primary source: U.S. Department of Justice, Bureau of Justice Statistics, *State and Local Police Departments, 1990*, Bulletin NCJ-133284 (Washington, D.C.: U.S. Department of Justice, February 1992), p. 3, Table 6. **Remarks**: Also included in source: average operating expenditures per department, per sworn officer, and per resident.

★312★

Law Enforcement

Operating Expenditures of Local Police Departments per Resident, 1990

Size of Population	Average Operating Expenditure per Resident
All sizes	$108
1,000,000 or more	144
500,000 to 999,999	114
250,000 to 499,999	128
100,000 to 249,000	108
50,000 to 99,999	95
25,000 to 49,999	93
10,000 to 24,999	96
2,500 to 9,999	85
Under 2,500	95

Source: U.S. Department of Justice. Bureau of Justice Statistics. *Sourcebook of Criminal Justice Statistics 1991*. Washington, D.C.: U.S. Department of Justice, 1992, p. 44. Primary source: U.S. Department of Justice, Bureau of Justice Statistics, *State and Local Police Departments, 1990*, Bulletin NCJ-133284 (Washington, D.C.: U.S. Department of Justice, February 1992), p. 3, Table 6. **Remarks**: Also included in source: average operating expenditures per department, per sworn officer, and per employee.

★313★

Law Enforcement

Operating Expenditures of Local Police Departments per Department, 1990

Size of Population	Average Operating Expenditure per Department
All sizes	$1,675,000
1,000,000 or more	334,542,000
500,000 to 999,999	79,567,000
250,000 to 499,999	46,500,000
100,000 to 249,000	16,268,000
50,000 to 99,999	6,528,000
25,000 to 49,999	3,225,000
10,000 to 24,999	1,513,000
2,500 to 9,999	442,000
Under 2,500	115,000

Source: U.S. Department of Justice. Bureau of Justice Statistics. *Sourcebook of Criminal Justice Statistics 1991*. Washington, D.C.: U.S. Department of Justice, 1992, p. 44. Primary source: U.S. Department of Justice, Bureau of Justice Statistics, *State and Local Police Departments, 1990*, Bulletin NCJ-133284 (Washington, D.C.: U.S. Department of Justice, February 1992), p. 3, Table 6. **Remarks**: Also included in source: average operating expenditures per sworn officer, per employee, and per resident.

★314★

Law Enforcement

Operating Expenditures of Sheriffs' Departments per Employee, 1990

Size of Population	Average Operating Expenditure per Employee
All sizes	$43,100
1,000,000 or more	55,400
500,000 to 999,999	43,400
250,000 to 499,999	40,600
100,000 to 249,000	52,400
50,000 to 99,999	33,200
25,000 to 49,999	27,800

[Continued]

★314★

Operating Expenditures of Sheriffs' Departments per Employee, 1990

[Continued]

Size of Population	Average Operating Expenditure per Employee
10,000 to 24,999	32,100
Under 10,000	32,700

Source: U.S. Department of Justice. Bureau of Justice Statistics. *Sourcebook of Criminal Justice Statistics 1991*. Washington, D.C.: U.S. Department of Justice, 1992, p. 44. Primary source: U.S. Department of Justice, Bureau of Justice Statistics, *Sheriffs' Departments 1990*, Bulletin NCJ-133283 (Washington, D.C.: U.S. Department of Justice, February 1992), p. 3, Table 6. **Remarks**: Also included in source: average operating expenditures per department, per sworn officer, and per resident.

★315★

Law Enforcement

Operating Expenditures of Sheriffs' Departments per Resident, 1990

Size of Population	Average Operating Expenditure per Resident
All sizes	$40
1,000,000 or more	38
500,000 to 999,999	34
250,000 to 499,999	34
100,000 to 249,000	48
50,000 to 99,999	34
25,000 to 49,999	28
10,000 to 24,999	38
Under 10,000	56

Source: U.S. Department of Justice. Bureau of Justice Statistics. *Sourcebook of Criminal Justice Statistics 1991*. Washington, D.C.: U.S. Department of Justice, 1992, p. 44. Primary source: U.S. Department of Justice, Bureau of Justice Statistics, *Sheriffs' Departments 1990*, Bulletin NCJ-133283 (Washington, D.C.: U.S. Department of Justice, February 1992), p. 3, Table 6. **Remarks**: Also included in source: average operating expenditures per department, per sworn officer, and per employee.

★316★

Law Enforcement

Number of Full-time employees in Local Police Departments, 1990

Size of Population	Average Number of Employees
1,000,000 or more	6,810
500,000 to 999,999	1,611
250,000 to 499,999	968
100,000 to 249,000	357
50,000 to 99,999	153
25,000 to 49,999	73
10,000 to 24,999	36
2,500 to 9,999	12
Under 2,500	3

Source: U.S. Department of Justice. Bureau of Justice Statistics. *Sourcebook of Criminal Justice Statistics 1991*. Washington, D.C.: U.S. Department of Justice, 1992, p. 40. Primary source: U.S. Department of Justice, Bureau of Justice Statistics, *State and Local Police Departments, 1990*, Bulletin NCJ-133284 (Washington, D.C.: U.S. Department of Justice, February 1992), p. 3, Table 4. **Remarks**: Also included in source: average number of full and part-time employees, sworn and civilian.

★317★

Law Enforcement

Number Employed in Police Protection, By State, 1990

State	Average no. of police per 10,000 population	State	Average no. of police per 10,000 population	State	Average no. of police per 10,000 population
United States average	28.1	Kentucky	20.1	North Dakota	20.5
Alabama	24.8	Louisiana	27.8	Ohio	24.6
Alaska	27.9	Maine	23.5	Oklahoma	27.1
Arizona	29.7	Maryland	30.6	Oregon	22.9
Arkansas	21.2	Massachusetts	29.2	Pennsylvania	24.9
California	28.7	Michigan	23.5	Rhode Island	29.8
Colorado	27.4	Minnesota	20.3	South Carolina	24.3
Connecticut	28.7	Mississippi	21.6	South Dakota	21.1
Delaware	28.1	Missouri	28.3	Tennessee	25.3
District of Columbia	92.1	Montana	24.2	Texas	25.8
Florida	33.8	Nebraska	24.0	Utah	21.6
Georgia	28.1	Nevada	36.1	Vermont	21.7
Hawaii	28.4	New Hampshire	27.0	Virginia	24.2
Idaho	25.8	New Jersey	41.0	Washington	21.5
Illinois	34.4	New Mexico	29.5	West Virginia	16.2
Indiana	23.2	New York	36.9	Wisconsin	25.8
Iowa	21.8	North Carolina	25.4	Wyoming	34.7
Kansas	26.8				

Source: "Police States," *American Demographics*, 14, April 1992, p. 20. Primary source: Bureau of the Census, *Public Employment: 1990*, Series GE-90-I.

★318★

Law Enforcement

Training Requirements for Local Police Departments, 1990

Size of Population	Average Number of Hours Required		
	Total	Classroom	Field
All sizes	543	402	141
1,000,000 or more	1,266	774	492
500,000 to 999,999	1,086	678	408
250,000 to 499,999	1,408	722	686
100,000 to 249,000	1,137	595	542
50,000 to 99,999	926	495	431
25,000 to 49,999	820	458	362
10,000 to 24,999	665	451	214

[Continued]

★318★

Training Requirements for Local Police Departments, 1990

[Continued]

Size of Population	Average Number of Hours Required		
	Total	Classroom	Field
2,500 to 9,999	536	413	123
Under 2,500	399	347	52

Source: U.S. Department of Justice. Bureau of Justice Statistics. *Sourcebook of Criminal Justice Statistics 1991*. Washington, D.C.: U.S. Department of Justice, 1992, p. 40. Primary source: U.S. Department of Justice, Bureau of Justice Statistics, *State and Local Police Departments, 1990*, Bulletin NCJ-133284 (Washington, D.C.: U.S. Department of Justice, February 1992), p. 3, Table 4. **Remarks**: Also included in source: average number of full and part-time employees, sworn and civilian.

★319★

Law Enforcement

Number of People per Lawyer, by State, 1988

State	Population per lawyer	State	Population per lawyer	State	Population per lawyer
United States	340	Kentucky	496	North Dakota	519
Alabama	565	Louisiana	383	Ohio	384
Alaska	270	Maine	412	Oklahoma	366
Arizona	394	Maryland	316	Oregon	357
Arkansas	608	Massachusetts	243	Pennsylvania	396
California	304	Michigan	427	Rhode Island	374
Colorado	269	Minnesota	339	South Carolina	633
Connecticut	246	Mississippi	598	South Dakota	515
Delaware	364	Missouri	382	Tennessee	514
District of Columbia	19	Montana	373	Texas	388
Florida	371	Nebraska	358	Utah	464
Georgia	414	Nevada	423	Vermont	362
Hawaii	351	New Hampshire	415	Virginia	403
Idaho	447	New Jersey	283	Washington	360
Illinois	285	New Mexico	430	West Virginia	622
Indiana	551	New York	219	Wisconsin	448
Iowa	460	North Carolina	634	Wyoming	410
Kansas	417				

Source: U.S. Bureau of the Census. *Statistical Abstract of the United States 1992*. 112th edition. Washington, D.C.: U.S. Department of Commerce, 1992, p. 192. Primary source: American Bar Foundation, Chicago, IL, *The 1971 Lawyer Statistical Report in the 1980s*. **Remarks**: Source also includes: total number of lawyers per state, 1970-1988.

★ 320 ★

Law Enforcement

Payments from Lawsuits Against Corporate Directors

According to a survey by Wyatt Co., a consulting firm in Washington, DC, the average payment arising from lawsuits against corporate directors was $1,710,000 in 1989, rising to $3,360,000 in 1992.

Source: "Odds and Ends," *The Wall Street Journal*, May 10, 1993, p. B1.

★ 321 ★

Law Enforcement

Libel Awards Won Against Publishing and Media Entities

The *Bulletin* of the Libel Defense Resource Center (LDRC) reported that the average libel award won against publishing and media entities was $9 million in the 1990s, compared with $1.5 million during the 1980s.

Source: "Publishers Support Your Local Justice!" Howard G. Zaharoff, *Publishers Weekly*, 239, April 19, 1993, p. 36.

★ 322 ★

Law Enforcement

Compensation from Injury Due to Childhood Vaccination

The average award to children who have suffered neurological damage as a result of adverse reactions to childhood vaccination is $1.1 million (from a federal vaccine injury compensation program). But the compensation to families of children who have died as a result of such complications is capped at $250,000.

Source: "Will It Hurt?" Elena Neuman, *Insight*, April 20, 1992.

Chapter 6
EDUCATION

College Costs

★ 323 ★

Student Loans

The federal government subsidizes 4.3 million college students with loans averaging $3,000 each and costing $6 billion per year. Banking fees to issue these loans are $1 billion per year.

Source: "New Schools of Thought on Paying for College," *Business Week*, 22, October 12, 1992, p. 58.

★ 324 ★

College Costs

Debt for College Graduates of Public Institutions

The average debt for college graduates of public institutions was $8,000 in 1985. It increased to $14,000 by 1990. As a consequence, student loans are $8.5 billion in default.

Source: Rukeyser, Louis. *Louis Rukeyser's Business Almanac*. New York: Simon & Shuster, 1991, p. 429.

★ 325 ★

College Costs

Four-Year Colleges Costs, 1991-1992

	Average Costs			
	Public		Private	
	Resident	Commuter	Resident	Commuter
Tuition	$2,137	$2,137	$10,017	$10,017
Books and supplies	485	485	508	508
Room & board	3,351	1,468	4,386	1,634
Transportation	464	793	470	795
Other	1,147	1,153	911	1,029
Total	7,584	6,036	16,292	13,983

Source: "Recession Blamed for Big Tuition Increase as Colleges Struggle to Offset Budget Cuts," *The Chronicle of Higher Education*, 38, October 23, 1991, p. A30. Primary source: The College Board. **Remarks**: Source also includes: average costs at two-year colleges and average annual increase in tuition and fees.

★ 326 ★

College Costs

Two-Year College Costs, 1991-1992

	Average Cost		
	Public	Private	
	Commuter	Resident	Commuter
Tuition	$1,022	$5,290	$5,290
Books and supplies	480	476	476
Room & board	1,543	3,734	1,529
Transportation	902	519	786
Other	966	895	925
Total	4,913	10,914	9,006

Source: "Recession Blamed for Big Tuition Increase as Colleges Struggle to Offset Budget Cuts," *The Chronicle of Higher Education*, 38, October 23, 1991, p. A30. Primary source: The College Board. **Remarks**: Source also includes: average costs at four-year colleges and average annual increase in tuition and fees.

★ 327 ★

College Costs

Annual Increase in Tuition and Fees at Four-Year Institutions

Year	Average Increase	
	Public	Private
1981-1982	16%	13%
1982-1983	20%	13%
1983-1984	12%	11%
1984-1985	8%	9%
1987-1988	6%	8%
1988-1989	5%	9%
1990-1991	7%	8%
1991-1992	12%	7%

Source: "Recession Blamed for Big Tuition Increase as Colleges Struggle to Offset Budget Cuts," *The Chronicle of Higher Education*, 38, October 23, 1991, p. A30. Primary source: The College Board. **Remarks**: Source also includes: average costs at four-year and two-year colleges and universities.

★ 328 ★

College Costs

College Expenditures on Athletics per Capita, 1987

Membership	Per capita expenditures($)	
	Women	Men
All AIAW Colleges	1,382	3,013
All AIAW & NCAA Div. I	2,156	5,257
NCAA Div. II	1,143	1,883
NCAA Div. III	218	315
All NCAA	1,686	3,755

Source: "Athletes, Budgets, and Per Capita Expenditures (PCE) at AIAW and NCAA Colleges," *Women and Public Polices*, 1987, Table 6, p. 103 (Princeton, NJ: Princeton University Press, 1987). Primary source: *AIAW Competitive Division Structure Implementation Study: Final Data Summary*, Association for Intercollegiate Athletics for Women, Fall 1978, Table XIV.

★ 329 ★

College Costs

Appropriations for Higher Education Operations by State, 1989

State	Appropriations for current operations per FTE student (dol.)	State	Appropriations for current operations per FTE student (dol.)	State	Appropriations for current operations per FTE student (dol.)
Alabama	3,865	Kentucky	3,643	North Dakota	2,803
Alaska	9,879	Louisiana	2,634	Ohio	3,392
Arizona	3,944	Maine	5,167	Oklahoma	3,698
Arkansas	3,767	Maryland	4,011	Oregon	3,796
California	4,907	Massachusetts	5,273	Pennsylvania	3,996
Colorado	2,853	Michigan	3,980	Rhode Island	4,402
Connecticut	6,345	Minnesota	4,231	South Carolina	3,790
Delaware	4,649	Mississippi	3,343	South Dakota	3,122
District of Columbia	7,695	Missouri	3,577	Tennessee	4,210
Florida	4,984	Montana	3,491	Texas	2,910
Georgia	4,235	Nebraska	3,191	Utah	4,079
Hawaii	7,229	Nevada	4,100	Vermont	2,337
Idaho	4,431	New Hampshire	2,782	Virginia	3,885
Illinois	3,650	New Jersey	5,648	Washington	4,307
Indiana	3,894	New Mexico	4,405	West Virginia	2,918
Iowa	4,155	New York	5,934	Wisconsin	4,028
Kansas	3,722	North Carolina	5,082	Wyoming	6,028

Source: U.S. Bureau of the Census. *Statistical Abstract of the United States 1992.* 112th edition. Washington, D.C.: U.S. Department of Commerce, 1992, p. 168. Primary source: Research Associates of Washington, DC, *State Profiles: Financing Public Higher Education,* annual. *Note:* 1. FTE = Full-time equivalent.

★ 330 ★

College Costs

Net Tuition Revenue for Higher Education Operations by State, 1989

State	Tuition per FTE student (dol.)	State	Tuition per FTE student (dol.)	State	Tuition per FTE student (dol.)
Alabama	1,495	Kentucky	1,361	North Dakota	1,445
Alaska	1,468	Louisiana	1,660	Ohio	2,079
Arizona	1,426	Maine	1,492	Oklahoma	1,113
Arkansas	1,389	Maryland	1,657	Oregon	1,340
California	518	Massachusetts	1,501	Pennsylvania	2,731
Colorado	2,027	Michigan	2,417	Rhode Island	1,912
Connecticut	1,595	Minnesota	1,339	South Carolina	1,572
Delaware	3,890	Mississippi	1,463	South Dakota	1,821
District of Columbia	653	Missouri	1,505	Tennessee	1,397
Florida	1,061	Montana	797	Texas	960
Georgia	1,229	Nebraska	1,157	Utah	1,210
Hawaii	717	Nevada	1,099	Vermont	5,491
Idaho	683	New Hampshire	3,432	Virginia	1,745
Illinois	1,044	New Jersey	1,553	Washington	1,075
Indiana	2,010	New Mexico	955	West Virginia	1,171
Iowa	1,911	New York	1,220	Wisconsin	1,771
Kansas	1,261	North Carolina	831	Wyoming	801

Source: U.S. Bureau of the Census. *Statistical Abstract of the United States 1992.* 112th edition. Washington, D.C.: U.S. Department of Commerce, 1992, p. 168. Primary source: Research Associates of Washington, DC, *State Profiles: Financing Public Higher Education,* annual. *Note:* FTE = Full-time equivalent.

★ 331 ★

College Costs

Per Capita State and Local Government Expenditures for Higher Education by State, 1988-1989

State	Expenditure per Capita ($)	State	Expenditure per Capita ($)	State	Expenditure per Capita ($)
United States	272	Kentucky	250	North Dakota	419
Alabama	297	Louisiana	215	Ohio	255
Alaska	458	Maine	260	Oklahoma	283
Arizona	345	Maryland	273	Oregon	331
Arkansas	217	Massachusetts	201	Pennsylvania	152
California	316	Michigan	365	Rhode Island	253
Colorado	286	Minnesota	334	South Carolina	260
Connecticut	196	Mississippi	279	South Dakota	204
Delaware	444	Missouri	219	Tennessee	260
District of Columbia	157	Montana	212	Texas	293
Florida	187	Nebraska	347	Utah	396
Georgia	213	Nevada	218	Vermont	397
Hawaii	330	New Hampshire	204	Virginia	303
Idaho	270	New Jersey	245	Washington	334
Illinois	233	New Mexico	385	West Virginia	227
Indiana	295	New York	231	Wisconsin	367
Iowa	395	North Carolina	338	Wyoming	426
Kansas	341				

Source: National Education Association. *Rankings of the States, 1991.* Washington, D.C.: National Education Association, 1991, p. 58. Primary source: Bureau of the Census, *Government Finances: 1988-1989*, p. 99.

★ 332 ★

College Costs

Per Capita State Expenditures for State Institutions of Higher Education by State, 1988-1989

State	Expenditure per Capita ($)	State	Expenditure per Capita ($)	State	Expenditure per Capita ($)
United States	249	Kentucky	250	North Dakota	422
Alabama	297	Louisiana	213	Ohio	244
Alaska	458	Maine	262	Oklahoma	283
Arizona	286	Maryland	220	Oregon	258
Arkansas	217	Massachusetts	200	Pennsylvania	137
California	300	Michigan	320	Rhode Island	253
Colorado	276	Minnesota	291	South Carolina	260
Connecticut	196	Mississippi	215	South Dakota	204
Delaware	444	Missouri	179	Tennessee	260
District of Columbia	NA	Montana	203	Texas	251
Florida	182	Nebraska	291	Utah	396
Georgia	211	Nevada	218	Vermont	397
Hawaii	330	New Hampshire	204	Virginia	303
Idaho	252	New Jersey	210	Washington	334
Illinois	180	New Mexico	370	West Virginia	227
Indiana	295	New York	192	Wisconsin	280
Iowa	342	North Carolina	316	Wyoming	357
Kansas	278				

Source: National Education Association. *Rankings of the States, 1991*. Washington, D.C.: National Education Association, 1991, p. 57. Primary source: Bureau of the Census, *Government Finances: 1988-1989*, p. 45.

★ 333 ★

College Costs

Saving for A Child's Education

Year Student Enters College	Average Amount to Save Each Year to Fund 4 Years	
	Public College	Private College
1994	$9,647	$25,386
1996	6,379	16,788
1998	5,015	13,197
2000	4,266	11,278
2002	3,847	10,124
2004	3,566	9,384
2006	3,380	8,896
2008	3,258	8,573
2010	3,179	8,366

Source: "Amount to Save Each Year to Fully Fund, by Matriculation, Four Years at an Average Public/Private College," *Consumers' Research*, 75, May 1992, p. 11. Primary source: College Savings Bank, April 1992.

Grade Point Average

★ 334 ★

Grade Point Averages of Working High School Students

Findings from a study of California and Wisconsin high-school students.

Hours Worked	Grade Point Average
0	3.0
1-10	3.03
11-15	2.93
16-20	2.86
21+	2.66

Source: "Too Old, Too Fast?" Steven Waldman and Karen Springen, *Newsweek*, 120, November 16, 1992, p. 80. Primary source: L. Steinberg, S.M. Dornbusch.

Language

★ 335 ★

Most Common English Words

The following are the most common words used by the average person who speaks English: the, of, and, to, a, in, that, is, I, it, for, as.

Source: Brandreth, Gyles. *Your Vital Statistics*. New York: Citadel, 1986, p. 50.

★336★

Language

Number of Languages Spoken per Student in Selected Countries

Country	Average Number of Foreign Languages
Luxembourg	2.7
Denmark	1.8
Netherlands	1.6
Great Britain	0.5
Ireland	0.4
Germany	0.4

Source: "European Topics," *International Herald Tribune*, November 19, 1992, p. 9.

★337★

Language

Vocabulary of High School Graduates

The vocabulary of the average high school graduate is said to consist of 50,000 words.

Source: "Spell Checkers: The Dictionaries of Tomorrow?" *Consumer Reports*, 56, October 1991, p. 672.

Libraries and Books

★ 338 ★

Public Library Expenditures per Capita by State, 1989

State	Expenditure per Capita ($)	State	Expenditure per Capita ($)	State	Expenditure per Capita ($)
Alabama	8.13	Kentucky	7.04	North Dakota	8.01
Alaska	23.64	Louisiana	11.53	Ohio	23.34
Arizona	14.53	Maine	12.66	Oklahoma	10.24
Arkansas	5.98	Maryland	24.45	Oregon	14.74
California	15.89	Massachusetts	19.18	Pennsylvania	10.99
Colorado	18.01	Michigan	13.83	Rhode Island	11.71
Connecticut	22.10	Minnesota	18.62	South Carolina	8.72
Delaware	8.22	Mississippi	6.58	South Dakota	13.09
District of Columbia	29.46	Missouri	13.37	Tennessee	7.70
Florida	11.93	Montana	7.56	Texas	8.94
Georgia	10.06	Nebraska	13.16	Utah	13.64
Hawaii	17.23	Nevada	9.65	Vermont	12.37
Idaho	9.84	New Hampshire	14.46	Virginia	16.29
Illinois	19.25	New Jersey	21.16	Washington	19.81
Indiana	17.95	New Mexico	11.34	West Virginia	7.46
Iowa	11.85	New York	29.48	Wisconsin	16.53
Kansas	15.85	North Carolina	10.55	Wyoming	19.83

Source: The World Almanac and Book of Facts 1992. New York: World Almanac, 1991, p. 216. Primary source: National Center for Educational Statistics, U.S. Dept. of Education.

★ 339 ★

Libraries and Books

Book and Serials Collections in Public Libraries per Capita, 1990

According to a 1990 study of 8,978 public libraries across the United States conducted by the National Center for Education Statistics (NCES), public libraries possess an average of 2.5 volumes per capita in the book and serials collections, or a total of 613 million volumes.

Source: "Government Releases Telling Data on U.S. Public Libraries," *Library Journal*, 117, October 15, 1992, p. 16. Primary source: "Public Libraries in the U.S.: 1990," National Center for Education Statistics. **Remarks**: Also included in source: visits per capita; circulation of material per capita; items checked out per capita; states with the highest and lowest expenditures per capita; and states with the highest and lowest circulation per capita.

★ 340 ★

Libraries and Books

Visits to Public Libraries per Capita, 1990

According to a study of 8,978 public libraries across the United States conducted by the National Center for Education Statistics (NCES), in 1990, there were an average of 3.13 visits per capita to public libraries, or 507 million visits nationwide, checking out an average of almost two items per trip.

Source: "Government Releases Telling Data on U.S. Public Libraries," *Library Journal*, 117, October 15, 1992, p. 16. Primary source: "Public Libraries in the U.S.: 1990," National Center for Education Statistics. **Remarks**: Also included in source: book and serials collections per capita; states with the highest and lowest expenditures per capita; and states with the highest and lowest circulation per capita.

★ 341 ★

Libraries and Books

States with the Highest Public Library Circulation per Capita, 1990

State	Circulation per Capita
Ohio	10.21
Maryland	9.63
Washington	9.46
Minnesota	9.0
Wyoming	8.72
Nationwide	5.75

Source: "Government Releases Telling Data on U.S. Public Libraries," *Library Journal*, 117, October 15, 1992, p. 17. Primary source: "Public Libraries in the U.S.: 1990," National Center for Education Statistics. **Remarks**: Also included in source: book and serials collections per capita; visits per capita; circulation of material per capita; items checked out per capita; states with the highest and lowest expenditures per capita; and states with the lowest circulation per capita.

★ 342 ★

Libraries and Books

States with the Lowest Public Library Circulation per Capita, 1990

State	Circulation per Capita
Mississippi	2.98
South Carolina	3.25
Tennessee	3.46
Georgia	3.68

[Continued]

★ 342 ★

States with the Lowest Public Library Circulation per Capita, 1990

[Continued]

State	Circulation per Capita
Nevada	3.75
Nationwide	5.75

Source: "Government Releases Telling Data on U.S. Public Libraries," *Library Journal*, 117, October 15, 1992, p. 17. Primary source: "Public Libraries in the U.S.: 1990," National Center for Education Statistics. **Remarks**: Also included in source: book and serials collections per capita; visits per capita; circulation of material per capita; items checked out per capita; states with the highest and lowest expenditures per capita; and states with highest circulation per capita.

★ 343 ★

Libraries and Books

Median per Capita Level of Support for Public Libraries

According to recent Gallup telephone polls, the public would be willing to double support for public libraries from the current $16 in per capita spending. 62% of those polled felt that an average of $36.94 per capita should go toward funding public libraries, and 38% thought the community should spend an average of $30.60 per capita.

Source: "Public Would Double Library Support, Says National Survey," *Library Journal*, 117, August, 1992, p. 16.

★ 344 ★

Libraries and Books

Circulation of Books from Public Libraries in Selected Countries

Country	Number of People	Average Number of Books	Year
Canada	1.0	5.3	1980
Chile	2.6	1.0	1980
Germany	1.0	2.7	1980
Great Britain	1.0	11.4	1977
Japan	1.0	1.0	1980
Netherlands	1.0	1.1	1979
New Zealand	1.0	9.4	1979
Nigeria	431.0	1.0	1979
Poland	1.0	4.1	1980
Sweden	1.0	9.3	1980

[Continued]

★ 344 ★

Circulation of Books from Public Libraries in Selected Countries

[Continued]

Country	Number of People	Average Number of Books	Year
United States	1.0	4.5	1978
Zambia	37.0	1.0	1977

Source: Brandreth, Gyles. *Your Vital Statistics*. New York: Citadel, 1986, p. 31.

★ 345 ★

Libraries and Books

Holdings in School Libraries by Size of School, 1984-1985

Size of School	Average Number			
	Books	Periodicals	Films	Audio
Less than 300	4,793	23	300	175
300-499	6,927	27	499	311
500-699	8,250	32	640	396
700-999	9,602	46	705	444
1,000-1,999	13,802	73	928	765
2,000-	18,082	115	912	846
ALL	7,668	34	540	353

Source: National Center for Education Statistics. *Digest of Education Statistics 1991*. Washington, D.C.: U.S. Department of Education, 1991, p. 409. Primary source: "National Survey of Public and Private School Libraries and Media Centers, 1985," U.S. Department of Education, National Center for Education Statistics. **Remarks**: Source also includes: holdings in elementary and high school libraries; expenditures per student in school libraries by size of library, in elementary schools and high schools.

★ 346 ★

Libraries and Books

Holdings in Elementary School Libraries, 1984-1985

Number of Students in Elementary School	Average Number			
	Books	Periodicals	Films	Audio
Less than 300	4,720	19	313	184
300-499	6,867	24	509	313
500-699	8,178	28	637	396
700-999	9,242	38	742	461

[Continued]

★346★

Holdings in Elementary School Libraries, 1984-1985

[Continued]

Number of Students in Elementary School	Average Number			
	Books	Periodicals	Films	Audio
1,000-1,999	11,214	45	888	991
2,000-	14,644	73	253	422

Source: National Center for Education Statistics. *Digest of Education Statistics 1991*. Washington, D.C.: U.S. Department of Education, 1991, p. 409. Primary source: "National Survey of Public and Private School Libraries and Media Centers, 1985," U.S. Department of Education, National Center for Education Statistics. **Remarks**: Source also includes: holdings in school libraries by size of school, and in high schools; expenditures per student in school libraries by size of library, in elementary schools and high schools.

★347★

Libraries and Books

Holdings in High School Libraries, 1984-1985

Number of Students in High School	Average Number			
	Books	Periodicals	Films	Audio
Less than 300	5,275	45	216	113
300-499	7,624	57	384	288
500-699	8,901	68	665	398
700-999	10,987	76	563	380
1,000-1,999	15,334	90	952	631
2,000-	18,247	117	943	866

Source: National Center for Education Statistics. *Digest of Education Statistics 1991*. Washington, D.C.: U.S. Department of Education, 1991, p. 409. Primary source: "National Survey of Public and Private School Libraries and Media Centers, 1985," U.S. Department of Education, National Center for Education Statistics. **Remarks**: Source also includes: holdings in school libraries by size of school and in elementary schools; expenditures per student in school libraries by size of library, in elementary schools and high schools.

★348★

Libraries and Books

Expenditures in School Libraries by Size of School, 1984-1985

Size of School	Average Expenditure per Student				
	Books	Periodical subscriptions	Audio-visual	Computer hardware	Computer software
Less than 300	$9.18	$2.52	$2.54	$5.96	$1.48
300-499	5.58	1.18	1.65	3.04	0.80
500-699	5.25	1.01	1.46	2.41	0.54
700-999	4.67	1.14	1.52	1.94	0.48
1,000-1,999	4.47	1.23	1.41	1.71	0.35

[Continued]

★ 348 ★

Expenditures in School Libraries by Size of School, 1984-1985

[Continued]

Size of School	Average Expenditure per Student				
	Books	Periodical subscriptions	Audio-visual	Computer hardware	Computer software
2,000-	4.53	1.14	1.15	0.94	0.13
ALL	6.24	1.49	1.80	3.41	0.84

Source: National Center for Education Statistics. *Digest of Education Statistics 1991*. Washington, D.C.: U.S. Department of Education, 1991, p. 409. Primary source: "National Survey of Public and Private School Libraries and Media Centers, 1985," U.S. Department of Education, National Center for Education Statistics. **Remarks**: Source also includes: holdings in school libraries by size of school, in elementary schools, and high schools; expenditures per student in elementary and high school libraries.

★ 349 ★

Libraries and Books

Expenditures in Elementary School Libraries, 1984-1985

Number of Students in Elementary School	Average Expenditure per Student				
	Books	Periodical subscriptions	Audio-visual	Computer hardware	Computer software
Less than 300	$8.63	$2.00	$2.12	$5.57	$1.53
300-499	5.29	1.00	1.62	3.08	0.80
500-699	5.07	0.86	1.44	2.36	0.55
700-999	4.44	0.90	1.45	2.09	0.48
1,000-1,999	3.81	0.70	1.44	1.88	0.36
2,000-	5.13	0.53	1.22	0.00	0.02

Source: National Center for Education Statistics. *Digest of Education Statistics 1991*. Washington, D.C.: U.S. Department of Education, 1991, p. 409. Primary source: "National Survey of Public and Private School Libraries and Media Centers, 1985," U.S. Department of Education, National Center for Education Statistics. **Remarks**: Source also includes: holdings in school libraries by size of school, in elementary schools, and high schools; expenditures per student in school libraries by size of library and in high schools.

★ 350 ★

Libraries and Books

Expenditures in High School Libraries, 1984-1985

Number of Students in High School	Average Expenditure per Student				
	Books	Periodical subscriptions	Audio-visual	Computer hardware	Computer software
Less than 300	$12.76	$5.94	$5.37	$8.55	$1.16
300-499	9.01	3.31	2.00	2.62	0.79
500-699	6.87	2.35	1.65	2.87	0.45
700-999	5.56	2.05	1.77	1.36	0.48

[Continued]

★ 350 ★

Expenditures in High School Libraries, 1984-1985

[Continued]

Number of Students in High School	Average Expenditure per Student				
	Books	Periodical subscriptions	Audio-visual	Computer hardware	Computer software
1,000-1,999	4.86	1.55	1.39	1.62	0.34
2,000-	4.50	1.17	1.15	0.98	0.14

Source: National Center for Education Statistics. *Digest of Education Statistics 1991*. Washington, D.C.: U.S. Department of Education, 1991, p. 409. Primary source: "National Survey of Public and Private School Libraries and Media Centers, 1985," U.S. Department of Education, National Center for Education Statistics. **Remarks**: Source also includes: holdings in school libraries by size of school, in elementary schools, and high schools; expenditures per student in school libraries by size of library and in elementary schools.

★ 351 ★

Libraries and Books

Videos in Library Collections, 1992

Type of Library	Average Number of Videos
Academic	506
Public	380
School	175
Special	87
Combined	282

Source: "ABC-CLIO Study Says Libraries Hold Line on Video Budgets," *Library Journal*, 118, June 1, 1993, p. 18. Primary source: *The Video Annual 1993*.

★ 352 ★

Libraries and Books

Library Spending for Videos, 1992

Type of Library	Average Amount Spent on Videos
Academic	$5,547
Public	3,267
Special	1,614
School	1,049
Combined	2,610

Source: "ABC-CLIO Study Says Libraries Hold Line on Video Budgets," *Library Journal*, 118, June 1, 1993, p. 18. Primary source: *The Video Annual 1993*.

★ 353 ★

Libraries and Books

A Day at the Library of Congress

At the Library of Congress on an average workday:

1. 730 books cataloged.
2. 1,900 inquiries answered in its Copyright Office.
3. 2,000 responses to research assignments for Congress handled in its Congressional Research Service Office.
4. $1.2 million of taxpayers money spent.

Source: "The Washington Post Looks at the Library," Linton Weeks, *Library of Congress Information Bulletin*, v50, n17, September 9, 1991, p. 327.

★ 354 ★

Libraries and Books

Cities with the Most Money Spent in Bookstores per Household, 1987

City	Average Household Spending	City	Average Household Spending
Austin, TX	$195.86	San Diego,CA	$85.00
Madison, WI	175.70	Portland, OR	84.50
San Jose, CA	164.25	New York, NY	82.93
Boston, MA	158.19	Raleigh-Durham, NC	82.55
San Francisco, CA	136.95	Nashville, TN	82.32
Lansing, MI	134.11	Nassau-Suffolk, NY	78.04
Washington, DC	109.43	Middlesex, NJ	75.25
Seattle, WA	106.91	Salt Lake City, UT	73.04
Oakland, CA	95.07	Chicago, IL	71.62
Columbus, OH	86.69	U.S.	57.17
Los Angeles, CA	85.76		

Source: Los Angeles Public Library/State of California, *Scan/Info*, 4, September 1992, p. 17, from *Los Angeles Times*, August 26, 1992, p. D2. Primary source: American Booksellers Association, from the U.S. Census. **Remarks**: Source also includes: average number of bookstores per 100,000 households in selected cities.

★355★

Libraries and Books

Cities with the Most Bookstores per Household, 1987

Cities	Average No. of Bookstores per 100,000 Households	Cities	Average No. of Bookstores per 100,000 Households
San Francisco, CA	2.43	San Jose, CA	1.53
Austin, TX	2.21	Charlotte, NC	1.50
Madison, WI	1.95	Oakland, CA	1.44
Raleigh-Durham, NC	1.92	Nassau-Suffolk, NY	1.43
Washington, DC	1.80	Columbus, OH	1.43
Portland, OR	1.70	Baltimore, MD	1.40
Boston, MA	1.67	Norfolk, VA	1.39
Nashville, TN	1.64	Denver, CO	1.39
San Diego, CA	1.57	Miami, FL	1.38
Seattle, WA	1.56	U.S.	1.24
Salt Lake City, UT	1.55		

Source: Los Angeles Public Library/State of California, *Scan/Info*, 4, September 1992, p. 17, from *Los Angeles Times*, August 26, 1992, p. D2. Primary source: American Booksellers Association, from the U.S. Census. **Remarks**: Also included in source: average household spending in bookstores per household.

★356★

Libraries and Books

Length of Visit to Children's Section in Bookstore

Type of Store	Average Length of Visit (in minutes)
Children's only	23.6
General/Specialty	17.1
National/Regional Chains	21.5
College/University	17.0

Source: "Customers' Average Length of Visit to Children's Section in Store," *Publishers Weekly*, 239, January 13, 1992, p. 28.

★357★

Libraries and Books

Amount Spent on Children's Books per Customer

Type of Store	Average Spent on Children's Books per Customer
Children's only	$23.40
General/Specialty	13.70
National/Regional Chains	13.70
College/University	12.90

Source: "Average Dollar Purchase of Children's Books per Customer," *Publishers Weekly*, 239, January 13, 1992, p. 28.

★358★

Libraries and Books

Number of Children's Books Purchased per Customer

Type of Store	Average No. of Children's Books Purchased per Customer
Children's only	3.5
General/Specialty	2.8
National/Regional Chains	3.1
College/University	2.3

Source: "Average Dollar Purchase of Children's Books per Customer," *Publishers Weekly*, 239, January 13, 1992, p. 28.

★359★

Libraries and Books

Cost of Hardcover Books by Subject, 1977-1992

Category	Average Price			
	1977	1990	1991	1992 Prelim.
Art	21.24	42.18	44.99	43.35
Biography	15.34	29.58	27.52	29.79
Business	18.00	45.48	43.38	44.11
Education	12.95	38.72	41.26	49.11
Fiction	10.09	19.83	21.30	20.34
History	17.12	36.43	39.87	38.58
Juveniles	6.65	13.01	16.64	14.51
Language	14.96	42.98	51.71	50.88
Law	25.04	60.78	64.89	75.84

[Continued]

★ 359 ★

Cost of Hardcover Books by Subject, 1977-1992

[Continued]

Category	Average Price			
	1977	1990	1991	1992 Prelim.
Literature	15.78	35.80	36.76	39.17
Medicine	24.00	72.24	71.44	75.00
Philosophy, Psychology	14.43	40.58	42.74	45.96
Poetry, Drama	13.63	32.19	33.29	36.88
Religion	12.26	31.31	32.33	35.66
Science	24.88	74.39	80.14	80.63
Sociology, Economics	29.88	42.10	48.43	45.69
Sports, Recreation	12.28	30.52	30.68	34.73
Technology	23.61	76.80	76.40	81.21
Travel	18.44	30.41	33.50	32.87
Total	19.22	42.12	44.17	45.25

Source: "Moving On Up?" Chandler Grannis, *Publishers Weekly*, 240, March 1, 1993, pp. S32-35.
Remarks: Source also includes: average price for mass market and trade paperbacks, 1977-1992.

★ 360 ★

Libraries and Books

Cost of Hardcover Books (less than $81.00) by Subject, 1977-1992

	Average Price			
	1987	1990	1991	1992 Prelim.
Art	33.07	37.31	37.92	36.99
Biography	22.75	25.47	25.23	26.47
Business	30.48	36.43	38.01	39.62
Education	29.37	36.35	36.51	39.87
Fiction	17.02	19.27	19.55	19.60
History	29.58	33.77	35.31	35.84
Juveniles	11.29	12.88	13.37	13.88
Language	33.26	34.74	40.32	38.99
Law	37.40	40.42	41.31	42.97
Literature	28.97	33.50	33.64	35.18
Medicine	38.33	41.53	40.19	41.38
Philosophy, Psychology	31.48	35.76	36.75	39.14
Poetry, Drama	24.99	29.93	31.32	32.35
Religion	23.09	29.30	29.70	31.22
Science	43.02	45.45	45.77	47.03
Sociology, Economics	31.65	37.78	39.19	40.38
Sports, Recreation	23.32	30.00	29.42	33.25
Technology	41.68	45.73	46.38	48.65

[Continued]

★ 360 ★

Cost of Hardcover Books (less than $81.00) by Subject, 1977-1992

[Continued]

	Average Price			
	1987	1990	1991	1992 Prelim.
Travel	24.45	29.12	29.00	30.08
Total	28.96	31.60	31.95	33.38

Source: "Moving On Up?" Chandler Grannis, *Publishers Weekly*, 240, March 1, 1993, pp. S32-35.
Remarks: Source also includes: average price for mass market and trade paperbacks, 1977-1992.

★ 361 ★

Libraries and Books

Cost of Mass Market Paperbacks by Subject, 1990-1992

	Average Price		
	1990	1991	1992 Prelim.
Art	13.40	10.13	13.00
Biography	7.24	6.18	6.19
Business	7.85	8.36	8.15
Education	8.08	5.98	7.28
Fiction	4.05	4.49	4.49
History	6.24	8.37	7.13
Juveniles	3.56	3.38	3.61
Language	5.36	7.26	6.59
Law	6.61	9.85	4.99
Literature	6.45	5.89	6.96
Medicine	8.24	5.88	9.49
Philosophy, Psychology	6.81	7.78	7.27
Poetry, Drama	6.08	6.59	6.36
Religion	6.95	7.79	6.33
Science	9.17	8.27	8.00
Sociology, Economics	6.88	7.56	6.18
Sports, Recreation	5.29	7.50	7.08
Technology	24.16	25.58	32.88
Travel	11.22	10.95	9.00
Total	4.57	5.08	5.10

Source: "Moving On Up?" Chandler Grannis, *Publishers Weekly*, 240, March 1, 1993, pp. S32-35.
Remarks: Source also includes: average price of trade paperbacks and hardcover books.

★ 362 ★

Libraries and Books

Cost of Trade Paperbacks by Subject, 1977-1992

Category	Average Price			
	1977	1990	1991	1992 Prelim.
Art	6.27	17.90	19.11	19.20
Biography	4.91	13.05	13.38	13.68
Business	7.09	19.61	21.20	22.31
Education	5.72	19.20	23.30	23.48
Fiction	4.20	11.32	12.17	13.69
History	5.81	17.39	16.99	17.38
Juveniles	2.68	6.78	7.60	7.34
Language	7.79	16.67	16.83	18.05
Law	10.66	25.15	24.62	23.64
Literature	5.18	15.54	15.31	16.89
Medicine	7.63	22.82	24.20	25.28
Philosophy, Psychology	5.57	15.29	16.54	17.20
Poetry, Drama	4.71	12.43	11.88	13.12
Religion	3.68	12.12	12.41	13.21
Science	8.81	28.01	28.31	27.98
Sociology, Economics	6.03	19.60	19.68	19.72
Sports, Recreation	4.87	13.67	15.03	15.49
Technology	7.97	30.06	33.39	28.72
Travel	5.21	13.32	14.06	14.64
Total	5.93	17.45	18.39	18.60

Source: "Moving On Up?" Chandler Grannis, *Publishers Weekly*, 240, March 1, 1993, pp. S32-35.
Remarks: Source also includes: average price of mass market paperbacks and hardcover books.

★ 363 ★

Libraries and Books

Cost of Periodical Subscriptions by Subject, 1989-1993

Subject	Average No. of Titles	Average Cost 1989	Average Cost 1990	Average Cost 1991	Average Cost 1992	Average Cost 1993	'89-'93 % of change
General works	78	$51.40	$55.64	$65.05	$63.28	$68.44	33.15
Philosophy & Religion	123	61.27	64.92	75.43	78.62	88.53	44.49
Psychology	132	99.09	109.06	122.49	136.98	149.29	50.66
History	202	53.95	57.20	63.49	69.10	74.16	37.47
Geography	61	141.36	155.85	193.25	218.88	236.79	67.51
Recreation	16	45.60	47.95	56.86	60.75	65.81	44.33
Business & Economics	247	119.56	131.08	159.79	175.55	196.86	64.65
Political Science	62	70.77	76.47	92.58	99.94	112.50	58.96
Law	85	55.02	58.92	68.86	73.98	83.86	52.44
Education	97	82.41	92.03	107.48	113.67	130.57	58.43
Art & Architecture	18	67.25	73.14	79.05	85.51	91.23	35.67
Language & Literature	292	49.32	51.36	59.02	62.08	67.57	37.01
General Science	66	180.49	190.34	234.15	300.35	348.18	92.91
Math & Computer Science	131	356.43	385.19	466.66	506.08	567.11	59.11
Physics	189	631.46	680.06	812.40	927.27	1090.34	72.67
Chemistry	193	608.29	641.26	789.09	915.75	1042.36	71.36
Geology	68	283.69	297.66	366.22	380.53	443.72	56.41
Biology	231	347.81	372.62	443.14	474.94	548.21	57.62

[Continued]

★ 363 ★

Cost of Periodical Subscriptions by Subject, 1989-1993

[Continued]

Subject	Average No. of Titles	Average Cost 1989	Average Cost 1990	Average Cost 1991	Average Cost 1992	Average Cost 1993	'89-'93 % of change
Engineering & Technology	386	298.67	323.34	415.76	480.38	530.49	77.62
Lib. & Information Science	62	82.65	91.52	106.11	119.82	127.08	53.76

Source: "The Art of Projecting: The Cost of Keeping Periodicals," *Library Journal*, 118, April 15, 1993, pp. 42-48. **Remarks**: Source also includes: percentage of change in the cost of periodicals, 1989-1993; cost of periodical subscription by country/continent; percentage of change in cost of periodicals by country/continent; cost of magazine article summaries, 1989-1993; and cost projections by broad subject.

★ 364 ★

Libraries and Books

Cost of Periodical Subscription by Country/Continent, 1989-1993

Continent/Country	Average No. of Titles	Average Cost 1989	Average Cost 1990	Average Cost 1991	Average Cost 1992	Average Cost 1993	'89-'93 % of Change
North America							
United States	2146	$141.89	$155.80	$173.87	$194.34	$216.43	52.53
Canada	104	55.96	60.46	76.37	82.55	82.11	46.73
Other	7	37.63	38.94	37.58	37.63	40.71	8.21
Total Avg. for all N.A.	2257	137.88	151.28	168.88	188.52	209.40	51.88
Europe							
France	148	138.21	155.35	186.55	192.57	218.88	58.37
Germany	311	376.99	407.32	483.87	477.60	582.19	54.60
Ireland	37	435.79	507.32	606.60	674.25	890.36	104.31
Italy	53	110.84	120.89	134.77	139.44	140.84	27.07
Netherlands	309	590.92	633.00	833.46	869.25	1066.37	80.46
Switzerland	128	448.23	461.71	638.48	634.72	714.71	59.45
United Kingdom	1021	249.86	276.74	340.07	404.29	427.39	71.05
Other	205	145.24	153.84	189.08	213.63	249.28	71.63
Total Avg. for all Europe	2212	308.07	333.87	418.95	463.58	530.88	72.33
Asia							
Japan	81	202.87	202.83	216.13	225.50	246.72	21.61
Other	43	80.66	83.45	102.47	119.99	127.33	57.86
Total Avg. for all Asia	124	159.80	159.25	175.47	189.76	207.89	30.10
Australia and New Zealand	70	96.81	106.31	119.13	125.32	128.22	32.44
South America	16	55.33	56.59	60.16	62.01	69.62	25.84
Africa	14	29.79	33.67	69.76	56.09	61.81	107.48

Source: "The Art of Projecting: The Cost of Keeping Periodicals," *Library Journal*, 118, April 15, 1993, pp. 42-48. **Remarks**: Source also includes: cost of periodical subscriptions by subject; percentage of change in the cost of periodicals, 1989-1993; percentage of change in cost of periodicals by country/continent; cost of magazine article summaries, 1989-1993; and cost projections by broad subject.

★ 365 ★

Libraries and Books

Books Read by Men and Women

According to a recent Gallup Poll, women who read, read an average of 18 books per year, but men readers, read just 12 books per year. Adults who read books today is double the number who read books in the 1950s.

Source: "Business Reports: Media: The 1990s Will Be Better for Reading," Judith Waldrop, *American Demographics*, 13, November 1991, p. 17.

★ 366 ★

Libraries and Books

Reading Habits of an Eleven-Year-Old

The average eleven-year-old reads only 11 pages of text per day.

Source: "New Computerization," *Advertising Marketing Bulletin*, 48, Spring 1993, p. 10.

School Enrollment

★ 367 ★

Public School Enrollment

According to the National Center for Education Statistics, an average of 500 students are enrolled per public school in the United States. Broken down into type of school: an average of 430 students are enrolled at elementary schools, 560 at middle schools, and 700 at high schools.

Source: "Flashcard," *The New York Times*, January 10, 1993, p. 4A.

★ 368 ★

School Enrollment

Average Daily Attendance in Public Schools by State, 1990-1991

State	Average number	State	Average number	State	Average number
United States	38,167,599	Kentucky	569,200	North Dakota	113,000
Alabama	681,865	Louisiana	715,814	Ohio	1,517,000
Alaska	101,150	Maine	198,758	Oklahoma	543,200
Arizona	598,499	Maryland	661,765	Oregon	448,200
Arkansas	409,535	Massachusetts	771,163	Pennsylvania	1,504,000
California	4,978,018	Michigan	1,455,735	Rhode Island	124,440
Colorado	521,899	Minnesota	708,973	South Carolina	573,138
Connecticut	449,700	Mississippi	474,109	South Dakota	120,205
Delaware	92,500	Missouri	731,650	Tennessee	766,337
District of Columbia	70,548	Montana	136,569	Texas	3,172,518
Florida	1,778,494	Nebraska	254,811	Utah	416,229
Georgia	1,055,184	Nevada	186,413	Vermont	88,266
Hawaii	159,093	New Hampshire	158,042	Virginia	932,143
Idaho	210,500	New Jersey	1,005,263	Washington	783,608
Illinois	1,572,455	New Mexico	254,794	West Virginia	297,309
Indiana	855,945	New York	2,306,000	Wisconsin	699,480
Iowa	454,280	North Carolina	1,010,040	Wyoming	92,300
Kansas	387,462				

Source: National Education Association. *Rankings of the States, 1991*. Washington, D.C.: National Education Association, 1991, p. 13.

★ 369 ★

School Enrollment

Average Daily Membership in Public Schools by State, 1990-1991

State	Average number	State	Average number	State	Average number
Alabama	718,859	Kentucky	599,658	North Dakota	116,730
Alaska	109,946	Louisiana	763,910	Ohio	1,687,900
Arizona	634,039	Maine	208,350	Oklahoma	573,300
Arkansas	429,586	Maryland	711,576	Oregon	479,900
California	NA	Massachusetts	830,928	Pennsylvania	1,630,000
Colorado	NA	Michigan	NA	Rhode Island	133,857
Connecticut	473,200	Minnesota	747,862	South Carolina	597,777
Delaware	98,650	Mississippi	497,614	South Dakota	125,832
District of Columbia	78,300	Missouri	NA	Tennessee	816,918
Florida	1,834,198	Montana	145,287	Texas	NA
Georgia	1,128,089	Nebraska	266,983	Utah	439,895
Hawaii	170,893	Nevada	197,264	Vermont	92,588
Idaho	NA	New Hampshire	167,240	Virginia	990,176
Illinois	1,692,041	New Jersey	1,080,928	Washington	836,294
Indiana	903,882	New Mexico	283,104	West Virginia	NA
Iowa	476,145	New York	2,522,000	Wisconsin	748,667
Kansas	410,982	North Carolina	1,070,850	Wyoming	97,200

Source: National Education Association. *Rankings of the States, 1991*. Washington, D.C.: National Education Association, 1991, p. 13.

★ 370 ★

School Enrollment

Students per Teacher Enrolled in Public Schools, by State, 1990-1991

State	Average number	State	Average number	State	Average number
United States	17.2	Kentucky	17.2	North Dakota	15.6
Alabama	18.1	Louisiana	18.0	Ohio	17.2
Alaska	17.0	Maine	14.5	Oklahoma	15.7
Arizona	18.4	Maryland	17.0	Oregon	18.5
Arkansas	17.0	Massachusetts	15.4	Pennsylvania	16.6
California	23.1	Michigan	19.6	Rhode Island	14.6
Colorado	17.8	Minnesota	17.4	South Carolina	17.2
Connecticut	13.4	Mississippi	18.0	South Dakota	15.4
Delaware	16.7	Missouri	15.8	Tennessee	19.1
District of Columbia	13.2	Montana	15.9	Texas	16.4
Florida	17.2	Nebraska	14.8	Utah	25.6
Georgia	16.7	Nevada	19.4	Vermont	13.5
Hawaii	17.5	New Hampshire	16.5	Virginia	15.6
Idaho	19.6	New Jersey	13.6	Washington	20.1
Illinois	16.8	New Mexico	17.4	West Virginia	15.0
Indiana	17.5	New York	14.0	Wisconsin	16.6
Iowa	15.5	North Carolina	17.0	Wyoming	15.0
Kansas	15.0				

Source: National Education Association. *Rankings of the States, 1991*. Washington, D.C.: National Education Association, 1991, p. 19.

★ 371 ★

School Enrollment

Students per Teacher in Average Daily Attendance in Public Schools, by State, 1990-1991

State	Average number	State	Average number	State	Average number
United States	15.9	Kentucky	15.5	North Dakota	15.0
Alabama	17.0	Louisiana	16.4	Ohio	14.7
Alaska	15.4	Maine	13.7	Oklahoma	14.8
Arizona	17.3	Maryland	15.7	Oregon	17.1
Arkansas	16.0	Massachusetts	14.3	Pennsylvania	15.0
California	23.2	Michigan	18.0	Rhode Island	13.2
Colorado	16.1	Minnesota	16.3	South Carolina	15.9
Connecticut	12.7	Mississippi	17.1	South Dakota	14.4
Delaware	15.5	Missouri	14.3	Tennessee	17.6
District of Columbia	11.5	Montana	14.2	Texas	15.4
Florida	16.5	Nebraska	13.8	Utah	23.9
Georgia	15.3	Nevada	18.0	Vermont	12.4
Hawaii	16.3	New Hampshire	15.1	Virginia	14.6
Idaho	18.7	New Jersey	12.6	Washington	18.7
Illinois	14.5	New Mexico	15.7	West Virginia	13.8
Indiana	15.7	New York	12.4	Wisconsin	14.7
Iowa	14.6	North Carolina	15.8	Wyoming	14.1
Kansas	13.3				

Source: National Education Association. *Rankings of the States, 1991*. Washington, D.C.: National Education Association, 1991, p. 20.

★ 372 ★

School Enrollment

Students per Teacher in Average Daily Membership by State, 1990-1991

State	Average number	State	Average number	State	Average number
Alabama	18.0	Louisiana	17.5	Ohio	16.4
Alaska	16.7	Maine	14.4	Oklahoma	15.6
Arizona	18.4	Maryland	16.9	Oregon	18.3
Arkansas	16.8	Massachusetts	15.4	Pennsylvania	16.3
Connecticut	13.4	Minnesota	17.2	Rhode Island	14.2
Delaware	16.6	Mississippi	17.9	South Carolina	16.5
District of Columbia	12.8	Montana	15.1	South Dakota	15.1
Florida	17.0	Nebraska	14.5	Tennessee	18.7
Georgia	16.4	Nevada	19.0	Utah	25.3
Hawaii	17.5	New Hampshire	16.0	Vermont	13.0
Illinois	15.6	New Jersey	13.5	Virginia	15.5
Indiana	16.6	New Mexico	17.4	Washington	20.0
Iowa	15.3	New York	13.6	Wisconsin	15.7
Kansas	14.1	North Carolina	16.8	Wyoming	14.8
Kentucky	16.4	North Dakota	15.5		

Source: National Education Association. *Rankings of the States, 1991*. Washington, D.C.: National Education Association, 1991, p. 20.

★ 373 ★

School Enrollment

Students per Teacher in All Schools in Selected Countries, 1970-1988

Country	1970	1980	1985	1988
Australia	-	16.0	13.9	14.5
Brazil	-	23.5	22.3	22.4
Canada	20.9	16.8	-	-
China	30.1	23.4	22.1	20.0
Egypt	33.2	-	26.2	24.0
France	20.0	21.4	17.9	15.6
Germany, Federal Republic of (former)	19.4	16.3	15.0	14.8
Indonesia	30.1	26.7	21.8	20.2
Iran, Islamic Republic of	32.8	-	-	24.1
Italy	15.6	12.3	10.8	10.1
Japan	21.8	20.9	20.5	19.4
Korea	49.9	43.5	35.4	31.1
Mexico	34.8	30.2	26.1	24.9
Nigeria	32.3	35.7	-	-
Pakistan	32.1	27.9	29.4	29.3
Philippines	29.4	31.3	31.2	33.1
Poland	17.5	16.6	16.5	14.0
Spain	28.5	23.9	22.6	22.3

[Continued]

★ 373 ★

Students per Teacher in All Schools in Selected Countries, 1970-1988

[Continued]

Country	1970	1980	1985	1988
Sweden	13.6	-	-	-
Thailand	30.5	31.1	19.4	18.2
United Kingdom	19.6	16.9	-	-
United States	22.4	18.6	17.6	17.0
Yugoslavia (former)	24.3	20.2	19.6	18.8

Source: National Center for Education Statistics. *Digest of Education Statistics 1991*. Washington, D.C.: U.S. Department of Education, 1991, p. 395. Primary source: United Nations Educational, Scientific, and Cultural Organization, Paris, *Statistical Yearbook*; and U.S. Department of Education, National Center for Education Statistics, Common Core of Data surveys and surveys of private schools.

★ 374 ★

School Enrollment

Students per Teacher in Elementary Schools in Selected Countries, 1970-1988

Country	1970	1980	1985	1988
Australia	28.0	18.8	15.9	16.6
Brazil	23.6	25.6	23.9	23.9
Canada	23.4	15.8	7.5	7.3
China	33.3	26.6	24.9	22.8
Egypt	38.0	-	31.9	29.9
France	26.0	24.0	20.5	21.0
Germany, Federal Republic of (former)	25.5	18.4	17.3	17.5
Indonesia	28.9	32.4	25.3	23.6
Iran, Islamic Republic of	32.4	27.4	21.9	26.1
Italy	21.6	16.2	13.5	12.5
Japan	26.2	25.1	23.9	22.2
Korea	56.9	47.5	37.9	36.3
Mexico	45.9	39.1	33.6	31.3
Nigeria	34.1	37.2	44.1	40.4
Pakistan	41.5	36.5	38.6	39.3
Philippines	28.6	30.4	30.9	33.0
Poland	23.0	19.7	20.7	15.6
Spain	34.0	28.3	25.3	24.7
Sweden	20.0	16.3	6.5	6.2
Thailand	34.7	24.7	20.1	18.7
United Kingdom	23.3	18.9	17.6	20.3
United States	24.6	20.1	19.2	17.9

[Continued]

★374★

Students per Teacher in Elementary Schools in Selected Countries, 1970-1988

[Continued]

Country	1970	1980	1985	1988
USSR (former)	10.9	9.1	9.4	8.5
Yugoslavia (former)	27.1	24.1	23.6	22.7

Source: National Center for Education Statistics. *Digest of Education Statistics 1991*. Washington, D.C.: U.S. Department of Education, 1991, p. 395. Primary source: United Nations Educational, Scientific, and Cultural Organization, Paris, *Statistical Yearbook*; and U.S. Department of Education, National Center for Education Statistics, Common Core of Data surveys and surveys of private schools.

★375★

School Enrollment

Students per Teacher in Secondary Schools in Selected Countries, 1970-1988

Country	1970	1980	1985	1988
Australia	-	12.9	12.1	12.6
Brazil	13.2	14.2	14.6	14.6
Canada	16.9	17.8	-	-
China	21.8	17.9	17.2	15.5
Egypt	25.0	24.0	20.4	18.0
France	15.8	19.6	16.3	13.0
Germany, Federal Republic of (former)	12.3	14.4	14.3	14.0
Indonesia	13.1	14.9	15.3	14.8
Iran, Islamic Republic of	34.2	-	-	20.9
Italy	11.5	10.2	9.5	9.1
Japan	18.4	17.2	17.9	17.5
Korea	36.5	39.1	33.2	26.9
Mexico	14.5	17.7	17.2	17.3
Nigeria	21.2	28.8	-	-
Pakistan	19.8	17.5	18.2	18.1
Philippines	33.1	34.1	32.3	33.4
Poland	10.2	12.0	10.6	10.8
Spain	21.5	20.9	21.0	20.9
Sweden	10.1	-	-	-
Thailand	15.5	-	17.6	16.8
United Kingdom	15.9	15.3	-	-
United States	19.6	16.6	15.5	15.8
Yugoslavia (former)	22.4	18.5	17.7	17.1

Source: National Center for Education Statistics. *Digest of Education Statistics 1991*. Washington, D.C.: U.S. Department of Education, 1991, p. 395. Primary source: United Nations Educational, Scientific, and Cultural Organization, Paris, *Statistical Yearbook*; and U.S. Department of Education, National Center for Education Statistics, Common Core of Data surveys and surveys of private schools.

School Expenditures

★ 376 ★

Expenditures per Student in Selected Countries, 1987-1988

Country	1987-88 Expenditure per Student
United States	$3,603
Australia	2,127
Austria	3,226
Belgium	2,867
Canada	4,207
Denmark	3,610
Finland	3,264
France	2,393
Germany, West (former)	2,300
Ireland	1,251
Italy	1,910
Japan	2,096
Netherlands	2,259
New Zealand	1,670
Norway	4,368
Portugal	980
Spain	855
Sweden	4,295
Switzerland	4,515
Turkey	639
United Kingdom	2,687

Source: National Center for Education Statistics. *Digest of Education Statistics 1991.* Washington, D.C.: U.S. Department of Education, 1991, Table 381. Primary source: United Nations Educational, Scientific, and Cultural Organization, Paris, *Statistical Yearbook*; and U.S. Department of Education, National Center for Education Statistics, Common Core of Data surveys and surveys of private schools.

★ 377 ★

School Expenditures

Expenditures per Capita for Public Schools, 1970-1990

Year	Per capita (dol.)	Year	Per capita (dol.)
1970	194	1986	640
1975	292	1987	679
1980	428	1988	725
1982	492	1989	783

[Continued]

★ 377 ★

Expenditures per Capita for Public Schools, 1970-1990

[Continued]

Year	Per capita (dol.)	Year	Per capita (dol.)
1983	514	1990	845
1984	543	1991, total	907
1985	591		

Source: U.S. Bureau of the Census. *Statistical Abstract of the United States 1992*. 112th edition. Washington, D.C.: U.S. Department of Commerce, 1992, p. 154. Primary source: National Education Association, Washington, DC, *Estimates of School Statistics*, annual. **Remarks**: Source also includes: per capita spending on public schools by state, and average expenditures per student in daily attendance by year and by state.

★ 378 ★

School Expenditures

Expenditures per Capita for Public Schools by State, 1991

State	Per capita (dol.)	State	Per capita (dol.)	State	Per capita (dol.)
U.S.	907	Kentucky	725	North Dakota	719
Alabama	694	Louisiana	758	Ohio	861
Alaska	1,626	Maine	1,062	Oklahoma	776
Arizona	892	Maryland	944	Oregon	925
Arkansas	668	Massachusetts	866	Pennsylvania	889
California	918	Michigan	926	Rhode Island	892
Colorado	841	Minnesota	990	South Carolina	835
Connecticut	1,151	Mississippi	1,726	South Dakota	716
Delaware	891	Missouri	742	Tennessee	618
District of Columbia	1,010	Montana	983	Texas	905
Florida	862	Nebraska	757	Utah	828
Georgia	859	Nevada	1,004	Vermont	992
Hawaii	784	New Hampshire	881	Virginia	941
Idaho	725	New Jersey	1,223	Washington	1,095
Illinois	788	New Mexico	915	West Virginia	883
Indiana	925	New York	1,186	Wisconsin	928
Iowa	846	North Carolina	813	Wyoming	1,178
Kansas	906				

Source: U.S. Bureau of the Census. *Statistical Abstract of the United States 1992*. 112th edition. Washington, D.C.: U.S. Department of Commerce, 1992, p. 154. Primary source: National Education Association, Washington, DC, *Estimates of School Statistics*, annual. **Remarks**: Source also includes: per capita spending on public schools by year, and average expenditures per student in daily attendance by year and by state.

★379★

School Expenditures

Expenditures per Student in Daily Public School Attendance, 1970-1990

Year	Expenditures per pupil in ADA (dol.)	Year	Expenditures per pupil in ADA (dol.)
1970	773	1986	3,764
1975	1,286	1987	3,996
1980	2,230	1988	4,280
1982	2,753	1989	4,620
1983	2,960	1990	4,975
1984	3,185	1991, total	5,261
1985	3,483		

Source: U.S. Bureau of the Census. *Statistical Abstract of the United States 1992*. 112th edition. Washington, D.C.: U.S. Department of Commerce, 1992, p. 154. Primary source: National Education Association, Washington, DC, *Estimates of School Statistics*, annual. **Remarks**: Source also includes: per capita spending on public schools by year and by state, and average expenditures per student in daily attendance by state. *Note:* ADA: Average Daily Attendance.

★ 380 ★

School Expenditures

Expenditures per Student in Daily Public School Attendance, by State, 1991

State	Expenditures per pupil in ADA (dol.)	State	Expenditures per pupil in ADA (dol.)	State	Expenditures per pupil in ADA (dol.)
U.S.	5,261	Kentucky	4,390	North Dakota	3,685
Alabama	3,648	Louisiana	4,012	Ohio	5,639
Alaska	7,887	Maine	5,894	Oklahoma	3,742
Arizona	4,231	Maryland	6,184	Oregon	5,291
Arkansas	3,334	Massachusetts	6,351	Pennsylvania	6,534
California	4,826	Michigan	5,257	Rhode Island	6,989
Colorado	4,809	Minnesota	5,260	South Carolina	4,327
Connecticut	7,914	Mississippi	3,322	South Dakota	3,730
Delaware	6,016	Missouri	4,415	Tennessee	3,707
District of Columbia	8,210	Montana	5,184	Texas	4,238
Florida	5,154	Nebraska	4,381	Utah	2,993
Georgia	4,860	Nevada	4,564	Vermont	5,740
Hawaii	5,008	New Hampshire	5,504	Virginia	5,360
Idaho	3,200	New Jersey	9,159	Washington	5,045
Illinois	5,062	New Mexico	4,446	West Virginia	5,046
Indiana	5,051	New York	8,500	Wisconsin	5,946
Iowa	4,839	North Carolina	4,802	Wyoming	5,255
Kansas	5,009				

Source: U.S. Bureau of the Census. *Statistical Abstract of the United States 1992*. 112th edition. Washington, D.C.: U.S. Department of Commerce, 1992, p. 154. Primary source: National Education Association, Washington, DC, *Estimates of School Statistics*, annual. **Remarks**: Source also includes: per capita spending on public schools by year and by state, and average expenditures per student in daily attendance by year. *Note:* ADA: Average Daily Attendance.

★ 381 ★

School Expenditures

Expenditures per Capita for Public School, an Historical Perspective, 1899-1989

Year	Expenditures per Capita
1899-1900	$2.83
1909-1910	4.71
1919-1920	9.91
1929-1930	19.03
1939-1940	17.91
1949-1950	39.00
1959-1960	87.00
1969-1970	200.00
1979-1980	424.00
1988-1989	772.00

Source: The World Almanac and Book of Facts 1992. New York: World Almanac, 1991, p. 213. Primary source: U.S. Department of Education, National Center for Education Statistics.

★ 382 ★

School Expenditures

Tuition at Private Schools in the United States, 1988

Orientation	Average Annual Tuition		
	Elementary	Secondary	Combined
Catholic	$1,005	$2,045	$3,382
Religious (non-Catholic)	1,619	3,592	2,052
Secular private	3,091	6,391	3,941

Source: U.S. Bureau of the Census. *Statistical Abstract of the United States 1992*. 112th edition. Washington, D.C.: U.S. Department of Commerce, 1992, p. 158. Primary source: U.S. National Center for Health Statistics, *Digest of Education Statistics*, 1990.

★ 383 ★

School Expenditures

Tuition and Boarding Fees in Great Britain

The average tuition and boarding expenses per year at first-rank schools in Great Britain is $17,000.

Source: "His School Days Now Would Tickle Tom Brown," William E. Schmidt, *The New York Times*, November 2, 1992, p. A4.

★ 384 ★

School Expenditures

Cost of Boarding Schools in the United States

The average cost per year for a five-day boarding school in the United States is $13,000, but for a seven-day boarding school the average cost is $16,000.

Source: "Boarding Schools Report Gains in 5-Day Programs," *The Wall Street Journal*, October 19, 1992, p. B1.

★ 385 ★

School Expenditures

Cost of School: Alternatives to Public Education

Type	Average Annual Cost per Student
Home-schools	$400-500
Catholic schools	
Elementary	$1,000
Secondary	$2,500

[Continued]

★ 385 ★

Cost of School: Alternatives to Public Education

[Continued]

Type	Average Annual Cost per Student
Private schools	$7,200
Boarding schools	$13,700

Source: "The Exodus," *U.S. News & World Report*, 113, December 9, 1991, pp. 66- 75.

★ 386 ★

School Expenditures

Median Custodial and Maintenance Cost of Schools per Student, 1991

According to the *American School & University*'s annual Maintenance and Operations Cost Study, the median maintenance and operations costs of schools per student was $489.45 in 1991.

Source: "Maintenance & Operations Costs Expressed in Dollars per Student," *AS&U*, 64, April 1992, p. 34.

★ 387 ★

School Expenditures

Median Custodial and Maintenance Cost of Schools per Square Foot, 1991

According to the *American School & University*'s annual Maintenance and Operations Cost Study, the median maintenance and operations costs of schools per square foot was $3.45 in 1991.

Source: "Maintenance & Operations Costs Expressed in Dollars per Square Foot," *AS & U*, 64, April 1992, p. 34.

★ 388 ★

School Expenditures

Cost of Educating Handicapped Students

The cost of educating handicapped students is 2.3 times more than for nonhandicapped students. The average expenditure is $4,000 to $6,000 per student for one year of regular education, and $10,000 to $14,000 for one year of education for a handicapped student.

Source: "Commentary: Every Child's Right: Literacy," Anne McGill-Franzen and Richard L. Allington, *Reading Teacher*, 45, October 1991, pp. 86-90.

Test Scores

★ 389 ★

1992 Scholastic Aptitude Test (SAT) Scores

Maximum score: 1,600.

	Average SAT scores		
	Verbal	Mathematics	Total
Big city schools	411	465	876
Rural schools	414	459	873
Other areas	432	485	917
Private schools	469	526	995
Public schools	420	475	895
Parochial schools	439	474	913
Female	418	456	874
Male	428	499	927
Blacks	352	385	737
Puerto Ricans	366	406	772
Mexican Americans	372	425	797
Native Americans	395	442	837
Asians	413	532	945
White	442	491	933
Foreign students	389	533	922
ALL	423	476	899

Source: "Test Scores Are Up Slightly for College-Bound Students," *The New York Times*, August 27, 1992, p. A18. Primary source: The College Board.

★ 390 ★

Test Scores

Scholastic Aptitude Test (SAT) Scores, 1972-1992

The 1992 average SAT score was 899 out of a possible 1,600 points. This is down from the 1972 average score of 937. 1972 was the first year that The College Board compiled average scores, and the scores haven't been as high since.

Source: "Putting Them to the Test," *USA Today*, August 27, 1992, p. D1. Primary source: The College Board. **Remarks**: Source also includes: average SAT scores by sex, by race/ethnicity, and by parents' education.

★ 391 ★

Test Scores

Scholastic Aptitude Test (SAT) Scores by Parents' Educational Level

Maximum score: 1,600.

Parents' Education	Average SAT Scores
Graduate degree	1006
Bachelor's degree	943
Associate's degree	864
High school diploma	837
No high school diploma	747

Source: "Putting Them to the Test," *USA Today*, August 27, 1992, p. D1. Primary source: The College Board. **Remarks**: Source also includes: average SAT scores, 1972-1992, and scores by sex, and race/ethnicity.

★ 392 ★

Test Scores

Scholastic Aptitude Test (SAT) Scores at Selected Universities

Maximum score: 1,600.

University	Average SAT Score
California Institute of Technology	1400
Massachusetts Institute of Technology	1375
Harvard University	1370
Yale University	1365
Stanford University	1365
Princeton University	1340
Rice University	1323
Johns Hoplkins University	1315
Dartmouth College	1310
Duke University	1305

Source: "U.S. News Top 25 National Universities," *U.S. News & World Report*, 112, September 30, 1991, p. 93. **Remarks**: Also included in source: average SAT scores at 25 colleges and universities.

★ 393 ★

Test Scores

American College Testing (ACT) Mean Scores, 1970-1990

Maximum score: 36.

	Mean ACT Score		
	English	Mathematics	Total
1970	18.5	20.0	19.9
Male	17.6	21.1	20.3
Female	19.4	18.8	19.4
1975	17.7	17.6	18.6
Male	17.1	19.3	19.5
Female	18.3	16.2	17.8
1980	17.9	17.4	18.5
Male	17.3	18.9	19.3
Female	18.3	16.2	17.9
1985	18.1	17.2	18.6
Male	17.6	18.6	19.4
Female	18.6	16.0	17.9
1990	20.5	19.9	20.6
Male	20.1	20.7	21.0
Female	20.9	19.3	20.3

Source: The World Almanac and Book of Facts 1992. New York: World Almanac, 1991. p. 219. Primary source: The American College Testing Program.

★ 394 ★

Test Scores

Mathematics Proficiency Test Scores of Eighth Graders by State, 1990

State	Average proficiency	State	Average proficiency	State	Average proficiency
Nationwide	261	Illinois	260	New York	261
Alabama	252	Indiana	267	North Carolina	250
Arizona	259	Iowa	278	North Dakota	281
Arkansas	256	Kentucky	256	Ohio	264
California	256	Louisiana	246	Oklahoma	263
Colorado	267	Maryland	260	Oregon	271
Connecticut	270	Michigan	264	Pennsylvania	266
Delaware	261	Minnesota	276	Rhode Island	260
District of Columbia	231	Montana	280	Texas	258
Florida	255	Nebraska	276	Virginia	264
Georgia	258	New Hampshire	273	West Virginia	256
Hawaii	251	New Jersey	269	Wisconsin	274
Idaho	272	New Mexico	256	Wyoming	272

Source: Mullis, Ina V.S., et al, *The State of Mathematics Achievement*. Washington, DC: National Center for Education Statistics, June, 1991. **Remarks**: Source also includes: percentage of students in each state at or above grade level in math proficiency.

★ 395 ★

Test Scores

Mean Scores of Mathematics Proficiency Test in Selected Countries

Based on a scale of 0 to 1,000, with a mean of 500.

Country	Mean Proficiency Score
Korea	567.8
Canada (Quebec–French)	543.0
Canada (British Columbia)	539.8
(Quebec–English)	535.8
(New Brunswick–English)	529.0
(Ontario–English)	516.1
(New Brunswick–French)	514.2
Spain	511.7
Great Britain	509.9
Ireland	504.3
Canada (Ontario–French)	481.5
United States	473.9

Source: Medrich, Elliott A. and Jeanne E. Griffith, *International Mathematics and Science Assessments: What Have We Learned,* Washington, DC: National Center for Education Statistics, 1992.

★ 396 ★

Test Scores

Mean Scores of Science Proficiency Test in Selected Countries

Based on a scale of 0 to 1,000, with a mean of 500.

Country	Mean Proficiency Score
Canada (British Columbia)	551.3
Korea	549.9
Great Britain	519.5
Canada (Quebec–English)	515.3
(Ontario–English)	514.7
(Quebec–French)	513.4
(New Brunswick–English)	510.5
Spain	503.9
United States	478.5
Ireland	469.3
Canada (Ontario–French)	468.3
(New Brunswick–French)	468.1

Source: Medrich, Elliott A. and Jeanne E. Griffith, *International Mathematics and Science Assessments: What Have We Learned,* Washington, DC: National Center for Education Statistics, 1992.

Miscellaneous

★ 397 ★

Recruiters on College Campuses, 1977-1993

Year	Average Number of Company Recruiters per Campus
1977	35.7
1978	38.5
1984	36.8
1985	37.8
1986	42.0
1992	26.0
1993	23.0

Source: "Job Prospects Sour for College Grads," *The Wall Street Journal*, May 20, 1993, p. B1. Primary source: College Placement Council.

★ 398 ★

Miscellaneous

Metropolitan Areas with the Highest Percentage of People 25 and Older with Advanced Degrees

Metropolitan Area	Percentage
Washington, D.C.	16.6%
Boston	12.0%
San Francisco	11.1%
New York	10.5%
Baltimore	9.2%
National metro average	8%

Source: "Demographics," *U.S. News & World Report*, 113, August 10, 1992, p. 15. Primary source: U.S. Census, 1990.

★ 399 ★

Miscellaneous

Years of School Completed by Sex and Race/ Ethnicity, 1970, 1991

	Median School Years Completed	
	1970	1991
White	12.1	12.8
Male	12.1	12.8
Female	12.1	12.7
Black	9.8	12.4
Male	9.4	12.4
Female	10.0	12.4
Hispanic	9.1	12.0
Male	9.3	12.1
Female	8.9	12.0

Source: U.S. Bureau of the Census. *Statistical Abstract of the United States 1992.* 112th edition. Washington, D.C.: U.S. Department of Commerce, 1992, p. 144. Primary source: U.S. Bureau of the Census, *U.S. Census of Population Reports*, series P-20, no. 462.

★ 400 ★

Miscellaneous

Immigrants' Level of Education

Immigrants to the United States from Central Europe around 1940 averaged three-fourths a year more schooling than the average American, but by 1980, recent immigrants averaged two-thirds of a year less than the average American.

Source: Los Angeles Public Library/State of California, *Scan/Info*, 4, August 1992, p. 16. Primary source: *San Francisco Chronicle*, August 14, 1992, p. A13.

★ 401 ★

Miscellaneous

Schools with the Smallest Freshmen and Sophomore Class Size, 1993

School	Average Freshmen and Sophomore Class Size
Bennington College	7
Wilson College	10
Prescott College	11
Sweet Briar College	11
Centenary College of LA	12

[Continued]

★ 401 ★

Schools with the Smallest Freshmen and Sophomore Class Size, 1993

[Continued]

School	Average Freshmen and Sophomore Class Size
Rosemont College	12
Wells College	12

Source: Money College Guide 1993, p. 98. **Remarks**: Source also includes: schools with the highest and lowest professors' salaries.

Chapter 7
EXPENDITURES

Automobile Expenses

★ 402 ★

New Car Price for a 1967 "Comparable Car"

1967 average consumer expenditure per new car is inflated to current dollars.

Year	Average New Car Price for a 1967 "Comparable Car"	
	With Added Safety Features	Without Added Safety Features
1970	$3,601	3,459
1980	6,863	5,764
1985	8,984	8,984
1989	10,248	7,825
1990	10,581	7,938

Source: U.S. Bureau of the Census. *Statistical Abstract of the United States 1992.* 112th edition. Washington, D.C.: U.S. Department of Commerce, 1992, p. 613. Primary source: Motor Vehicle Manufacturers Association of the United States, Inc., Detroit, MI, *Motor Vehicle Facts and Figures*, annual. **Remarks**: Source also includes: average price of a new car and number of weeks of median family income to equal average cost of new car, all by year.

★ 403 ★

Automobile Expenses

Cost of Owning and Operating an Automobile per Mile, 1975-1990

Year	Average Cost per Mile (cents)	Average Cost per 100,000 Miles (dollars)
1975	18.31	$1,831
1980	27.95	2,795
1985	27.20	2,720
1986	29.59	2,959
1987	32.64	3,264
1988	33.40	3,341
1989	38.20	3,820
1990	41.00	4,100

Source: U.S. Bureau of the Census. *Statistical Abstract of the United States 1992.* 112th edition. Washington, D.C.: U.S. Department of Commerce, 1992, p. 613. Primary source: Motor Vehicle Manufacturers Association of the United States, Inc., Detroit, MI, *Motor Vehicle Facts and Figures*, annual.

★ 404 ★

Automobile Expenses

Cost of Owning an Automobile, 1975-1990

Year	Cost for a New Car		
	Foreign	Domestic	Average
1975	$3,708	$2,648	$3,542
1980	7,609	7,482	7,574
1985	11,733	12,022	12,022
1989	14,959	15,127	15,292
1990	15,641	17,010	16,012

Source: U.S. Bureau of the Census. *Statistical Abstract of the United States 1992.* 112th edition. Washington, D.C.: U.S. Department of Commerce, 1992, p. 613. Primary source: Motor Vehicle Manufacturers Association of the United States Inc., Detroit, MI, *Motor Vehicle Facts and Figures*, annual. **Remarks**: Source also includes: estimated average new car price for a 1967 "comparable car," and number of weeks of median family income to equal average cost of new car, all by year.

★ 405 ★

Automobile Expenses

Weeks of Family Income to Equal Price of New Car

Year	Weeks of Income for Average New Car
1970	18.7
1980	18.7
1985	23.0
1989	24.5
1990	24.5

Source: U.S. Bureau of the Census. *Statistical Abstract of the United States 1992.* 112th edition. Washington, D.C.: U.S. Department of Commerce, 1992, p. 613. Primary source: Motor Vehicle Manufacturers Association of the United States, Inc., Detroit, MI, *Motor Vehicle Facts and Figures*, annual. **Remarks**: Source also includes: average price of a new car and estimated average new car price for a 1967 "comparable car," all by year.

★ 406 ★

Automobile Expenses

Cost per Mile of Operating an Automobile, by Type of Car

	Average Cost (cents) per Mile				
	Large	Intermediate	Compact	Sub-compact	Van
Depreciation	9.6	8.6	7.3	5.9	10.7
Maintenance	6.0	5.2	4.6	5.1	6.9
Gas & oil	7.0	5.7	4.6	4.4	9.1
Parking & tolls	0.9	0.9	0.9	0.9	0.9
Insurance	4.9	5.6	4.3	5.0	8.9
Taxes	2.2	1.8	1.6	1.4	2.7
Total	30.6	27.8	23.3	22.7	39.2

Source: Edmond's Car Savvy. Alhambra, CA: Edmund Publications, 1991, p. 23.

★ 407 ★

Automobile Expenses

Cost of Owning and Operating an Automobile by Size, 1990-1991

Size	Average Annual Cost		Average Cost per Mile (cents)	
	1990	1991	1990	1991
Subcompact	$3,300	$3,580	33.00	35.83
Mid-size	4,100	4,364	41.00	43.64
Full-size	4,400	5,559	44.00	55.59

Source: Los Angeles Public Library/State of California, *Scan/Info*, 4, January 1992, p. 5, from *Chicago Tribune*, April 14, 1991, pt. 17, 1:1. Primary source: American Automobile Association.

★ 408 ★

Automobile Expenses

Automobiles with the Highest Fuel Cost by Size of Car, 1992

Size	Automobile	Average Annual Fuel Cost
2-seater	Geo Metro LSi Convertible	$419
Minicomp.	Nissan NX	562
Subcomp.	Geo Metro XFi	328
Compact	Ford Escort FS	515
Mid-size	Mazda 626/MX-6	666
Large	Saab 9000	819
Sm. wagon	Ford Escort Wagon	545
Mid-size wagon	Eagle Summit Wagon	720

Source: The World Almanac and Book of Facts 1992. New York: World Almanac, 1991, p. 715. Primary source: U.S. Department of Energy, Environmental Protection Agency. **Remarks**: Source also includes: automobiles with the lowest fuel costs, 1992.

★ 409 ★

Automobile Expenses

Automobiles with the Lowest Fuel Cost by Size of Car, 1992

Size	Automobile	Average Annual Fuel Cost
2-seater	Cadillac Allante	$1,191
Minicomp.	Porsche 911 Carrera	1,126
Subcomp.	Aston-Martin Virage Saloon	1,557
Compact	BMW 535i	1,001

[Continued]

★409★

Automobiles with the Lowest Fuel Cost by Size of Car, 1992

[Continued]

Size	Automobile	Average Annual Fuel Cost
Mid-size	Rolls-Royce Bentley	1,557
Mid-size wagon	Mercedes-Benz 300TE	1,126

Source: The World Almanac and Book of Facts 1992. New York: World Almanac, 1991, p. 715. Primary source: U.S. Department of Energy, Environmental Protection Agency. **Remarks**: Source also includes: automobiles with the highest fuel costs, 1992.

★410★

Automobile Expenses

Monthly Car Payment, 1975 and 1992

According to Dean Witter Reynolds Inc., a New York investment banking firm, in 1992, the average car payment was $308 with a 4 1/2 year loan period. In 1975, the average car payment was $133 and had a three year loan payment.

Source: "Cautious Customers," Joann Muller, *Detroit Free Press*, August 24, 1992, pp. 8F-9F.

★411★

Automobile Expenses

Annual Cost for Operating an Automobile in Selected Cities

Cost is based on a car driven 15,000 miles annually, at 50 cents per mile.

City	Average Annual Cost
Most Expensive	
Los Angeles, CA	$7,529
Least Expensive	
Sioux Falls, SD	4,626

Source: Los Angeles Public Library/State of California, *Scan/Info*, 4, June 1992, p. 15, from *Los Angeles Times*, May 27, 1992, p. E2. Primary source: Runzheimer International.

★ 412 ★

Automobile Expenses

Cost of Renting an Automobile

Car	Average Cost per Day (at 50 cents per mile)
Lamborghini Diablo	$1,500[1]
Rolls-Royce Corniche	750
Porsche 911 Turbo	350
Mazda Miata	129
Alfa Romeo Spider	80
Volkswagen Beetle	50
Ford Escort	50
Nissan Sentra	50
Toyota Corolla	50

Source: "Price Tag: Renting Exotic Automobiles," *The New York Times*, August 8, 1992, p. 50. Primary source: Budget Rent-A-Car, Car Collection division; Hertz Rent-A-Car; Avis Rent-A-Car; *Trend* magazine; *Car and Driver* magazine. *Note:* 1. $10 per mile.

★ 413 ★

Automobile Expenses

Cost of Automobile by Race/Ethnicity and Sex

According to a study by the American Bar Foundation, black men pay an average of $1,150 more for a new car than white men: an average of 15 percent markup for white men compared with 24 percent for black men. Black women pay $450 more than white men and white women pay $200 more than white men.

Source: "White Men Get Cars Cheapest," Patricia Braus, *American Demographics*, 15, June 1993, pp. 20-21.

★ 414 ★

Automobile Expenses

Expenditure for New Automobile Buyers, by Type of Household, 1989

Type of Household	Average Expenditure Toward New Vehicle
All	$1,218
Married couples	1,651
without children	1,376
with children	1,789
oldest under 6	1,332
oldest 6-17	1,797
oldest 18-	2,120
Single parents	734

[Continued]

★ 414 ★

Expenditure for New Automobile Buyers, by Type of Household, 1989

[Continued]

Type of Household	Average Expenditure Toward New Vehicle
Single persons	600
Others	877

Source: "Married and Motoring," *American Demographics*, 13, October 1991, p. 6. Primary source: U.S. Bureau of Labor Statistics, 1989 Consumer Expenditure Survey.

★ 415 ★

Automobile Expenses

New Car Loan by Finance Companies

Year	Average Interest	Average Time (months)	Average Financed	Average Payment
1979	13.5%	44.2	$6,037	$174.10
1980	14.8	45.0	6,322	183.91
1981	16.0	45.3	7,339	216.50
1982	15.9	46.0	8,178	238.16
1983	12.7	45.9	8,787	242.25
1984	14.6	48.3	9,337	256.78
1985	12.0	51.1	9,883	248.44
1986	9.4	50.0	10,665	258.61
1988	12.6	56.2	11,663	275.80
1990	12.5	54.6	12,071	291.31

Source: MVMA Motor Vehicle Facts and Figures '91. Detroit, MI: Motor Vehicle Manufacturing Association, 1991, p. 61. **Remarks**: Source also includes: average interest rate, maturity, amount and payment of used car loans by finance companies.

★416★

Automobile Expenses

Used Car Loan by Finance Companies

Year	Average Interest	Average Time (months)	Average Financed	Average Payment
1979	18.1%	34.4	$3,549	$132.82
1980	19.1	34.8	3,810	143.44
1981	20.0	35.8	4,339	162.13
1982	20.8	37.0	4,746	174.84
1983	18.8	38.0	5,033	176.77
1984	17.8	39.7	5,691	191.00
1985	17.6	41.2	6,091	198.03
1986	16.0	42.6	6,555	202.68
1988	15.1	46.7	7,824	222.58
1990	16.0	46.1	8,289	242.07

Source: MVMA Motor Vehicle Facts and Figures '91. Detroit, MI: Motor Vehicle Manufacturing Association, 1991, p. 61. **Remarks**: Source also includes: average interest rate, maturity, amount and payment of new car loans by finance companies.

Charitable Giving

★417★

Charitable Giving per Household, 1989, 1991

According to Independent Sector, a non-profit coalition in Washington, DC, the average charitable gift from 72% of American households in 1991 was $899, down from $978 from 75% of households in 1989.

Source: "Recession Hurts Giving of Cash, Not Volunteering," Pamela Sebastian, *The Wall Street Journal*, October 16, 1992, p. B3A. **Remarks**: Source also includes: charitable giving by race/ethnicity, and average amount of time Americans spend in volunteer work.

★418★

Charitable Giving

Charitable Giving by Teenagers, 1991

According to Independent Sector, a non-profit coalition in Washington, DC, of teens who gave money to charitable causes, the average gift was $56 in 1991, compared to $46 in 1989.

Source: "Teenagers Volunteered More in '91 Than in '89," *The Wall Street Journal*, December 4, 1992, p. A5B.

★ 419 ★

Charitable Giving

Charitable Giving per Household by Age, 1989

Age	Average Amount per Household	Percent of Household Income
18-24 yrs.	$484	1.5%
25-34 yrs.	893	2.6
35-44 yrs.	956	2.6
45-54 yrs.	1,098	2.8
55-64 yrs.	1,420	4.7
65-74 yrs.	1,070	3.7
75 & up	698	4.5

Source: U.S. Bureau of the Census. *Statistical Abstract of the United States 1992*. 112th edition. Washington, D.C.: U.S. Department of Commerce, 1992, p. 375. Primary source: Virginia Hodgkinson and Murray Weitzman, and the Gallup Organization, Inc., *Giving and Volunteering in the United States: 1990*. **Remarks**: Source also includes: charitable giving per household by income.

★ 420 ★

Charitable Giving

Charitable Giving per Household by Income, 1989

Income	Average Amount per Household	Percent of Household Income
Under $10,000	$379	5.5%
$10,000-19,000	485	3.2
$20,000-29,000	728	2.9
$30,000-39,000	894	2.6
$40,000-49,000	831	1.8
$50,000-59,000	941	1.7
$60,000-74,999	1,250	1.8
$75,000-99,999	2,793	3.2
$100,000 & up	2,893	2.9

Source: U.S. Bureau of the Census. *Statistical Abstract of the United States 1992*. 112th edition. Washington, D.C.: U.S. Department of Commerce, 1992, p. 375. Primary source: Virginia Hodgkinson and Murray Weitzman, and the Gallup Organization, Inc., *Giving and Volunteering in the United States: 1990*. **Remarks**: Source also includes: charitable giving per household by age.

★ 421 ★

Charitable Giving

Charitable Giving per Household by Type of Charity, 1989

Type of Charity	Average Contribution per Household
Arts, culture, humanities	$193
Education	291
Environment	88
Health	143
Human services	263
International	202
Private, community foundations	116
Public, societal benefit	120
Recreation - adults	135
Religion	896
Youth	129

Source: U.S. Bureau of the Census. *Statistical Abstract of the United States 1992.* 112th edition. Washington, D.C.: U.S. Department of Commerce, 1992, p. 375. Primary source: Virginia Hodgkinson and Murray Weitzman, and the Gallup Organization, Inc., *Giving and Volunteering in the United States: 1990.*

Cost of Child Care

★ 422 ★

Most Expensive Cities in the United States for Child Day-Care

City	Average Weekly Cost
New York, NY	$104
Boston, MA	101
Minneapolis, MN	86
Anchorage, AK	84
Manchester, NH	84
San Francisco, CA	82
Portland, ME	80
Washington, DC	77
Hartford, CT	76
Philadelphia, PA	76

Source: Krantz, Les. *The Best and Worst of Everything.* New York: Prentice Hall General Reference, 1991, p. 74. Primary source: Runzheimer International; Donnelley Marketing Information Services. **Remarks**: Source also includes: least expensive cities in the United States for child day-care.

★ 423 ★

Cost of Child Care

Least Expensive Cities in the United States for Child Day-Care

City	Average Weekly Cost
Ogden, UT	$39
Jackson, MI	43
Columbia, SC	45
Huntington, WV	45
Mobile, AL	46
Tampa, FL	46
New Orleans, LA	47
Cheyenne, WY	48
Grand Island, NE	48
Tucson, AZ	49

Source: Krantz, Les. *The Best and Worst of Everything.* New York: Prentice Hall General Reference, 1991, p. 74. Primary source: Runzheimer International; Donnelley Marketing Information Services. **Remarks**: Source also includes: most expensive cities in the United States for child day-care.

★ 424 ★

Cost of Child Care

Cost of Child Care, 1988

In 1988, the average cost of child care was $54 per week or $2,800 per year, an expense incurred by only 40% of working mothers.

Source: "Most Working Mothers Have Free Child Care," *The Wall Street Journal*, September 29, 1992, p. B1.

★ 425 ★

Cost of Child Care

Cost of Child Care per Hour by Type of Care, 1990

Type	Average Hourly Rate
In-home care	$2.30
Day care center	1.67
Family day care	1.35
Relative	1.11

Source: "Who's Minding the Children?" *The New York Times*, January 28, 1993, p. C6. Primary source: The Urban Institute, from the National Child Care Survey, 1965 and 1990. **Remarks**: Source also includes: child care by family income, 1989.

★ 426 ★

Cost of Child Care

Cost of Child Care as a Percentage of Family Income, 1990

Income	Average Percentage of Family Income
Under $15,000	23%
$15,000-$24,999	12%
$25,000-$34,999	8%
$35,000-$49,999	7%
$50,000 & up	6%

Source: "Who's Minding the Children?" *The New York Times*, January 28, 1993, p. C6. Primary source: The Urban Institute, from the National Child Care Survey, 1965 and 1990. **Remarks**: Source also includes: average cost of child care per hour by type of care, 1990.

★ 427 ★

Cost of Child Care

Cost of Raising a Child by Family Income

Family Income	Average Annual Cost of Raising Children, birth-2yrs.
Above $48,300	$8,770
Below $29,900	4,330

Source: "Spending Money: Rich Children Are Different," Thomas Exter, *American Demographics*, 13, August 1991, p. 6.

★ 428 ★

Cost of Child Care

Cost of Raising a Child From Birth to Age 18

According to a government study done by the Agriculture Department's Family Economics Research Group, the average cost of raising a child from birth to age 18 in 1992 dollars is $128,670, not including college.

The average annual cost by the age of the child is:

Age	Average Annual Cost
Under 3 yrs.	$6,610
3-5 yrs.	7,010
6-8 yrs.	6,960
9-11 yrs.	6,770

[Continued]

★ 428 ★

Cost of Raising a Child From Birth to Age 18

[Continued]

Age	Average Annual Cost
12-14 yrs.	7,540
15-17 yrs.	8,000

Source: "Just Think of It All As Another Mortgage," *The Wall Street Journal*, June 2, 1993, p. B1.

Cost of Diapers

★ 429 ★

Per-Diaper Cost

Diaper	Average Cost
Cloth	7 to 9 cents
Diaper Service	13 to 17 cents
Disposable	25 cents

Source: "Cotton vs. Disposables: What's the Damage? William Rathje and Cullen Murphy, *Garbage*, 4, October/November 1992, pp. 29-30.

★ 430 ★

Cost of Diapers

Diaper Cost for Two Years

These figures are based on the following: Disposables—50 disposable diapers per week at a cost of 26 cents each; Diaper service—$11.40 for weekly delivery of 60 diapers and $82 for overwraps and setup fee; Cloth diapers—home laundering one-time cost for four dozen cloth diapers and one dozen overwraps is $200; use of electric water heater, top-loading washer, and electric dryer (electricity cost of 8.2 cents per kilowatt-hour) for three washes a week; and detergent cost of 40 cents per load.

Type	Average Cost
Cloth washed at home	$526
Cloth from service	1,268
Disposables	1,352

Source: "Two Years of Diapering," *Consumer Reports*, 56, August 1991, p. 552.

Cost of Energy

★ 431 ★

Lifetime Energy Costs of Household Appliances

Electric Appliance	Average Energy Cost	
	Average Model	Efficient Model
Air conditioning, Central	$4,560	$2,280
Dishwasher	706	454
Dryer	983	772
Light	56	13
Refrigerator	1,824	274
Television, Color	252	101
Washer	1,264	632
Water heater	3,237	1,618

Source: World Watch, May/June 1988, p. 30. Primary source: American Council for an Energy-Efficient Economy.

★ 432 ★

Cost of Energy

Annual Energy Cost of Household Appliances

Data include electric appliances only and assume an average cost of 7 cents per kilowatt hour.

Household appliance/item	Actual energy cost (dollars)
Space heating	350
Refrigerator/freezer	239
Water heating	154
Air-conditioning	109
Lighting	84
Washer/dryer	77
Furnace fan	46
Range/oven	42
Color TV	22
Dishwasher	12
Microwave	7
Other appliances	22

Source: Changing by Degrees: Steps to Reduce Greenhouse Gases, U.S. Congress, Office of Technology Assessment, OTA-O-482, Washington, D.C., February 1991, p. 117, from Office of Technology Assessment, 1991, adapted from U.S. Department of Energy, Energy Information Administration, *Household Energy Consumption and Expenditure* (Washington, DC: 1989). Lighting data provided by A. Mejer, Lawrence Berkeley Laboratory.

★ 433 ★

Cost of Energy

Cost of Energy by Income, 1990

Average Income	Average Energy Cost	Percent of Income
$5,637	$1,226	21.7%
14,115	1,659	11.8
24,500	2,143	8.7
38,376	2,532	6.6
76,660	3,147	4.1

Source: "Who Pays What for Energy," *The Wall Street Journal*, February 2, 1993, p. A16. Primary source: U.S. Bureau of Labor Statistics, *Consumer Expenditures Survey 1990*.

★ 434 ★

Cost of Energy

Monthly Cost of Electricity in Selected Cities

City	Average Monthly Cost (750 kwh)
Wichita, KS	$69.48
Boston, MA	69.65
Miami, FL	70.47
Toledo, OH	71.68
Riverside, CA	72.66
Philadelphia, PA	72.99
Pittsburgh, PA	73.98
San Diego, CA	74.98
Newark, NJ	75.19
New York City, NY	86.37

Source: Krantz, Les. *The Best and Worst of Everything*. New York: Prentice Hall General Reference, 1991, p. 76. Primary source: U.S. Bureau of the Census.

★ 435 ★

Cost of Energy

Annual Household Energy Costs by Year, 1980-1987

Year	Average Cost of Energy per Household
1980	$815
1981	917
1982	1,022
1983	1,048
1985	1,123
1987	1,080

Source: U.S. Bureau of the Census. *Statistical Abstract of the United States 1992*. 112th edition. Washington, D.C.: U.S. Department of Commerce, 1992, p. 567. Primary source: U.S. Energy Information Administration, *Residential Energy Consumption Survey*. **Remarks**: Source also includes: cost of energy by geographical region of the United States.

★ 436 ★

Cost of Energy

Annual Household Energy Costs by Geographical Region of the United States, 1987

Region	Average Cost of Energy per Household
Northeast	$1,276
Midwest	1,124
South	1,081
West	891

Source: U.S. Bureau of the Census. *Statistical Abstract of the United States 1992*. 112th edition. Washington, D.C.: U.S. Department of Commerce, 1992, p. 567. Primary source: U.S. Energy Information Administration, *Residential Energy Consumption Survey*. **Remarks**: Source also includes: cost of energy by year, 1980-1987.

★ 437 ★

Cost of Energy

Annual Household Energy Costs by Type of Housing, 1987

Housing	Average Cost of Energy per Household
Single-family detached	$1,226
Single-family attached	1,135
Two-to four-unit bldg.	905
Five or more unit bldg.	681
Mobile home	948

Source: U.S. Bureau of the Census. *Statistical Abstract of the United States 1992*. 112th edition. Washington, D.C.: U.S. Department of Commerce, 1992, p. 567. Primary source: U.S. Energy Information Administration, *Residential Energy Consumption Survey*. **Remarks**: Source also includes: cost of energy by year house built, and by family income; consumption of energy by type of housing, year built and family income.

★ 438 ★

Cost of Energy

Annual Household Energy Costs by the Year House Was Built, 1987

Year House Built	Average Cost of Energy per Household
1939 or earlier	$1,105
1940-1949	1,042
1950-1959	1,121
1960-1969	1,060
1970-1974	1,062
1975-1979	1,131
1980-1984	984
1985 or later	1,049

Source: U.S. Bureau of the Census. *Statistical Abstract of the United States 1992*. 112th edition. Washington, D.C.: U.S. Department of Commerce, 1992, p. 567. Primary source: U.S. Energy Information Administration, *Residential Energy Consumption Survey*. **Remarks**: Source also includes: cost of energy by type of housing and by family income; consumption of energy by type of housing, year house built and family income.

★ 439 ★

Cost of Energy

Cost of Fluorescent Lighting Vs. Incandescent Lighting

The average cost of an incandescent light bulb is 80 cents, while a fluorescent light bulb costs $20. But fluorescent lighting is more energy-efficient and lasts 10 times as long. At the national average electricity rate of 8.25 cents per kilowatt-hour, the total cost for 10,000 hrs. for a 60-watt incandescent bulb is $57.50, and for a Sylvania F13TT/27K fluorescent bulb, the cost is $33.04. The savings for using fluorescent lighting is $24.46.

Source: "Fluorescent Vs. Incandescent," *Consumer Reports*, 57, October 1992, p. 668.

Cost of Food

★ 440 ★

Cost of the Federal Price Support Program for Farms, 1991

In 1991, the average cost of the federal price support program for farms, in terms of increased food costs to consumers, was $210 per household, or $19.6 billion total, according to the Organization of Economic Cooperation and Development (OECD) which monitors agriculture worldwide.

Source: "Why You May Pay More for Less Food," Chris Warden, *Consumers' Research*, 76, April 1993, p. 15.

★ 441 ★

Cost of Food

Weekly Expenditures for Fruits and Vegetables, 1982, 1989

Product	Average Weekly Expenditures	
	1982	1989
Canned fruits & vegetables	$2.42	$3.23
Frozen fruits & vegetables	1.28	1.84
Fresh fruits & vegetables	3.54	4.87
Total	7.24	9.94

Source: U.S. Department of Commerce. International Trade Administration. *U.S. Industrial Outlook '92*. Washington, D.C.: U.S. Department of Commerce, 1992, p. 32-10. Primary source: U.S. Department of Labor, Bureau of Labor Statistics.

★ 442 ★

Cost of Food

Weekly Expenditures for Meat, 1982-1989

Product	Average Weekly Expenditures			
	1982/83	1985	1987	1989
Beef	$4.10	$3.71	$3.68	$3.89
Pork	2.29	2.35	2.22	2.24
Poultry	1.50	1.60	1.68	1.92
Seafood	1.14	1.24	1.27	1.36
Total	9.03	8.91	8.85	9.41

Source: U.S. Department of Commerce. International Trade Administration. *U.S. Industrial Outlook '92*. Washington, D.C.: U.S. Department of Commerce, 1992, p. 32-4. Primary source: U.S. Department of Labor, Bureau of Labor Statistics, *Consumer Expenditure Surveys, 1982/83-1989.*

★ 443 ★

Cost of Food

Weekly Expenditures for Nonalcoholic Drinks, 1985-1989

Product	Average Weekly Expenditures per Household				
	1985	1986	1987	1988	1989
Soft drinks	$2.01	$1.99	$2.23	$2.34	$2.52
Cola	1.45	1.44	1.64	1.72	1.85
Other	0.55	0.56	0.58	0.61	0.67
Fruit drinks	0.27	0.31	0.26	0.30	0.36
Juices					
Frozen	0.39	0.34	0.32	0.37	0.35
Canned, bottled	0.77	0.72	0.82	0.87	0.98
Coffee & tea	1.19	1.25	1.07	0.99	1.02

Source: U.S. Department of Commerce. International Trade Administration. *U.S. Industrial Outlook '92*. Washington, D.C.: U.S. Department of Commerce, 1992, p. 32-37. Primary source: U.S. Department of Labor, Bureau of Labor Statistics, *Consumer Expenditure Surveys, 1985-1989.*

★ 444 ★

Cost of Food

Weekly Expenditures for Meals Away-from-home, 1985-1989

Item	Average Weekly Expenditures				
	1985	1986	1987	1988	1989
Breakfast	$1.39	$1.52	$1.51	$1.54	$1.65
Lunch	7.95	8.02	9.10	9.48	10.25
Dinner	9.80	9.68	10.43	10.56	11.71
Snack	2.14	2.42	2.55	2.53	2.50
Alcohol	2.31	2.00	2.17	2.13	2.19
Total	23.59	23.64	25.76	26.24	28.30

Source: U.S. Department of Commerce. International Trade Administration. *U.S. Industrial Outlook '92*. Washington, D.C.: U.S. Department of Commerce, 1992, p. 39-7. Primary source: U.S. Department of Labor, Bureau of Labor Statistics, *Consumer Expenditure Surveys, 1985-1989.*

★ 445 ★

Cost of Food

Weekly Expenditures for Selected Food Products, 1985-1989

Item	Average Weekly Expenditures				
	1985	1986	1987	1988	1989
Cereal	$1.71	$1.78	$2.01	$2.10	$2.51
Bakery	3.74	3.51	3.74	3.90	4.38
Eggs	0.58	0.57	0.55	0.53	0.63
Dairy	5.12	4.82	5.28	5.27	5.85
Sugar & sweets	1.48	1.42	1.43	1.50	1.65
Fats, oils	1.12	0.99	0.97	1.05	1.13

Source: U.S. Department of Commerce. International Trade Administration. *U.S. Industrial Outlook '92*. Washington, D.C.: U.S. Department of Commerce, 1992, p. 39-7. Primary source: U.S. Department of Labor, Bureau of Labor Statistics, *Consumer Expenditure Surveys, 1985-1989.* **Remarks**: Source also includes: at-home food expenditures for meat, poultry, fish, fresh and processed fruits and vegetables.

★ 446 ★

Cost of Food

Household Expenditures for Non-essential Food Items, 1990

Item	Average Spending
Condiments & seasonings	$74
Fats & oils	68
Sugar & sweets, including candy, gum, jam	94
Potato chips, nuts, other snacks	67
Miscellaneous prepared food	146
Coffee, tea, soda	213

Source: "Households Bring Most Bacon Home," Diane Crispell, *The Wall Street Journal*, February 21, 1992, p. B1. Primary source: Bureau of Labor Statistics.

★ 447 ★

Cost of Food

Spending for At-home Food Vs. Away-from-home Food, 1990

According to the Bureau of Labor Statistics, spending for food eaten at home has grown faster than spending for food eaten away from home. In 1990, at-home food averaged $2,485, an increase of 5.3% over 1989; while food away from home averaged $1,811, a 1.7% increase from the previous year.

Source: "Households Bring Most Bacon Home," Diane Crispell, *The Wall Street Journal*, February 21, 1992, p. B1. Primary source: Bureau of Labor Statistics.

★ 448 ★

Cost of Food

Annual Household Food Expenditures in Urban and Rural Areas

Food expenditures for rural households are higher than for urban households. The average household spending for food eaten at home is $2,548 in rural areas, and $2,364 in cities. This is due in part to the urban lifestyle: 34% more is spent on food away from home in city households than in rural ones.

Source: "Spending Money: Old Bucks in the Woods," Thomas Exter, *American Demographics*, 13, May 1991, p. 6.

★ 449 ★

Cost of Food

Cost of Groceries by Region of the United States, 1992/1993

Area	Average Cost of 36 Selected Grocery Items	Price Change From Year Earlier
Northeast	$84.24	-$0.91
Mid-south	82.02	-$0.86
Southeast	78.84	-$1.51
Great Lakes	81.73	-$1.94
Plains	82.07	-$1.47
South Central	81.94	-$2.11
West	84.60	-$0.80
California	88.77	+$0.44
U.S.	82.20	-$1.01

Source: "Supermarket Chains Chop Prices to Keep Rivals From Cutting Into Their Business," Kathleen Deveny, *The Wall Street Journal*, March 25, 1993, p. B1. Primary source: Information Resources Inc.

★ 450 ★

Cost of Food

Cost of Thanksgiving Dinner

According to the American Farm Bureau, Department of Information and Public Relations' annual survey conducted in November, the average cost of a homemade Thanksgiving meal for ten people is $25.95.

Source: The American Farm Bureau, Department of Information and Public Relations, Park Ridge, Ill.

Cost of Gift Giving

★ 451 ★

Holiday Spending per Household, 1991

The average household spent $400 on holiday gifts, 1992, compared with $375 in 1991. The biggest spenders were those aged 45-54, whose average spending was $467. One in three households spent less than $200.

Source: "A Big Ho Ho Ho for America's Retailers?" *U.S. News & World Report*, 113, December 7, 1992, p. 16.

★ 452 ★

Cost of Gift Giving

Parents' Holiday Spending per Child, 1992

According to Gallup's "North Pole Poll," parents spend an average of $213 per child aged 7 to 16 years for Christmas gifts in 1992; compared with $209, the amount children thought parents should spend.

Source: "Bottom Line: How Parents Stack Up," Neely Tucker, *Detroit Free Press*, December 30, 1992, p. 1D.

★ 453 ★

Cost of Gift Giving

Grandparents' Gift Giving per Grandchild

According to the Data Group, grandparents spend an average of $82 per grandchild for a holiday gift, $42 for a birthday gift, $74 for a special occasion such as a graduation; and $19 for other occasions like Easter or Valentines Day.

Source: "Growing Importance of Grandparents," Joe Schwartz and Judith Waldrop, *American Demographics*, 14, February 1992, pp. 10-11.

Cost of Owning a Pet

★ 454 ★

Cost of Owning a Dog

Note: Figures are based on an average life of 11 years.

Type of Cost	Average cost		
	Initial	Annual	Lifetime
Flea & pest treatment	---	$80	$880
Food	---	400	4,400
Grooming supplies	$130	100	1,230
Leashes, collars, toys	100	100	1,200
Training	50-100	50-200	600-2,300
Veterinarian fees, licenses	150	290	3,340
Total	480	1,170	13,350

Source: "The Cost of That Dog in the Window," *The New York Times*, June 13, 1992, p. 48. Primary source: American Kennel Club.

★ 455 ★

Cost of Owning a Pet

Annual Veterinary Cost by Type of Pet

Type	Average Annual Cost	
	1983	1987
Dog	$66.87	$82.86
Cat	36.83	54.26
Bird	----	4.47
Horse	----	120.75

Source: U.S. Bureau of the Census. *Statistical Abstract of the United States 1992.* 112th edition. Washington, D.C.: U.S. Department of Commerce, 1992, p. 238. Primary source: American Veterinary Medical Association, Schaumburg, IL, *The Veterinary Service Market for Companion Animals, July 1983* and *August 1988.*

★ 456 ★

Cost of Owning a Pet

Purchase Price of a Pet, by Type, 1991

Kind	Average Price
Bird	$6 to $1000
Cat	$10 to $800
Dog	$30 to $2,000
Fish	25 cents to $250
Gerbil	$5 to $10
Guinea pig	$15 to $20
Hamster	$5 to $10
Horse	$250 to $2,000
Rabbit	$15 to $25
Turtles	$5 to $10

Source: "Price Tag: The Top 10 Pets," *The New York Times*, November 12, 1992, p. C2. Primary source: Humane Society of the United States; American Veterinary Medical Association; American Pet Products Manufacturers Association; *World Almanac and Book of Facts 1992.*

Cost of Telephones

★ 457 ★

Cost of Residential Telephone Service

In 1990, the national average for residential local telephone service with unlimited calling was $12.40. The average cost of local service including subscriber line charges and taxes was $17.78, compared with $17.53 for 1989. In 1989 the average household spent $567 for telephone service, compared with $325 for 1980. Telephone service has accounted for two percent of the average household budget every year for the last 15 years.

Source: U.S. Department of Commerce. International Trade Administration. *U.S. Industrial Outlook '92*. Washington, D.C.: U.S. Department of Commerce, 1992, p. 28-11.

★ 458 ★

Cost of Telephones

Cost of Cellular Phones

According to market research firm, Herschel Shosteck Associates, mobile or car phones have fallen in price from an average of $1,000 in 1988 to $320 in 1992. Cellular phones, in soft or hard cases that can be used in or out of cars, have also dropped in price from an average of $1,100 to $340. Prices are even dropping for portable hand-held phones, from $1,400 to $600.

Source: "An Eye on Phones," Edward C. Baig, *U.S. News & World Report*, 113, June 22, 1992, pp. 78-82. **Remarks**: Source also includes: cost of cellular phones per minute.

★ 459 ★

Cost of Telephones

Cost of Cellular Phones per Month, 1987-1991

Year	Average Monthly Cost
1987	$96.83
1988	98.02
1989	89.30
1990	80.90
1991	72.74

Source: U.S. Bureau of the Census. *Statistical Abstract of the United States 1992*. 112th edition. Washington, D.C.: U.S. Department of Commerce, 1992, p. 553. Primary source: Cellular Telecommunications Industry Association, Washington, DC, *State of the Cellular Industry*, annual. **Remarks**: Source also includes: average length of cellular phone call.

Cost of Water

★ 460 ★

Cost of Water in Selected Countries

Country	Average Cost of Water per Cubic Meter
Australia	$1.06
Belgium	.96
Canada	.27
Finland	.72
France	.92
Germany	1.17
Great Britain	.81
Ireland	.47
The Netherlands	.87
Norway	.32
Sweden	.68
United States	.36

Source: "More Money Going Down the Drain," *Parade* Magazine, February 7, 1993, p. 20.

★ 461 ★

Cost of Water

Cost of Water in New York City

In New York City the cost of water is $2.62 per 100 cubic feet. There are 748 gallons in 100 cubic feet.

Activity	Average Amount	Average Cost
Shower	50 gallons for 5 min.	18 cents
Flush toilet	5 gallons	2 cents
Washer, per load	49 gallons for 18 lbs.	17 cents

Source: "Water: How Much It Costs," *The New York Times*, October 8, 1992, p. B4. Primary source: New York City Department of Environmental Protection.

Credit Cards

★ 462 ★

Outstanding Balance of Credit Cards

The average balance of private-label credit cards (those which can only be used in stores operated by a particular retailer) is about $250, while all- purpose credit cards, like Visa and Mastercard, have an average balance of $1,400.

Source: "More Major Marketers Are Now Offering Their Own General-Purpose Credit Cards," Peter Pae, *The Wall Street Journal*, August 26, 1992, B1.

★ 463 ★

Credit Cards

Credit Card Debt by Educational Level

Education level	Average Debt per Month
College graduates	$391
Some college education	159
High school graduates	83
Some high school education	51

Source: Rukeyser, Louis. *Louis Rukeyser's Business Almanac.* New York: Simon & Shuster, 1991, p. 410. Primary source: Federal Reserve and *American Demographics* magazine. **Remarks**: Source also includes: credit card usage during the Christmas buying season and consumer installment credit as a percentage of disposable income.

★ 464 ★

Credit Cards

Credit Card Usage

The average American has 6.9 credit cards. During the Christmas buying season, Visa credit cards alone are used an average of 5,340 times every minute. Consumer installment credit averages 19.4% of disposable income.

Source: Rukeyser, Louis. *Louis Rukeyser's Business Almanac.* New York: Simon & Shuster, 1991, p. 410. Primary source: Federal Reserve and *American Demographics* magazine. **Remarks**: Source also includes: credit card debt by educational level.

Household Expenses (general)

★ 465 ★

Annual Household Expenditures, 1984-1990

Item	1984	1985	1988	1989	1990
All consumer units (1,000)	$90,233	$91,564	$94,862	$95,818	$96,968
Total expenditures	21,975	23,490	25,892	27,810	28,369
Food, total	3,290	3,477	3,748	4,152	4,296
Food at home, total	1,970	2,037	2,136	2,390	2,485
Food away from home	1,320	1,441	1,612	1,762	1,811
Alcoholic beverages	275	306	269	284	293
Housing total	6,674	7,087	8,079	8,609	8,886
Shelter	3,489	3,833	4,493	4,835	5,032
Fuels, utilities, and public services	1,638	1,648	1,747	1,835	1,890
Household operations and furnishings	1,241	1,282	1,477	1,546	1,557
Housekeeping supplies	307	325	361	394	406
Apparel and services	1,319	1,420	1,489	1,582	1,617
Transportation, total	4,304	4,587	5,093	5,187	5,122
Vehicles	1,813	2,043	2,361	2,291	2,129
Gasoline and motor oil	1,058	1,035	932	985	1,047
Other transportation	1,433	1,509	1,800	1,911	1,946
Health care	1,049	1,108	1,298	1,407	1,480
Life insurance	300	278	314	346	345
Pensions and Social Security	1,598	1,738	1,935	2,125	2,248
Other expenditures[1]	2,936	3,269	3,426	3,857	3,806

Source: U.S. Bureau of the Census. *Statistical Abstract of the United States 1992*. 112th edition. Washington, D.C.: U.S. Department of Commerce, 1992, p. 444. Primary source: U.S. Bureau of Labor Statistics, *Consumer Expenditures in 1990*. *Notes:* 1. Includes entertainment, personal care, reading, education, cash contributions, and miscellaneous expenditures.

★ 466 ★

Household Expenses (general)

Annual Household Expenditures by Age of Householder, 1990

Age	Total expenditures	Food at home	Food away from home	Housing Total	Apparel and services	Transportation: Vehicle purchases	Transportation: Gasoline motor oil	Transportation: All other transportation	Health care	Personal insurance and pensions	Personal taxes
All consumer units	$28,369	$2,485	$1,811	$8,886	$1,617	$2,129	$1,047	$1,946	$1,480	$2,592	$2,952
Under 25 years old	16,518	1,285	1,476	4,845	1,034	1,591	722	1,185	403	972	843
25 to 34 years old	28,107	2,340	1,760	9,349	1,571	2,421	1,080	1,914	981	2,761	2,954
35 to 44 years old	35,579	3,134	2,246	11,354	2,310	2,523	1,245	2,313	1,415	3,700	4,471
45 to 54 years old	36,996	3,008	2,482	10,719	2,165	2,967	1,391	2,692	1,597	3,847	4,070
55 to 64 years old	29,244	2,601	1,830	8,610	1,557	2,014	1,134	2,151	1,791	2,958	3,507

[Continued]

★ 466 ★

Annual Household Expenditures by Age of Householder, 1990

[Continued]

Age	Total expen-ditures	Food at home	Food away from home	Housing Total	Apparel and services	Transportation			Health care	Personal insurance and pen-sions	Per-sonal taxes
						Vehi-cle pur-chases	Gaso-line motor oil	All other trans-porta-tion			
65 to 74 years old	20,895	2,106	1,199	6,591	972	1,163	792	1,511	2,197	1,071	1,378
75 years old and over	15,448	1,654	752	5,527	489	921	396	815	2,223	261	829

Source: U.S. Bureau of the Census. *Statistical Abstract of the United States 1992*. 112th edition. Washington, D.C.: U.S. Department of Commerce, 1992, p. 442. Primary source: U.S. Bureau of Labor Statistics, *Consumer Expenditures in 1990*. **Remarks**: Source also includes: 1990 average annual household expenditures, by region of residence, by size of household, and by occupation of householder.

★ 467 ★

Household Expenses (general)

Annual Household Expenditures by Region of Residence, 1990

Region	Total expen-ditures	Food at home	Food away from home	Housing Total	Apparel and serv-ices	Transportation			Health care	Per-sonal insur-ance and pen-sions	Per-sonal taxes
						Vehi-cle pur-chases	Gaso-line, motor oil	All other trans-porta-tation			
All consumer units	$28,369	$2,485	$1,811	$8,886	$1,617	$2,129	$1,047	$1,946	$1,480	$2,592	$2,952
Northeast	29,489	2,599	2,024	9,789	1,808	1,920	867	2,026	1,396	2,690	3,094
Midwest	25,919	2,313	1,709	7,837	1,357	2,068	1,000	1,726	1,336	2,408	2,426
South	27,011	2,381	1,696	8,000	1,549	2,275	1,152	1,884	1,600	2,395	2,715
West	32,445	2,749	1,909	10,699	1,852	2,174	1,115	2,240	1,544	3,042	3,701

Source: U.S. Bureau of the Census. *Statistical Abstract of the United States 1992*. 112th edition. Washington, D.C.: U.S. Department of Commerce, 1992, p. 442. Primary source: U.S. Bureau of Labor Statistics, *Consumer Expenditures in 1990*. **Remarks**: Source also includes: 1990 average annual household expenditures, by age of household, by size of household, and by occupation of householder.

★ 468 ★

Household Expenses (general)

Annual Household Expenditures by Occupation of Householder, 1990

Occupation	Total expen-ditures	Food at home	Food away from home	Hous-ing total	Apparel and ser-vices	Transportation			Health care	Per-sonal insur-ance and pen-sions	Per-sonal taxes
						Vehi-pur-chases	Gasoline, motor oil	All other trans-porta-tation			
All consumer units	$28,369	$2,485	$1,811	$8,886	$1,617	$2,129	$1,047	$1,946	$1,480	$2,592	$2,952
Self employed workers	35,795	2,745	2,338	10,673	1,921	2,584	1,192	2,317	2,192	3,867	2,825
Wage and salary earners:											
Managers and professionals	41,901	2,796	2,775	13,372	2,557	2,990	1,289	2,886	1,685	4,747	5,907
Technical, sales, and clerical	29,047	2,453	1,972	9,254	1,659	1,910	1,076	2,192	1,276	2,942	3,271
Service workers	21,253	2,177	1,347	6,518	1,343	1,613	878	1,558	903	1,809	1,639

[Continued]

★ 468 ★

Annual Household Expenditures by Occupation of Householder, 1990

[Continued]

Occupation	Total expen-ditures	Food at home	Food away from home	Hous-ing total	Apparel and ser-vices	Transportation			Health care	Per-sonal insur-ance and pen-sions	Per-sonal taxes
						Vehi-pur-chases	Gasoline, motor oil	All other trans-porta-tation			
Construction workers/mechanics	28,100	2,571	1,660	8,460	1,514	2,576	1,466	1,900	1,116	2,766	2,900
Operators, fabricators and laborers	25,465	2,652	1,546	7,311	1,348	2,455	1,233	1,750	1,030	2,402	2,176
Retired	18,144	2,007	961	5,993	709	1,307	640	1,199	2,058	436	1,027

Source: U.S. Bureau of the Census. *Statistical Abstract of the United States 1992.* 112th edition. Washington, D.C.: U.S. Department of Commerce, 1992, p. 442. Primary source: U.S. Bureau of Labor Statistics, *Consumer Expenditures in 1990.* **Remarks**: Source also includes: 1990 average annual household expenditures, by age of householder, by region of residence, and by size of household.

★ 469 ★

Household Expenses (general)

Annual Household Expenditures of Husband-Wife Families, 1990

	Husband and wife only	Oldest child under 6	Oldest child 6 to 17	Oldest child 18 or older
Income before taxes	$36,196	$40,687	$44,628	$50,200
Persons per household	2.0	3.4	4.2	3.9
Annual expenditures	$31,509	$35,009	$38,779	$42,785
Food, total	$4,567	$4,660	$6,187	$6,887
Food at home	2,507	2,972	3,660	4,034
Food away from home	2,061	1,687	2,527	2,853
Alcoholic beverages	337	266	289	341
Housing	9,593	12,723	11,672	11,171
Apparel	1,657	1,881	2,373	2,450
Transportation	5,748	6,229	7,163	8,993
Health care	2,091	1,551	1,597	1,969
Entertainment	1,668	1,613	2,216	1,992
Personal-care products	418	370	499	620
Reading	186	184	193	200
Education	253	255	555	1,247
Tobacco and supplies	269	249	324	441
Miscellaneous	650	699	832	873
Cash contributions	1,197	535	868	1,374
Personal insurance and pensions	2,875	3,795	4,011	4,226

Source: "How Children Spend Their Parents' Money," *American Demographics*, 14, February 1992, p. 6. Primary source: 1990 Consumer Expenditure Survey. **Remarks**: Source also includes: 1990 annual household expenditures for one-parent families.

★ 470 ★

Household Expenses (general)

Annual Household Expenditures of One-Parent Families, 1990

	One parent, children under 18
Consumer units	6,074
Income before taxes	$17,415
Persons per household	2.9
Annual expenditures	$19,230
Food, total	$3,539
Food at home	2,397
Food away from home	1,142
Alcoholic beverages	164
Housing	7,007
Apparel	1,502
Transportation	2,708
Health care	629
Entertainment	880
Personal-care products	270
Reading	79
Education	261
Tobacco and supplies	208
Miscellaneous	475
Cash contributions	351
Personal insurance and pensions	1,157

Source: "How Children Spend Their Parents' Money," *American Demographics*, 14, February 1992, p. 6. Primary source: 1990 Consumer Expenditure Survey. **Remarks**: Source also includes: 1990 annual household expenditures for husband-wife families.

★ 471 ★

Household Expenses (general)

Household Spending as a Percentage of Income, 1990

This is spending for two-earner family with two dependent children.

Category	Average Spending	Percentage of Income
Family income	$54,926	100.0%
Federal taxes	16,608	30.2
State & local taxes	5,218	9.5
After tax expenditures	33,100	60.3
Housing & household operations	6,927	16.3
Food	6,061	11.0
Health & personal care	4,853	8.8

[Continued]

★ 471 ★

Household Spending as a Percentage of Income, 1990

[Continued]

Category	Average Spending	Percentage of Income
Transportation	4,012	7.3
Recreation	2,545	4.6
Clothing	2,456	4.5
Insurance & pensions	1,255	2.3
Other	2,991	5.5

Source: "Average Family's Budget," *Consumers' Research*, 75, April 1992, p. 21. Primary source: Tax Foundation.

Household Expenses for Miscellaneous Items

★ 472 ★

Weekly Expenditures for Selected Non-food Products, 1985-1989

Item	Average Weekly Consumer Expenditures				
	1985	1986	1987	1988	1989
Tobacco	$3.78	$3.60	$3.71	$3.81	$3.81
Pet food	1.22	1.09	1.15	1.20	1.36
Personal care	2.65	2.72	2.93	3.22	3.28
Non-prescription drugs	0.96	1.01	1.10	1.19	1.23
Housekeeping supplies	6.24	6.07	6.56	6.94	7.57
Total	14.85	14.49	15.45	16.36	17.25

Source: U.S. Department of Commerce. International Trade Administration. *U.S. Industrial Outlook '92*. Washington, D.C.: U.S. Department of Commerce, 1992, p. 39-7. Primary source: U.S. Department of Labor, Bureau of Labor Statistics, *Consumer Expenditure Surveys, 1985-1989*.

★ 473 ★

Household Expenses for Miscellaneous Items

Annual Household Expenditures for Women's Clothing, 1984-1988

Year	Average Cost of Women's Clothing per Household
1984	$440
1985	500
1986	470
1987	520
1988	490

Source: "The Fashion Harvest," Judith Waldrop, *American Demographics*, 12, August 1990, p. 4. Primary source: U.S. Department of Labor, Bureau of Labor Statistics, *Consumer Expenditure Surveys, 1985-1989.*

★ 474 ★

Household Expenses for Miscellaneous Items

Household Expenses in Selected Metropolitan Areas, 1990

Metropolitan Area	Total expenditures[1]	Food	Housing		Apparel and services	Transportation			Health care
			Total[1]	Shelter		Total[1]	Vehicle purchases	Gasoline and motor oil	
Anchorage, AK	$43,434	$5,554	$13,396	$8,529	$2,016	$8,610	$4,221	$1,195	$1,671
Atlanta, GA	32,760	4,158	10,775	6,401	2,108	5,543	2,109	1,058	1,798
Baltimore, MD	30,768	4,529	9,895	6,074	1,978	5,100	2,059	1,005	1,453
Boston-Lawrence-Salem, MA-NH	30,518	4,036	11,461	7,624	1,659	4,896	1,762	884	1,165
Chicago-Gary-Lake County, IL-IN-WI	32,890	5,151	10,728	6,487	2,184	5,301	2,180	986	1,256
Cleveland-Akron-Lorain, OH	26,357	4,166	7,468	3,853	2,043	4,502	1,623	874	1,340
Dallas-Fort Worth, TX	34,534	4,530	10,296	5,453	1,982	6,948	3,085	1,328	1,552
Detroit-Ann Arbor, MI	28,658	3,760	9,521	5,644	1,262	5,883	2,483	1,160	1,157
Honolulu, HI	33,320	5,455	10,359	6,770	1,477	5,200	2,002	858	1,691
Houston-Galveston-Brazoria, TX	30,217	4,195	9,063	5,077	1,589	6,278	2,483	1,231	1,469
Los Angeles-Long Beach, CA	36,061	5,236	12,656	7,995	2,404	6,004	2,520	1,087	1,633
Miami-Fort Lauderdale, FL	33,205	4,926	10,470	6,204	1,587	6,839	3,125	954	1,742
Minneapolis-St. Paul, MN-WI	32,827	4,621	10,340	6,543	1,816	5,295	1,727	1,196	1,374
New York-Northern New Jersey-Long Island, NY-NJ-CT	32,680	5,120	11,169	6,968	2,290	4,866	1,771	766	1,490
Philadelphia-Wilmington-Trenton, PA-NJ-DE-MD	31,429	4,619	10,173	5,562	2,037	5,223	1,940	847	1,646
San Diego, CA	32,024	4,351	11,642	7,179	1,849	5,019	1,491	1,182	1,285
San Francisco-Oakland-San Jose, CA	38,927	5,292	13,727	9,449	2,556	6,470	2,570	1,024	1,348
Seattle-Tacoma, WA	33,426	4,750	10,759	6,721	1,667	5,761	2,179	1,009	1,578
St. Louis-East St. Louis-Alton, MO-IL	27,491	3,813	8,793	4,386	1,340	4,888	1,932	1,013	1,428
Washington, DC-MD-VA	37,505	4,825	12,905	8,172	2,469	5,922	2,377	1,014	1,918

Source: U.S. Bureau of the Census. *Statistical Abstract of the United States 1992.* 112th edition. Washington, D.C.: U.S. Department of Commerce, 1992, p. 444. Primary source: U.S. Bureau of Labor Statistics, *Consumer Expenditures in 1990. Note:* 1. Includes other items not shown separately.

★ 475 ★

Household Expenses for Miscellaneous Items

Cost of Furniture

Householders under age 35 spend an average of $310 per year on furniture, according to a consumer expenditure survey done by the U.S. Bureau of Labor Statistics.

Source: "Spending Money: Nice Niches for Furniture," Thomas Exter, *American Demographics*, 13, March 1991, p. 6. Primary source: U.S. Department of Labor, Bureau of Labor Statistics, *Consumer Expenditure Surveys, 1985-1989.*

★ 476 ★

Household Expenses for Miscellaneous Items

Cost of Household Footwear per Year, 1984-1991

Year	Average Cost of Footwear per Household
1984	$185
1985	184
1986	167
1987	184
1988	196
1989	189
1990	225
1991	242

Source: U.S. Department of Labor, Bureau of Labor Statistics, *Consumer Expenditure Surveys, 1984-1981.*

★ 477 ★

Household Expenses for Miscellaneous Items

Annual Cost of Deodorant in Selected Cities

City	Average Annual Cost
Denver, CO	$4.93
Portland, ME	5.00
Hartford, CT	5.12
Houston, TX	5.12
Dallas/Ft. Worth, TX	5.24
Seattle, WA	5.60
Boston, MA	5.73
Salt Lake City, UT	6.70
Grand Rapids, MI	7.45
Pittsburgh, PA	8.04

Source: Krantz, Les. *The Best and Worst of Everything.* New York: Prentice Hall General Reference, 1991, p. 95. Primary source: Arbitron/SAMI.

★ 478 ★

Household Expenses for Miscellaneous Items

Cost for a Night at Home in Selected Cities

City	Average Cost of a McDonald's Quarter Pounder, Six-Pack of Beer, New Monopoly Game
Cleveland	$17.57
Orlando	17.97
Tulsa	18.42
Anchorage	26.69
New York City	27.44
286 other cities	18.53

Source: "A Burger, a Cold One, and Thou," *Business Week*, 22, August 31, 1992, p. 34. Primary source: ACCRA.

Housing

★ 479 ★

Public Housing

Tenants in public housing with an average income of $7,700, pay about 30% of their rent, or an average of $170 per month.

Source: "The Larger Context: Barefoot Capitalism," Michael Novak, *Forbes*, 17, May 1991, p. 90.

★ 480 ★

Housing

Retirement Communities

According to the American Association of Homes for the Aging (AAHA), in the 700 continuing-care retirement communities (CCRC) in the United States, the median entrance fee for a one-bedroom unit is $58,637, and the median monthly fee is $871.

Source: "Personal Business: Planning Before You Settle on a Retirement Community, Susan B. Garland, *Business Week*, 20, May 1991, pp. 150-151.

★ 481 ★

Housing

Nursing Homes

In 1990, the average cost of a nursing home was $31,000 per year, or $86 per day. This is double the 1980 cost. The price ranges from $60,000 in large cities to $20,000 in rural areas.

Source: "Policies for Old-Age Care," Virginia Wilson, *Newsweek*, 120, April 20, 1992, p. 61.

★ 482 ★

Housing

Cost of Mobile Homes Vs. On-Site Homes

Type	Average Cost
Single section mobile	$19,800
Double section mobile	36,000
On-site homes	149,800

Source: "The Ready-Made Home Market," *The New York Times*, September 14, 1992, p. D3. Primary source: Manufactured Housing institute.

★ 483 ★

Housing

Housing Affordability, 1993 Vs. 1983

	1993	1983
Average family income	$32,000	$21,500
Average house price	$100,000	$55,000
Interest rate	7.5%	13%
Average monthly payment	$625	$540
Share of income to housing	23%	30%

Source: "Housing Affordability Today Vs. '83," *USA Today*, March 19-21, 1993, p. 1A.

★ 484 ★

Housing

Share of Income for Housing in Selected Cities

City	Average Share of Income for Housing
Youngstown, OH	8.9%
Oklahoma City, OK	9.2
Louisville, KY	9.3
Mobile, AL	10.1
Houston, TX	10.5

[Continued]

★ 484 ★

Share of Income for Housing in Selected Cities

[Continued]

City	Average Share of Income for Housing
Jacksonville, FL	10.6
Salt Lake City, UT	10.9
Indianapolis, IN	11.1
Akron, OH	11.2
Tucson, AZ	11.9
Austin, TX	12.3
Denver, CO	12.4
Greenville, NC	12.6
Albuquerque, NM	13.2
Milwaukee, WI	14.1
Middlesex/Somerset, NJ	14.5
Raleigh, NC	14.9
Tacoma, WA	16.2

Source: "The Paycheck Pinch Test," *U.S. News & World Report*, 113, April 5, 1993, p. 98. Primary source: The WEFA Group, HSH Associates.

★ 485 ★

Housing

1993 Median Price for Homes in Selected Cities

City	Median Price
Expensive:	
Middlesex/Somerset, NJ	$184,461
Raleigh, NC	115,510
Tacoma, WA	115,432
Milwaukee, WI	106,204
Portland, OR	104,493
Moderate:	
Indianapolis, IN	91,213
Austin, TX	90,104
Houston, TX	89,748
Dayton, OH	89,705
Akron, OH	86,233
Inexpensive:	
Tucson, AZ	78,591
Louisville, KY	75,355
Mobile, AL	71,042

[Continued]

★ 485 ★

1993 Median Price for Homes in Selected Cities

[Continued]

City	Median Price
Oklahoma City, OK	68,080
Youngstown, OH	62,229

Source: "The Paycheck Pinch Test," *U.S. News & World Report*, 113, April 5, 1993, p. 98. Primary source: The WEFA Group.

★ 486 ★

Housing

Cities With the Biggest Rise in the Median Price of Homes, 1992

City	Median 1992 Price	Percentage Change ('91-'92)
Akron, OH	$79,262	10.8%
Oklahoma City, OK	61,950	10.1
Tacoma, WA	107,138	9.7
Portland, OR	97,275	9.4
Houston, TX	81,012	8.9
Gary-Hammond, IN	75,938	8.9
San Antonio, TX	71,438	8.8
Milwaukee, WI	98,138	8.0
Tucson, AZ	71,218	7.6
Greenville-Spartanburg, SC	83,862	7.6

Source: "Big Winners in '92," *U.S. News & World Report*, 114, April 5, 1993, p. 75. Primary source: National Association of Realtors, the WEFA Group.

★ 487 ★

Housing

Monthly Mortgage Payment, 1976-1990

Year	Average Monthly Payment	Percent of Income
1976	$329	24.0%
1980	599	32.4
1984	868	30.3
1985	896	30.0
1986	852	28.6
1987	939	29.3
1988	1,008	32.8

[Continued]

★ 487 ★

Monthly Mortgage Payment, 1976-1990

[Continued]

Year	Average Monthly Payment	Percent of Income
1989	1,054	31.8
1990	1,127	32.8

Source: U.S. Bureau of the Census. *Statistical Abstract of the United States 1992*. 112th edition. Washington, D.C.: U.S. Department of Commerce, 1992, p. 724. Primary source: Chicago Title Insurance Company, Chicago, IL, *The Guarantor*, bimonthly. **Remarks**: Source also includes: median purchase price; average age of first-time and repeat buyer; and down payment of first-time and repeat buyers, all for 1976-1990.

★ 488 ★

Housing

Median Purchase Price for First-time Buyers Vs. Repeat Buyers, 1976-1989

Year	Median Price of Home	
	First-time Buyer	Repeat Buyer
1976	$37,670	$50,090
1980	61,450	75,750
1984	81,500	100,400
1985	75,100	106,200
1986	74,700	114,860
1987	84,730	115,430
1988	97,100	141,400
1989	105,200	144,700

Source: U.S. Bureau of the Census. *Statistical Abstract of the United States 1992*. 112th edition. Washington, D.C.: U.S. Department of Commerce, 1992, p. 724. Primary source: Chicago Title Insurance Company, Chicago, IL, *The Guarantor*, bimonthly. **Remarks**: Source also includes: average monthly payment; average age of first-time and repeat buyer; and down payment of first-time and repeat buyers, all for 1976-1990.

★ 489 ★

Housing

Median Purchase Price for First-time Buyers Vs. Repeat Buyers, 1990-1992

Year	Median Price of Home	
	First-time Buyer	Repeat Buyer
1990	$106,000	$149,400
1991	118,700	152,500
1992	122,400	158,000

Source: Chicago Title and Trust Family of Title Insurers, Chicago, IL, *Who's Buying Homes in America*, 1993, p. 3. **Remarks**: Source also includes: average price of homes, average monthly payment, monthly payment as a percentage of income, average age of buyer, average number of houses looked at, average number of months buyers looked before deciding, all for 1990-1992.

★ 490 ★

Housing

Average Monthly Payment for First-time Buyers Vs. Repeat Buyers, 1990-1992

Year	Average Monthly Payment	
	First-time Buyer	Repeat Buyer
1990	$1,010	$1,221
1991	1,046	1,230
1992	968	1,156

Source: Chicago Title and Trust Family of Title Insurers, Chicago, IL, *Who's Buying Homes in America*, 1993, p. 3. **Remarks**: Source also includes: median price of homes, average price of homes, monthly payment as a percentage of income, average age of buyer, average number of houses looked at, average number of months buyers looked before deciding, all for 1990-1992.

★ 491 ★

Housing

Average Monthly Payment by Household Income, 1990-1992

Income	1990	1991	1992
Less than $30,000	$692	$748	$619
30,000-$40,000	772	869	766
$41,000-$50,000	950	1,030	881
$51,000-$60,000	1,164	1,036	1,027
$61,000-	1,515	1,484	1,402

Source: Chicago Title and Trust Family of Title Insurers, Chicago, IL, *Who's Buying Homes in America*, 1993, p. 3. **Remarks**: Source also includes: median price of homes, average price of homes, average monthly payment, monthly payment as a percentage of income, average age of buyer, average number of houses looked at, average number of months buyers looked before deciding, all for 1990-1992.

★ 492 ★

Housing

Median Price of a Home by Household Income, 1990-1992

Income	First-time Buyer			Repeat Buyer		
	1990	1991	1992	1990	1991	1992
30,000-$40,000	$78,600	$89,000	$88,300	$97,500	$87,800	$98,800
$41,000-$50,000	102,100	111,400	106,500	118,200	125,500	128,300
$51,000-$60,000	133,100	119,600	120,800	155,900	132,100	127,400
$61,000-	154,000	154,000	175,700	208,500	233,400	209,100

Source: Chicago Title and Trust Family of Title Insurers, Chicago, IL, *Who's Buying Homes in America*, 1993, p. 7.

★ 493 ★

Housing

Median Value of Owner-Occupied Housing, by State, 1990

State	Median (dol.)	State	Median (dol.)	State	Median (dol.)
United States	79,100	Kentucky	50,500	North Dakota	50,800
Alabama	53,700	Louisiana	58,500	Ohio	63,500
Alaska	94,400	Maine	87,400	Oklahoma	48,100
Arizona	80,100	Maryland	116,500	Oregon	67,100
Arkansas	46,300	Massachusetts	162,800	Pennsylvania	69,700
California	195,500	Michigan	60,600	Rhode Island	133,500
Colorado	82,700	Minnesota	74,000	South Carolina	61,100
Connecticut	177,800	Mississippi	45,600	South Dakota	45,200
Delaware	100,100	Missouri	59,800	Tennessee	58,400
District of Columbia	123,900	Montana	56,600	Texas	59,600
Florida	77,100	Nebraska	50,400	Utah	68,900
Georgia	71,300	Nevada	95,700	Vermont	95,500
Hawaii	245,300	New Hampshire	129,400	Virginia	91,000
Idaho	58,200	New Jersey	162,300	Washington	93,400
Illinois	80,900	New Mexico	70,100	West Virginia	47,900
Indiana	53,900	New York	131,600	Wisconsin	62,500
Iowa	45,900	North Carolina	65,800	Wyoming	61,600
Kansas	52,200				

Source: U.S. Bureau of the Census. *Statistical Abstract of the United States 1992*. 112th edition. Washington, D.C.: U.S. Department of Commerce, 1992, p. 718. Primary source: U.S. Bureau of the Census, *1990 Census of Housing, General Housing Characteristics*, series CH-1, and *Census of Population and Housing, 1990: Summary Tape File 1C*. **Remarks**: Source also includes: median rent by state, 1990.

★ 494 ★

Housing

Median Rental Payments, by State, 1990

State	Median (dol.)	State	Median (dol.)	State	Median (dol.)
United States	374	Kentucky	250	North Dakota	266
Alabama	229	Louisiana	260	Ohio	296
Alaska	503	Maine	358	Oklahoma	259
Arizona	370	Maryland	473	Oregon	344
Arkansas	230	Massachusetts	506	Pennsylvania	322
California	561	Michigan	343	Rhode Island	416
Colorado	362	Minnesota	384	South Carolina	276
Connecticut	510	Mississippi	215	South Dakota	242
Delaware	425	Missouri	282	Tennessee	273
District of Columbia	441	Montana	251	Texas	328
Florida	402	Nebraska	282	Utah	300
Georgia	344	Nevada	445	Vermont	378
Hawaii	599	New Hampshire	479	Virginia	411
Idaho	261	New Jersey	521	Washington	383
Illinois	369	New Mexico	312	West Virginia	221
Indiana	291	New York	428	Wisconsin	331
Iowa	261	North Carolina	284	Wyoming	270
Kansas	285				

Source: U.S. Bureau of the Census. *Statistical Abstract of the United States 1992.* 112th edition. Washington, D.C.: U.S. Department of Commerce, 1992, p. 719. Primary source: U.S. Bureau of the Census, *1990 Census of Housing, General Housing Characteristics*, series CH-1, and *Census of Population and Housing, 1990: Summary Tape File 1C*. **Remarks**: Source also includes: median value of owner- occupied housing, 1990.

★ 495 ★

Housing

Median Price of New Homes by Region of the United States, 1970-1991

Year	Northeast	Midwest	South	West
1970	$30,300	$24,400	$20,300	$24,000
1971	30,600	27,200	22,500	25,500
1972	31,400	29,300	25,800	27,500
1973	37,100	32,900	30,900	32,400
1974	40,100	36,100	34,500	35,800
1975	44,000	39,600	37,300	40,600
1976	47,300	44,800	40,500	47,200
1977	51,600	51,500	44,100	53,500
1978	58,100	59,200	50,300	61,300
1980	69,500	63,400	59,600	72,300
1981	76,000	65,900	64,400	77,800
1982	78,200	68,900	66,100	75,000
1983	82,200	79,500	70,900	80,100
1984	88,600	85,400	72,000	87,300
1985	103,300	80,300	75,000	92,600

[Continued]

★ 495 ★

Median Price of New Homes by Region of the United States, 1970-1991

[Continued]

Year	Northeast	Midwest	South	West
1986	125,000	88,300	80,200	95,700
1987	140,000	95,000	88,000	111,000
1988	149,000	101,600	92,000	126,500
1990	159,000	107,900	99,000	147,500
1991	155,400	110,000	100,000	142,300

Source: U.S. Bureau of the Census. *Statistical Abstract of the United States 1992*. 112th edition. Washington, D.C.: U.S. Department of Commerce, 1992, p. 712. Primary source: U.S. Bureau of the Census and U.S. Department of Housing and Urban Development, *Current Construction Reports*, series C25, *Characteristics of New Housing*, annual; and *One-Family Houses Sold*, monthly.

★ 496 ★

Housing

Median Price of Existing Homes by Region of the United States, 1970-1991

Year	Median sales price (dol.)				
	Total	Northeast	Midwest	South	West
1970	23,000	25,200	20,100	22,200	24,300
1971	24,800	27,100	22,100	24,300	26,500
1972	26,700	29,800	23,900	26,400	28,400
1973	28,900	32,800	25,300	29,000	31,000
1974	32,000	35,800	27,700	32,300	34,800
1975	35,300	39,300	30,100	34,800	39,600
1976	38,100	41,800	32,900	36,500	46,100
1977	42,900	44,400	36,700	39,800	57,300
1978	48,700	47,900	42,200	45,100	66,700
1980	62,200	60,800	51,900	58,300	89,300
1981	66,400	63,700	54,300	64,400	96,200
1982	67,800	63,500	55,100	67,100	98,900
1983	70,300	72,200	56,600	69,200	94,900
1984	72,400	78,700	57,100	71,300	95,800
1985	75,500	88,900	58,900	75,200	95,400
1986	80,300	104,800	63,500	78,200	100,900
1987	85,600	133,300	66,000	80,400	113,200
1988	89,300	143,000	68,400	82,200	124,900
1990	95,500	141,200	74,000	85,900	139,600
1991	100,300	141,900	77,800	88,900	147,200

Source: U.S. Bureau of the Census. *Statistical Abstract of the United States 1992*. 112th edition. Washington, D.C.: U.S. Department of Commerce, 1992, p. 712. Primary source: National Association of Realtors, Washington, DC, *Home Sales*, monthly, and *Home Sales Yearbook 1990*.

★ 497 ★

Housing

Price of New Mobile Homes by Region of the United States, 1975-1991

Year	Average Sales Price (dol.)				
	U.S.	Northeast	Midwest	South	West
1975	10,600	10,500	10,700	9,000	13,600
1980	19,800	18,500	18,600	18,200	25,400
1981	19,900	19,000	18,900	18,400	25,600
1982	19,700	19,800	20,000	18,500	24,700
1983	21,000	21,400	20,400	19,700	27,000
1984	21,500	22,200	21,100	20,200	27,400
1985	21,800	22,700	21,500	20,400	28,700
1986	22,400	24,400	21,800	20,700	29,900
1987	23,700	25,600	23,700	21,900	31,000
1988	25,100	27,000	24,600	22,700	33,900
1989	27,200	30,200	26,700	24,100	37,800
1990	27,800	30,000	27,000	24,500	39,300
1991, prel.	27,800	30,500	27,600	24,500	38,600

Source: U.S. Bureau of the Census. *Statistical Abstract of the United States 1992.* 112th edition. Washington, D.C.: U.S. Department of Commerce, 1992, p. 712. Primary source: U.S. Bureau of the Census, *Current Construction Reports*, series C20.

★ 498 ★

Housing

Counties (Affluent Suburbs) With High Cost and High Ownership Housing, 1990

County	percent owner-occupied	median value of owner-occupied dwelling units
Hunterdon County, NJ (Middlesex-Somerset-Hunterdon)	80.5%	$209,892
Putnam County, NY (New York)	81.9	194,943
Sussex County, NJ (Newark)	82.3	156,336
Calvert County, MD (Washington)	85.0	136,084
Ocean County, NJ (Monmouth-Ocean)	82.9	125,959
Stafford County, VA (Washington)	81.9	125,389
Douglas County, CO (Denver)	85.2	119,454
Queen Annes County, MD (Baltimore)	81.0	117,975
Pike County, PA	83.3	117,702
Fayette County, GA (Atlanta)	86.1	116,709
Poquoson City, VA (Norfolk-Virginia Beach-Newport News)	82.7	113,636
Geauga County, OH (Cleveland)	85.7	107,687
Spotsylvania County, VA	81.9	103,982
Lincoln County, ME	83.2	102,984
Livingston County, MI (Detroit)	84.5	97,287
Forsyth County, GA (Atlanta)	81.9	96,203
Washington County, MN (Minneapolis-St. Paul)	83.9	94,220

[Continued]

★ 498 ★

Counties (Affluent Suburbs) With High Cost and High Ownership Housing, 1990

[Continued]

County	percent owner-occupied	median value of owner-occupied dwelling units
Hanover County, VA (Richmond-Petersburg)	83.5	91,308
Scott County, MN (Minneapolis-St. Paul)	81.9	90,943

Source: "Four Ways to Find a Home," William Dunn, *American Demographics*, 14, July 1992, pp. 56-61. Primary source: 1990 Census. **Remarks**: Source also includes: low cost and high ownership housing, high cost & low ownership housing, and low cost & low ownership housing, 1990.

★ 499 ★

Housing

Counties (Rural) With Low Cost and High Ownership Housing, 1990

County	percent owner-occupied	median value of owner-occupied dwelling units	County	percent owner-occupied	median value of owner-occupied dwelling units
Bollinger County, MO	81.7%	$29,384	Jackson County, KS	81.4%	$34,241
Grundy County, TN	81.4	29,793	Perry County, MS	82.5	34,243
Newton County, TX	83.2	30,442	Washington County, AL	87.2	34,250
Catahoula Parish, LA	81.5	31,555	La Salle Parish, LA	82.1	34,256
Wayne County, TN	83.7	32,839	Marshall County, MN	82.1	34,552
Edmonson County, KY	85.6	32,967	Leake County, MS	84.8	34,654
Greene County, MS	86.4	33,750	Decatur County, TN	80.5	34,706
Calhoun County, MS	80.8	33,914	Lowndes County, AL	80.5	34,760
Fulton County, AR	81.7	34,016			

Source: "Four Ways to Find a Home," William Dunn, *American Demographics*, 14, July 1992, pp. 56-61. Primary source: 1990 Census. **Remarks**: Source also includes: high cost & high ownership housing, high cost & low ownership housing, and low cost & low ownership housing, 1990.

★ 500 ★

Housing

Counties (Affluent and Urban) With High Cost and Low Ownership Housing, 1990

County	percent owner-occupied	median value of owner-occupied dwelling units
New York County, NY (New York)	17.9%	$486,842
Pitkin County, CO	52.4	451,968
San Francisco County, CA (San Francisco)	34.5	298,866
Honolulu County, HI (Honolulu)	52.0	283,596
Santa Barbara County, CA (Santa Barbara-Santa Maria-Lompoc)	54.7	249,944
Arlington County, VA (Washington)	44.6	230,980
Alexandria city, VA (Washington)	40.5	228,602
Alameda County, CA (Oakland)	53.3	227,186
Los Angeles County, CA (Los Angeles-Long Beach)	48.2	226,446
Maui County, HI	57.6	202,059
Monterey County, CA (Salinas-Seaside-Monterey)	50.6	198,241
Kings County, NY (New York)	25.9	196,147
Essex County, NJ (Newark)	45.3	196,064
Queens County, NY (New York)	42.4	190,990
San Diego, CA (San Diego)	53.8	186,698
Passaic County, NJ (Bergen-Passaic)	55.8	185,492
Bronx County, NY (New York)	17.9	173,919
Suffolk County, MA (Boston)	32.5	162,082
Hudson County, NJ (Jersey City)	32.5	156,962
Yolo County, CA (Sacramento)	51.9	137,761
Eagle County, CO	57.5	135,840
Sacramento, CA (Sacramento)	56.6	129,806
Providence County, RI	53.5	127,402
District of Columbia (Washington)	38.9	123,938
San Joaquin County, CA (Stockton)	57.6	121,654
Summit County, CO	48.2	121,401

Source: "Four Ways to Find a Home," William Dunn, *American Demographics*, 14, July 1992, pp. 56-61. Primary source: 1990 Census. **Remarks**: Source also includes: high cost & high ownership housing, low cost & high ownership housing, and low cost & low ownership housing, 1990.

★ 501 ★

Housing

Counties (of College Students and/or the Elderly) With Low Cost and Low Ownership Housing, 1990

County	percent owner-occupied	median value of owner-occupied dwelling units	County	percent owner-occupied	median value of owner-occupied dwelling units
Pemiscot County, MO	56.9%	$28,803	Desha County, AR	66.0%	$36,685
Madison Parish, LA	64.3	30,091	Coahoma County, MS	56.7	36,717
Lee County, AR	62.8	31,527	Phillips County, AR	53.7	36,853
Mississippi County, MO	64.9	32,273	Nodaway County, MO	65.4	37,063
Monroe County, AR	63.1	32,322	Sunflower County, MS	60.1	37,825
New Madrid County, MO	64.0	32,774	Davison County, SD	60.5	38,353
Beadle County, SD	65.8	34,164	Cameron County, TX (Brownsville-Harlingen)	64.4	38,436
Bell County, KY	65.8	34,199	Yazoo County, MS	66.3	38,442
Poinsett County, AR	65.2	34,467	St. Francis County, AR	61.1	38,563
McDonough County, IL	62.2	36,022			
Humphreys County, MS	58.7	36,604			

Source: "Four Ways to Find a Home," William Dunn, *American Demographics*, 14, July 1992, pp. 56-61. Primary source: 1990 Census. **Remarks**: Source also includes: high cost & high ownership housing, low cost and high ownership housing, and high cost & low ownership housing, 1990.

★ 502 ★

Housing

Counties With the Highest-Priced Housing, 1990

County	median value of owner-occupied dwelling units
New York County, NY (New York)	$486,842
Pitkin County, CO	451,968
Marin County, CA (San Francisco)	354,150
San Mateo County, CA (San Francisco)	343,938
San Francisco County, CA (San Francisco)	298,866
Santa Clara County, CA (San Jose)	289,430
Honolulu County, HI (Honolulu)	283,596
Westchester County, NY (New York)	283,462
Santa Cruz County, CA (Santa Cruz)	256,098
Orange County, CA (Anaheim-Santa Ana)	252,662
Santa Barbara County, CA (Santa Barbara-Santa Maria-Lompoc)	249,944
Fairfield County, CT (Bridgeport-Stamford, Norwalk-Danbury)	249,829
Ventura County, CA (Oxnard-Ventura)	245,263
Arlington County, VA (Washington)	230,980
Alexandria City, VA (Washington)	228,602
Bergen County, NJ (Bergen-Passaic)	227,661
Alameda County, CA (Oakland)	227,186

[Continued]

★ 502 ★

Counties With the Highest-Priced Housing, 1990

[Continued]

County	median value of owner-occupied dwelling units
Los Angeles County, CA (Los Angeles-Long Beach)	226,446
Contra Costa County, CA (Oakland)	219,407
Morris County, NJ (Newark)	217,334

Source: "The Highest-Priced Housing," *American Demographics*, 14, October 1992, p. 25. Primary source: 1990 Census.

★ 503 ★

Housing

Counties in the United States With the Highest Rent, 1990

County	Average Monthly Rent
Marin County, CA	$763
Fairfax County, VA	748
Orange County, CA	728
Santa Clara County, CA	715
San Mateo County, CA	711
Montgomery County, MD	698
Suffolk County, NY	696
Ventura County, CA	695
Loudoun County, VA	682
Nassau County, NY	678
Arlington County, VA	678
Putnam County, NY	672
Alexandria City, VA	667
Pitkin County, CO	663
Morris County, NJ	659
Maui County, HI	658

Source: "The Highest Rents," *American Demographics*, 14, October 1992, p. 26. Primary source: 1990 Census.

Loans and Insurance

★504★

Interest Rates, 1992

Type	Average Loan Rate
Mortgage (30-year fixed)	8.12% (19-year low)
Home equity loan	7.83% (all-time low)
Auto loan	9.39% (20-year low)
Credit card	18.45% (10-year low)

Source: "How to Profit From Low Loan Rates," Elizabeth MacDonald, *Money*, 21, September 1992, p. 27.

★505★

Loans and Insurance

Debt of Medical Students

According to the American Association of Medical Colleges, the average debt of the 81% of medical school graduates who were in debt in 1989 was $42,374, up from $19,700 in 1981. The debt for minority and underprivileged students to complete four years of medical school training was even greater: 91% had an average debt of $48,168 and 41% had debts exceeding $50,000.

Source: "Statements on Introduced Bills and Joint Resolutions," William S. Cohen, *Congressional Record*, January 14, 1991, pS747.

★506★

Loans and Insurance

Cost of Household Auto Insurance Vs. Claims Filed

During the last ten years, the average household spent $8910 on auto insurance but filed only one claim averaging $600.

Source: "How to Choose the Right Company," *Consumer Reports*, 57, August 1992, pp. 489-499.
Remarks: Source also includes: response of 63,000 consumers to a questionnaire rating 49 insurance companies.

★ 507 ★

Loans and Insurance

Life Insurance per Household, 1991

In 1991, for those households with life insurance, the average amount was $126,000.

Source: American Council of Life Insurance, Washington DC.

Support Payments

★ 508 ★

Recipients of Child Support Payments, 1990

Recipient	Average Child Support Payment
Divorced or separated	$3,268
Never married	1,888
Black mothers	2,263
Hispanic mothers	2,965
White mothers	3,132
ALL	2,995

Source: Krantz, Les. *What the Odds Are: A to Z Odds on Everything You Hoped or Feared Could Happen.* New York: Prentice Hall General Reference, 1992, p. 59.

★ 509 ★

Support Payments

Support Payments to Relative Outside Provider's Home, 1988

Relationship to Provider	Average Annual Payment
Children, all	$1,699
Children, 21 yrs. & up	3,085
Parents	1,330
Ex-spouses	4,855
Other relatives	1,737
ALL	1,925

Source: "Who's Helping Out?" *American Demographics*, 14, August 1992, p. 24. Primary source: Census Bureau.

Wedding and Funeral Expenses

★510★

Cost of a Wedding and Reception, 1988

According to a survey of 3,700 readers of *Modern Bride* magazine, the average wedding and reception for 200 guests in 1988 was $13,370, up from $10,500 in 1987.

Source: Bluman, Allan. *Elementary Statistics*. Dubuque, IA: William C. Brown Publishers, 1992, p. 77.

★511★

Wedding and Funeral Expenses

Cost of a Formal Wedding, 1990

According to a survey of the readers of *Bride's* magazine, mostly middle class and upper middle class, the average formal wedding cost in 1990, including the honeymoon, was $19,344.

Based on 200 guests and five attendants, an itemized list of the cost is:

Item	Average Cost
Attendants' dresses	$745+
Attendants' gifts	238
Attendants' suits	333
Bride's dress & veil	963
Bride's mother's dress	236
Clergy & church	166
Engagement ring	2,285
Flowers	478
Groom's suit (rented)	82
Honeymoon	3,200
Invitations	286
Limousine	201
Music	882
Photographer	908
Reception	5,900
Rehearsal dinner	501
Trousseau clothes	936
Wedding rings	1,004

Source: "It's One Party Even the Recession Can't Spoil," Rachel Powell, *The New York Times*, June 23, 1991, p. C10.

★512★

Wedding and Funeral Expenses

Cost of a Japanese Shinto Wedding

The average Shinto wedding service costs $24,000, but can easily run as high as $40,000. Some of the expenses incurred are: the average rental fee for a Japanese bridal dress is $1,600; the average cost per wedding guest is $350; and the Japanese custom of serving seven-course meals, with lobster or steak, costs an average of $130 per plate.

Source: "Japanese Pay for Tradition," David J. Morrow, *Detroit Free Press*, February 14, 1993, p. 1Q, 4Q.

★513★

Wedding and Funeral Expenses

Cost of a Japanese Funeral

An average funeral service in Tokyo, Japan cost $20,000, but many run as high as $75,000. This includes mortician and temple fees, flowers, altars, and food and drink for guests. Because of the scarcity of land, most Japanese are cremated, but even a small temple lot can cost $28,000.

Source: "International Business: Japan: Rest in Peace...With Lasers, Smoke, and Synthesizer Music," Karen Lowry Miller, *Business Week*, 21, September 16, 1991, p. 48.

★514★

Wedding and Funeral Expenses

Cost of an American Funeral

According to a 1989 survey by the Federal Trade Commission (FTC), the average funeral-burial package for Americans costs about $3,800; cremation-based funerals cost an average of $1,500.

Source: "Changing Styles Bring Cremation Industry to Life," *American Demographics*, 14, December 1992, p. 25.

Miscellaneous Spending

★ 515 ★

Annual Safety Deposit Box Fees, 1986, 1992

NOTE: Boxes are normally 18 to 24 inches deep.

Size	Average Annual Fee	
	1986	1992
3-inch by 5-inch	$13.70	$15.92
10-inch by 10-inch	56.90	67.14

Source: "The Price of Safety," *The New York Times*, April 24, 1993, p. 35. Primary source: Sheshunoff Information Services.

★ 516 ★

Miscellaneous Spending

Cost of a High School Prom, 1993

According to *Your Prom* magazine, the average cost of a high school prom in 1993 is $1,058.

An itemized list of the costs:

His		Hers	
Dinner	$54	Boutonniere	$14
Flowers	20	Casual clothes	93
Limousine	161	Cosmetics	24
Shoes & accessories	43	Gown	200
Tickets	43	Shoes & accessories	169
Tux rental	89	Visit to beauty salon	38
Other	63	Other	47
Total	473		585

Source: "Prom's Price," *USA Weekend*, May 14-16, 1993, p. 14.

★ 517 ★

Miscellaneous Spending

Cost of Magazines

According to ABC figures, the 1990 average price of an annual magazine subscription was $27.11, up 4.4 percent from 1989; while the average newsstand price was $2.65, up 8.6 percent from 1989.

Source: U.S. Department of Commerce. International Trade Administration. *U.S. Industrial Outlook '92*. Washington, D.C.: U.S. Department of Commerce, 1992, p. 25-7.

★518★

Miscellaneous Spending

Cost of Selected Weight Loss Programs

The average cost per pound of weight loss for a 12-week diet program:

Program	Average Cost
HMR	$7.94
Medifast	6.35
Optifast	9.98
United Weight Control	11.56
Diet Center	4.08
Jenny Craig	10.43
Nutrisystems	8.62
Registered Dietitian	6.80
Weight Watchers	1.10
TOPS	0.03
Overeaters Anonymous	0.00

Source: "The Cost of Losing Weight," *The New York Times*, November 24, 1992, p. C11. Primary source: *Journal of the American College of Nutrition.*

★519★

Miscellaneous Spending

Cost of Cigarettes

The average smoker spends $750 per year on cigarettes.

Source: "Smoking Out the Elusive Smoker," Walecia Konrad and Christopher Power, *Business Week*, 22, March 16, 1992, pp. 62-63.

★520★

Miscellaneous Spending

Lottery Wagers

Americans annually wager $90 per capita on lotteries. With 32 states and the District of Columbia allowing lotteries, $200 billion is bet annually in the United States.

Source: "A Global Framework for Analyzing Gaming Business Units," Lawrence Dandurand and others, *Nevada Review of Business & Economics*, 15, Spring- Summer 1991, pp. 2-13.

★ 521 ★

Miscellaneous Spending

States With the Most Lottery Spending per Year

State	Average per Capita Lottery Spending
Massachusetts	$235
District of Columbia	197
Maryland	185
Connecticut	162
New Jersey	155
Michigan	132
Ohio	128
Pennsylvania	121
Illinois	113
Delaware	89

Source: Krantz, Les. *The Best and Worst of Everything.* New York: Prentice Hall General Reference, 1991, p. 173. Primary source: *Gaming & Wagering Business.*

★ 522 ★

Miscellaneous Spending

College Students' Discretionary Spending

College students spend an average of $135 per month on snack foods, compact discs, movie tickets, and other discretionary items.

Source: "The Media Business: Advertising; New Marketing Specialists Tap Collegiate Consumers," Eben Shapiro, *The New York Times,* February 27, 1992, p. D16.

★ 523 ★

Miscellaneous Spending

Banking Fees by Type of Transaction, 1990-1991

Type	Average Fee		
	1990	1991	Change
Monthly checking account	$2.91	$3.75	$0.84
Stop-payment	11.21	12.35	1.14
Insufficient funds	13.00	14.17	1.17
Overdrafts	12.49	13.87	1.38
Minimum balance to avoid fee	124.84	165.36	40.52

Source: "Charges for Accounts Grow More Expensive," Michael Quint, *The New York Times,* December 12, 1992, p. 37. Primary source: Federal Reserve Board. **Remarks**: Source also includes: average cost per check written.

★ 524 ★

Miscellaneous Spending

Cost per Written Check

Each check written costs banks an average of 68 cents.

Source: "Charges for Accounts Grow More Expensive," Michael Quint, *The New York Times*, December 12, 1992, p. 37. Primary source: Federal Reserve Board. **Remarks**: Source also includes: average fees for various banking transactions.

Chapter 8
GOVERNMENT TAXING AND SPENDING

Spending

★525★

Newly Awarded Monthly Social Security Benefit by Type of Beneficiary, 1989

Beneficiary	Average Monthly Benefit ($)
Retired workers and dependents	
Workers	541
Spouses	268
Children	235
Disabled workers and dependents	
Workers	566
Spouses	145
Children	149
Widows and widowers (nondisabled)	527
Disabled widows and widowers	375
Widowed mothers and fathers	376
Surviving children	386

Source: U.S. Department of Health & Human Resources. *Fast Facts & Figures About Social Security*. Washington D.C.: U.S. Department of Health & Human Resources, 1990, p. 13.

★ 526 ★

Spending

Monthly Social Security Family Benefit by Type of Beneficiary, 1989

Beneficiary	Average Monthly Benefit ($)
Retired worker, alone	552
Retired worker and wife	966
Nondisabled aged widow or widower	522
Widowed mother or father, 2 children	1,120
Disabled worker, alone	539
Disabled worker, wife, 2 or more children	972

Source: U.S. Department of Health & Human Resources. *Fast Facts & Figures About Social Security*. Washington D.C.: U.S. Department of Health & Human Resources, 1990, p. 16.

★ 527 ★

Spending

Monthly Social Security Benefit by Type of Beneficiary, 1989

Beneficiary	Average Monthly Benefit ($)
Retired workers	567
Spouses	293
Children	242
Disabled workers	556
Spouses	144
Children	157
Survivors	
Widows and widowers (nondisabled)	522
Disabled widows and widowers	367
Mothers and fathers	388
Children	385

Source: U.S. Department of Health and Human Resources. *Fast Facts & Figures About Social Security*. Washington D.C.: U.S. Department of Health & Human Resources, 1990, p. 15.

★ 528 ★

Spending

Monthly Social Security Benefit by Sex, 1989

Beneficiary	Average Monthly Benefit	
	Men ($)	Women ($)
Total	627	458
Retired workers	639	488
Spouses	189	294
Disabled workers	617	438
Spouses	91	145
Survivors		
Widows and widowers (nondisabled)	382	523
Disabled widows and widowers	224	369
Mothers and fathers	247	395

Source: U.S. Department of Health & Human Resources. *Fast Facts & Figures About Social Security*. Washington D.C.: U.S. Department of Health & Human Resources, 1990, p. 21.

★ 529 ★

Spending

Monthly Social Security Benefit by Type of Beneficiary, 1970-1990

Type of beneficiary	1970	1980	1983	1984	1985	1986	1987	1988	1989	1990
Average monthly benefit, current dollars										
Retired workers	118	341	441	461	479	489	513	537	567	603
Retired worker and wife	199	567	743	780	814	831	873	914	966	1,027
Disabled workers	131	371	456	471	484	488	508	530	556	587
Wives and husbands	59	164	217	227	236	241	253	265	281	298
Children of retired workers	45	140	176	186	198	204	216	228	242	259
Children of deceased workers	82	240	298	314	330	337	352	368	385	406
Children of disabled workers	39	110	136	139	142	141	146	151	157	164
Widowed mothers	87	246	309	322	332	338	353	368	388	409
Widows and widowers, nondisabled	102	311	396	415	433	444	468	493	522	557
Parents	103	276	350	364	378	386	407	428	454	482
Special benefits	45	105	129	134	138	140	145	151	158	167
Average monthly benefit constant (1990) dollars										
Retired workers	397	529	582	586	586	591	595	596	602	603
Retired worker and wife	669	879	981	991	996	1,006	1,012	1,015	1,025	1,027
Disabled workers	440	575	602	598	592	591	589	588	590	587
Wives and husbands	198	254	287	288	289	292	293	294	298	298
Children of deceased workers	276	372	394	399	405	408	409	409	409	406
Widowed mothers	292	381	408	409	406	409	409	409	412	409
Widows and widowers, nondisabled	343	482	523	527	530	538	543	547	554	556

Source: U.S. Bureau of the Census. *Statistical Abstract of the United States 1992*. 112th edition. Washington, D.C.: U.S. Department of Commerce, 1992, p. 361. Primary source: U.S. Social Security Administration, *Annual Statistical Supplement to the Social Security Bulletin*.

★ 530 ★

Spending

Monthly Social Security Benefit for Retired Workers by State, 1990

State	Average Monthly Benefit Retired workers ($)	State	Average Monthly Benefit Retired workers ($)	State	Average Monthly Benefit Retired workers ($)
Alabama	555	Kentucky	553	North Dakota	567
Alaska	602	Louisiana	560	Ohio	618
Arizona	610	Maine	555	Oklahoma	575
Arkansas	539	Maryland	601	Oregon	614
California	615	Massachusetts	605	Pennsylvania	621
Colorado	588	Michigan	643	Rhode Island	601
Connecticut	661	Minnesota	587	South Carolina	561
Delaware	627	Mississippi	520	South Dakota	557
District of Columbia	515	Missouri	589	Tennessee	561
Florida	602	Montana	588	Texas	583
Georgia	560	Nebraska	595	Utah	611
Hawaii	593	Nevada	604	Vermont	589
Idaho	586	New Hampshire	605	Virginia	567
Illinois	641	New Jersey	659	Washington	624
Indiana	628	New Mexico	568	West Virginia	595
Iowa	605	New York	645	Wisconsin	618
Kansas	617	North Carolina	561	Wyoming	604

Source: U.S. Bureau of the Census. *Statistical Abstract of the United States 1992*. 112th edition. Washington, D.C.: U.S. Department of Commerce, 1992, p. 362. Primary source: U.S. Social Security Administration, *Social Security Bulletin*. **Remarks**: Source also includes: social security benefits for disabled workers and widows and widowers, by state, 1990.

★ 531 ★

Spending

Monthly Social Security Benefit for Disabled Workers by State, 1990

State	Average Monthly Benefit Disabled workers ($)	State	Average Monthly Benefit Disabled workers ($)	State	Average Monthly Benefit Disabled workers ($)
Alabama	560	Kentucky	584	North Dakota	556
Alaska	584	Louisiana	595	Ohio	613
Arizona	619	Maine	534	Oklahoma	570
Arkansas	553	Maryland	606	Oregon	600
California	591	Massachusetts	577	Pennsylvania	608
Colorado	582	Michigan	644	Rhode Island	555
Connecticut	590	Minnesota	572	South Carolina	559
Delaware	603	Mississippi	538	South Dakota	533
District of Columbia	525	Missouri	576	Tennessee	556
Florida	598	Montana	602	Texas	579
Georgia	559	Nebraska	569	Utah	575
Hawaii	582	Nevada	616	Vermont	578
Idaho	593	New Hampshire	583	Virginia	572
Illinois	612	New Jersey	610	Washington	602
Indiana	606	New Mexico	572	West Virginia	627
Iowa	579	New York	613	Wisconsin	594
Kansas	570	North Carolina	551	Wyoming	580

Source: U.S. Bureau of the Census. *Statistical Abstract of the United States 1992*. 112th edition. Washington, D.C.: U.S. Department of Commerce, 1992, p. 362. Primary source: U.S. Social Security Administration, *Social Security Bulletin*. **Remarks**: Source also includes: social security benefits for retired workers and widows and widowers, by state, 1990.

★ 532 ★

Spending

Monthly Social Security Benefit for Widows and Widowers by State, 1990

State	Average Monthly Benefit Widows and widowers ($)	State	Average Monthly Benefit Widows and widowers ($)	State	Average Monthly Benefit Widows and widowers ($)
Alabama	480	Kentucky	496	North Dakota	524
Alaska	537	Louisiana	509	Ohio	585
Arizona	581	Maine	526	Oklahoma	529
Arkansas	477	Maryland	568	Oregon	581
California	581	Massachusetts	586	Pennsylvania	587
Colorado	559	Michigan	600	Rhode Island	578
Connecticut	621	Minnesota	556	South Carolina	482
Delaware	585	Mississippi	449	South Dakota	524
District of Columbia	474	Missouri	548	Tennessee	497
Florida	574	Montana	559	Texas	538
Georgia	492	Nebraska	570	Utah	581
Hawaii	530	Nevada	579	Vermont	561
Idaho	558	New Hampshire	578	Virginia	516
Illinois	601	New Jersey	612	Washington	591
Indiana	589	New Mexico	518	West Virginia	527
Iowa	567	New York	598	Wisconsin	584
Kansas	579	North Carolina	487	Wyoming	568

Source: U.S. Bureau of the Census. *Statistical Abstract of the United States 1992*. 112th edition. Washington, D.C.: U.S. Department of Commerce, 1992, p. 362. Primary source: U.S. Social Security Administration, *Social Security Bulletin*. **Remarks**: Source also includes: social security benefits for retired and disabled workers, by state, 1990.

★ 533 ★

Spending

Monthly Social Security Benefits for Women, 1989

	White Women ($)	African-American Women ($)	Other Women ($)
Average monthly benefit	442	369	373
Retired workers	469	396	424
Their spouses	282	220	225
Disabled workers	424	396	387
Their spouses	144	118	106
Widows	498	381	387
Disabled widows	363	304	290

Source: "Heading for Hardship: The Future of Older Women in America," Fran Leonard and Laura Loeb, *USA Today*, January 1992, p. 20.

★534★

Spending

Monthly Payment for Aid to Families with Dependent Children (AFDC) by State, 1990

State	Average monthly payment per family 1990 ($)	State	Average monthly payment per family 1990 ($)	State	Average monthly payment per family 1990 ($)
U.S.	396	Kentucky	219	North Dakota	377
Alabama	121	Louisiana	169	Ohio	323
Alaska	720	Maine	425	Oklahoma	299
Arizona	264	Maryland	371	Oregon	368
Arkansas	193	Massachusetts	510	Pennsylvania	391
California	640	Michigan	500	Rhode Island	508
Colorado	316	Minnesota	496	South Carolina	209
Connecticut	593	Mississippi	124	South Dakota	290
Delaware	293	Missouri	275	Tennessee	193
District of Columbia	398	Montana	343	Texas	166
Florida	269	Nebraska	329	Utah	353
Georgia	260	Nevada	275	Vermont	518
Hawaii	590	New Hampshire	415	Virginia	267
Idaho	282	New Jersey	364	Washington	449
Illinois	329	New Mexico	264	West Virginia	243
Indiana	271	New York	530	Wisconsin	459
Iowa	378	North Carolina	242	Wyoming	352
Kansas	328				

Source: U.S. Bureau of the Census. *Statistical Abstract of the United States 1992.* 112th edition. Washington, D.C.: U.S. Department of Commerce,1992, p. 371. Primary source: U.S. Administration for Children and Families, *Quarterly Public Assistance Statistics*, annual. **Remarks**: Source also includes: the total number of recipients and total payments for the years, 1980-1990, by state; supplemental social security income—number of recipients and total payments for the years, 1980-1990, all by state.

★535★

Spending

Monthly Payment for Aid to Families with Dependent Children (AFDC), 1960-1989

Year	Average monthly payment	
	Family ($)	Recipient ($)
1960	443	116
1970	585	153
1980	421	145
1989 (est.)	384	133

Source: U.S. Department of Health & Human Resources. *Fast Facts & Figures About Social Security.* Washington D.C.: U.S. Department of Health & Human Resources, 1990, p. 38. Primary source: *Annual Statistical Supplement 1989*, table 9.G1, and Office of Family Assistance, Family Support Administration.

★ 536 ★

Spending

Federal Aid per Capita to State and Local Government by State, 1990

State	Federal Aid Per capita ($)	State	Federal Aid Per capita ($)	State	Federal Aid Per capita ($)
Alabama	520	Kentucky	555	North Dakota	737
Alaska	1,303	Louisiana	630	Ohio	497
Arizona	442	Maine	621	Oklahoma	498
Arkansas	532	Maryland	491	Oregon	601
California	468	Massachusetts	641	Pennsylvania	515
Colorado	434	Michigan	511	Rhode Island	770
Connecticut	600	Minnesota	541	South Carolina	542
Delaware	470	Mississippi	620	South Dakota	734
District of Columbia	2,831	Missouri	425	Tennessee	557
Florida	354	Montana	740	Texas	406
Georgia	484	Nebraska	494	Utah	487
Hawaii	540	Nevada	368	Vermont	670
Idaho	565	New Hampshire	385	Virginia	361
Illinois	462	New Jersey	514	Washington	528
Indiana	437	New Mexico	633	West Virginia	562
Iowa	464	New York	876	Wisconsin	519
Kansas	412	North Carolina	444	Wyoming	1,253

Source: U.S. Bureau of the Census. *Statistical Abstract of the United States 1992.* 112th edition. Washington, D.C.: U.S. Department of Commerce, 1992, p. 283. Primary source: U.S. Bureau of the Census, *Federal Expenditures by State for Fiscal Year 1990.* **Remarks**: Source also includes: federal aid to state and local governments, by selected programs, by state.

★ 537 ★

Spending

Federal Funds Distribution per Capita by State, 1990

State	Federal funds Per capita ($)	State	Federal funds Per capita ($)	State	Federal funds Per capita ($)
Alabama	4,272	Kentucky	3,670	North Dakota	4,555
Alaska	5,867	Louisiana	3,582	Ohio	3,496
Arizona	4,112	Maine	4,011	Oklahoma	3,753
Arkansas	3,509	Maryland	5,671	Oregon	3,457
California	3,891	Massachusetts	4,949	Pennsylvania	3,823
Colorado	4,428	Michigan	3,142	Rhode Island	4,303
Connecticut	4,484	Minnesota	3,445	South Carolina	3,919
Delaware	3,226	Mississippi	3,912	South Dakota	4,114
District of Columbia	28,592	Missouri	4,741	Tennessee	3,701
Florida	3,970	Montana	4,186	Texas	3,428
Georgia	3,265	Nebraska	3,448	Utah	3,779
Hawaii	4,927	Nevada	3,448	Vermont	3,148
Idaho	3,862	New Hampshire	3,209	Virginia	5,874
Illinois	3,210	New Jersey	3,664	Washington	4,140
Indiana	3,051	New Mexico	5,703	West Virginia	3,685
Iowa	3,587	New York	3,918	Wisconsin	3,052
Kansas	3,850	North Carolina	3,043	Wyoming	4,089

Source: U.S. Bureau of the Census. *Statistical Abstract of the United States 1992*. 112th edition. Washington, D.C.: U.S. Department of Commerce, 1992, p. 324. Primary source: U.S. Bureau of the Census, *Federal Expenditures by State for Fiscal Year*, annual. **Remarks**: Source also includes: federal funds distribution for defense, procurement, grants, and salaries and wages, by state.

★ 538 ★

Spending

Federal Grants to State and Local Governments for Highway Trust Fund by State, 1990

State	Highway Trust Fund Per capita ($)	State	Highway Trust Fund Per capita ($)	State	Highway Trust Fund Per capita ($)
Alabama	72.3	Kentucky	43.6	North Dakota	110.2
Alaska	291.1	Louisiana	40.0	Ohio	37.2
Arizona	50.8	Maine	53.9	Oklahoma	55.6
Arkansas	54.6	Maryland	76.3	Oregon	55.9
California	45.2	Massachusetts	30.1	Pennsylvania	63.9
Colorado	75.9	Michigan	38.3	Rhode Island	125.8
Connecticut	119.0	Minnesota	64.1	South Carolina	46.1
Delaware	60.1	Mississippi	48.3	South Dakota	136.8
District of Columbia	110.5	Missouri	53.1	Tennessee	55.5
Florida	37.0	Montana	120.0	Texas	60.4
Georgia	59.7	Nebraska	65.6	Utah	70.3
Hawaii	112.5	Nevada	74.9	Vermont	97.4
Idaho	125.2	New Hampshire	50.2	Virginia	51.1
Illinois	44.0	New Jersey	45.8	Washington	82.3
Indiana	46.8	New Mexico	63.3	West Virginia	60.5
Iowa	67.1	New York	33.6	Wisconsin	45.3
Kansas	65.3	North Carolina	58.2	Wyoming	245.7

Source: U.S. Bureau of the Census. *Statistical Abstract of the United States 1992*. 112th edition. Washington, D.C.: U.S. Department of Commerce, 1992, p. 604. Primary source: U.S. Bureau of the Census, *Federal Expenditures by State for Fiscal Year*, annual. **Remarks**: Source also includes: federal grants to state and local governments for mass transportation, by state, 1990.

★ 539 ★

Spending

Federal Grants to State and Local Governments for Urban Mass Transportation Administration (UMTA) by State, 1990

State	UMTA Per capita ($)	State	UMTA Per capita ($)	State	UMTA Per capita ($)
U.S.	15.0	North Carolina	2.6	North Dakota	2.9
Alabama	3.6	Kentucky	3.1	Ohio	10.1
Alaska	2.8	Louisiana	-	Oklahoma	4.4
Arizona	7.1	Maine	8.9	Oregon	10.1
Arkansas	2.8	Maryland	20.1	Pennsylvania	15.1
California	16.4	Massachusetts	22.3	Rhode Island	8.3
Colorado	8.6	Michigan	6.2	South Carolina	3.3
Connecticut	14.0	Minnesota	7.7	South Dakota	2.6
Delaware	6.4	Mississippi	2.8	Tennessee	3.6
District of Columbia	534.2	Missouri	13.1	Texas	6.4
Florida	7.2	Montana	3.0	Utah	12.7
Georgia	13.5	Nebraska	5.4	Vermont	3.1
Hawaii	13.1	Nevada	4.6	Virginia	5.2
Idaho	2.8	New Hampshire	2.8	Washington	14.6
Illinois	28.4	New Jersey	28.7	West Virginia	3.6
Indiana	2.0	New Mexico	7.2	Wisconsin	6.8
Iowa	6.0	New York	46.8	Wyoming	4.9
Kansas	2.7				

Source: U.S. Bureau of the Census. *Statistical Abstract of the United States 1992*. 112th edition. Washington, D.C.: U.S. Department of Commerce, 1992, p. 604. Primary source: U.S. Bureau of the Census, *Federal Expenditures by State for Fiscal Year*, annual. **Remarks:** Source also includes: federal grants to state and local governments for the Highway Trust Fund, by state, 1990.

★ 540 ★

Spending

State Legislative Appropriations for State Arts Agencies by State, 1991

State	1991 Per capita ($)	State	1991 Per capita ($)	State	1991 Per capita ($)
Alabama	0.39	Kentucky	0.90	North Dakota	0.43
Alaska	2.60	Louisiana	0.22	Ohio	1.12
Arizona	0.56	Maine	0.61	Oklahoma	1.02
Arkansas	0.41	Maryland	1.56	Oregon	0.54
California	0.57	Massachusetts	2.10	Pennsylvania	0.99
Colorado	0.47	Michigan	0.98	Rhode Island	1.01
Connecticut	0.67	Minnesota	0.96	South Carolina	1.04
Delaware	1.97	Mississippi	0.20	South Dakota	0.58
District of Columbia	5.27	Missouri	0.88	Tennessee	0.88
Florida	1.81	Montana	0.98	Texas	0.20
Georgia	0.52	Nebraska	0.66	Utah	2.49
Hawaii	10.89	Nevada	0.30	Vermont	0.85
Idaho	0.66	New Hampshire	0.47	Virginia	0.65
Illinois	0.91	New Jersey	1.51	Washington	0.49
Indiana	0.51	New Mexico	0.74	West Virginia	1.33
Iowa	0.45	New York	2.83	Wisconsin	0.50
Kansas	0.43	North Carolina	0.82	Wyoming	0.77

Source: U.S. Bureau of the Census. *Statistical Abstract of the United States 1992.* 112th edition. Washington, D.C.: U.S. Department of Commerce, 1992, p. 244. Primary source: National Assembly of State Arts Agencies, Washington, DC, unpublished data. **Remarks**: Source also includes: legislative appropriations for state arts agencies, 1988-1990.

★ 541 ★

Spending

Veterans Compensation Benefits, By Period of Service, 1980-1990

Period of Service and Veteran Status	Average Payment				
	1980	1985	1988	1989	1990
Total	2,370	3,505	3,949	4,111	4,335
Living veterans	2,600	3,666	4,003	4,126	4,320
Service connected	2,669	3,692	3,967	4,078	4,250
Nonservice connected	2,428	3,581	4,133	4,308	4,591
Deceased veterans	1,863	3,066	3,787	4,062	4,382
Service connected	3,801	5,836	6,649	6,992	7,349
Nonservice connected	1,228	1,809	2,224	2,358	2,548
Prior to World War I	1,432	1,855	2,226	2,388	2,616
Living	2,634	4,436	9,149	8,411	10,502
World War I	1,683	2,461	2,986	3,181	3,435
Living	2,669	4,439	5,759	6,316	6,922
World War II	2,307	3,317	3,724	3,851	4,052
Living	2,462	3,460	3,823	3,936	4,123
Korean conflict	2,691	4,114	4,618	4,836	5,105
Living	2,977	4,260	4,658	4,852	5,103
Peace-time	3,080	3,973	4,023	4,042	4,132

[Continued]

★ 541 ★

Veterans Compensation Benefits, By Period of Service, 1980-1990

[Continued]

Period of Service and Veteran Status	Average Payment				
	1980	1985	1988	1989	1990
Living	2,828	3,589	3,613	3,621	3,709
Vietnam era	2,795	4,021	4,463	4,683	4,945
Living	2,709	3,849	4,229	4,416	4,671

Source: U.S. Bureau of the Census. *Statistical Abstract of the United States 1992*. 112th edition. Washington, D.C.: U.S. Department of Commerce, 1992, p. 349. Primary source: U.S. Dept. of Veteran Affairs, *Annual Report of the Secretary of Veterans Affairs*. **Remarks**: Source also includes: the number of veterans on the rolls, 1980-1990.

★ 542 ★

Spending

Number of Recipients with Limited Income Receiving Monthly Benefits, 1990

In thousands. Average Monthly Recipients.

Program	1990	Program	1990
MEDICAL CARE		Interest reduction payments	531
Medicaid	25,255	Rural rental housing loans	16
Veterans	585		
Indian Health Services	1,100	EDUCATION AID	
Community health centers	5,350	Stafford loans	3,624
		Pell grants	3,434
CASH AID		Head Start	541
A.F.D.C.	11,439	College Work-Study Program	835
Supplemental Security Income	4,913	Supplemental Educational	
Earned income tax credit	33,693	Opportunity Grants	633
Pensions for needy veterans	1,080		
General assistance	1,205	JOBS AND TRAINING	
		Training for disadvantaged	
FOOD BENEFITS		adults and youth	416
Food stamps	21,500	Job Corps	40
School lunch program	11,600	Summer youth employment	
Women, infants and children	4,500	program	625
Nutrition program for elderly	3,548	Work incentive program	444
		Senior community service	
HOUSING BENEFITS		employment program	65
Lower-income housing asst.	2,500		
Low-rent public housing	1,405	ENERGY AID	
Rural housing loans	25	Low-income energy	
		assistance	5,800

Source: U.S. Bureau of the Census. *Statistical Abstract of the United States 1992*. 112th edition. Washington, D.C.: U.S. Department of Commerce, 1992, p. 356. Primary source: U.S. Dept. of Veterans Affairs, *Annual Report of the Secretary of Veterans Affairs*. **Remarks**: Source also includes: number of recipients with limited income receiving monthly benefits, 1985 and 1989.

★ 543 ★

Spending

Cost of Government Social-Welfare Programs per Person by Type of Program

Government social-welfare programs on the federal, state, and local levels spend $3,364 per person. This includes $112 on veterans' benefits; $192 on health and medical care; $447 on public aid; $826 on education; and $1,671 on social insurance.

Source: Rukeyser, Louis. *Louis Rukeyser's Business Almanac.* New York: Simon & Shuster, 1991, p. 133.

★ 544 ★

Spending

Government Expenditure per Person, 1950-1992

Federal government expenditures per person averaged $5,086 in 1990, up from $280 in 1950. State and local government spending has increased from $184 in 1950 for every individual, to $3,015 in 1990.

Source: Rukeyser, Louis. *Louis Rukeyser's Business Almanac.* New York: Simon & Shuster, 1991, p. 129.

★ 545 ★

Spending

Welfare Benefits for Immigrants, 1950s-1980s

According to a study by Trejo and George Borjas, a University of California San Diego economist, immigrants who arrived in the late 1950s drew an average of $7,177 in household welfare, compared to $7,907 for other Americans. Welfare benefits increased to $13,552 for immigrants who arrived between 1975 and 1980.

Source: Los Angeles Public Library/State of California, *Scan/Info*, 4, August 1992, pp. 15-16. Primary source: *San Francisco Chronicle*, August 8, 1992, p. A13.

★ 546 ★

Spending

Allowances per House of Representatives Member, 1991

In 1991 each member of the House of Representatives received an average allowance of $176,900 for office expenses (varying because of the district's rental rates and distance from Washington) and an average of $200,000 for mailing expenses. Each member spent an average of $154,800 for office expenses. These expenses included an average of: $46,150 for equipment and supplies; $23,700 for travel and food; $14,900 for printing and constituent communications; $29,700 for telecommunications and nonfranked mail; $28,500 for rent and office upkeep; and $4,900 for publications and subscriptions. In addition, the average member's payroll was $479,900 and franked mail expenses were $100,000. Total spent on office expenses, payroll and franked mail was an average of $734,700 per member.

Source: "Lawmakers' Pay, Perks Add to Total," Anne Willette, *USA Today*, September 28, 1992, p. 1A-5A.

★ 547 ★

Spending

The National Debt

The national debt is $13,000 per capita or $52,000 for the average family of 4. Based on an average household income of $48,840, the national debt is 71% of annual personal income.

Source: American Citizens Committee on Reducing Debt. *In Accord*, Winter 1990- 1991.

★ 548 ★

Spending

The National Debt, 1900 to the Present

The national debt per capita was only $16.60 in 1900; now it is around $18,000 per person.

Source: U.S. News and World Report, 113, February 1, 1993, p. 13. **Remarks**: Source also includes: federal debt inherited by Carter, Reagan, Bush and Clinton.

★ 549 ★

Spending

Military Contracts in Selected Cities

In 1989 in New London, CT and its suburbs, military contracts averaged $9,785 per person, more than three times what was spent in Dallas-Fort Worth, the nation's second most military-dependent city, which received $2,852 per person. Long Island received $1,236 per capita in military contracts. The national average in 1990 was $544 per capita.

Source: "Winning the Cold War and Losing a Job," Kirk Johnson, *New York Times*, February 9, 1993, pp. A1, B4. **Remarks**: Source also includes: per capita spending for military contracts in selected states.

★ 550 ★

Spending

Military Contracts in Selected States

In 1990 the national per capita average for military spending was $544. The states where military contracts exceeded the national average were.

State	Average per Capita
Massachusetts	$1,318
Connecticut	1,293
Virginia	1,266
Missouri	1,178
Colorado	991
Arizona	907
Maryland	903
Alaska	802
California	738
Maine	690
Rhode Island	548
Mississippi	547

Source: Johnson, Kirk, "Winning the Cold War and Losing a Job," *New York Times*, February 9, 1993, p. A1, B4. Primary source: Defense Department and Census Bureau Information. **Remarks**: Source also includes: per capita spending for military contracts in selected cities.

★ 551 ★

Spending

Reduction in Armed Services Personnel

Due to defense cuts, an average of 350,000 personnel are leaving the armed services each year.

Source: "From Soldiers to McVets," Perri Capell, *American Demographics*, 14, November 1992, p. 58. **Remarks**: Source also includes: cuts in military personnel for 1992, 1993, 1994.

★552★

Spending

Food Stamps

The Federal food stamp program which began in 1964 has the highest number ever enrolled today: an average monthly enrollment of 25 million people in 1992, or 10.4% of the population. $173 was the average family benefit in 1992; the maximum benefit was $370.

Source: "Food Stamp Users up Sharply in Sign of Weak Recovery," Jason De Parle, *New York Times*, March 2, 1993, pp. A1, A18.

★553★

Spending

Cost of the Persian Gulf War

The cost for the United States to engage in air combat only during the Persian Gulf War averaged $175 million per day; the cost to engage in combined air/ground combat averaged $500 million per day. The total cost to the United States alone was $61 billion.

Source: Center for Defense Information, Washington, D.C.

★554★

Spending

HUD Spending in the Poorest Cities

Cities ranked by percent of population below poverty line.

City	Average Spending per Person	Cities	Average Spending per Person
1. Laredo, TX	$55.72	11. Newark, NJ	$75.17
2. Detroit, MI	86.35	12. Baton Rouge, LA	45.38
3. New Orleans, LA	61.56	13. Buffalo, NY	100.59
4. Miami, FL	57.56	14. El Paso, TX	37.51
5. Flint, MI	65.91	15. Shreveport, LA	34.08
6. Gary, IN	67.55	16. Birmingham, AL	53.26
7. Cleveland, OH	98.27	17. St. Louis, MO	109.10
8. Hartford, CT	58.57	18. Macon, GA	33.74
9. Atlanta, GA	49.90	19. Cincinnati, OH	70.42
10. Dayton, OH	71.01	20. Fresno, CA	31.21

Source: "The Ways of Washington," *U.S. News & World Report*, 114, March 29, 1993, p. 26. Primary source: U.S. Census Bureau, U.S. Department of Housing and Urban Development.

★ 555 ★

Spending

Money Spent on Local Parks

A study by Pennsylvania State University, which examined the use of local parks across the United States, found that three-fourths of the population used them occasionally or regularly, and said they received their money's worth from the average of $45 per person a year cities and towns spent on them.

Source: "Have Backyards Supplanted Parks? Survey Says No," Trish Hall, *The New York Times*, August 26, 1992, p. C1.

Taxes

★ 556 ★

Highest Taxes Per Year in Selected Countries

Country	Average Tax Per Person
Austria	$5,302
Denmark	$8,151
France	$5,802
Finland	$5,499
Germany	$6,540
Netherlands	$5,483
Norway	$8,346
Sweden	$8,385
Switzerland	$6,707
United States	$4,944

Source: Krantz, Les. *The Best and Worst of Everything.* New York: Prentice Hall General Reference, 1991, pp. 145-146. Primary source: Tax Foundation, Washington, D.C.

★ 557 ★

Taxes

Highest Property Taxes in Selected United States' Cities

City	Property Tax, Per Person
Anchorage, AK	$492
Baltimore, MD	$400
Boston, MA	$611
Jersey City, NJ	$371
New York, NY	$588

[Continued]

★557★

Highest Property Taxes in Selected United States' Cities

[Continued]

City	Property Tax, Per Person
Richmond, VA	$489
Rochester, NY	$473
San Francisco, CA	$383
Virginia Beach, VA	$312
Washington, DC	$727

Source: Krantz, Les. *The Best and Worst of Everything.* New York: Prentice Hall General Reference, 1991, p. 100. Primary source: U.S. Bureau of the Census. **Remarks**: Source also includes: lowest property taxes in selected United States cities.

★558★

Taxes

Lowest Property Taxes in Selected United States' Cities

City	Property Tax, Per Person
Akron, OH	$49.00
Aurora, CO	$52.00
Colorado Springs, CO	$42.00
Columbus, OH	$24.00
Mesa, AZ	$8.00
Mobile, AL	$25.00
Riverside, CA	$48.00
Toledo, OH	$32.00
Tucson, AZ	$34.00
Tulsa, OK	$45.00

Source: Krantz, Les. *The Best and Worst of Everything.* New York: Prentice Hall General Reference, 1991, p. 100. Primary source: U.S. Bureau of the Census. **Remarks**: Source also includes: lowest property taxes in selected United States cities.

★559★

Taxes

Federal, State, and Local Tax per Person

In 1960 in the United States every man, woman, and child paid an average of $629.00 in federal, state, and local taxes. In 1990 the average was $7,592.

Source: Rukeyser, Louis. *Louis Rukeyser's Business Almanac.* New York: Simon & Shuster, 1991, p. 166.

★ 560 ★

Taxes

Social Security Taxes Per Earner, 1990

The average American earner pays $1,652.20 per year in social security taxes.

Source: U.S. Department of Health & Human Resources. *Fast Facts & Figures About Social Security.* Washington D.C.: U.S. Department of Health & Human Resources, 1990, p. 1. **Remarks**: Source also includes: maximum earnings subject to social security tax, and minimum taxes paid to social security.

★ 561 ★

Taxes

Number of Days Spent Working to Pay Taxes, 1950-1991

Number of days the average American worked to pay all federal, state, and local taxes.

Year	Number of Days	
	From...	Days
1950	January 1-April 3	93 days
1955	January 1-April 9	99 days
1960	January 1-April 17	107 days
1965	January 1-April 15	105 days
1970	January 1-April 28	118 days
1975	January 1-April 28	118 days
1980	January 1-May 1	121 days
1985	January 1-May 1	121 days
1990	January 1-May 5	125 days
1991	January 1-May 8	128 days

Source: Rukeyser, Louis. *Louis Rukeyser's Business Almanac.* New York: Simon & Shuuster, 1991, p. 166. Primary source: Tax Foundation.

★ 562 ★

Taxes

Number of Minutes Spent Working to Pay Taxes

The average American worker spends 1 hr. 46 min. of each working day to pay federal taxes, and 59 min. to pay all other taxes.

Source: "Tax Facts," *USA Today*, March 22, 1993, p. 3B. Primary source: Tax Foundation.

★ 563 ★

Taxes

Percent of income spent on taxes

The median U.S. family pays 40% of its income for federal, state, and local taxes, or 2 out of every 5 dollars.

Source: International Herald Tribune, February 3, 1993, p. 1F.

★ 564 ★

Taxes

IRS Delinquency Payments per Return, 1991

The 3.2 million delinquency penalties from the IRS amounted to an average of $363 each in 1991.

Source: "Tax Penalty Ambush," Greg Anrig, Jr. and Elizabeth M. Macdonald, *Money*, 21, June 1992, pp. 157-164. **Remarks**: Source also includes: other IRS fines and information on how to appeal them.

★ 565 ★

Taxes

Federal Individual Income Tax per Return, 1980-1989

Year	Average Amount
1980	$3,387
1985	$3,932
1988	$4,705
1989	$4,800

Source: U.S. Bureau of the Census. *Statistical Abstract of the United States 1992*. 112th edition. Washington, D.C.: U.S. Department of Commerce, 1992, p. 326. Primary source: U.S. Internal Revenue Service, *Statistics of Income Bulletin*, and *Statistics of Income, Individual Income Tax Returns*, annual. **Remarks**: Source also includes: the number of returns and the total amount obtained from income tax returns, 1980-1989.

★ 566 ★

Taxes

Most Frequently Claimed Deductions on Federal Income Tax Forms, 1991

In 1991 the most frequently claimed deductions on schedule A of Form 1040 were: interest expenses, with an average of $7,090; state and local taxes, with an average of $4,300; and charitable contribution, with an average of $420.

Source: "Tax Facts," *USA Today*, March 22, 1993, p. 3B. **Remarks**: Source also includes: other income tax facts are also included, e.g. personal income tax paid, 1985-1991, time spent each day earning money to pay taxes.

★ 567 ★

Taxes

Taxes Collected From Taxpayers Who Don't File Returns

When taxpayers don't file returns, the IRS constructs substitute returns and sends out bills. An average of 400,000 substitute returns are issued each year, paying out $2 billion, or an average of $5,200 per return.

Source: "Tax Penalty Ambush," Greg Anrig, Jr. and Elizabeth M. Macdonald, *Money*, 21, June 1992, pp. 157-164. **Remarks**: Source also includes: IRS fines and information on how to appeal them.

★ 568 ★

Taxes

Fines Collected From Taxpayers for Negligence Penalties

1991, the IRS exacted 2.4 million negligence penalties, which averaged $283 each.

Source: "Tax Penalty Ambush," Greg Anrig, Jr. and Elizabeth M. Macdonald, *Money*, 21, June 1992, pp. 157-164. **Remarks**: Source also includes: IRS fines and information on how to appeal them.

★ 569 ★

Taxes

Fines Collected From Taxpayers for Underpayment of Taxes

When a taxpayer underpays taxes by $5,000 or 10% of the correct amount, whichever is greater, a penalty of 20% of the underpayment is owed. In 1991, the IRS exacted 35,000 of these penalties, which averaged $4,361 each.

Source: "Tax Penalty Ambush," Greg Anrig, Jr. and Elizabeth M. Macdonald, *Money*, 21, June 1992, pp. 157-164. **Remarks**: Source also includes: other IRS fines and information on how to appeal them.

★570★

Taxes

Income Tax Refunds

According to the IRS, most Americans allow too much money withheld from their income for taxes. The average refund is $900.

Source: "Pocket Guide to Money," *Consumer Reports*, 57, May 1992, p. 333.

★571★

Taxes

Federal Individual Income Tax per Capita, by State, 1989

State	Per capita (dol.)	State	Per capita (dol.)	State	Per capita (dol.)
Alabama	1,252	Kentucky	1,246	North Dakota	1,211
Alaska	2,470	Louisiana	1,145	Ohio	1,626
Arizona	1,490	Maine	1,480	Oklahoma	1,244
Arkansas	1,041	Maryland	2,284	Oregon	1,534
California	2,087	Massachusetts	2,391	Pennsylvania	1,780
Colorado	1,740	Michigan	1,812	Rhode Island	1,889
Connecticut	3,139	Minnesota	1,756	South Carolina	1,256
Delaware	2,009	Mississippi	927	South Dakota	1,201
District of Columbia	2,507	Missouri	1,575	Tennessee	1,461
Florida	1,976	Montana	1,216	Texas	1,629
Georgia	1,572	Nebraska	1,459	Utah	1,131
Hawaii	1,958	Nevada	2,306	Vermont	1,651
Idaho	1,184	New Hampshire	2,226	Virginia	1,978
Illinois	2,073	New Jersey	2,685	Washington	1,964
Indiana	1,563	New Mexico	1,159	West Virginia	1,075
Iowa	1,399	New York	2,246	Wisconsin	1,571
Kansas	1,653	North Carolina	1,522	Wyoming	1,555

Source: U.S. Bureau of the Census. *Statistical Abstract of the United States 1992*. 112th edition. Washington, D.C.: U.S. Department of Commerce, 1992, p. 328. Primary source: U.S. Internal Revenue Service, *Statistics of Income Bulletin*, spring 1990. **Remarks**: Source also includes: the number of returns, the adjusted gross income and the total obtained from income tax returns by state, 1989.

★ 572 ★

Taxes

State Taxes Per Capita by State, 1990

State	Average Per capita taxes ($)	State	Average Per capita taxes ($)	State	Average Per capita taxes ($)
Alabama	945	Louisiana	968	Ohio	1,054
Alaska	2,811	Maine	1,271	Oklahoma	1,105
Arizona	1,194	Maryland	1,349	Oregon	980
Arkansas	961	Massachusetts	1,557	Pennsylvania	1,112
California	1,458	Michigan	1,220	Rhode Island	1,229
Colorado	931	Minnesota	1,558	South Carolina	1,128
Connecticut	1,602	Mississippi	931	South Dakota	718
Delaware	1,696	Missouri	965	Tennessee	870
Florida	1,027	Montana	1,073	Texas	866
Georgia	1,092	Nebraska	958	Utah	1,026
Hawaii	2,107	Nevada	1,317	Vermont	1,182
Idaho	1,130	New Hampshire	536	Virginia	1,066
Illinois	1,127	New Jersey	1,349	Washington	1,525
Indiana	1,100	New Mexico	1,329	West Virginia	1,243
Iowa	1,193	New York	1,590	Wisconsin	1,340
Kansas	1,077	North Carolina	1,186	Wyoming	1,347
Kentucky	1,156	North Dakota	1,059	United States	1,211

Source: The World Almanac and Book of Facts 1992. New York: World Almanac, 1991, p. 154. Primary source: Census Bureau, U.S. Dept. of Commerce. **Remarks**: Source also includes: state revenues, state debts, and U.S. aid, by state.

★ 573 ★

Taxes

U.S. Aid Per Capita by State, 1990

State	Average Per capita U.S. aid ($)	State	Average Per capita U.S. aid ($)	State	Average Per capita U.S. aid ($)
Alabama	509	Louisiana	568	Ohio	439
Alaska	1,131	Maine	563	Oklahoma	427
Arizona	352	Maryland	438	Oregon	576
Arkansas	486	Massachusetts	549	Pennsylvania	443
California	539	Michigan	449	Rhode Island	665
Colorado	413	Minnesota	519	South Carolina	495
Connecticut	533	Mississippi	579	South Dakota	647
Delaware	460	Missouri	352	Tennessee	508
Florida	309	Montana	685	Texas	382
Georgia	444	Nebraska	431	Utah	560
Hawaii	550	Nevada	304	Vermont	699
Idaho	490	New Hampshire	336	Virginia	340
Illinois	392	New Jersey	471	Washington	461
Indiana	406	New Mexico	514	West Virginia	525
Iowa	464	New York	717	Wisconsin	483
Kansas	386	North Carolina	399	Wyoming	1,110
Kentucky	486	North Dakota	693	United States	477

Source: The World Almanac and Book of Facts 1992. New York: World Almanac, 1991, p. 154. Primary source: Census Bureau, U.S. Dept. of Commerce. **Remarks**: Source also includes: state revenues, state expenditures, state debts, and state taxes, by state.

★574★

Taxes

Federal Tax Liability by State, 1989

State	Average Federal Tax Liability ($)	State	Average Federal Tax Liability ($)	State	Average Federal Tax Liability ($)
United States	4,820	Kentucky	3,855	North Dakota	3,410
Alabama	3,946	Louisiana	4,007	Ohio	4,182
Alaska	4,555	Maine	3,728	Oklahoma	3,880
Arizona	4,218	Maryland	5,309	Oregon	4,140
Arkansas	3,378	Massachusetts	5,374	Pennsylvania	4,591
California	5,526	Michigan	4,817	Rhode Island	4,544
Colorado	4,483	Minnesota	4,381	South Carolina	3,659
Connecticut	6,756	Mississippi	3,358	South Dakota	3,446
Delaware	4,840	Missouri	4,306	Tennessee	4,194
District of Columbia	5,641	Montana	3,525	Texas	4,907
Florida	5,121	Nebraska	3,856	Utah	3,559
Georgia	4,405	Nevada	5,239	Vermont	4,045
Hawaii	4,658	New Hampshire	5,053	Virginia	4,896
Idaho	3,602	New Jersey	6,162	Washington	4,913
Illinois	5,406	New Mexico	3,633	West Virginia	3,619
Indiana	4,225	New York	5,730	Wisconsin	3,998
Iowa	3,760	North Carolina	4,037	Wyoming	4,471
Kansas	4,492				

Source: Almanac of the 50 states 1992. Burlington, VT: Information Publications, 1991, p. 442.

★575★

Taxes

State Government Revenue, Debt, and Expenditures, 1970-1990

Year	General Revenue per capita ($)	Debt Outstanding per capita ($)	General Expenditures per capita ($)
1970	384	207	383
1980	1,034	540	1,010
1985	1,536	89	1,449
1986	1,636	1,030	1,565
1987	1,728	1,094	1,664
1988	1,815	1,129	1,763
1989	1,948	1,194	1,895
1990	2,086	1,283	2,047

Source: U.S. Bureau of the Census. *Statistical Abstract of the United States 1992.* 112th edition. Washington, D.C.: U.S. Department of Commerce, 1992, pp. 286-287. Primary source: U.S. Census Bureau, *State Government Finances*, series GF, No. 3; *Census of Governments: 1977 and 1987* vol. 6; *Historical Statistics on Governmental Finances and Employment.* **Remarks**: Source also includes: state government revenue, debt, and expenditures, by state.

★576★

Taxes

State Government Revenue Per Capita by State, 1990

State	General Revenue Per capita ($)	State	General Revenue Per capita ($)	State	General Revenue Per capita ($)
Alabama	879	Louisiana	2,115	Ohio	1,841
Alaska	8,804	Maine	2,319	Oklahoma	1,900
Arizona	1,905	Maryland	2,185	Oregon	2,051
Arkansas	1,725	Massachusetts	2,622	Pennsylvania	1,847
California	2,327	Michigan	2,120	Rhode Island	2,568
Colorado	1,756	Minnesota	2,512	South Carolina	1,987
Connecticut	2,627	Mississippi	1,758	South Dakota	1,882
Delaware	3,159	Missouri	1,566	Tennessee	1,638
Florida	1,594	Montana	2,289	Texas	1,528
Georgia	1,727	Nebraska	1,810	Utah	2,049
Hawaii	3,467	Nevada	1,889	Vermont	2,545
Idaho	1,980	New Hampshire	1,410	Virginia	1,880
Illinois	1,823	New Jersey	2,364	Washington	2,362
Indiana	1,930	New Mexico	2,688	West Virginia	2,127
Iowa	2,099	New York	2,915	Wisconsin	2,244
Kansas	1,770	North Carolina	1,862	Wyoming	3,649
Kentucky	1,994	North Dakota	2,549		

Source: U.S. Bureau of the Census. *Statistical Abstract of the United States 1992*. 112th edition. Washington, D.C.: U.S. Department of Commerce, 1992, p. 286. Primary source: U.S. Census Bureau, *State Government Finances*, series GF, No. 3; *Census of Governments: 1977 and 1987* vol. 6. **Remarks**: Source also includes: state government debt and expenditure by state, and state government revenue, debt, and expenditure, 1970-1989.

★ 577 ★

Taxes

State Government Outstanding Debt per Capita by State, 1990

State	Debt Outstanding Per capita ($)	State	Debt Outstanding Per capita ($)	State	Debt Outstanding Per capita ($)
Alabama	985	Louisiana	3,026	Ohio	1,033
Alaska	10,065	Maine	1,731	Oklahoma	1,181
Arizona	598	Maryland	1,390	Oregon	2,308
Arkansas	743	Massachusetts	3,111	Pennsylvania	920
California	970	Michigan	987	Rhode Island	3,605
Colorado	735	Minnesota	860	South Carolina	1,117
Connecticut	3,343	Mississippi	522	South Dakota	2,568
Delaware	4,471	Missouri	1,026	Tennessee	537
Florida	769	Montana	1,747	Texas	463
Georgia	481	Nebraska	863	Utah	1,039
Hawaii	3,065	Nevada	1,308	Vermont	2,236
Idaho	970	New Hampshire	2,793	Virginia	983
Illinois	1,335	New Jersey	2,446	Washington	1,168
Indiana	747	New Mexico	1,208	West Virginia	1,378
Iowa	675	New York	2,587	Wisconsin	1,251
Kansas	124	North Carolina	463	Wyoming	2,067
Kentucky	1,437	North Dakota	1,365		

Source: U.S. Bureau of the Census. *Statistical Abstract of the United States 1992*. 112th edition. Washington, D.C.: U.S. Department of Commerce, 1992, pp. 286-287. Primary source: U.S. Census Bureau, *State Government Finances*, series GF, No. 3; *Census of Governments: 1977 and 1987* vol. 6. **Remarks**: Source also includes: state government revenue and expenditure by state, and state government revenue, debt, and expenditure, 1970-1989.

★ 578 ★

Taxes

State Government General Expenditures per Capita by State, 1990

State	General Expenditures Per capita ($)	State	General Expenditures Per capita ($)	State	General Expenditures Per capita ($)
Alabama	1,831	Louisiana	2,020	Ohio	1,889
Alaska	7,790	Maine	2,234	Oklahoma	1,784
Arizona	2,056	Maryland	2,057	Oregon	1,957
Arkansas	1,672	Massachusetts	2,832	Pennsylvania	1,787
California	2,359	Michigan	2,104	Rhode Island	2,741
Colorado	1,708	Minnesota	2,379	South Carolina	1,943
Connecticut	2,702	Mississippi	1,708	South Dakota	1,841
Delaware	2,994	Missouri	1,505	Tennessee	1,616
Florida	1,589	Montana	2,066	Texas	1,391
Georgia	1,759	Nebraska	1,784	Utah	2,014
Hawaii	3,201	Nevada	1,968	Vermont	2,603
Idaho	1,818	New Hampshire	1,512	Virginia	1,920
Illinois	1,754	New Jersey	2,334	Washington	2,340
Indiana	1,802	New Mexico	2,568	West Virginia	1,969
Iowa	2,137	New York	2,763	Wisconsin	2,146
Kansas	1,747	North Carolina	1,894	Wyoming	3,270
Kentucky	1,927	North Dakota	2,483		

Source: U.S. Bureau of the Census. *Statistical Abstract of the United States 1992.* 112th edition. Washington, D.C.: U.S. Department of Commerce, 1992, pp. 286-287. Primary source: U.S. Census Bureau, *State Government Finances*, series GF, No. 3; *Census of Governments: 1977 and 1987* vol. 6. **Remarks**: Source also includes: state government revenue and debt by state, and state government revenue, debt, and expenditure, 1970-1989.

★ 579 ★

Taxes

City Government General Revenue per Capita in Selected Cities, 1990

Cities	General Revenue Per capita ($)
New York City, NY	4,360
Los Angeles, CA	1,050
Chicago, IL	1,019
Houston, TX	763
Philadelphia, PA	1,571
San Diego, CA	1,016
Detroit, MI	1,465
Dallas, TX	822
Phoenix, AZ	1,036
San Antonio, TX	566
San Jose, CA	819
Baltimore, MD	2,203
Indianapolis, IN	1,191

[Continued]

★ 579 ★

City Government General Revenue per Capita in Selected Cities, 1990

[Continued]

Cities	General Revenue Per capita ($)
San Francisco, CA	3,202
Jacksonville, FL	1,060
Columbus, OH	835
Milwaukee, WI	932
Memphis, TN	1,162
Washington, DC	6,612
Boston, MA	2,775

Source: U.S. Bureau of the Census. *Statistical Abstract of the United States 1992*. 112th edition. Washington, D.C.: U.S. Department of Commerce, 1992, p. 299. Primary source: U.S. Census Bureau, *City Government Finances*, series GF, No. 4, annual. **Remarks**: Source also includes: city government taxes, expenditure, and debt for 75 of the largest cities in the United States.

★ 580 ★

Taxes

Per Capita City Government General Expenditures in Selected Cities, 1990

Cities	General Expenditure Per capita ($)
New York City, NY	4,256
Los Angeles, CA	963
Chicago, IL	962
Houston, TX	817
Philadelphia, PA	1,566
San Diego, CA	811
Detroit, MI	1,408
Dallas, TX	795
Phoenix, AZ	1,246
San Antonio, TX	730
San Jose, CA	921
Baltimore, MD	1,989
Indianapolis, IN	1,237
San Francisco, CA	2,577
Jacksonville, FL	1,094
Columbus, OH	853
Milwaukee, WI	920
Memphis, TN	1,112

[Continued]

★ 580 ★

Per Capita City Government General Expenditures in Selected Cities, 1990
[Continued]

Cities	General Expenditure Per capita ($)
Washington, DC	6,713
Boston, MA	2,593

Source: U.S. Bureau of the Census. *Statistical Abstract of the United States 1992*. 112th edition. Washington, D.C.: U.S. Department of Commerce, 1992, p. 299. Primary source: U.S. Census Bureau, *City Government Finances*, series GF, No. 4, annual. **Remarks**: Source also includes: city government revenue, taxes, and debt for 75 of the largest cities in the United States.

★ 581 ★

Taxes

Per Capita City Government Gross Outstanding Debt in Selected Cities, 1990

Cities	Gross Debt Outstanding Per capita ($)
New York City, NY	3,537
Los Angeles, CA	1,873
Chicago, IL	1,443
Houston, TX	2,126
Philadelphia, PA	2,305
San Diego, CA	1,314
Detroit, MI	1,400
Dallas, TX	1,590
Phoenix, AZ	2,228
San Antonio, TX	4,278
San Jose, CA	1,412
Baltimore, MD	1,673
Indianapolis, IN	1,311
San Francisco, CA	3,269
Jacksonville, FL	6,768
Columbus, OH	1,882
Milwaukee, WI	860
Memphis, TN	1,141
Washington, DC	5,548
Boston, MA	1,478

Source: U.S. Bureau of the Census. *Statistical Abstract of the United States 1992*. 112th edition. Washington, D.C.: U.S. Department of Commerce, 1992, p. 299. Primary source: U.S. Census Bureau, *City Government Finances*, series GF, No. 4, annual. **Remarks**: Source also includes: city government revenue, taxes, and expenditure for 75 of the largest cities in the United States.

★ 582 ★

Taxes

Per Capita City Government Revenue, 1980-1990

Item	Per Capita ($)		
	1980	1985	1990
Revenue	672	1,007	1,323
General revenue	539	781	1,035
Intergovernmental revenue	200	244	296
From State governments only	113	157	224
Taxes	222	325	450
Property	119	160	229
Percent of total taxes	(X)	(X)	(X)
Sales and gross receipts	58	95	125
General	36	58	76
Selective	22	36	49
Income, licenses and other	44	70	95
Charges and miscellaneous	117	212	289
Current charges only	70	113	165
Utility and liquor store revenue	111	180	217
Water system	35	58	76
Electric power system	57	91	113
Gas supply system	10	19	15
Transit system	7	11	12
Liquor stores	2	2	2
Insurance trust revenue	22	45	71

Source: U.S. Bureau of the Census. *Statistical Abstract of the United States 1992*. 112th edition. Washington, D.C.: U.S. Department of Commerce, 1992, p. 298. Primary source: U.S. Census Bureau, *City Government Finances*, series GF, No. 4, annual. **Remarks:** Source also includes: total city government revenue and percent distribution, 1980-1990.

★ 583 ★

Taxes

Per Capita City Government Expenditures, 1980-1990

Item	Per Capita ($)		
	1980	1985	1990
Expenditure	664	954	1,299
General expenditure	513	720	1,004
Police protection	58	86	119
Fire protection	32	46	62
Highways	42	59	79
Sewerage and other sanitation	56	78	108
Public welfare	27	39	52
Education	66	78	114
Libraries	6	9	12

[Continued]

★583★

Per Capita City Government Expenditures, 1980-1990

[Continued]

Item	Per Capita ($)		
	1980	1985	1990
Health and hospitals	32	45	60
Parks and recreation	24	34	50
Housing and community develop	25	35	50
Airports	8	12	19
Financial administration	13	18	25
General control	14	23	34
General public buildings	8	10	12
Interest on general debt	22	47	74
Other and unallocable	81	101	136
Utility and liquor store expenditure	130	201	251
Water system	42	65	89
Electric power system	61	93	114
Gas supply system	9	19	15
Transit system	16	23	32
Liquor stores	2	2	2
Insurance trust expenditure	21	33	44

Source: U.S. Bureau of the Census. *Statistical Abstract of the United States 1992*. 112th edition. Washington, D.C.: U.S. Department of Commerce, 1992, p. 298. Primary source: U.S. Census Bureau, *City Government Finances*, series GF, No. 4, annual. **Remarks**: Source also includes: total city government revenue and percent distribution, 1980-1990.

★584★

Taxes

Per Capita City Government Outstanding Debt, 1980-1990

Item	Per Capita ($)		
	1980	1985	1990
Debt outstanding, year end	610	961	1,392
Long-term	584	926	1,356
Full faith and credit	280	316	448
Nonguaranteed	303	611	908
Short-term	26	36	36
Net long-term debt outstanding	514	849	909
Long-term debt issued	89	175	183
Long-term debt retired	39	78	118

Source: U.S. Bureau of the Census. *Statistical Abstract of the United States 1992*. 112th edition. Washington, D.C.: U.S. Department of Commerce, 1992, p. 298. Primary source: U.S. Census Bureau, *City Government Finances*, series GF, No. 4, annual. **Remarks**: Source also includes: total city government outstanding debt and percent distribution, 1980-1990.

★ 585 ★

Taxes

Per Capita State and Local Government Revenue From the Federal Government, 1980-1990

Source of revenue	Per capita ($)		
	1980	1985	1990
Revenue	1,993	3,015	4,150
Federal government	366	445	550
Public welfare	110	164	241
Highways	40	54	58
Education	64	70	93
Employment security administration	9	11	12
Revenue sharing	30	19	-
Other and unallocable	114	126	146

Source: U.S. Bureau of the Census. *Statistical Abstract of the United States 1992*. 112th edition. Washington, D.C.: U.S. Department of Commerce, 1992, p. 284 Primary source: U.S. Census Bureau, *Census of Governments: 1982*; *Historical Statistics on Governmental Finances and Employment* (GC 82(6)- 4); *State Government Finances*, series GF, No. 5. **Remarks**: Source also includes: the total state and local government revenue from the federal government and percent distribution; total and per capita revenue from state and local sources; and state and local government expenditures by function, 1980- 1990.

★ 586 ★

Taxes

Per Capita State and Local Government Revenue From State and Local Sources, 1980-1990

Source of revenue	Per capita ($)		
	1980	1985	1990
Revenue	1,993	3,015	4,150
From state and local sources	1,627	2,570	3,600
General, net intergovernmental	1,321	2,059	2,866
Taxes	986	1,465	2,017
Property	302	435	626
Sales and gross receipts	353	529	715
Individual income	186	294	425
Corporation income	59	80	95
Other	87	128	156
Charges and miscellaneous	335	594	849
Utility and liquor stores	113	188	236
Water supply system	30	50	71
Electric power system	50	91	118
Transit system	11	18	21
Gas supply system	8	15	12
Liquor stores	14	14	14
Insurance trust revenue	193	323	498

[Continued]

★ 586 ★

Per Capita State and Local Government Revenue From State and Local Sources, 1980-1990

[Continued]

Source of revenue	Per capita ($)		
	1980	1985	1990
Employee retirement	112	223	379
Unemployment compensation	60	74	74

Source: U.S. Bureau of the Census. *Statistical Abstract of the United States 1992.* 112th edition. Washington, D.C.: U.S. Department of Commerce, 1992, p. 284 Primary source: U.S. Census Bureau, *Census of Governments: 1982*; *Historical Statistics on Governmental Finances and Employment* (GC 82(6)- 4); *State Government Finances*, series GF, No. 5. **Remarks**: Source also includes: the total and per capita revenue for state and local government from the federal government and percent distribution; total and per capita revenue from state and local sources; and state and local government expenditures by function, 1980-1990.

★ 587 ★

Taxes

City Government Taxes per Capita in Selected Cities, 1990

Cities	Taxes Per capita (dol.)
New York City, NY	2,063
Los Angeles, CA	519
Chicago, IL	536
Houston, TX	408
Philadelphia, PA	964
San Diego, CA	325
Detroit, MI	503
Dallas, TX	481
Phoenix, AZ	343
San Antonio, TX	214
San Jose, CA	418
Baltimore, MD	852
Indianapolis, IN	533
San Francisco, CA	1,185
Jacksonville, FL	352
Columbus, OH	459
Milwaukee, WI	258
Memphis, TN	239
Washington, DC	3,744
Boston, MA	996

Source: U.S. Bureau of the Census. *Statistical Abstract of the United States 1992.* 112th edition. Washington, D.C.: U.S. Department of Commerce, 1992, p. 299. Primary source: U.S. Census Bureau, *City Government Finances*, series GF, No. 4, annual. **Remarks**: Source also includes: city government revenue, expenditure, and debt for 75 of the largest cities in the United States.

★588★

Taxes

Deductions Claimed on Personal Income Tax Returns, 1991

Prepared by the Research Institute of America, New York, from the total deductions for each type of deduction in each income group, divided by the number of actual claims. Itemized deductions were claimed on 28.2% of all 1991 returns.

Income	Averages for Itemized Deductions			
	Medical	Tax	Gift	Interest
$25,000-$30,000	$3,084	$2,115	$1,244	$5,116
$30,000-$40,000	3,319	2,540	1,315	5,675
$40,000-$50,000	3,273	3,141	1,399	6,119
$50,000-$75,000	4,864	4,161	1,683	7,220
$75,000-100,000	4,031	6,012	2,247	9,261
$100,000-$200,000	14,430	9,616	3,501	13,536
$200,000-$500,000	34,467	21,374	8,169	21,835
500,000-1,000,000	57,740	48,289	20,272	34,566
$1,000,000	86,034	157,694	95,047	78,834

Source: "A Special Summary and Forecast of Federal and State Tax Developments," *The Wall Street Journal*, June 16, 1993, p. A1.

★589★

Taxes

Tax Rate by Age

The average lifetime tax rate for "Baby Boomers" is 31 percent, compared to 24 percent for the parents of "Baby Boomers." But if current policies on tax and spending continues, the grandchildren of this generation would pay 71 percent of their income in taxes.

Source: "The Spend-Now, Tax-Later Orgy," Sylvia Nasar, *The New York Times*, January 14, 1993, p. D2.

★590★

Taxes

Estate Taxes, 1987-1992

Year	Average Estate Tax per Return
1987	$119,000
1988	140,000
1989	150,000
1990	162,000

[Continued]

★ 590 ★

Estate Taxes, 1987-1992

[Continued]

Year	Average Estate Tax per Return
1991	160,000
1992	158,000

Source: "Tax Advisors Say Time for Substantial Gifts May be Now," Jan M. Rosen, *The New York Times*, December 19, 1992, p. 35.

★ 591 ★

Taxes

Cost per Household for Daily Import and Export Controls

The average household pays $100 per year for dairy import and export controls, along with price supports and other regulations.

Source: "Udder Insanity," *Consumer Reports*, 57, May 1992, pp. 330-332.

★ 592 ★

Taxes

Government Spending by Age

According to the Congressional Budget Office (CBO), in 1990, the United States spent an average of $11,300 apiece on its 31 million elderly, compared to $4,500 on each of its 64 million children.

Source: "Cover Story: Is Uncle Sam Shortchanging Young Americans?" Aaron Berstein, *Business Week*, 21, August 19, 1991, p. 85.

★ 593 ★

Taxes

Personal Exemptions, 1950 and 1989

In 1950, when the median family income was $3,319, the personal exemption was $600 per year. This meant that an average family of four, even with no other exemptions, could deduct about 72% of its income from taxation. But in 1989, when the median family income was $34,213, and the personal exemption was $2,050, an average family of four with no other exemptions could deduct only 24% of its income from taxation. In 1950, the $600 exemption represented 18.1% of the median family income; in 1989, it was less than 6% of the family income.

Source: "Personal Exemption Remarks," Dan Coats, *Congressional Record*, February 6, 1991, pA1659.

Chapter 9
HEALTH AND MEDICINE

Cost of Health Care (General)

★594★

Health Care Expenditures per Capita in Selected Countries, 1970-1989

Country	Per-capita health expenditures in U.S. dollars			
	1970	1980	1985	1990
Australia	$204	$595	$952	$1,151
Canada	274	806	1,315	1,795
Denmark	209	571	770	963
France	192	656	991	1,379
Germany	199	740	1,047	1,287
Greece	62	196	292	406
Italy	147	541	761	1,113
Japan	126	515	785	1,113
Norway	153	624	899	1,281
Spain	82	322	437	730
Sweden	274	859	1,125	1,421
Switzerland	247	734	1,102	1,389
United Kingdom	144	445	642	909
United States	346	1,063	1,711	2,566

Source: "The High Cost of Health," *The Wall Street Journal*, January 20, 1993, p. R12. Primary source: Organization for Economic Cooperation and Development, Paris, France, *OECD Health Data*, 1991 and *OECD Health Systems: Facts and Trends*, 1992.

★595★

Cost of Health Care (General)

Personal Health Care Expenditures per Capita, by State, 1966-1982

State	1966	1982	State	1966	1982	State	1966	1982
United States	$201	$1,220						
Alabama	145	1,033	Kentucky	$155	$957	North Dakota	$197	$1,325
Alaska	227	1,187	Louisiana	156	1,106	Ohio	195	1,247
Arizona	190	1,112	Maine	173	1,091	Oklahoma	183	1,086
Arkansas	142	994	Maryland	190	1,232	Oregon	197	1,165
California	242	1,451	Massachusetts	253	1,508	Pennsylvania	201	1,273
Colorado	233	1,209	Michigan	211	1,281	Rhode Island	231	1,351
Connecticut	236	1,348	Minnesota	216	1,229	South Carolina	125	857
Delaware	209	1,153	Mississippi	115	897	South Dakota	181	1,154
District of Columbia	430	2,838	Missouri	198	1,285	Tennessee	166	1,144
Florida	184	1,228	Montana	175	1,036	Texas	177	1,110
Georgia	150	1,048	Nebraska	195	1,216	Utah	158	896
Hawaii	208	1,228	Nevada	196	1,380	Vermont	197	978
Idaho	153	868	New Hampshire	188	986	Virginia	151	1,054
Illinois	220	1,308	New Jersey	192	1,115	Washington	219	1,165
Indiana	182	1,101	New Mexico	157	904	West Virginia	161	1,057
Iowa	197	1,176	New York	258	1,417	Wisconsin	192	1,219
Kansas	195	1,271	North Carolina	143	931	Wyoming	200	873

Source: National Center for Health Statistics. *Health United States 1991*. Washington, D.C.: U.S. Department of Health and Human Services, 1992, p. 285. Primary source: *Office of the Actuary: Personal health care expenditures by State, selected years 1966-1982*, by K.R. Levit. Health Care Financing Review. HCFA Pub. No. 03199. Health Care Financing Administration. Washington, DC. U.S. Government Printing Office, Summer 1985.

★596★

Cost of Health Care (General)

National Health Expenditures from Public Funds per Capita, 1929-1990

Year	Public funds, amount per capita	Year	Public funds, amount per capita
1929	$4	1975	$245
1935	4	1980	447
1940	6	1985	707
1950	21	1986	762
1955	26	1987	827
1960	35	1988	893
1965	50	1989	983
1970	129	1990	1,089

Source: National Center for Health Statistics. *Health United States 1991*. Washington, D.C.: U.S. Department of Health and Human Services, 1992, p. 274. Primary source: Office of National Health Statistics, Office of the Actuary: National Health Expenditures, 1990. Health Care Financing Review, Vol 13, No. 1. **Remarks**: Source also includes: national health expenditures from public funds per capita, 1929-1990.

★ 597 ★

Cost of Health Care (General)

National Health Expenditures from Private Funds per Capita, 1929-1990

Year	Private funds, amount per capita	Year	Private funds, amount per capita
1929	$25	1975	$346
1935	18	1980	616
1940	23	1985	1,003
1950	58	1986	1,060
1955	75	1987	1,134
1960	108	1988	1,253
1965	154	1989	1,363
1970	217	1990	1,478

Source: National Center for Health Statistics. *Health United States 1991*. Washington, D.C.: U.S. Department of Health and Human Services, 1992, p. 274. Primary source: Office of National Health Statistics, Office of the Actuary: National Health Expenditures, 1990. *Health Care Financing Review*, Vol 13, No. 1. **Remarks**: Source also includes: national health expenditures from private funds per capita, 1929-1990.

★ 598 ★

Cost of Health Care (General)

Personal Health Care Expenditures per Capita, 1929-1990

Year	Per capita	Year	Per capita
1929	$26	1977	$653
1935	21	1978	726
1940	26	1979	812
1950	70	1980	933
1955	93	1981	1,073
1960	126	1982	1,193
1965	175	1983	1,299
1970	302	1984	1,394
1971	328	1985	1,496
1972	362	1986	1,606
1973	401	1987	1,743
1974	456	1988	1,897
1975	519	1989	2,062
1976	586	1990	2,255

Source: National Center for Health Statistics. *Health United States 1991*. Washington, D.C.: U.S. Department of Health and Human Services, 1992, p. 285. Primary source: *Office of the Actuary: Personal health care expenditures by State, selected years 1966-1982*, by K.R. Levit. Health Care Financing Review. HCFA Pub. No. 03199. Health Care Financing Administration. Washington, DC. U.S. Government Printing Office, Summer 1985.

★ 599 ★

Cost of Health Care (General)

Health Care Cost per Family, 1991

In 1991, the average family's health care costs were $6,535, of which the family share was $4,296 and the employer's share was $2,239.

Source: "The Health Care Dilemma," *USA Today*, March 22, 1993, p. 11A. **Remarks**: Source also includes: cost of health care to society in general.

★ 600 ★

Cost of Health Care (General)

Cost of HMO Health Care

According to the Health Insurance Association of America, in 1990, the average monthly premium per family was $311 for HMOs (Health Maintenance Organization) that provided care through health centers, and $316 for those plans that provided care in doctors' offices. Health insurance with managed-care features averaged $319 for monthly benefits per family.

Source: "Are HMOs the Answer?" *Consumer Reports*, 57, August 1992, pp. 519-527.

★ 601 ★

Cost of Health Care (General)

Hidden Health Care Costs per Household

Because of hidden costs, the average expenditure on health care per household is $8,000. This includes an average of : $2,300 spent directly by members of the household; $1,580 out-of-pocket payment per household to health-care providers; $590 deduction from paychecks for employee contribution to group insurance; $130 deduction from social security checks as the employee-paid premium for federal supplemental medical insurance; $3,930 per household for higher rates of federal income, social security, state income, state sales, and other taxes to support federal, state and local health efforts. In addition, employers contribute an average of $1,580 per household for health benefits.

Source: "Hidden Factors Boost Annual Cost to $8,000 for Average Household," *The Christian Science Monitor*, November 18, 1991, p. 10. **Remarks**: Source also includes: amount the government spends on health care compared to its spending for defense and education.

★ 602 ★

Cost of Health Care (General)

Health Care Cost per Vehicle at Ford Motor Company

In 1993, the cost of health care for employees at the Ford Motor Company averaged $800 per automobile. This is double the per vehicle cost of steel.

Source: "Costing the Big Three Billions," *The Detroit News and Free Press*, March 21, 1993, p. 1A. **Remarks**: Source also includes: total cost (in billions) for health care at General Motors, Ford and Chrysler.

★ 603 ★

Cost of Health Care (General)

Health Insurance Costs per Employee by Size of Company, 1992

Size of company	Average Cost Per Employee
Fewer than 500	$3,500
500-999	3,628
1,000-2,499	3,599
2,500-4,999	3,579
5,000-9,999	3,714
10,000-19,000	3,847
20,000-39,000	4,003
40,000-	3,775

Source: "Company Health Costs Soar," *USA Today*, April 27, 1993, p. 2A. Primary source: Employee Benefit Research Institute; A. Foster Higgins & Co, Inc. **Remarks**: Source also includes: average health insurance costs per employee, 1984-1992 and percentage of companies offering health insurance to full- time employees by size of company.

★ 604 ★

Cost of Health Care (General)

Health Care Costs as a Percentage of a Company's Earnings

According to a survey by Foster Higgins & Co., a benefits consulting firm, from a survey of 1,955 business and government employers, health care costs consumed 26% of the average company's net earnings.

Source: "Health Care a Growing Burden," Milt Freudenheim, *The New York Times*, January 29, 1991, p. D1. **Remarks**: Source also includes: average amount businesses pay for health care per employee, and percentage change from previous years' costs.

★ 605 ★

Cost of Health Care (General)

Administrative Costs of Health Care

The administrative and billing costs of managing health care average 24 cents of every dollar spent on health care.

Source: "How to End the Health Care Crisis," Marty Russo, *USA Today* (magazine), March 1992, p. 20. **Remarks**: Source also includes: total cost United States spends on health care compared to other countries.

★ 606 ★

Cost of Health Care (General)

Health Care Costs per Family Headed by a 65-Year-Old, 1961-1991

	Average Out-of-Pocket Costs per Family		
	1961	1972	1991
Hospital	$228	$175	$90
Doctor	316	184	408
Nursing home	287	582	1,194
Private insurance	304	309	653
Medicare	---	214	320
Other	454	390	640

Source: "Health Costs," *The Wall Street Journal*, February 26, 1992, p. B6. Primary source: Families USA.

★ 607 ★

Cost of Health Care (General)

Health Care Costs per Person, 65-year-old or Older, 1991-2031

Year of retirement at age 65	Average Out-of-Pocket Medical Expenses (not covered by Medicare or other insurance)
1991	$1,680
2001	2,436
2011	3,261
2021	4,066
2031	4,826

Source: "Effect of Normal Medical Expenses on Single Person," *Management Review*, 80, April 1992, p. 33. Primary source: Northwestern Life Insurance Co., November 1991. **Remarks:** Also included in source: age that assets would be depleted.

★ 608 ★

Cost of Health Care (General)

Health Budget in Developing Countries

In developing countries the average annual health budget is only $2 to $3 per person. Compare this to the average cost of $1 to test a sample of donated blood for HIV infection. Five years ago it cost $30 to test donated blood in a country where the average health care was only $1 per person for a year. The World Health Organization (WHO) estimates that 1% to 5% of new HIV infections in developing countries are due to tainted blood transfusions.

Source: "The Cost of Clean Blood," Sharon Kingman, *New Scientist*, 136, September 1992, p. 20.

★ 609 ★

Cost of Health Care (General)

Health Care Cost for Japanese Companies

According to Hewitt Associates, a benefits consultant, Japanese companies spend an average of $700 per employee on health care premiums per year, compared with $3,452 spent by American companies.

Source: "Japan's Health System Provides Effective Though Spartan Care," *The New York Times*, December 28, 1992, p. A8.

★ 610 ★

Cost of Health Care (General)

Hidden Costs of Insurance Claims

For each insurance claim submitted, physicians spend an average of $8; another $8 is spent for every check written by employers and insurance companies. Investigating the legitimacy of a claim involves another hidden cost. Thus, $25 in administrative costs can be added to a $25 doctor's visit, doubling the cost of medical care.

Source: "How to Solve the Health Care Crisis," John C. Goodman and Gerald L. Musgrave, *Consumer's Research*, 75, March 1992, pp. 10-14.

Cost of Illness or Injury

★ 611 ★

Arthritis Sufferers

Arthritis costs an average of $150,000 per person for care and treatment, while federal investment in research on arthritis and related disorders averages $3 per person.

Source: Debra Lappin, (Chairperson, Government Affairs Committee, The Arthritis Foundation), U.S. House Committee on Appropriations, Depts. of Labor, Health and Human Services, Education... Approp. for 1991, Hearing, Washington, DC, 1990.

★ 612 ★

Cost of Illness or Injury

Cost of AIDS

Each year in the United States treatment of AIDS, involving multiple hospitalizations with intensive care and outpatient expenses for drugs and services, costs an average of $100,000 per patient.

Source: "Experts See a Long Battle," *The Christian Science Monitor*, July 31, 1992.

★613★

Cost of Illness or Injury

Coronary Bypass Surgery

Coronary bypass surgery can cost $30,000 for one operation. Cardiovascular surgeons bring in the most revenue to hospitals of any specialty: an average of $10,942 per inpatient hospital admission in 1989, more than twice the average doctor's rate. This according to an annual survey by Jackson-Coker, an Atlanta physician-recruiting firm.

Source: "The Cardiac Money Machine," *Consumer Reports*, 57, July 1992, p. 446.

★614★

Cost of Illness or Injury

Cost of Heart Transplants, 1988

Health Status of Patient	Average Cost	Average Hospital Stay (days)	Survival Rate for 1 year
Life support or	$127,989	37	76.7%
intensive care			83.0%
In hospital	88,153	28	83.6%
Not hospitalized	76,319	16	
Homebound			84.0%
Not homebound			88.0%
ALL	91,570	23	82.4%

Source: "On Waiting List Before Birth: A Case of Transplant Ethics," Gina Kolata, *The New York Times*, January 20, 1993. Primary source: Dr. Roger W. Evans, Mayo Clinic. **Remarks**: Source also includes: figures for liver transplants.

★615★

Cost of Illness or Injury

Cost of Liver Transplants, 1988

Health Status of Patient	Average Cost	Average Hospital Stay (days)	Survival Rate for 1 year
Life support or	$211,711	50	36.4%
intensive care			64.3%
In hospital	154,077	42	69.6%
Not hospitalized	114,797	25	
Homebound			73.9%

[Continued]

★615★

Cost of Liver Transplants, 1988

[Continued]

Health Status of Patient	Average Cost	Average Hospital Stay (days)	Survival Rate for 1 year
Not homebound			82.9%
ALL	145,795	33	63.8%

Source: "On Waiting List Before Birth: A Case of Transplant Ethics," Gina Kolata, *The New York Times*, January 20, 1993. Primary source: Dr. Roger W. Evans, Mayo Clinic. **Remarks**: Source also includes figures for heart transplants.

★616★

Cost of Illness or Injury

Cost of Tuberculosis

According to Dr. George DiFerdinando, medical director for New York State's Department of Tuberculosis Control, the average hospitalization cost for each tuberculosis patient in 1990 was $15,000. In New York City alone, 90% of TB patients were hospitalized, financed by taxpayers through Medicaid.

Source: "TB Returns to Haunt the Poor," Patricia Braus, *American Demographics*, 14, February 1992, p. 20.

★617★

Cost of Illness or Injury

Cost of Cancer

According to Martin Brown, of the National Cancer Institute, the overall costs of cancer in the United States in 1990 were estimated to have been $104 billion, or an average of $416 per person.

Some average costs relating to cancer detection are:

Type of Cancer-Detection Procedure	Average Cost
Mammographic exam	$120 per person
Pap test	$10 per lab fee
Fecal-occult blood test	$5 per test
Sigmoidscopy	$100 per procedure

Source: "The National Economic Burden of Cancer: An Update," Martin L. Brown, *Journal of the National Cancer Institute*, 82, December 5, 1990, pp. 1811-1814.

★ 618 ★

Cost of Illness or Injury

Cost of Knee Replacement

Due to injuries, each year an average of one million patients have part or all of the cartilage in the knee removed, sometimes to the extent of having a knee replaced with an artificial joint. In the United States in 1992, about 187,000 such operations were performed at an average cost of $40,000.

Source: "Cartilage Transplants in Knees Show Promise," *The Wall Street Journal*, June 9, 1993, B1.

★ 619 ★

Cost of Illness or Injury

Consumers' Perceptions of Medical Care Costs

Type of Care	Average National Fee	Low Estimate	High Estimate	Average Estimate
Broken arm	$418	$10	$3,800	$583
Delivery of baby				
Caesarean	2,636	200	70,000	3,661
normal	1,999	45	80,000	2,595
Hysterectomy	2,493	100	45,000	3,048
Mammogram	88	10	4,000	315
Office visit	47	3	200	151
Open-heart surgery	7,280	10	99,000	12,804
Pap smear & exam	105	8	2,000	320
Physical exam	111	8	2,500	270
Tonsillectomy & adenoidectomy	726	50	12,000	1,437

Source: "Firms Use Financial Incentives to Make Employees Seek Lower Health-Care Fees," Glenn Ruffenach, *The Wall Street Journal*, February 9, 1993, p. B10. Primary source: Medrisk, Inc.

★ 620 ★

Cost of Illness or Injury

Cost of Treatment for AIDS, Breast Cancer, and Kidney Dialysis

Illness	Average Cost
Kidney dialysis	$185,000
AIDS	102,000
Breast cancer	52,000

Source: "The Cost of Care," *Newsweek*, 120, November 23, 1992, p. 48. Primary source: Agency for Health Care Policy and Research.

★ 621 ★

Cost of Illness or Injury

Cost of Substance Abuse

Substance abuse costs $177 billion annually for direct, indirect and related charges. This translates into an average of $750 for every man, woman, and child in the United States.

Source: Jones, Richard L. (Child Welfare League of America). U.S. House. Committee on Ways and Means, "Impact of Substance Abuse on State and Local Child Welfare Systems." Hearing, April 30, Washington, GPO, 1991.

★ 622 ★

Cost of Illness or Injury

Prostate Surgery

The average prostate operation costs $8,000-$12,000. Americans spend $3-$4 billion on surgery for benign prostate enlargement every year.

Source: "The Prostate Puzzle," *Consumer Reports*, 57, July 1993, p. 459.

By Type of Illness or Injury

★ 623 ★

Colds per Year by Age

Age (years)	Average Number of Colds per Year
0-4	5
5-19	3
20-39	2
40-	2

Source: "Colds Target the Young," *USA Today*, February 5, 1993, p. 1A. Primary source: University of Michigan.

★ 624 ★

By Type of Illness or Injury

Deaths from the Flu

There are an average of 20,000 deaths per year due to flu-related complications.

Source: "Database," *U.S. News & World Report*, 113, January 20, 1992, p. 14.

★ 625 ★

By Type of Illness or Injury

Risk of Stroke for Men by State, 1987-1989

Average annual death rate from stroke for men in the United States is 31.9 per 100,000 men.

Higher Than Average	About Average	Lower Than Average
Alabama	California	Alaska
Arkansas	Idaho	Arizona
Georgia	Illinois	Colorado
Indiana	Michigan	Connecticut
Kentucky	Missouri	Delaware
Louisiana	Ohio	Florida
Mississippi	Oregon	Hawaii
North Carolina	Pennsylvania	Iowa
Oklahoma	Texas	Maine
South Carolina	West Virginia	Maryland
Tennessee	Wisconsin	Massachusetts
Virginia	Wyoming	Minnesota
Washington, DC		Montana
		Nebraska
		Nevada
		New Hampshire
		New Jersey
		New Mexico
		New York
		North Dakota
		Rhode Island
		South Dakota
		Utah
		Vermont
		Washington

Source: "The Risk of Stroke, State by State," *The New York Times*, July 29, 1992, p. C12. **Remarks**: Source also includes: risk of stroke for women by state, 1987-1989.

★ 626 ★

By Type of Illness or Injury

Risk of Stroke for Women by State, 1987-1989

Average annual death rate from stroke for women in the United States is 27.3 per 100,000 women.

Higher Than Average	About Average	Lower Than Average
Alabama	Idaho	Alaska
Arkansas	Illinois	Arizona
California	Missouri	Colorado
Georgia	Ohio	Connecticut
Indiana	Oregon	Delaware

[Continued]

★ 626 ★

Risk of Stroke for Women by State, 1987-1989

[Continued]

Kentucky	Texas	Florida
Louisiana	Utah	Hawaii
Michigan	West Virginia	Iowa
Mississippi	Washington	Massachusetts
North Carolina	Wyoming	Maine
Oklahoma		Maryland
South Carolina		Minnesota
Tennessee		Montana
Virginia		Nebraska
Washington, DC		Nevada
		New Hampshire
		New Jersey
		New Mexico
		New York
		North Dakota
		Pennsylvania
		Rhode Island
		South Dakota
		Vermont
		Wisconsin

Source: "The Risk of Stroke, State by State," *The New York Times*, July 29, 1992, p. C12. **Remarks**: Source also includes: risk of stroke for men by state, 1987-1989.

★ 627 ★

By Type of Illness or Injury

Deaths From Cardiovascular Diseases

According to the American Heart Association, on average, every 34 seconds an American dies from heart and blood vessel diseases, accounting for nearly one million deaths per year. Cardiovascular diseases are the No. 1 killer in America, killing more people than cancer or AIDS.

Source: "Away From Politics," *International Herald Tribune*, January 19, 1993, p. 3.

★ 628 ★

By Type of Illness or Injury

Deaths From Cancer

According to the American Cancer Society, an average of 1,401 Americans die every day from cancer. One out of every five deaths in the United States is due to cancer. An estimated 526,000 people died of this disease in 1993.

Source: American Cancer Society, *Cancer Facts and Figures, 1993.*

★ 629 ★

By Type of Illness or Injury

Deaths From Bee Stings

An average of 18-20 people who are very allergic die from bee stings every year in the United States.

Source: "Abuzz Over Bee Keeping," Rif S. El-Makkakh, et. al., *Science Teacher*, 58, September 1991, pp. 26-31.

★ 630 ★

By Type of Illness or Injury

Melanoma Survival Time

Victims of advanced melanoma, a deadly skin cancer, who receive standard treatments have an average survival time of 7.3 months. However, a new vaccine being tested has given more than half of those receiving it a survival time of 23 months.

Source: "Early Data Finds Promise in New Cancer Therapy," Natalie Angier, *The New York Times*, July 31, 1992, p. A10.

★ 631 ★

By Type of Illness or Injury

Brain Injuries From Automobile Crashes

According to The Society of Automotive Engineers (SAE), an average of 80 brain injuries per day, or 29,000 per year, are the result of automobile crashes. Of these, 13,000 or 35 per day are severe and 2,000 or five per day result in a nonresponsive state.

Source: Joan Claybrook (President, Public Citizen) U.S. Senate Committee on Commerce, Science, and Transportation, "Reauthorization of the National Highway Traffic Safety Administration," Hearing, April 11, 1991, Washington, DC, GPO, 1991.

★ 632 ★

By Type of Illness or Injury

Arthritis Sufferers

Arthritis affects one American every 33 seconds, or about one million more every year. One in two people over the age of 65 have arthritis, and one in seven people in the American population as a whole. An average of one in three families is affected by this disease.

Source: Debra, Lappin, (Chairperson, Government Affairs Committee, The Arthritis Foundation), U.S. House Committee on Appropriations, Depts. of Labor, Health and Human Services, Education... Approp. for 1991, Hearing, Washington, DC, 1990.

★ 633 ★

By Type of Illness or Injury

Lou Gehrig's Disease

Once Lou Gehrig's disease (Amyotropic Lateral Sclerosis or ALS) has been diagnosed, the average life expectancy is two to three years. ALS strikes an average of 13 people each day, 91 per week, 5,000 every year. 300,000 people from today's population will die from this disease.

Source: "National Amyotrophic Lateral Sclerosis Awareness Month," Dante B. Fascell, *Congressional Record*, August 1, 1991, pE2814.

★ 634 ★

By Type of Illness or Injury

AIDS

Some statistics on AIDS: 1.) every 54 seconds there is a new HIV infection in the United States; 2.) every nine minutes there is one AIDS death; 3.) every day there are 267 new cases; 4.) every month there are 8,000 new cases; 5.) an average of one in every 1,000 college students are infected with AIDS.

Source: "We Have Lost the War Against AIDS," Larry Kramer, *USA Today* (magazine), May 1992, p. 72.

★ 635 ★

By Type of Illness or Injury

AIDS Life Expectancy

According to the Federal Centers for Disease Control and Prevention, the average life expectancy for adults who contract AIDS is 18 months to two years after diagnosis.

Source: "Insurers Accused of Discrimination in AIDS Coverage," Milt Freudenheim, *The New York Times*, June 1, 1993, p. D2.

★ 636 ★

By Type of Illness or Injury

Hearing-Impaired Children

According to child hearing experts, three years old is the average age for identifying deaf or severely hearing-impaired children in the United States. A panel of experts assembled by the National Institute of Health recommended that all infants be tested in the first three months of life by an otocoustic emission test, costing an average of $25.

Source: "U.S. Panel Backs Testing All Babies to Uncover Hearing Losses Early," Warren Leary, *The New York Times*, March 10, 1993, p. C12.

★637★

By Type of Illness or Injury

Lead Exposure and Children

According to the findings of an Australian study (and similar results from a Boston study), the mental development of children from middle class families are just as likely to be damaged from lead exposure (mainly from house paint) as children from poor families. An average deficit of eight I.Q. points was discovered when children's average lifetime blood lead concentrations rose from 10 to 35 micrograms per deciliter. In this study an average of 10 micrograms per deciliter of lead exposure was linked to significant I.Q. deficits in 10-year-olds.

Source: "Study Documents Lead-Exposure Damage in Middle-Class Children," Jane E. Brody, *The New York Times*, October 29, 1992, p. A20.

★638★

By Type of Illness or Injury

Toxicity of Lead

Testing of municipal water supplies by the Environmental Protection Agency concluded that lead in drinking water exceeds federally permissible levels in 20% of the nation's large cities. Because lead is so toxic, 15 parts per billion, or the average of 15 raindrops in an Olympic-sized pool, can be dangerous, especially if infants drink it in their formulas.

Source: "Lead Levels Excessive in Water, E.P.A. Says," Michael Specter, *The New York Times*, October 21, 1992, p. B8.

★639★

By Type of Illness or Injury

Migraine Headaches

According to Dr. Seymour Solomon, director of the headache unit at Montefiore Medical Center in New York City, migraine patients have an average of two to three headaches per month, but many have them several times per week. An estimated 24 million Americans afflicted with migraines lose more than 157 million work days per year.

Source: "Migraine Misery: Whose Head Hurts Most," Patricia Braus, *American Demographics*, 14, September 1992, p. 25.

★ 640 ★

By Type of Illness or Injury

Sunburn

An average fair-skinned person can prevent sunburn for about five hours in most parts of the United States by applying SPF 15 sunscreen.

Place	Average time (minutes) for unprotected, fair skinned person to redden
Northern California, Kansas City, New York City, Indianapolis	35-40
Brownsville, TX; Miami; Riyadh Hong Kong	20-24
Panama City; Caracas; Nairobi; Singapore	20

Source: "Sunscreens," *Consumer Reports*, 56, June 1991, p. 400. Primary source: The Skin Cancer Foundation and Madhu A. Pathak, Dept of Dermatology, Harvard Medical School.

★ 641 ★

By Type of Illness or Injury

Cancer in California by Race/Ethnicity, 1988-1990

Type	Cases of cancer per 100,000 people each year			
	Asian/ Native Amer.	Black	Hispanic	White
Breast	32.3	53.9	25.5	47.0
Colo rectal	33.2	76.7	27.5	65.1
Lung/bronchus	58.9	99.4	58.6	120.3
Prostate	45.4	150.6	66.3	108.8
ALL cancers	273.3	415.9	235.9	401.9

Source: Los Angeles Public Library/State of California, *Scan/Info*, 5, May 1993, p. 13, in *Los Angeles Times*, May 14, 1993, p. A3. Primary source: California Department of Health Services.

Hospitals

★ 642 ★

Daily Hospital Census by State, 1990

State	Average daily census (1,000)	State	Average daily census (1,000)	State	Average daily census (1,000)
U.S.	843.7	Kentucky	12.2	North Dakota	3.4
Alabama	15.6	Louisiana	14.4	Ohio	35.1
Alaska	1.0	Maine	4.2	Oklahoma	9.0
Arizona	8.5	Maryland	15.5	Oregon	6.5
Arkansas	8.3	Massachusetts	26.2	Pennsylvania	51.5
California	71.2	Michigan	27.4	Rhode Island	3.7
Colorado	9.2	Minnesota	16.8	South Carolina	10.4
Connecticut	11.6	Mississippi	11.1	South Dakota	3.5
Delaware	2.2	Missouri	19.2	Tennessee	19.9
District of Columbia	6.4	Montana	3.1	Texas	47.7
Florida	40.0	Nebraska	6.5	Utah	3.4
Georgia	23.6	Nevada	2.5	Vermont	1.6
Hawaii	3.4	New Hampshire	3.4	Virginia	21.2
Idaho	2.4	New Jersey	30.9	Washington	10.3
Illinois	40.0	New Mexico	4.2	West Virginia	6.6
Indiana	16.8	New York	90.5	Wisconsin	16.3
Iowa	11.3	North Carolina	22.6	Wyoming	1.7
Kansas	9.9				

Source: U.S. Bureau of the Census. *Statistical Abstract of the United States 1992*. 112th edition. Washington, D.C.: U.S. Department of Commerce, 1992, p. 113. Primary source: American Hospital Association, Chicago, IL, *Hospital Statistics, 1991-1992*. **Remarks**: Also included in source: total number of hospitals, hospital beds, patients admitted to hospitals, occupancy rate, and outpatients visits, all by state, 1990.

★ 643 ★

Hospitals

Days Hospitalized per Year by Race/Ethnicity, Age, and Income

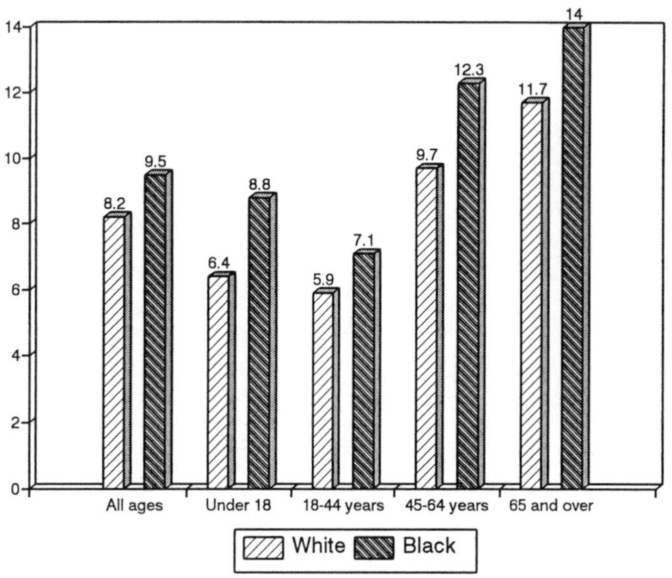

Days per person.

Characteristic	White	Black
All family incomes[1]		
All ages	8.2	9.5
Under 18 years	6.4	8.8
18-44 years	5.9	7.1
45-64 years	9.7	12.3
65 years and over	11.7	14.0
Less than $20,000		
All ages	9.5	10.0
Under 18 years	6.5	9.4
18-44 years	6.6	7.4
45-64 years	11.7	13.5
65 years and over	12.1	13.5
$20,000 or more		
All ages	6.9	7.7
Under 18 years	6.2	7.4
18-44 years	5.4	6.2

[Continued]

★ 643 ★

Days Hospitalized per Year by Race/Ethnicity, Age, and Income

[Continued]

Characteristic	White	Black
45-64 years	8.4	10.6
65 years and over	10.5	13.7

Source: "Average Annual Number of Days Hospitalized per Person Hospitalized during the Year Preceding Interview by Race, Age, and family Income: United States, 1985-87," *Health of Black and White Americans, 1985-87*, 1990, p. 21. Primary source: National Health Interview Survey, U.S. Department of Health and Human Services, Public Health Service, Centers for Disease Control, Division of Health Statistics, Hyattsville, MD, January 1990. *Note:* 1. Includes unknown family income.

★ 644 ★

Hospitals

Hospital Stay by Age, Sex, Race/Ethnicity, 1964-1990

Characteristic	Average length of stay		
	1964	1985	1990
		Number of days	
Total	8.9	7.1	6.7
Age			
Under 15 years	6.0	5.3	5.8
Under 5 years	7.8	5.1	6.2
5-14 years	4.3	5.5	5.2
15-44 years	7.6	5.6	5.4
45-64 years	10.7	8.0	6.7
65 years and over	12.1	8.7	8.4
65-74 years	11.9	8.4	8.0
75 years and over	12.4	9.1	8.9
Sex			
Male	9.7	7.4	6.8
Female	8.2	6.8	6.5
Race			
White	8.6	6.9	6.5
Black	12.7	8.5	7.8

Source: National Center for Health Statistics. *Health United States 1991*. Washington, D.C.: U.S. Department of Health and Human Services, 1992, p. 224. Primary source: Division of Health Interview Statistics, National Center for Health Statistics: Data from the National Health Interview Survey.

★ 645 ★

Hospitals

Hospital Stay by Sex and Age, 1970-1989

Selected Characteristics	Average stay (days)		
	Total	Male	Female
1970	8.0	8.7	7.6
1980	7.3	7.7	7.0
1983	6.9	7.4	6.6
1984	6.6	7.0	6.3
1985	6.5	6.9	6.2
1986	6.4	6.8	6.1
1987	6.4	6.9	6.1
1988	6.5	7.1	6.2
1989	6.5	7.0	6.1
Age			
Under 1 year old	6.0	5.8	6.4
1 to 4 years old	3.5	3.5	3.5
5 to 14 years old	5.0	5.3	4.7
15 to 24 years old	4.1	5.7	3.5
25 to 34 years old	4.5	6.6	3.9
35 to 44 years old	5.6	6.2	5.1
45 to 64 years old	6.7	6.7	6.6
65 to 74 years old	8.2	7.9	8.5
75 years old and over	9.4	7.3	9.5

Source: U.S. Bureau of the Census. *Statistical Abstract of the United States 1992*. 112th edition. Washington, D.C.: U.S. Department of Commerce, 1992, p. 115. Primary source: U.S. National Center for Health Statistics, *Vital Health Statistics*, series 13 and unpublished data. **Remarks**: Also included in source: patients' discharged from hospitals, days of hospital care, and hospital beds used per day.

★ 646 ★

Hospitals

Hospital Beds Used per Day, 1970-1989

Year	Beds used per day[1]	Age	Beds used per day[1]
1970	321	Under 1 year old	332
1980	337	1 to 4 years old	53
1983	316	5 to 14 years old	39
1984	286	15 to 24 years old	112
1985	261	25 to 34 years old	143
1986	250	35 to 44 years old	138
1987	244	45 to 64 years old	248

[Continued]

★ 646 ★

Hospital Beds Used per Day, 1970-1989

[Continued]

Year	Beds used per day[1]	Age	Beds used per day[1]
1988	228	65 to 74 years old	580
1989, total	223	75 years old and over	1,120

Source: U.S. Bureau of the Census. *Statistical Abstract of the United States 1992.* 112th edition. Washington, D.C.: U.S. Department of Commerce, 1992, p. 115. Primary source: U.S. National Center for Health Statistics, *Vital Health Statistics*, series 13 and unpublished data. **Remarks**: Also included in source: patients' discharged from hospitals, days of hospital care, and average hospital stay. *Notes:* 1. Average daily number of beds occupied per 100,000 civilian population.

★ 647 ★

Hospitals

Population per Hospital Bed in Selected Countries

Country	Year	Average Number of People per Bed
Australia	1977	85
Brazil	1976	245
Canada	1978	165
Chile	1980	292
Egypt	1981	500
France	1977	82
Germany, West (former)	1980	87
England & Wales	1974	117
India	1981	1,265
Indonesia	1979	1,787
Japan	1981	86
Madagascar	1978	189
Nigeria	1980	1,251
Poland	1979	132
Saudi Arabia	1981	645
Sweden	1980	68
United States	1979	165
USSR (former)	1978	82
Zambia	1981	289

Source: Brandreth, Gyles. *Your Vital Statistics.* New York: Citadel, 1986; p. 62. **Remarks**: Source also includes: total number of hospitals and hospital beds in 22 countries.

★ 648 ★

Hospitals

Community Hospital Care Costs, 1970-1990

Type of expense	1970	1975	1980	1983	1985	1987	1988	1989	1990
Average cost per day, total	$74	$134	$245	$369	$460	$539	$586	$637	$687
Average cost per stay, total	605	1,030	1,851	2,789	3,245	3,850	4,207	4,588	4,947

Source: U.S. Bureau of the Census. *Statistical Abstract of the United States 1992*. 112th edition. Washington, D.C.: U.S. Department of Commerce, 1992, p. 114. Primary source: American Hospital Association, Chicago, IL, *Hospital Statistics, 1991-1992*.

★ 649 ★

Hospitals

Community Hospital Care Costs by State, 1990

In dollars.

State	Average Cost per day 1990	Average Cost per stay 1990	State	Average Cost per day 1990	Average Cost per stay 1990	State	Average Cost per day 1990	Average Cost per stay 1990
United States	$687	$4,947	Kentucky	$563	$3,762	North Dakota	$427	$4,468
Alabama	588	4,175	Louisiana	701	4,575	Ohio	720	4,801
Alaska	1,070	6,249	Maine	574	4,604	Oklahoma	632	4,302
Arizona	867	4,877	Maryland	678	4,640	Oregon	800	4,432
Arkansas	534	3,730	Massachusetts	788	5,709	Pennsylvania	662	5,120
California	939	5,709	Michigan	716	5,358	Rhode Island	663	4,839
Colorado	725	5,209	Minnesota	536	4,782	South Carolina	590	4,168
Connecticut	825	6,238	Mississippi	439	3,116	South Dakota	391	3,905
Delaware	771	5,112	Missouri	679	5,022	Tennessee	633	4,340
District of Columbia	995	7,876	Montana	405	3,973	Texas	752	4,663
Florida	769	5,312	Nebraska	490	4,675	Utah	832	4,409
Georgia	630	4,303	Nevada	854	5,511	Vermont	598	4,343
Hawaii	638	6,048	New Hampshire	671	4,544	Virginia	635	4,408
Idaho	547	3,701	New Jersey	613	4,573	Washington	817	4,519
Illinois	717	5,253	New Mexico	734	4,172	West Virginia	565	3,918
Indiana	667	4,390	New York	641	6,397	Wisconsin	554	4,083
Iowa	495	4,135	North Carolina	595	4,408	Wyoming	462	3,990
Kansas	532	4,161						

Source: U.S. Bureau of the Census. *Statistical Abstract of the United States 1992*. 112th edition. Washington, D.C.: U.S. Department of Commerce, 1992, p. 114. Primary source: American Hospital Association, Chicago, IL, *Hospital Statistics, 1991-1992*.

★ 650 ★

Hospitals

Cost of Hospitals per Day, 1971-1989

Year	Expenses for inpatient care		
	Per inpatient day	Year	Per inpatient day
1971	$83		
1972	95	1981	$284
1973	102	1982	327
1974	113	1983	368
1975	133	1984	410
1976	152	1985	460
1977	173	1986	499
1978	194	1987	537
1979	216	1988	581
1980	244	1989	631

Source: National Center for Health Statistics. *Health United States 1991*. Washington, D.C.: U.S. Department of Health and Human Services, 1992, p. 272. Primary source: American Hospital Association: *Hospital Statistics*, 1990. **Remarks**: Also included in source: cost of hospitals per stay, 1971-1989.

★ 651 ★

Hospitals

Cost of Hospitals per Stay, 1971-1989

Year	Expenses for inpatient care		
	Per inpatient stay	Year	Per inpatient stay
1971	$667		
1972	747	1981	$2,168
1973	794	1982	2,493
1974	883	1983	2,776
1975	1,025	1984	2,984
1976	1,172	1985	3,239
1977	1,317	1986	3,530
1978	1,470	1987	3,849
1979	1,631	1988	4,194
1980	1,844	1989	4,572

Source: National Center for Health Statistics. *Health United States 1991*. Washington, D.C.: U.S. Department of Health and Human Services, 1992, p. 272. Primary source: American Hospital Association: *Hospital Statistics*, 1990. **Remarks**: Also included in source: cost of hospitals per day, 1971-1989.

★ 652 ★

Hospitals

Hospital Care Expenditures per Capita by State, 1966-1982

State	Amount per capita		State	Amount per capita		State	Amount per capita	
	1966	1982		1966	1982		1966	1982
United States	$80	$57	Kentucky	$60	$433	North Dakota	$83	$624
Alabama	61	541	Louisiana	63	549	Ohio	74	599
Alaska	149	552	Maine	74	517	Oklahoma	63	498
Arizona	78	498	Maryland	84	606	Oregon	66	468
Arkansas	56	443	Massachusetts	116	810	Pennsylvania	82	675
California	88	626	Michigan	90	628	Rhode Island	101	623
Colorado	100	557	Minnesota	89	540	South Carolina	51	397
Connecticut	91	578	Mississippi	48	431	South Dakota	75	530
Delaware	91	552	Missouri	81	679	Tennessee	67	578
District of Columbia	192	2,021	Montana	67	445	Texas	69	495
Florida	66	569	Nebraska	75	568	Utah	58	399
Georgia	56	492	Nevada	68	630	Vermont	86	443
Hawaii	79	479	New Hampshire	73	458	Virginia	63	506
Idaho	50	335	New Jersey	71	498	Washington	72	434
Illinois	90	700	New Mexico	69	449	West Virginia	70	564
Indiana	63	512	New York	110	679	Wisconsin	76	539
Iowa	69	536	North Carolina	57	428	Wyoming	85	398
Kansas	76	593						

Source: National Center for Health Statistics. *Health United States 1991*. Washington, D.C.: U.S. Department of Health and Human Services, 1992, p. 287. Primary source: *Office of the Actuary: Personal health care expenditures by State, selected years 1966-1982*, by K.R. Levit. Health Care Financing Review. HCFA Pub. No. 03199. Health Care Financing Administration. Washington, DC. U.S. Government Printing Office, Summer 1985.

★ 653 ★

Hospitals

Hospital Stay for Medicare Enrollees by Geographic Region, 1980-1989

Region	Average length of stay		
	1980	1985	1989
New England	12.1	10.1	10.5
Middle Atlantic	13.4	8.5	11.6
East North Central	11.2	8.6	8.7
West North Central	9.9	7.7	7.9
South Atlantic	10.3	8.5	9.0
East South Central	9.6	8.0	8.2
West South Central	9.1	7.5	8.1
Mountain	8.7	7.0	7.1
Pacific	8.7	7.2	7.4
U.S.	10.6	8.2	8.9

Source: National Center for Health Statistics. *Health United States 1991*. Washington, D.C.: U.S. Department of Health and Human Services, 1992, p. 296. Primary source: Bureau of Data Management and Strategy, Health Care Financing Administration: Unpublished data. **Remarks**: Also included in source: number of hospital discharges, average days of care, average total charges, and medicare payment per enrollee, all for the aged and disabled.

★ 654 ★

Hospitals

Cost of Hospital Care per Day for Medicare Enrollees by Geographic Region, 1980-1989

Region	Average Charge per Day		
	1980	1985	1989
New England	$295	$559	$838
Middle Atlantic	304	559	825
East North Central	298	623	971
West North Central	246	580	915
South Atlantic	277	613	963
East South Central	249	561	754
West South Central	259	599	1,003
Mountain	310	706	1,173
Pacific	424	907	1,480
U.S.	296	623	980

Source: National Center for Health Statistics. *Health United States 1991*. Washington, D.C.: U.S. Department of Health and Human Services, 1992, p. 296. Primary source: Bureau of Data Management and Strategy, Health Care Financing Administration: Unpublished data. **Remarks**: Also included in source: number of hospital discharges, average length of stay, average total charges, and medicare payment per enrollee, all for the aged and disabled.

★ 655 ★

Hospitals

Hospital Medicare and Supplementary Medical Insurance Payments per Enrollee by Geographic Region, 1980-1989

Geographic division	Benefit payments					
	Medicare			Supplementary medical insurance		
	1980	1985	1989	1980	1985	1989
	Amount per enrollee					
United States	$909	$1,585	$1,848	$390	$770	$1,197
New England	978	1,661	1,932	402	769	1,141
Middle Atlantic	965	1,792	2,094	428	893	1,372
East North Central	1,008	1,603	1,849	370	706	1,142
West North Central	888	1,476	1,581	304	643	889
South Atlantic	818	1,486	1,736	384	771	1,254
East South Central	754	1,413	1,905	281	544	1,026
West South Central	798	1,488	1,871	352	653	1,224
Mountain	782	1,309	1,616	368	667	1,095
Pacific	1,003	1,713	1,865	509	1,008	1,285

Source: National Center for Health Statistics. *Health United States 1991*. Washington, D.C.: U.S. Department of Health and Human Services, 1992, p. 296. Primary source: Bureau of Data Management and Strategy, Health Care Financing Administration: Unpublished data. **Remarks:** Also included in source: number of hospital discharges, average length of stay, and average total charges, all for the aged and disabled.

★ 656 ★

Hospitals

Outpatient Visits per Year

There is an increase in the demand for services at the hospitals which comprise the National Association of Public Hospitals (NAPH). Outpatient visits at member hospitals, which averaged 180,000 in 1986, have jumped to an average of 212,000 visits in 1989.

Source: David Ramsey, Dept of Health and Hospitals, State of Louisiana. U.S. House Committee on Ways and Means, "Health Insurance for the Unemployed," Hearing, March 21, 1991, Washington, DC, 1991, pp. 84-85.

★ 657 ★

Hospitals

Inpatient and Outpatient Hospital Charges, 1990

According to the American Hospital Association, in 1990, the average charge for an outpatient hospital visit was $223, while the average cost for an inpatient hospital day was $1,019.

Source: "Poor Access to Health Care May Result in More Hospitalization, Study Finds," Ron Winslow, *The Wall Street Journal*, November 4, 1992, p. B9.

★ 658 ★

Hospitals

Hospital Costs Not Reimbursed by Insurance

According to a 1988 survey of 60 public hospitals around the country, public hospitals lost an average of $40 million to care that was not reimbursed by private insurance, Medicaid, or Medicare. Even after receiving subsidies from local governments, their operating costs fell short by an average of $9 million.

Source: "Bed at a Public Hospital Can Take Days to Get," Philip Hilts, *The New York Times*, January 30, 1991, p. A20. **Remarks**: Also included in source: average emergency room wait, number of cocaine-addicted babies born in hospitals, drug-related emergency room visits, and average wait for admission to hospital from emergency room.

★ 659 ★

Hospitals

Addicted Babies Born in Hospitals

According to a 1988 survey of 60 public hospitals around the country, an average of 104 cocaine-addicted babies were born at public hospitals, with predictions that the numbers will keep rising.

Source: "Bed at a Public Hospital Can Take Days to Get," Philip Hilts, *The New York Times*, January 30, 1991, p. A20. **Remarks**: Also included in source: average emergency room wait, average hospital costs not reimbursed by insurance, drug-related emergency room visits, and average wait for admission to hospital from emergency room.

★ 660 ★

Hospitals

Drug-Related Emergency Room Visits

According to a 1988 survey of 60 public hospitals around the country, there were an average of 19,000 drug-related emergency room visits at public hospitals, almost 30 percent of the emergency department visits.

Source: "Bed at a Public Hospital Can Take Days to Get," Philip Hilts, *The New York Times*, January 30, 1991, p. A20. **Remarks**: Also included in source: average emergency room wait, average hospital costs not reimbursed by insurance, number of cocaine-addicted babies born in hospitals, and average wait for admission to hospital from emergency room.

★ 661 ★

Hospitals

Wait for Admission to Hospital from Emergency Room

According to a 1988 survey of 60 public hospitals around the country, the average wait for admission to a public hospital from the emergency room was 5.6 hours. The average of the longest wait at all the hospitals was 31.6 hours, while waits of 3-4 days were common.

Source: "Bed at a Public Hospital Can Take Days to Get," Philip Hilts, *The New York Times*, January 30, 1991, p. A20. **Remarks**: Also included in source: average emergency room wait, average hospital costs not reimbursed by insurance, number of cocaine-addicted babies born in hospitals, and drug-related emergency room visits.

★ 662 ★

Hospitals

Rural Hospitals

The average charge at rural hospitals is $850 per day, while urban hospitals, which offer a wide array of specialties, charge $1,350 a day.

Source: "150 Miles Away, the Doctor is Examining Your Tonsils," Ronald Smothers, *The New York Times*, September 16, 1992, p. C14.

★ 663 ★

Hospitals

Hospital Overcharges

The General Accounting Office estimates that 99 percent of hospital bills contain overcharges. A study done by a private insurer estimates that the average hospital bill contains unnecessary charges of $1,400.

Source: "Confusion and Error are Rife in Hospital Billing Practices," Elisabeth Rosenthal, *The New York Times*, January 27, 1993, p. C16. **Remarks**: Also included in source: a discussion of common errors in hospital billing practices.

★ 664 ★

Hospitals

Hospital Charges for Seat-Belted Victims of Automobile Crashes

According to Network of Employers for Traffic Safety (NETS), a coalition in Washington, the average hospital bill for seat-belted victims of automobile crashes is $800, while someone not wearing a safety belt has an average hospital bill of $2,500.

Source: "Seat Belts," *Business Week*, 22, November 2, 1992, p. 37.

★ 665 ★

Hospitals

Hospital Billing, Canada and the United States

The average Canadian hospital spends eight percent of its health care dollars on billing the single payer, the provincial government, and other administrative tasks, with an average of six employees. On the other hand, the average hospital in the United States spends 18 percent of its health care dollars on billing and administration, using 50 employees and one million dollars worth of equipment.

Source: "MediScare," Judy Haiven, *Mother Jones*, 16, March-April 1991, p. 50-53, 67-69.

★ 666 ★

Hospitals

Time and Cost of Physicians' Paperwork, Canada and the United States

The average Canadian doctor spends 30 seconds per patient on paperwork, costing one dollar on billing per patient. On the other hand, physicians in the United States spend an average of seven minutes on paperwork per patient, with another seven dollars going to a billing staff.

Source: "MediScare," Judy Haiven, *Mother Jones*, 16, March-April 1991, p. 50-53, 67-69.

★ 667 ★

Hospitals

Computers and the Reduction of Health Care Costs

According to researchers at Indiana University, charges to patients treated by doctors who use computers to order drugs and tests were 13% lower per hospital admission than charges for patients treated by doctors who used paper. Total charges averaged $900 less for patients whose doctors use computers compared with those whose doctors use traditional charts. One drawback: doctors using computers spend an average of five minutes more per patient to write orders than those doctors using charts.

Source: "Computers Helping Doctors Match Care With Costs Can Lower Bills, Study Says," *The Wall Street Journal*, January 20, 1993, p. B6.

Dental

★ 668 ★

Cost of Fillings

Type of filling	Average Cost	Average Lifespan (yrs.)
Amalgam	$51	10-20
Cosposite	64	3-10
Ceramic	404	10
Gold	361	20 or more

Source: "How Dental Fillings Compare," *Consumer Reports*, 56, May 1991, p. 319.

★ 669 ★

Dental

Dental Visits per Person by Selected Characteristics, 1987

Characteristic	Dental visits Number per person			Characteristic	Dental visits Number per person		
	1964	1983	1989		1964	1983	1989
Total	1.6	1.9	2.1	Race			
				White	1.7	2.0	2.3
Age				Black	0.8	1.2	1.2
2-14 years	1.3	2.0	2.1				
2-4 years	0.3	0.7	0.9	Family income			
5-14 years	1.9	2.5	2.5	Less than $14,000	0.9	1.2	1.3
15-44 years	1.9	1.9	2.0	$14,000-$24,999	0.9	1.5	1.6
45-64 years	1.7	2.0	2.4	$25,000-$34,999	1.4	2.2	2.2
65 years and over	0.8	1.5	2.0	$35,000-$49,999	1.9	2.5	2.7
65-74 years	0.9	1.8	2.2	$50,000 or more	2.7	2.9	3.1
75 years and over	0.6	1.1	1.8				
				Geographic region			
Sex				Northeast	2.1	2.4	2.2
Male	1.4	1.7	2.0	Midwest	1.6	1.9	2.1
Female	1.7	2.1	2.3	South	1.2	1.6	1.8
				West	1.7	2.0	2.4

Source: National Center for Health Statistics. *Health United States 1991*. Washington, D.C.: U.S. Department of Health and Human Services, 1992, p. 223. Primary source: Division of Health Interview Statistics, National Center for Health Statistics: Data from the National Health Interview Survey.

Doctor Visits

★ 670 ★

Visits to Physicians per Person Per Year, 1970-1989

Year	Sex		Race		Age (years)					
	Male	Female	White	Black	Under 6	6 to 16	17 to 24	25 to 44	45 to 64	65 and over
1970	4.1	5.1	4.8	3.9	5.9	2.9	4.6	4.6	5.2	6.3
1980	4.0	5.4	4.8	4.5	6.7	3.2	4.0	4.6	5.1	6.4
1985	4.4	6.1	5.4	4.7	6.3	3.1	4.2	4.9	6.1	8.3
1987	4.5	6.2	5.5	4.9	6.7	3.3	4.4	4.8	6.4	8.9
1988	4.5	6.2	5.6	4.6	7.0	3.4	3.8	5.1	6.1	8.7
1989	4.7	6.1	5.6	4.7	6.7	3.5	3.9	5.1	6.1	8.9

Source: U.S. Bureau of the Census. *Statistical Abstract of the United States 1992*. 112th edition. Washington, D.C.: U.S. Department of Commerce, 1992, p. 110. Primary source: U.S. National Center for Health Statistics, *Vital and Health Statistics*, series 10, and unpublished data. **Remarks**: Source also includes: dental visits per person per year.

★ 671 ★

Doctor Visits

Visits to Physicians per Person Per Year, 1984-1989

	Physician Contacts Per Person (Number)	
	1984	1989
Total	5.0	5.3
Sex:		
Male	4.4	4.8
Female	5.6	5.9
Race:		
White	5.1	5.5
Black	4.8	4.9
Income:		
Less than $14,000	5.8	6.3
$14,000 to $24,999	4.8	5.2
$25,000 to $34,999	4.9	5.5
$35,000 to $49,999	5.2	5.2
$50,000 or more	5.5	6.0

Source: U.S. Bureau of the Census. *Statistical Abstract of the United States 1992.* 112th edition. Washington, D.C.: U.S. Department of Commerce, 1992, p. 110. Primary source: U.S. National Center for Health Statistics, *Health USA*, 1990. **Remarks**: Source also includes: the percentage of visits by place of contact.

★ 672 ★

Doctor Visits

Mexican Doctors' Office Visit Charges

According to a report commissioned by Families USA Foundation in which 242 Mexican doctors were surveyed, Mexican doctors charged an average of $25.00 for an office visit. American doctors charged 2 1/2 to three times as much for a similar visit.

Source: "Mexican Health Care Draws Americans," Philip J Hilts, *The Detroit Free Press*, November 26, 1992, p. 23A. **Remarks**: Source also includes: the average California and Mexican prices for five drugs.

★ 673 ★

Doctor Visits

German Doctors' Charges by Type of Service Rendered

A.O.K. Sickness Fund, a German health insurance plan, pays doctors the following:.

Type of Service	Average Charge
Consultation	
Office or phone	$5.00
Night	12.50
House call	17.80
At night	45.00
Diagnosis of pregnancy	20.00
Attendance at normal birth	93.75
Attendance at breech birth	143.75
Examination of a baby	22.50
Electrocardiogram	15.60
Chest X-ray	28.13

Source: "How Much a Sickness Fund Pays," *New York Times*, January 23, 1993, p. 4. **Remarks**: Source also includes: average costs of German and American health care per person.

Drugs

★ 674 ★

Cost of Drug Testing

A new drug takes an average of 12 years and $230 million to develop. For every successful drug that reaches the pharmacy, 10,000 compounds are duds.

Source: "Penicillin From a Screen," *Newsweek*, 120, September 14, 1992, p. 58.

★ 675 ★

Drugs

Cost of Immunization

Type of Disease	Average Cost to Doctors of drugs for Childhood Immunization
Diptheria/tetanus/pertussis	$49.85
Oral polio	39.64
Measles/mumps/rubella	50.58
Hemophilus influenza	58.20
Hepatitis B	32.12
Total	230.39

Source: "Kids, Shots, and Drug Research. Let's Get Some Answers," *Business Week*, 23, February 15, 1992, p. 38. Primary source: Children's Defense Fund, Centers for Disease Control.

★ 676 ★

Drugs

Cost of Drugs

Due to inflation, a drug today costing the average American $53.76, cost only $20 in 1980, a 268 percent increase. At the current rate of inflation, that same drug will cost $77.06 in 1995 (365% increase), and $120.88 in the year 2000 (604% increase).

Source: "Communism Collapses...but Nothing Changes the Pharmaceutical Industry's Skyrocketing Pricing Practices," David Pryor, *Congressional Record*, September 10, 1991, pS.12618.

★ 677 ★

Drugs

The Price of Drugs, California and Mexico

Drug	Average price	
	California	Mexico
Cardizem	$24.04	$6.12
Ceclor	35.78	8.78
Mevacor	63.87	46.94
Prozac	34.09	23.72
Tagamet	89.70	21.82
Zantac	102.00	22.35

Source: "International Price Gap," *The Wall Street Journal*, June 29, 1993, p. B1. Primary source: Families USA Foundation.

Home Health Care

★ 678 ★

Cost of Home Health Care

Because of the increasing number of older Americans and the lower cost of nursing care at home, home health care has been growing at a rate of 20% for the last few years. The average cost for routine skilled nursing care at home is $750 a month, compared with $2,000 in an institution.

Source: U.S. Department of Commerce. International Trade Administration. *U.S. Industrial Outlook '92.* Washington, D.C.: U.S. Department of Commerce, 1992, p. 43-3.

Laughter

★ 679 ★

Laughter and Health

According to Joan Coggin, M.D., a cardiologist at Loma Linda University School of Medicine in Loma Linda, California, laughter has proven health benefits—it leads to stress reduction and relaxation. Adults laugh an average of only 15 times per day, while children laugh an average of 400 times daily.

Source: "Laugh—It's Healthy," *Glamour*, December 1992, p. 33.

Lifeline--Flying the Sick to Health

★ 680 ★

Cost per Lifeline Mission

Lifeline Pilots is a group of 300 volunteer pilots who fly critically ill patients on an average of 20 missions per month, in their own private planes, mostly in the Midwest. $600 is the average cost of a Lifeline mission, which is supported by donations from individuals and companies. Most passengers who use Lifeline do so after they have gone bankrupt after years of medical expenses.

Source: "Missions of Mercy: Pilots Ferry Patients to Hope for Healthy Life," Mark Shuman, *Chicago Tribune*, November 24, 1991, section 18, p. 10.

Mental Health

★ 681 ★

Mental Health Expenditures per Capita by Type of Organization, 1969-1988

Type of organization	Amount per capita					
	1969	1975	1979	1983	1986	1988
All organizations	$17	$31	$40	$62	$77	$95
State and county mental hospitals	9	15	17	24	26	29
Private psychiatric hospitals	1	2	3	7	11	19
Non-Federal general hospitals with separate psychiatric services	2	3	3	9	12	15
Veterans Administration medical centers	2	3	4	6	6	5
Federally funded community mental health centers	1	4	7	-	-	-
Residential treatment centers for emotionally disturbed children	1	1	2	2	4	5
Freestanding psychiatric outpatient clinics	1	2	3	2	2	3
All other organizations	0	1	1	12	16	19

Source: National Center for Health Statistics. *Health United States 1991*. Washington, D.C.: U.S. Department of Health and Human Services, 1992, p. 301. Primary source: Survey and Reports Branch, Division of Biometry and Applied Sciences, National Institute of Mental Health: R.W. Manderscheid and S.A. Barrett: Mental Health, United States, 1987.; unpublished data. **Remarks**: Source also includes: total mental health expenditures by type of organization.

★ 682 ★

Mental Health

Per Capita Expenditures of State Mental Health Agencies by State, 1987

State	Amount per capita 1987	State	Amount per capita 1987	State	Amount per capita 1987
United States	$38	Kentucky	$23	North Dakota	$42
Alabama	29	Louisiana	25	Ohio	33
Alaska	50	Maine	42	Oklahoma	30
Arizona	16	Maryland	49	Oregon	28
Arkansas	23	Massachusetts	62	Pennsylvania	50
California	30	Michigan	61	Rhode Island	41
Colorado	30	Minnesota	42	South Carolina	45
Connecticut	56	Mississippi	22	South Dakota	27
Delaware	41	Missouri	31	Tennessee	24
District of Columbia	130	Montana	28	Texas	18
Florida	25	Nebraska	21	Utah	19
Georgia	32	Nevada	28	Vermont	44
Hawaii	26	New Hampshire	36	Virginia	35
Idaho	17	New Jersey	43	Washington	37
Illinois	25	New Mexico	24	West Virginia	23
Indiana	31	New York	99	Wisconsin	31
Iowa	12	North Carolina	41	Wyoming	30
Kansas	28				

Source: National center for Health Statistics. *Health United States 1991*. Washington, D.C.: U.S. Department of Health and Human Services, 1992, p. 302. Primary source: National Association of State Mental Health Program Directors and the National Association of State Mental Health Program Directors Institute, Inc.: Final Report: Funding Sources and Expenditures of State Mental Health Agencies: Revenue/Expenditures Study Results, Fiscal Year 1987, Apr. 1990. **Remarks**: Source also includes: Mental Health Agencies Expenditures by state for the years 1981-1985.

★ 683 ★

Mental Health

Cost of Psychotherapy, 1991

Type of Therapy	Average Cost per Visit
Group	$50 per person
Family	150 per family
Individual	100 per person

Source: "Price Tag: Psychotherapy," *The New York Times*, February 4, 1993. Primary source: American Psychiatric Association; American Psychoanalytic Association. **Remarks**: Also included in source:Average charge per psychotherapy session by state, 1992.

★ 684 ★

Mental Health

Charge per Psychotherapy Session by State, 1992

Average Charge per Psychotherapy Session.

$62-$74	$75-$87	$88-$100
Alabama	Colorado	Alaska
Arkansas	Connecticut	Arizona
Delaware	Indiana	California
Georgia	Louisiana	Florida
Hawaii	Massachusetts	Illinois
Idaho	North Carolina	Maryland
Iowa	Ohio	New Hampshire
Kansas	Oregon	New Jersey
Kentucky	Pennsylvania	New Mexico
Maine	Rhode Island	New York
Michigan	South Dakota	Texas
Minnesota	Tennessee	
Mississippi	Utah	
Missouri	Virginia	
Montana	Wisconsin	
Nebraska		
North Dakota		
Oklahoma		
South Carolina		
Vermont		
Washington		
West Virginia		
Wyoming		

Source: "Price Tag: Psychotherapy," *The New York Times*, February 4, 1993. Primary source: American Psychiatric Association; American Psychoanalytic Association. **Remarks**: Also included in source: average cost of psychotherapy by type of session.

★ 685 ★

Mental Health

Length of Therapy Treatment

According to the findings of a 1982-83 nationwide survey of mental-health-care professionals conducted by Dr. David Knesper, a psychiatrist at the University of Michigan, patients of psychiatrists are in treatment for an average of nine months, while patients of psychologists are treated for five months. Psychiatric patients have an average of 25 visits, and psychologists' patients, 16 visits.

Source: "The Psyche Market," Paula Mergenbagen DeWitt, *American Demographics*, 14, September 1992, pp. 62-65.

★ 686 ★

Mental Health

Homeless, Mentally Ill Population in the United States

According to the findings of recent studies summarized by Dr. Ron Jemelka and his associates at the University of Washington, on a typical day an average of 250,000, or a quarter of a million, seriously mentally ill persons, such as schizophrenics and manic-depressives, are living on the streets, in public shelters, or in prison. On a typical day, an average of 30,700 of these mentally ill people can be found in jail, having no charges against them, but are merely waiting for the availability of space in a state psychiatric hospital.

Source: "The Mental Health Mess," E Fuller Torrey, *National Review*, 44, December 28, 1992, pp. 22-23.

Nursing Homes

★ 687 ★

Nursing Home Care per Capita by State, 1966 and 1982

State	Amount per capita		State	Amount per capita		State	Amount per capita	
	1966	1982		1966	1982		1966	1982
United States	$12	$114	Kentucky	$9	$104	North Dakota	$19	$154
Alabama	8	79	Louisiana	8	89	Ohio	12	143
Alaska	1	26	Maine	15	176	Oklahoma	19	111
Arizona	8	53	Maryland	9	102	Oregon	17	113
Arkansas	13	112	Massachusetts	22	192	Pennsylvania	12	116
California	11	91	Michigan	10	106	Rhode Island	15	214
Colorado	15	104	Minnesota	22	235	South Carolina	6	76
Connecticut	19	206	Mississippi	4	90	South Dakota	18	165
Delaware	8	86	Missouri	12	139	Tennessee	6	76
District of Columbia	6	55	Montana	12	92	Texas	11	88
Florida	11	65	Nebraska	17	140	Utah	9	63
Georgia	8	79	Nevada	7	82	Vermont	19	149
Hawaii	6	63	New Hampshire	16	90	Virginia	6	85
Idaho	12	84	New Jersey	10	97	Washington	16	137
Illinois	13	109	New Mexico	5	49	West Virginia	3	62
Indiana	12	129	New York	16	184	Wisconsin	14	150
Iowa	22	168	North Carolina	6	75	Wyoming	6	49
Kansas	18	163						

Source: National Center for Health Statistics. *Health United States 1991*. Washington, D.C.: U.S. Department of Health and Human Services, 1992, p. 290. Primary source: *Office of the Actuary: Personal health care expenditures by State, selected years 1966-1982*, by K.R. Levit. Health Care Financing Review. HCFA Pub. No. 03199. Health Care Financing Administration. Washington, DC. U.S. Government Printing Office, Summer 1985. **Remarks**: Source also includes: nursing home care per capita by state, 1969-1980.

★ 688 ★

Nursing Homes

Nursing Home Monthly Charges per Resident by Type of Facility, 1964-1985

Facility characteristic	Average monthly charge			
	1964	1973-74	1977	1985
All facilities	$186	$479	$689	$1,456
Ownership				
Proprietary	205	489	670	1,379
Nonprofit and government	145	456	732	1,624
Certification				
Skilled nursing facility	...	566	880	1,905
Skilled nursing and intermediate facility	...	514	762	1,571
Intermediate facility	...	376	556	1,179
Not certified	...	329	390	875
Bed size				
Less than 50 beds	---	397	546	1,036
50-90 beds	---	448	643	1,335
100-199 beds	---	502	706	1,478
200 beds or more	---	576	837	1,759
Geographic region				
Northeast	213	651	918	1,781
Midwest	171	433	640	1,399
South	161	410	585	1,256
West	204	454	653	1,458

Source: National Center for Health Statistics. *Health United States 1991*. Washington, D.C.: U.S. Department of Health and Human Services, 1992, p. 280. Primary source: National Center for Health Statistics. **Remarks**: Source also includes: monthly charges per nursing home resident by resident characteristics, 1964-1985.

★ 689 ★

Nursing Homes

Nursing Home Monthly Charges per Resident by Type of Resident, 1964-1985

Resident	Average monthly charge			
	1964	1973-74	1977	1985
All residents	186	479	689	1,456
Age:				
Under 65 years	155	434	585	1,379
65-74 years	184	473	669	1,372
75-84 years	191	488	710	1,468
85 years and older	194	485	719	1,497

[Continued]

★ 689 ★

Nursing Home Monthly Charges per Resident by Type of Resident, 1964-1985

[Continued]

Resident	Average monthly charge			
	1964	1973-74	1977	1985
Sex:				
Male	171	466	652	1,438
Female	194	484	705	1,463

Source: National Center for Health Statistics. *Health United States 1991*. Washington, D.C.: U.S. Department of Health and Human Services, 1992, p. 280. Primary source: National Center for Health Statistics. **Remarks**: Source also includes: monthly charges per nursing home resident by type of facility, 1964-1985.

★ 690 ★

Nursing Homes

Nursing Home Monthly Charges per Resident from Medicare, 1977-1985

Facility characteristic	Average monthly charge Medicare	
	1977	1985
All facilities	$1,167	$2,141
Ownership		
Proprietary	1,048	2,058
Nonprofit and government	1,325	2,456
Certification		
Skilled nursing facility	1,136	2,315
Skilled nursing and intermediate facility	1,195	2,156
Bed size		
Less than 50 beds	869	1,348
50-99 beds	1,141	1,760
100-199 beds	1,242	2,192
200 beds or more	1,179	2,767
Geographic location		
Northeast	1,369	2,109
Midwest	1,160	2,745
South	1,096	2,033
West	868	1,838

Source: National Center for Health Statistics. *Health United States 1991*. Washington, D.C.: U.S. Department of Health and Human Services, 1992, p. 279. Primary source: National Center for Health Statistics: The National Nursing Home Survey, 1977 Summary for the United States, and 1985 Summary of the United States. **Remarks**: Source also includes: nursing home monthly charges, according to residents' own income, Medicaid, and Public Assistance/Welfare.

★ 691 ★

Nursing Homes

Nursing Home Monthly Charges per Resident from Medicaid, 1977-1985

Facility characteristic	Average monthly charge Medicaid	
	1977	1985
All facilities	$720	$1,504
Ownership		
Proprietary	677	1,363
Nonprofit and government	825	1,851
Certification		
Skilled nursing facility	955	2,000
Skilled nursing and intermediate facility	739	1,509
Intermediate facility	563	1,150
Bed size		
Less than 50 beds	663	1,335
50-99 beds	634	1,323
100-199 beds	691	1,413
200 beds or more	925	1,919
Geographic region		
Northeast	975	2,035
Midwest	639	1,382
South	619	1,200
West	663	1,501

Source: National Center for Health Statistics. *Health United States 1991*. Washington, D.C.: U.S. Department of Health and Human Services, 1992, p. 279. Primary source: National Center for Health Statistics: The National Nursing Home Survey, 1977 Summary for the United States, and 1985 Summary of the United States. **Remarks**: Source also includes: nursing home monthly charges, according to residents' own income, Medicare, and Public Assistance/Welfare.

★ 692 ★

Nursing Homes

Nursing Home Monthly Charges per Resident from Public Assistance/Welfare, 1977-1985

Facility characteristic	Average monthly charge Public assistance welfare	
	1977	1985
All facilities	$508	$863
Ownership		
Proprietary	501	763
Nonprofit and government	534	1,237

[Continued]

★ 692 ★

Nursing Home Monthly Charges per Resident from Public Assistance/Welfare, 1977-1985

[Continued]

Facility characteristic	Average monthly charge Public assistance welfare	
	1977	1985
Certification		
Skilled nursing facility	575	1,338
Skilled nursing and intermediate facility	623	1,215
Intermediate facility	479	900
Not certified	401	664
Bed size		
Less than 50 beds	394	835
50-99 beds	493	774
100-199 beds	573	855
200 beds or more	602	1,071
Geographic region		
Northeast	511	738
Midwest	537	1,241
South	452	727
West	564	837

Source: National Center for Health Statistics. *Health United States 1991.* Washington, D.C.: U.S. Department of Health and Human Services, 1992, p. 279. Primary source: National Center for Health Statistics: The National Nursing Home Survey, 1977 Summary for the United States, and 1985 Summary of the United States. **Remarks**: Source also includes: nursing home monthly charges, according to residents' own income, Medicare, and Medicaid.

★ 693 ★

Nursing Homes

Number of Beds in Nursing Homes, 1986

Characteristic	Average number of beds	Characteristic	Average number of beds
Total	92	Government	126
Northeast	108	Proprietary	87
Midwest	94	Nonprofit	101
South	93		
West	72	3 to 9 beds	5
		10 to 24 beds	17
Certification:		25 to 49 beds	38
Medicare or Medicaid		50 to 74 beds	60
(skilled nursing facility)	122	75 to 99 beds	88
Medicaid		100 to 199 beds	131

[Continued]

★ 693 ★

Number of Beds in Nursing Homes, 1986

[Continued]

Characteristic	Average number of beds	Characteristic	Average number of beds
(intermediate care facility)	77	200 to 299 beds	231
Not certified nursing home	38	300 to 499 beds	356
		500 beds or more	659

Source: U.S. Bureau of the Census. *Statistical Abstract of the United States 1992.* 112th edition. Washington, D.C.: U.S. Department of Commerce, 1992, p. 119. Primary source: U.S. National Center for Health Statistics, *Vital and Health Statistics, Advance Data From Vital and Health Statistics*, no. 147, and unpublished data. **Remarks**: Source also includes: total number of nursing homes and residential facilities and occupancy rate, 1986.

★ 694 ★

Nursing Homes

Nursing Home Monthly Charges from Residents' Own Income, 1977-1985

Facility characteristic	Average monthly charge Own income or family support	
	1977	1985
All facilities	$690	$1,450
Ownership		
Proprietary	686	1,444
Nonprofit and government	698	1,462
Certification		
Skilled nursing facility	866	1,797
Skilled nursing and intermediate facility	800	1,643
Intermediate facility	567	1,222
Not certified	447	999
Bed size		
Less than 50 beds	516	886
50-99 beds	686	1,388
100-199 beds	721	1,567
200 beds or more	823	1,701
Geographic location		
Northeast	909	1,645
Midwest	652	1,398

[Continued]

★ 694 ★

Nursing Home Monthly Charges from Residents' Own Income, 1977-1985

[Continued]

Facility characteristic	Average monthly charge Own income or family support	
	1977	1985
South	585	1,359
West	663	1,489

Source: National Center for Health Statistics. *Health United States 1991*. Washington, D.C.: U.S. Department of Health and Human Services, 1992, p. 279. Primary source: National Center for Health Statistics: The National Nursing Home Survey, 1977 Summary for the United States, and 1985 Summary of the United States. **Remarks**: Source also includes: nursing home monthly charges per resident, according to Medicare, Medicaid, and Public Assistance/Welfare.

Personal Hygiene

★ 695 ★

Teens' Use of Grooming Aids

Grooming Aid	Average Number of Times per Week	
	Girls	Boys
Deordorant	7.4	6.5
Shampoo	5.7	5.8
Acne remedy	3.2	2.5

Source: "Teens Preen: Boys Vs. Girls," *USA Today*, May 26, 1993, p. 1D. Primary source: Teenage Research Unlimited.

★ 696 ★

Personal Hygiene

Shaving Habits of Men and Women

According to Gillette Company, men shave an average of 48 square inches of skin, compared with women who shave 412 square inches. But men shave more often and buy more blades than women: an average of 30 per year for men, 10 for women.

Source: "Gillette's New Sensor Picks Up a Big Edge in Razors for Women," *The Wall Street Journal*, December 17, 1992, p. B10.

★ 697 ★

Personal Hygiene

Time Women Spend Shaving

According to a survey by Gillette Company, women shave an average of 2.8 times per week. During the winter, only 8% of women shave daily, compared with 29% in the summer. Almost 36% shave in the morning; 50% in the evening.

Source: "Gillette Notes Shaving Habits of Women," *The Christian Science Monitor*, February 28, 1992, p. 8.

Smoking

★ 698 ★

Death Rate Among Smokers

Worldwide, almost three million people die annually from smoking-related causes, or an average of one death every ten seconds. 100,000 in Africa and Latin America; 200,000 in Asia (not counting China and India); China, 300,000; and the industrialized world, 1.5 million, of this number there an estimated 400,000 deaths each in the USA and former USSR.

Source: "Inter-Health Fights Life-Style Diseases," Nikolai Khaltaev, *World Health*, May-June 1991, pp. 18-20.

★ 699 ★

Smoking

Weight Gain of Smokers Who Quit the Habit

According to the results of two surveys of 9,004 people conducted in the 1970s and 1980s by Dr. David F. Williamson, an epidemiologist at the Federal Centers for Disease Control in Atlanta, six to eight pounds is the average weight gain (over a period of five years) for those who quit smoking. But a weight gain of 30 pounds occurs in one out of ten people who quit the habit, especially in those people who are underweight.

Source: "Modest Weight Gain Found in Those Who Quit Smoking," *The New York Times*, March 15, 1991, p. A21.

Women's Health

★ 700 ★

Ovarian Cancer

According to researcher, Dr. Marilyn Schapira, Medical College of Wisconsin, Milwaukee, and her colleagues, mass screening for ovarian cancer would only add hours to women's lives: eight hours would be added to the average life expectancy of a 40-year-old woman.

Source: "Mass Ovarian Cancer Testing Impractical," Doug Levy, *USA Today*, June 1, 1993, p. D1.

★ 701 ★

Women's Health

Pap Smear Testing

According to a study by Marianne C. Fahs and colleagues at Mount Sinai Medical Center and Memorial Sloan-Kettering Cancer Center, New York, pap smear testing every three years for women over 65 years of age would help reduce death from cervical cancer by 73% at an average cost of $2,254 per life saved each year. Compare this to pneumonia and flu vaccination programs which cost $1,934 to $3,415 per life saved each year.

Source: "Pap Smears for Some Women Over 65 are Cost-Effective, Study Says," Ron Winslow, *The Wall Street Journal*, September 15, 1992, p. B6.

★ 702 ★

Women's Health

Hysterectomy Surgery

Cost for a hysterectomy varies depending on the locale. According to the Health Insurance Association of America, in New York City the cost averaged $4,165, and in Atlanta, $1,885.

Source: The Universal Almanac 1992. Kansas City, MO: Andrews and McMeel, 1991, p. 262.

★ 703 ★

Women's Health

Childbirth Costs

Average total cost of maternity care, doctor's and hospital charges.

	1986	1991
Normal delivery		
Midwest	$2,460	$4,541
United States	2,550	4,720
Caesarean		
Midwest	4,220	7,334
United States	4,270	7,826

Source: "Costs of Having a Baby," *USA Today*, March 21, 1993, p. 15A. Primary source: *Source Book of Health Insurance Data, 1992*, Health Insurance Association of America.

★ 704 ★

Women's Health

AIDS and Women

According to a study done by the United Nations, an average of 3,000 women contract the AIDS virus and 500 die every day worldwide. Seventy percent of these women are between the ages of 15 and 25 years.

Source: "UN Says AIDS Hits Younger Women Harder," *Detroit Free Press*, July 29, 1993, p. 2A.

Alternative Therapy

★ 705 ★

Visits to Alternative Therapists

A 1990 national survey conducted by Dr. David M. Eisenberg of Beth Israel Hospital and Harvard Medical School, found that Americans spent $13.7 billion on alternative medicine, such as chiropractic manipulation of the spine, relaxation techniques, spiritual healing, herbal medicine, and homeopathy. Those who relied on alternative therapists averaged 19 visits during the year. The problem that most frequently cause people to seek unconventional therapy are backaches, anxiety, allergies, and chronic pain. Those most likely to use it are well-educated, middle-income whites from 25-49 years of age, living in the Western states.

Source: "Unusual Therapy Gains Popularity," Natalie Angier, *The New York Times*, January 26, 1993, p. A12.

Blood Pressure

★706★

Weight Loss and Lower Blood Pressure

According to the Trials of Hypertension Prevention, a multicenter study initiated by the National Heart, Lung, and Blood Institute, on nondrug therapies in 2000 people with high blood pressure, weight-loss was found to be the most effective way to reduce blood pressure. An average drop of nearly 4mmHg in diastolic and 2.5 mmHg in systolic pressure, resulted from an average weight loss of 15 pounds over a six months time period. After two years, of a nondrug regime (sodium and alcohol restrictions, exercise, and weight loss), people lowered their diastolic blood pressure by an average of 9 mmHg and their systolic pressure pressure by 11 mmHg.

Source: "How to lower Blood Pressure," *Consumer Reports*, 57, May 1992, pp. 300-301.

Miscellaneous

★707★

Exercise and Health

According to RAND Corporation, an average of 21 minutes of life would be gained and an average of 24 cents would be saved (in medical and other costs), for each additional mile walked or run by a sedentary person.

Source: Wellness Letter. University of California at Berkeley, School of Public Health, vol. 8, issue 8, May 1992, p. 1.

Chapter 10
HUMAN BODY

Blood

★ 708 ★

Blood Pressure

Higher readings than those shown are classified as hypertension.

Average Diastolic Blood Pressure	Average Systolic Blood Pressure		
	Less than 120	120-129	130-139
Less than 80	Optimal	Normal	High normal
80-84	Normal	Normal	High normal
85-89	High normal	High normal	High normal

Source: "A New View of Blood Pressure," *New York Times*, October 31, 1992, p. 5. Primary source: National High Blood Pressure Education Program. **Remarks**: Also included in source: systolic pressure, the higher number of a blood pressure reading, is the pressure as the heart pumps; diastolic pressure is the pressure when the heart relaxes between beats. The original chart includes blood pressure readings for the four new classifications of hypertension.

★ 709 ★

Blood

Replacement of Red Blood Cells

The red marrow of our bones produces the body's red blood cells. These are the most abundant, and are created and destroyed at an average rate of 2,000,000 per second to be replaced by the same number. Red blood cells have an average life-span of 120 days, after which they are removed by the spleen.

Source: Brandreth, Gyles. *Your Vital Statistics*. New York: Citadel, 1986, p. 60.

★ 710 ★

Blood

Quantity of Blood in the Human Body

A man weighing 154 pounds would have an average of 5.5 quarts of blood. A woman weighing 110 pounds would have an average of 3.5 quarts. In other words, the average-sized adult has 7-10 pints of blood.

Source: McAleer, Neil. *The Body Almanac.* New York: Doubleday, 1985, p. 143.

★ 711 ★

Blood

Blood Pumped Through the Kidneys

About 2.25 pints of blood are pumped through the kidneys every minute (or an average of 400 gallons per day) to produce an average of 2 to 3 pints of urine each day. The result is that most of our blood will have been filtered by our kidneys once every five minutes. This equals one-quarter of the heart's output when the body is at rest.

Source: Brandreth, Gyles. *Your Vital Statistics.* New York: Citadel, 1986, p. 102.

Bones

★ 712 ★

Number of Bones

Although babies are born with about 350 bones, an average adult has 206 bones because many of them fuse together between birth and maturity. Half of these are found in the hands and feet. The longest bone is the femur and makes up 27 percent of the height of the average person.

Source: Brandreth, Gyles. *Your Vital Statistics.* New York: 1986, p. 14.

★ 713 ★

Bones

Spine

The spine in the average male is 28 inches and in the average female, 24 inches. There is surprisingly little variation from individual to individual since the variations in height are due mostly to the length of the bones in the legs.

Source: Asimov, Isaac. *The Human Body.* New rev. ed. New York: A Mentor Book, 1992, p. 64.

Brain

★714★

Weight of the Brain

The average brain weighs 3 lbs. and contains 10,000 million nerve cells, which have 25,000 potential interconnections with other cells.

Source: The Guinness Book of Answers. 8th ed. Enfield, Eng.: Guinness Publishing, 1991, p. 148.

Cells

★715★

Number of Cells

The average person is made up of 50,000,000,000,000, or 50 trillion cells. Contrast this with the average elephant which has 6,500,000,000,000,000, or 6 1/2 quadrillion cells, or the average shrew with 7,000,000,000, or 7 billion cells.

Source: Asimov, Isaac. *The Human Body*. New rev. ed. New York: A Mentor Book, 1992, p. 312.

★716★

Cells

Life of a Cell

Cell	Average Lifespan
Blood cells	
Red blood cells	120 days
Lymphocytes	over 1 year
Other white cells	10 hours
Platelets	10 days
Brain cells	lifetime
Bone cells	25-30 years
Colon cells	3-4 days
Skin cells	19-34 days
Spermatozoa	2-3 days
Stomach cells	2 days

Source: Diagram Group. Comparisons. New York: St. Martin's Press, 1980, p. 175; *Guiness Book of Records 1992*. New York: Bantam Book, 1992, p. 180.

Chemicals

★717★

Chemical Content

The chemical content of the average person: oxygen, 65%; carbon 18.5%; hydrogen, 9.5%; calcium, 1.5%; phosphorus, 1%; potassium, sulphur, sodium, chlorine, magnesium all 0.35%, and traces of iron, iodine, zinc, and fluorine. Translated into commercial uses, the average human has enough: iron to make a 3 inch nail; sulphur to kill all the fleas on the average dog; carbon to equal 900 pencils; potassium to fire a toy cannon; fat to make 7 bars of soap; phosphorus to make 2,200 match heads; and water to fill a 10 gallon tank.

Source: Brandreth, Gyles. *Your Vital Statistics*. New York: Citadel, 1986, p. 14.

Eyes

★718★

Eye Movement

The muscles operating the lenses in the eye move an average of 100,000 times per day focusing on various objects. These muscles work harder than any others in the body: to give leg muscles the equivalent amount of work would require walking 50 miles per day.

Source: Brandreth, Gyles. *Your Vital Statistics*. New York: Citadel, 1986, p. 26.

★719★

Eyes

Blinking

The rate of blinking varies but on the average, the eye blinks once every five seconds or 17 thousand times each day or 6 1/4 million times a year.

Source: Brandreth, Gyles. *Your Vital Statistics*. New York: Citadel, 1986, p. 26.

Hair

★ 720 ★

Life of Human Hair

Beneath the human scalp, a follicle grows one hair for two to six years at an average rate of half an inch per month. At that point, the follicle shrinks toward the skin's surface. After resting for about six months, the follicle then sheds the hair. The average adult sheds 100 hairs a day.

Source: "What is Hair," *Consumer Reports*, 57, July 1992, p. 398.

★ 721 ★

Hair

Amount of Hair on a Human Head

There are about 100 thousand hairs on the average adult's scalp. Most redheads have about 90 thousand hairs, blonds have about 140 thousand and brunettes fall in between these two figures.

Source: Margo. *Growing New Hair*. Brookline, MA: Autumn Press, 1980, p. 40.

★ 722 ★

Hair

Male Facial Hair

On the face of the average adult male, every 24 hours about 25,000 hairs grow up to half a millimeter.

Source: How in the World. Pleasantville, NY: The Reader's Digest Association, 1990, p. 29.

★ 723 ★

Hair

Weight Human Hair Can Support

The average hair on the human head can support a weight of 2.82 ounces, so for example, a person who weighs 176 lbs. could hang from 1,000 hairs, or 1 percent of the average covering, assuming the hairs were not pulled out by the roots.

Source: Brandreth, Gyles. *Your Vital Statistics*. New York: Citadel, 1986, p. 22.

Heart

★ 724 ★

Heart Beat

In an average lifetime, the human heart beats 3000 million times, pumping 48 million gallons of blood around the body. In a regular rhythm, it beats an average of 72 times per minute.

Source: Brandreth, Gyles. *Your Vital Statistics*. New York: Citadel, 1986, p. 293.

Height and Weight

★ 725 ★

Height and Weight of Men and Women, 1960-1990

The average female is 5 feet 3 3/4 inches tall and weighs 135 lbs. The average male is 5 feet 9 inches tall and weighs 162 lbs. Between 1960 and 1990 the average American male grew two inches taller and 27 lbs. heavier, while the average American woman grew two inches taller and 1 lb. heavier.

Source: Asimov, Isaac. *The Human Body*. New rev. ed. New York: A Mentor Book, 1992, p. 314; *Diagram Group*. Comparisons. New York: St. Martin's Press, 1980, pp. 72-73.

★ 726 ★

Height and Weight

Height and Weight of Children

Average height and weight of children.

Age	Girls		Age	Boys	
	Height	Weight lbs.		Height	Weight lbs.
birth	1 ft. 8 in.	7 1/2	birth	1 ft. 6 in.	7 1/2
6 mos.	2 ft. 2 in.	16	6 mos.	2 ft. 2 in.	17
1	2 ft. 5 in.	20	1	2 ft. 5 in.	21
2	2 ft. 9 in.	25	2	2 ft. 9 in.	26
3	3 ft. 0 in.	30	3	3 ft. 0 in.	31
4	3 ft. 3 in.	33	4	3 ft. 3 in.	34
5	3 ft. 5 in.	38	5	3 ft. 6 in.	39
6	3 ft. 8 in.	45	6	3 ft. 9 in.	46
7	3 ft. 11 in.	49	7	3 ft. 11 in.	51
8	4 ft. 2 in.	56	8	4 ft. 2 in.	57
9	4 ft. 4 in.	62	9	4 ft. 4 in.	63
10	4 ft. 6 in.	69	10	4 ft. 6 in.	69
11	4 ft. 8 in.	77	11	4 ft. 8 in.	77
12	4 ft. 10 in.	86	12	4 ft. 10 in.	83

Source: The World Almanac and Book of Facts 1992. New York: World Almanac, 1991, p. 956. Primary source: *Physicians Handbook*, 1983. **Remarks**: Also included in source: height in centimeters and weight in kilograms.

★ 727 ★

Height and Weight

Height and Weight of Men and Women

Average height and weight of American men and women.

Height	20-24	25-29	30-39	40-49	50-59	60-69
Men						
5'2"	130	134	138	140	141	140
5'3"	136	140	143	144	145	144
5'4"	139	143	147	149	150	149
5'5"	143	147	151	154	155	153
5'6"	148	152	156	158	159	158
5'7"	153	156	160	163	164	163
5'8"	157	161	165	167	168	167
5'9"	163	166	170	172	173	172
5'10"	167	171	174	176	177	176
5'11"	171	175	179	181	182	181
6'0"	176	181	184	186	187	186
6'1"	182	186	190	192	193	191
6'2"	187	191	195	197	198	196
6'3"	193	197	201	203	204	200
6'4"	198	202	206	208	209	207
Women						
4'10"	105	110	113	118	121	123

[Continued]

★ 727 ★

Height and Weight of Men and Women

[Continued]

Height	20-24	25-29	30-39	40-49	50-59	60-69
4'11"	110	112	115	121	125	127
5'0"	112	114	118	123	127	130
5'1"	116	119	121	127	131	133
5'2"	120	121	124	129	133	136
5'3"	124	125	128	133	137	140
5'4"	127	128	131	136	141	143
5'5"	130	132	134	139	144	147
5'6"	133	134	137	143	147	150
5'7"	137	138	141	147	152	155
5'8"	141	142	145	150	156	158
5'9"	146	148	150	155	159	161
5'10"	149	150	153	158	162	163
5'11"	155	156	159	162	166	167
6'0"	157	159	164	168	171	172

Source: The World Almanac and Book of Facts 1992. New York: World Almanac, 1991, p. 956. Primary source: Society of Actuaries, *Build and Blood Pressure Study*, 1979. **Remarks**: The figure for weight includes indoor clothing and shoes.

★ 728 ★

Height and Weight

Weight Lost and Regained from Dieting

According to a survey conducted by *Consumer Reports* from the experiences of 95,000 people who have tried to lose weight, in addition to information from the commercial diet industry itself, the average dieter stayed on a diet program for about six months and lost about 10-20 percent of starting weight. But on the other hand, the average dieter gained back almost half of that weight in six months after stopping the program, and more than two- thirds of it after two years. The worst maintainers regained all the weight lost or more in nine months, while the best maintainers kept off two-thirds of their weight after two years.

Source: "Rating the Diets," *Consumer Reports*, 58, June 1993, pp. 353-357. **Remarks**: Source also includes: an evaluation of each of the major commercial diet programs.

★ 729 ★

Height and Weight

Miss America's Weight

Between 1954 and 1980, the weight of the average Miss America dropped from 132 pounds to 117 pounds.

Source: "'The Famine Within' Probes Women's Pursuit of Thinness," Kerry O'Neil, *Christian Science Monitor*, August 31, 1992, p. 11.

★ 730 ★

Height and Weight

Height and Weight of Americans Vs. What They Wish They Were

According to the Gallup Organization's "national body image survey," the average American man thinks women want a man to be six feet tall, 173 pounds, and muscular. But in fact, women want men the way they want themselves, i.e., 5-foot-11, 171 pounds, and muscular. In fact, the average American man is 5-foot-10, 172 pounds, with a lean build.

The average American woman thinks men want a woman to be 5-foot-4, 118 pounds, and wear a size 8. Whereas, the average American woman wants to be 5-foot-4 1/2, weigh 123 pounds, and wear a size 6-8. In fact the average American woman is 5-foot-3 1/2, weighs 134 pounds, and wears size 10-12.

Source: Bluman, Allan. *Elementary Statistics*. Dubuque, IA: William C. Brown Publishers, 1992, p. 66.

Lungs

★ 731 ★

Breathing

The average human breathes 5,000 gallons of air per day, or 13 million cubic feet of air in a lifetime.

Source: Omni, 15, October 1992, p. 39.

★ 732 ★

Lungs

Air Intake

At rest the average person breathes in air at a rate of 12 times per minute through a pair of lungs which weigh about 2.5 pounds.

Source: Brandreth, Gyles. *Your Vital Statistics*. New York: Citadel, 1986, p. 64.

★ 733 ★

Lungs

Capacity of the Human Lungs

1.7 gallons is the average maximum capactity of the lungs. However, since air can only be absorbed at the surface, what counts is not the volume of the lungs, but the area of surface that it presents. The surface area exposed by the lungs is 25 times the total skin surface of the body.

Source: Asimov, Isaac. *The Human Body.* New rev. ed. New York: A Mentor Book, 1992, p. 145-147.

★ 734 ★

Lungs

Oxygen and Carbon Dioxide Content of Air Breathed

On the average, the air humans breathe in is about .03 percent carbon dioxide and 21 percent oxygen; the air breathed out is 5.6 percent carbon dioxide and 14 percent oxygen.

Source: Asimov, Isaac. *The Human Body.* New rev. ed. New York: A Mentor Book, 1992, p. 148.

Mouth

★ 735 ★

Taste Buds

The average adult mouth has 10,000 taste buds in the throat, on the tongue and the roof of the mouth; the average child slightly more. Each bud, composed of about 40 cells, cluster together in tiny bumps called papillae. The average lifespan of these taste buds is about ten days (about the same as that of fireflies). There are several taste hairs protruding from the tip of each taste cell which provide the receptor surface for taste. The average mouth has one million of these taste hairs.

Source: McAleer, Neil. *The Body Almanac.* New York: Doubleday, 1985, pp. 62-63.

★736★

Mouth

Saliva

In the average person there are 45 fluid ounces of saliva flowing in a 24- hour period. But that rate can vary depending on circumstances. The flow is low (6/1,000 of a fluid ounce) when there is no food in the mouth or thoughts of food in the mind. But once food is introduced in the mouth or mind, this flow increases up to 1/8 of a fluid ounce per minute. In the average lifetime, about 10,000 gallons, or more than 1.2 million fluid ounces of saliva are produced—enough to fill a good-sized swimming pool.

Source: McAleer, Neil. *The Body Almanac.* New York: Doubleday, 1985, p. 63.

Muscles

★737★

Number of Muscles

In the average human body there are about 656 muscles. An average of 43 muscles are needed for a frown; 17 for a smile.

Source: Brandreth, Gyles. *Your Vital Statistics.* New York: Citadel, 1986, p. 26.

★738★

Muscles

Weight of Muscles

In the average man muscles make up 40 percent of body weight, or about 69 pounds. In the average woman muscles make up about 30 percent of body weight, or about 43 pounds.

Source: Asimov, Isaac. *The Human Body.* New rev. ed. New York: A Mentor Book, 1992, p. 118.

Skin

★ 739 ★

Weight of Human Skin

The skin of the average adult weighs about 5.9 lbs., and medically speaking, is considered to be the heaviest organ.

Source: Guinness Book of Records 1992. New York: Bantam Book, 1992, p. 183.

★ 740 ★

Skin

Area of Human Skin

The average woman has about 17 square feet of skin, and the average man has 20 square feet. The outer skin is replaced about once a month.

Source: Brandreth, Gyles. *Your Vital Statistics.* New York: Citadel, 1986, p. 16.

★ 741 ★

Skin

Contents of Human Skin

On average, 12 feet of nerves, 3 feet of blood vessels, 10 hair follicles, 100 sweat glands, and hundreds of nerve endings are in each square centimeter of skin. During the average lifetime 40 lbs. of dead skin are shed.

Source: Brandreth, Gyles. *Your Vital Statistics.* New York: Citadel, 1986, p. 16.

Sperm

★ 742 ★

Sperm Count

Average sperm concentrations have fallen to 66 million spermatozoa per milliliter in 1990, according to a study in the latest issue of *British Medical Journal*. On average, men have about half the sperm counts of their grandfathers' generation.

Source: "Human Sperm Count Study Shows World-Wide Decline," *Wall Street Journal*, September, 11, 1992, p. B13.

Temperature

★ 743 ★

Mean Body Temperature

In a new study, researchers have found that the average body temperature of healthy adults is 98.2 F degrees, with slightly lower readings in the morning and higher readings in the evening. Women's average temperature at 98.4 F degrees is slightly higher than men's average temperature of 98.1 F degrees. These findings were the result of a study which set out to test the 124-year-old standard established by Carl Wunderlich, who arrived at 98.6 F degrees as the mean body temperature.

Source: "Body Temperature of 98.2 Degrees Found to be Normal," Elyse Tanouye, *Wall Street Journal*, September 23, 1992, p. B4.

Calories

★ 744 ★

Energy Requirements

Age (years)	Males		Females	
	Average Weight (lbs)	Energy Needs (Calories)	Average Weight (lbs)	Energy Needs (Calories)
1	24	1,100	24	1,100
2-3	31	1,300	31	1,300
4-6	40	1,800	40	1,800
7-9	55	2,200	55	2,200
10-12	75	2,500	79	2,200
13-15	110	2,800	106	2,200
16-18	136	3,200	117	2,100
19-24	156	3,000	128	2,100
25-49	163	2,700	130	1,900
50-74	161	2,300	139	1,800
75+	152	2,000	141	1,500

Source: Adapted from National Research Council, Nutrition Board, and other sources.

★ 745 ★

Calories

Calorie Consumption in the Office

Working quietly in his office, an executive uses an average of 105 calories per hour, while a secretary burns up fewer, an average of 88.

Source: Brandreth, Gyles. *Your Vital Statistics*. New York: Citadel, 1986, p. 113.

★ 746 ★

Calories

Calorie Consumption for Selected Activities

Activity	Average Number of Calories Used Per Hour	Activity	Average Number of Calories Used Per Hour
Basketball	500	Sitting	100
Bicycling (5 1/2 mph)	210	Skiing (cross	
(13 mph)	660	country)	600-660
Bowling	220-270	(downhill)	570
Calisthenics	300	Snow shoveling	420-480
Digging	360-420	Square dancing	350
Gardening	200	Standing	140
Golfing (using power cart)	150-220	Swimming	
(pulling cart)	240-300	moderately	500-700
(carrying clubs)	300-360	Tennis	
Football	500	(doubles)	300-360
Handball (social)	600-660	(singles)	420-480
(competitive)	660	Vacuuming	240-300
Hoeing	300-360	Volleyball	350
Housework	180	Walking	
Jogging (5-10 mph)	500-800	(2 mph)	150-240
Lawn mowing (power)	250	(3.5 mph)	240-300
(hand)	420-480	(4 mph)	300-400
Raking leaves	300-360	(5 mph)	420-480

Source: Agress, Clarence M. *Energetics*. New York: Grosset & Dunlap, 1978, p. 35; Brody, Jane E. *Jane Brody's Good Food Book*. New York: W.W. Norton, 1985 p. 226; *New York Public Library Desk Reference*. New York: Webster's New World, 1989 p. 648.

★ 747 ★

Calories

Daily Calorie Consumption of College Students

The average college student weighing 115 pounds expends 2,093 calories a day. The total is derived from the total number of calories used per pound (listed below) multiplied by the weight of the student.

Activity	Hours spent in activity	Calories per pound per hour	Total calories used per pound
Asleep	8	0.4	3.2
Lying still, awake	1	0.5	0.5
Dressing and undressing	1	0.9	0.9
Studying in class, eating, studying, talking	8	0.7	5.6
Walking	1	1.5	1.5
Standing	1	0.8	0.8
Driving a car	1	1.0	1.0
Running	1/2	4.0	2.0
Playing ping-pong	1/2	2.7	1.3
Writing	2	0.7	1.4
Total	24	-	18.2

Source: Sherman, Irwin W. *Biology: A Human Approach*. 4th ed. New York: Oxford University Press, 1989, p. 80.

Chapter 11
INCOME

By Age

★ 748 ★

Weekly Income of Children

Results of a study of the purchasing power of children aged 4 to 12 conducted by James McNeal, professor of Marketing at Texas A&M University, and Dr. Chyon-Hwa Yea, a statistical consultant of Proctor & Gamble, showed that the average weekly allowance in 1991 was $8.87 for boys and $7.66 for girls. But girls save more, or an average of 43% of their allowance, while boys save 38%. Overall, children spent an average of $3.09 a week in 1989, and $4.87 in 1991.

Source: "Growing Up in the Market," James U. McNeal, *American Demographics*, 14, October 1992, p. 46.

★ 749 ★

By Age

Median Household Income of the Elderly by Age, 1991

Average median household income by age of household head.

55 & older	55-64	65-76	75-84	85 & older
$21,500	$32,365	$20,314	$13,740	$10,800

Source: "Retirement Funds," *American Demographics*, 14, April 1992, p. 31. Primary source: Current Population Survey, March 1991. **Remarks**: Source also includes: possible sources of income itemized.

★ 750 ★

By Age

Median Pensions, 1990

65 and Older	Annual Median Pension	
	Private Firms	Government
Men	$5,099	$11,684
Women	2,297	5,947

Source: "Benefits Vary," *The Detroit News*, December 20, 1992, p. 3F. Primary source: U.S. Census Bureau, Social Security Administration and National Taxpayers Union.

★ 751 ★

By Age

Military Retirement

Military retirement typically occurs to men (99% are men) in their early 40s after 20 years of service. The average monthly military retirement income is $1,300.

Source: "How to Reach Military Retirees," Frederick Day and Charles Jackson, *American Demographics*, 13, April 1991, p. 41.

★ 752 ★

By Age

Public Employee Retirement, 1970-1989

Year	Unit	Total	Level of Government		Type of Benefit		
			Federal Total	State and local	Age and service	Disability	Survivor Monthly
Average annual benefit per annuitant, current dollars							
1970	Dollars	3,005	3,440	2,413	3,271	3,136	1,512
1980	Dollars	7,190	9,181	4,859	7,719	7,588	3,855
1985	Dollars	9,137	11,659	6,512	9,592	10,232	5,556
1988	Dollars	10,395	12,771	8,016	10,877	12,001	6,330
Average annual benefit per annuitant, constant (1988) dollars							
1970	Dollars	8,700	9,959	6,986	9,470	9,079	4,377
1980	Dollars	10,312	13,167	6,969	11,070	10,883	5,529
1985	Dollars	10,169	12,975	7,247	10,675	11,387	6,183
1988	Dollars	10,395	12,771	8,016	10,877	12,001	6,330

Source: U.S. Bureau of the Census. *Statistical Abstract of the United States 1992*. 112th edition. Washington, D.C.: U.S. Department of Commerce, 1992, p. 364. Primary source: U.S. Social Security Administration, *Social Security Bulletin*, June 1990 and unpublished data.

★ 753 ★

By Age

Federal Civil Service Retirement, 1970-1990

Item	Unit	1970	1980	1984	1985	1986	1987	1988	1989	1990
Average monthly benefit										
Age and service	Dollars	362	992	1,149	1,189	1,197	1,267	1,263	1,310	1,369
Disability	Dollars	221	723	851	881	881	893	930	966	1,003
Survivors	Dollars	116	392	501	528	536	552	583	616	653

Source: U.S. Bureau of the Census. *Statistical Abstract of the United States 1992*. 112th edition. Washington, D.C.: U.S. Department of Commerce, 1992, p. 364. Primary source: U.S. Office of Personnel Management, *Compensation Report*, annual.

★ 754 ★

By Age

Discretionary Dollars by Age

Age	Average Discretionary Money
15-24	$7,790
25-29	9,130
30-34	10,920
35-39	12,400
45-49	14,450
50-54	13,550
55-59	14,580
60-64	14,360
65-69	12,920

Source: Krantz, Les. *The Best and Worst of Everything*. New York: Prentice Hall General Reference, 1991, p. 197. Primary source: U.S. Bureau of the Census.

By City or State

★ 755 ★

Hourly Wage in Selected Cities Worldwide

City	Average Hourly Wage (U.S. $)
Copenhagen	$10.10
Oslo	10.20
Los Angeles	10.40
New York	10.50
Helsinki	10.70

[Continued]

★ 755 ★

Hourly Wage in Selected Cities Worldwide

[Continued]

City	Average Hourly Wage (U.S. $)
Tokyo	10.90
Luxembourg	11.40
Geneva	14.50
Zurich	16.00

Source: "Average Earnings," *Newsweek*, 120, February 17, 1992, p. 6. Primary source: Union Bank of Switzerland.

★ 756 ★

By City or State

Per Capita Income by State, 1991

State	Income	State	Income	State	Income
U.S. Avg.	$19,092	Kentucky	$15,626	North Dakota	$15,605
Alabama	15,518	Louisiana	15,046	Ohio	17,770
Alaska	21,067	Maine	17,454	Oklahoma	15,541
Arizona	16,579	Maryland	22,189	Oregon	17,575
Arkansas	14,629	Massachusetts	23,003	Pennsylvania	19,306
California	20,847	Michigan	18,655	Rhode Island	19,207
Colorado	19,358	Minnesota	19,125	South Carolina	15,467
Connecticut	26,022	Mississippi	13,328	South Dakota	16,071
Delaware	20,816	Missouri	17,928	Tennessee	16,486
District of Columbia	24,063	Montana	15,675	Texas	17,230
Florida	18,992	Nebraska	17,718	Utah	14,625
Georgia	17,436	Nevada	19,783	Vermont	17,997
Hawaii	21,190	New Hampshire	21,760	Virginia	20,082
Idaho	15,333	New Jersey	25,666	Washington	19,484
Illinois	20,731	New Mexico	14,644	West Virginia	14,301
Indiana	17,179	New York	22,471	Wisconsin	17,939
Iowa	17,296	North Carolina	16,853	Wyoming	16,937
Kansas	18,322				

Source: Los Angeles Public Library/State of California, *Scan/Info*, 4, September 1992, p. 19. Primary source: *Los Angeles Times*, September 3, 1992, p. A22. From the Commerce Department.

★ 757 ★

By City or State

Per Capita Money Income by State, 1969-1987

State	Per capita Income (dol.)		State	Per Capita Income (dol.)		State	Per Capita Income (dol.)	
	1969	1987		1969	1987		1969	1987
U.S.	3,119	11,923	Kentucky	2,425	9,380	North Dakota	2,469	9,641
Alabama	2,317	9,615	Louisiana	2,330	8,961	Ohio	3,199	11,323
Alaska	3,725	13,263	Maine	2,548	10,478	Oklahoma	2,694	9,927
Arizona	2,937	11,521	Maryland	3,512	14,697	Oregon	3,148	11,045
Arkansas	2,142	9,061	Massachusetts	3,408	14,389	Pennsylvania	3,066	11,544
California	3,614	13,197	Michigan	3,357	11,973	Rhode Island	3,121	12,351
Colorado	3,106	12,271	Minnesota	3,038	12,281	South Carolina	2,303	9,967
Connecticut	3,885	16,094	Mississippi	1,925	8,088	South Dakota	2,387	8,910
Delaware	3,265	12,785	Missouri	2,952	11,203	Tennessee	2,464	10,448
District of Columbia	3,842	14,778	Montana	2,696	9,322	Texas	2,792	10,645
Florida	3,058	12,456	Nebraska	2,797	11,139	Utah	2,697	9,288
Georgia	2,640	11,406	Nevada	3,554	12,603	Vermont	2,772	11,234
Hawaii	3,373	12,290	New Hampshire	2,985	13,529	Virginia	2,996	13,658
Idaho	2,644	9,159	New Jersey	3,674	15,028	Washington	3,357	12,184
Illinois	3,495	12,437	New Mexico	2,437	9,434	West Virginia	2,333	8,980
Indiana	3,070	11,078	New York	3,608	13,167	Wisconsin	3,032	11,417
Iowa	2,884	11,198	North Carolina	2,474	10,856	Wyoming	2,895	9,826
Kansas	2,929	11,520						

Source: U.S. Bureau of the Census. *Statistical Abstract of the United States 1992.* 112th edition. Washington, D.C.: U.S. Department of Commerce, 1992, p. 455. Primary source: U.S. Bureau of the Census, *1970 Census of Population*, vol. 1; *1980 Census of Population*, Vol. 1, chap C (PC 80-1C), and *Current Population Reports*, series P-26, No. 88. **Remarks**: Source also includes: median family income by state, 1969 and 1979.

★ 758 ★

By City or State

Median Family Income by State, 1969 and 1979

State	Median Family Income (dol.)		State	Median Family Income (dol.)		State	Median Family Income (dol.)	
	1969	1979		1969	1979		1969	1979
U.S.	9,586	19,917	Kentucky	7,439	16,444	North Dakota	7,836	18,023
Alabama	7,263	16,347	Louisiana	7,527	18,088	Ohio	10,309	20,909
Alaska	12,441	28,395	Maine	8,205	16,167	Oklahoma	7,720	17,688
Arizona	9,185	19,017	Maryland	11,057	23,112	Oregon	9,487	20,027
Arkansas	6,271	14,641	Massachusetts	10,833	21,166	Pennsylvania	9,554	19,995
California	10,729	21,537	Michigan	11,029	22,107	Rhode Island	9,733	19,448
Colorado	9,552	21,279	Minnesota	9,928	21,185	South Carolina	7,620	16,978
Connecticut	11,808	23,149	Mississippi	6,068	14,591	South Dakota	7,490	15,993
Delaware	10,209	20,817	Missouri	8,908	18,784	Tennessee	7,446	16,564
District of Columbia	9,576	19,099	Montana	8,509	18,413	Texas	8,486	19,618
Florida	8,261	17,280	Nebraska	8,562	19,122	Utah	9,320	20,024
Georgia	8,165	17,414	Nevada	10,687	21,311	Vermont	8,928	17,205
Hawaii	11,552	22,750	New Hampshire	9,682	19,723	Virginia	9,044	20,018
Idaho	8,380	17,492	New Jersey	11,403	22,906	Washington	10,404	21,696
Illinois	10,957	22,746	New Mexico	7,845	16,928	West Virginia	7,414	17,308
Indiana	9,966	20,535	New York	10,609	20,180	Wisconsin	10,065	20,915
Iowa	9,016	20,052	North Carolina	7,770	16,792	Wyoming	8,944	22,430
Kansas	8,690	19,707						

Source: U.S. Bureau of the Census. *Statistical Abstract of the United States 1992.* 112th edition. Washington, D.C.: U.S. Department of Commerce, 1992, p. 455. Primary source: U.S. Bureau of the Census, *1970 Census of Population*, vol. 1; *1980 Census of Population*, Vol. 1, chap C (PC 80-1C), and *Current Population Reports*, series P-26, No. 88. **Remarks**: Source also includes: per capita family income by state, 1969 - 1987.

★ 759 ★

By City or State

Annual Pay by State, 1990

State	Average Annual Pay 1990	State	Average Annual Pay 1990	State	Average Annual Pay 1990
U.S.	23,602	Kentucky	19,947	North Dakota	17,626
Alabama	20,468	Louisiana	20,646	Ohio	22,843
Alaska	29,946	Maine	20,154	Oklahoma	20,288
Arizona	21,443	Maryland	24,730	Oregon	21,332
Arkansas	18,204	Massachusetts	26,689	Pennsylvania	23,457
California	26,180	Michigan	25,376	Rhode Island	22,388
Colorado	22,908	Minnesota	23,126	South Carolina	19,669
Connecticut	28,995	Mississippi	17,718	South Dakota	16,430
Delaware	24,423	Missouri	21,716	Tennessee	20,611
District of Columbia	33,717	Montana	17,895	Texas	22,700
Florida	21,032	Nebraska	18,577	Utah	20,074
Georgia	22,114	Nevada	22,358	Vermont	20,532
Hawaii	23,167	New Hampshire	22,609	Virginia	22,750
Idaho	18,991	New Jersey	28,449	Washington	22,646
Illinois	25,312	New Mexico	19,347	West Virginia	20,715
Indiana	21,699	New York	28,873	Wisconsin	21,101
Iowa	19,224	North Carolina	20,220	Wyoming	20,049
Kansas	20,238				

Source: U.S. Bureau of the Census. *Statistical Abstract of the United States 1992*. 112th edition. Washington, D.C.: U.S. Department of Commerce, 1992, p. 411. Primary source: U.S. Bureau of Labor Statistics, *Employment and Wages, Annual Averages 1990, Bulletin 2393*, and USDL New Release 91-390, *Average Annual Pay by State and Industry, 1990*. **Remarks**: Source also includes: average annual pay by state for 1988 and 1989.

★ 760 ★

By City or State

Per Capita Income in the Largest U.S. Cities, 1987

Cities	1987 (dol.)	Cities	1987 (dol.)
New York, NY	12,926	Columbus, OH	10,811
Los Angeles, CA	13,592	New Orleans, LA	9,340
Chicago, IL	10,806	Cleveland, OH	8,690
Houston, TX	12,007	El Paso, TX	8,027
Philadelphia, PA	10,002	Seattle, WA	14,438
San Diego, CA	12,978	Denver, CO	12,980
Detroit, MI	9,662	Nashville, TN	12,583
Dallas, TX	13,489	Austin, TX	11,860
San Antonio, TX	8,779	Kansas City, MO	12,077
Phoenix, AZ	12,375	Oklahoma City, OK	11,547
Honolulu, HI	12,734	Fort Worth, TX	11,082
Baltimore, MD	9,989	Atlanta, GA	11,689
San Jose, CA	13,711	Portland, OR	11,830
San Francisco, CA	15,137	Long Beach, CA	12,947
Indianapolis, IN	12,111	St. Louis, MO	9,718
Memphis, TN	10,347	Tucson, AZ	10,204
Jacksonville, FL	11,514	Albuquerque, NM	11,988
Washington, DC	14,778	Pittsburgh, PA	10,988
Milwaukee, WI	10,593	Miami, FL	9,830
Boston, MA	12,984	Cincinnati, OH	11,223

Source: U.S. Bureau of the Census. *Statistical Abstract of the United States 1992*. 112th edition. Washington, D.C.: U.S. Department of Commerce, 1992, p. 455. Primary source: U.S. Bureau of the Census, *1980 Census of Population*, vol. 1, chapter C (PC80-1-C) and *Current Population Reports*, series P-26, No. 88.

★ 761 ★

By City or State

Counties in the United States with Low per Capita Income, 1989

County	Per capita income 1989	County	Per capita income 1989
Starr, TX	$4,549	Owsley, KY	$7,334
Shannon, SD	5,294	Zapata, TX	7,334
Maverick, TX	6,155	Apache, AZ	7,406
Dimmit, TX	6,576	Clay, WV	7,427
Zavala, TX	6,739	Jackson, KY	7,479
McCreary, KY	6,834	Perry, AL	7,643
Sioux, ND	7,052	Buffalo, SD	7,695
Todd, SD	7,053	Corson, SD	7,731
Madison, LA	7,096	Magoffin, KY	7,772
Menitee, KY	7,107	St. Helena, LA	7,777
Elliott, KY	7,181	Summers, WV	7,791
Willacy, TX	7,199	Jefferson, MS	7,813
Jackson, SD	7,303		

Source: The Universal Almanac 1992. Kansas City, Mo: Andrews and McMeel, 1991, p. 224. Primary source: U.S. Bureau of Economic Analysis. *Survey of Current Business* (April 1991). **Remarks**: Source also includes: Counties with the highest per capita income.

★ 762 ★

By City or State

Counties in the United States with High per Capita Income, 1989

County	Per capita income 1989	County	Per capita income 1989
Haines Borrough, AK	$37,548	Hunterdon, NJ	30,301
Loving, TX	37,122	Morris, NJ	29,981
New York, NY	35,193	Bristol Bay Borrough, AK	29,755
Marin, CA	34,983	Montgomery, MD	29,639
Pitkin, CO	33,108	Nassau, NY	28,678
Arlington, VA	33,039	Fairfax, Fairfax City and Falls Church, VA	28,366
Wheeler, NE	32,706	Howard, MD	28,252
Somerset, NJ	32,469	Valdez-Cordova, AK	27,727
Fairfield, CT	31,438	San Mateo, CA	27,659
Alexandria, VA	31,264	Kiowa, CO	26,916
Westchester, NY	31,188	Sherman, TX	26,900
Cheyenne, CO	31,117	Nantucket, MA	26,652
Bergen, NJ	30,967		

Source: The Universal Almanac 1992. Kansas City, Mo: Andrews and McMeel, 1991, p. 224. Primary source: U.S. Bureau of Economic Analysis. *Survey of Current Business* (April 1991). **Remarks**: Source also includes: Counties with the lowest per capita income.

★ 763 ★

By City or State

Median Household Income by State, 1990

State	1990 (dol.)	State	1990 (dol.)	State	1990 (dol.)
U.S.	29,943	Kentucky	24,780	North Dakota	25,264
Alabama	23,357	Louisiana	22,405	Ohio	30,013
Alaska	39,298	Maine	27,464	Oklahoma	24,384
Arizona	29,224	Maryland	38,857	Oregon	29,281
Arkansas	22,786	Massachusetts	36,247	Pennsylvania	29,005
California	33,290	Michigan	29,937	Rhode Island	31,968
Colorado	30,733	Minnesota	31,465	South Carolina	28,735
Connecticut	38,870	Mississippi	20,178	South Dakota	24,571
Delaware	30,804	Missouri	27,332	Tennessee	22,592
District of Columbia	27,392	Montana	23,375	Texas	28,228
Florida	26,685	Nebraska	27,482	Utah	30,142
Georgia	27,561	Nevada	32,023	Vermont	31,098
Hawaii	38,921	New Hampshire	40,805	Virginia	35,073
Idaho	25,305	New Jersey	38,734	Washington	32,112
Illinois	32,542	New Mexico	25,039	West Virginia	22,137
Indiana	26,928	New York	31,591	Wisconsin	30,711
Iowa	27,288	North Carolina	26,329	Wyoming	29,460
Kansas	29,917				

Source: U.S. Bureau of the Census. *Statistical Abstract of the United States 1992*. 112th edition. Washington, D.C.: U.S. Department of Commerce, 1992, p. 448. Primary source: U.S. Bureau of the Census, *Census Population Reports*, series P-60, No. 174. **Remarks**: Source also includes: median household income by state, 1984-1989.

★ 764 ★

By City or State

Hourly Earnings of Production Workers in Manufacturing Industries by State, 1990

State	1990 in dollars	State	1990 in dollars	State	1990 in dollars
U.S.	10.83	Kentucky	10.70	North Dakota	9.27
Alabama	9.39	Louisiana	11.61	Ohio	12.64
Alaska	12.46	Maine	10.59	Oklahoma	10.73
Arizona	10.21	Maryland	11.57	Oregon	11.15
Arkansas	8.51	Massachusetts	11.39	Pennsylvania	11.04
California	11.48	Michigan	13.86	Rhode Island	9.45
Colorado	10.94	Minnesota	11.23	South Carolina	8.84
Connecticut	11.53	Mississippi	8.37	South Dakota	8.48
Delaware	12.39	Missouri	10.74	Tennessee	9.55
District of Columbia	12.51	Montana	11.51	Texas	10.47
Florida	8.98	Nebraska	9.66	Utah	10.32
Georgia	9.17	Nevada	11.05	Vermont	10.52
Hawaii	10.99	New Hampshire	10.83	Virginia	10.07
Idaho	10.60	New Jersey	11.76	Washington	12.61
Illinois	11.44	New Mexico	9.04	West Virginia	11.53
Indiana	12.03	New York	11.11	Wisconsin	11.11
Iowa	11.27	North Carolina	8.79	Wyoming	10.83
Kansas	10.94				

Source: U.S. Bureau of the Census. *Statistical Abstract of the United States 1992.* 112th edition. Washington, D.C.: U.S. Department of Commerce, 1992, p. 742. Primary source: U.S. Bureau of the Census, *Census Population Reports*, series P-60, No. 174. **Remarks**: Source also includes: average hourly earnings of production workers in manufacturing industries by state, 1980-1989.

By Country

★ 765 ★

Hourly Wage in the Manufacturing Industry in Selected Countries, 1991

Country	Average Hourly Wage	Average Hourly Benefits
Austria	$9.49	$9.42
France	8.49	7.61
Germany, West (former)	13.09	11.30
Great Britain	9.59	4.12
Greece	4.08	2.63
Italy	9.52	9.99
Japan	13.60	4.25
Netherlands	10.66	8.69

[Continued]

★ 765 ★

Hourly Wage in the Manufacturing Industry in Selected Countries, 1991

[Continued]

Country	Average Hourly Wage	Average Hourly Benefits
Portugal	2.70	2.05
Sweden	12.72	9.58
Switzerland	15.51	7.88
United States	11.40	4.22

Source: "Expensive Deutschland," *The Wall Street Journal*, January 23, 1993, p. A10. Primary source: Institute on the German Economy, Cologne. Based in part on preliminary figures.

★ 766 ★

By Country

Low Hourly Wages in Selected Countries, 1990

A study by the Department of Labor reports that among countries with low hourly wages, the average hourly wage is $1.85 in Mexico, $2.64 in Brazil, and $3.20 in Hong Kong.

Source: "Industrialized Nation's Labor Costs," *The Wall Street Journal*, January 6, 1992, p. A10.

★ 767 ★

By Country

Compensation Package for Employees in Selected Countries, 1992

Compensation includes cash pay, stock options, benefits, and perquisites.

Country	Average Total Compensation			
	Manufacturing	White Collar	Manager	CEO
Canada	$34,935	$47,231	$132,877	$416,066
France	30,019	62,279	190,354	479,772
Germany	36,857	59,916	145,627	390,933
Great Britain	26,084	74,761	162,190	439,441
Italy	31,537	58,263	219,573	463,009
Japan	34,263	59,916	185,437	390,723
United States	27,606	57,675	159,575	717,237

Source: "Pay Through the Ranks," *Wall Street Journal*, October 12, 1992, p. B1. Primary source: Towers Perrin.

★ 768 ★

By Country

Gross National Product (GNP) per Capita in Selected Countries

Country	GNP ($ per capita)	Country	GNP ($ per capita)	Country	GNP ($ per capita)
Albania	1,200	Gabon	2,770	Niger	290
Algeria	2,170	Gambia, The	230	Nigeria	250
Angola	620	Germany	18,366	Norway	21,850
Argentina	2,160	Ghana	380	Oman	5,220
Australia	14,440	Greece	5,340	Pakistan	370
Austria	17,360	Guatemala	920	Panama	1,780
Bahrain	6,360	Guinea	430	Papua New Guinea	900
Bangladesh	180	Guinea-Bissau	180	Paraguay	1,030
Barbados	6,370	Guyana	340	Peru	1,090
Belgium	16,390	Haiti	400	Philippines	700
Belize	1,600	Honduras	900	Poland	1,760
Benin	380	Hungary	2,560	Portugal	4,260
Bhutan	190	Iceland	21,240	Qatar	9,920
Bolivia	600	India	350	Romania	3,445
Botswana	940	Indonesia	490	Rwanda	310
Brazil	2,550	Iran Islamic Rep	1,800	Saudi Arabia	6,230
Bulgaria	2,320	Iraq	1,940	Senegal	650
Burkina Faso	310	Ireland	8,500	Sierra Leone	200
Burundi	220	Israel	9,750	Singapore	10,450
Cameroon	1,010	Italy	15,150	Solomon Islands	570
Canada	19,020	Jamaica	1,260	Somalia	170
Cape Verde	760	Japan	23,730	South Africa	2,460
Central African Rep	390	Jordan	1,730	Spain	9,150
Chad	190	Kenya	380	Sri Lanka	430
Chile	1,770	Korea, North	1,240	Sudan	540
China	360	Korea, Rep	4,400	Surinam	3,020
Colombia	1,190	Kuwait	16,380	Swaziland	900
Comoros	460	Laos	170	Sweden	21,710
Congo	930	Lesotho	470	Switzerland	30,270
Costa Rica	1,790	Libya	5,410	Syrian Arab Rep	1,020
Cote d'Ivoire	790	Luxembourg	24,860	Tanzania	120
Cuba	2,000	Madagascar	230	Thailand	1,170
Cyprus	7,050	Malawi	180	Togo	390
Czechoslovakia (former)	7,878	Malaysia	2,130	Trinidad & Tobago	3,160
Denmark	20,510	Mali	260	Tunisia	1,260
Djibouti	1,070	Malta	5,820	Turkey	1,360
Dominican Rep	790	Mauritania	490	Uganda	250
Ecuador	1,040	Mauritius	1,950	U. Arab Emirates	18,430
Egypt	630	Mexico	1,990	United Kingdom	14,570
El Salvador	1,040	Morocco	900	U.S.S.R. (former)	9,211
Equatorial Guinea	430	Mozambique	80	United States	21,100
Ethiopia	120	Namibia	1,245	Uruguay	2,620
Fiji	1,640	Nepal	170	Venezuela	2,450
Finland	22,060	Netherlands	16,010	Vietnam	215
France	17,830	New Zealand	11,800	Yemen	605

[Continued]

★ 768 ★

Gross National Product (GNP) per Capita in Selected Countries

[Continued]

Country	GNP ($ per capita)	Country	GNP ($ per capita)	Country	GNP ($ per capita)
Yugoslavia (former)	2,490	Zambia	390	Zimbabwe	640
Zaire	260				

Source: The 1992 Information Please Environmental Almanac. Boston, MA: Houghton Mifflin Company, 1992, pp. 300-301. Primary source: World Bank.

★ 769 ★

By Country

Gross National Product (GNP) per capita in the Czech Republic and Slovakia

The gross national product (GNP) for former Czechoslovakia, now the Czech Republic and Slovakia, is $2,250 for the Czech Republic and $1,840 for Slovakia.

Source: Los Angeles Public Library/State of California, *Scan/Info*, 5, January 1993, p. 18. Primary source: *New York Times*, January 1, 1993, p. A1.

★ 770 ★

By Country

Annual Income of Farmers Vs. Urban Dwellers in China

Official statistics in China indicate that farmers have annual earnings averaging only two-fifths of urban dwellers': $135 for farmers, $315 for city residents.

Source: "Discontent Is Growing on China's Farms," *The Wall Street Journal*, February 17, 1993, p. A10.

★ 771 ★

By Country

Monthly Income of Chinese Workers

Since the 1990 economic reforms in China, the income of Chinese workers has risen from only $12 (U.S. dollars) per month to $55.00 a month.

Source: "Learning the Business Basics About China," William G. Perry, Jr., *Management Review*, 80, May 1992, pp. 50-52.

By Education

★ 772 ★

Mean Money Earnings By Educational Level and Sex

Age and Sex	Total (Dol.)	Elementary, 8 years or less (Dol.)	High school			College			
			Total (Dol.)	1-3 years (Dol.)	4 years (Dol.)	Total (Dol.)	1-3 years (Dol.)	4 years (Dol.)	5 or more years (Dol.)
Male, total	34,886	19,188	27,131	22,564	28,043	43,217	34,188	44,554	55,831
25 to 34 years old	27,743	15,887	23,355	19,453	24,038	33,003	28,298	35,534	39,833
35 to 44 years old	37,958	18,379	28,205	23,621	28,927	45,819	36,180	47,401	58,542
45 to 54 years old	40,231	19,686	31,235	24,133	32,862	50,545	39,953	50,718	62,902
55 to 64 years old	37,469	22,379	29,460	25,280	30,779	50,585	36,954	55,518	61,647
65 years old and over	33,145	17,028	24,003	19,530	25,516	44,424	34,323	43,092	52,149
Female, total	22,768	13,322	18,469	15,381	18,954	27,493	22,654	28,911	35,827
25 to 34 years old	21,337	11,832	16,673	13,385	17,076	25,194	20,872	27,210	32,563
35 to 44 years old	24,453	13,714	19,344	15,695	19,886	29,287	23,307	31,631	37,599
45 to 54 years old	23,429	13,490	19,500	16,651	19,986	29,334	24,608	29,242	38,307
55 to 64 years old	21,388	13,941	18,607	15,202	19,382	26,930	23,364	27,975	33,383
65 years old and over	19,194	-	18,281	-	18,285	23,277	-	-	-

Source: U.S. Bureau of the Census. *Statistical Abstract of the United States 1992*. 112th edition. Washington, D.C.: U.S. Department of Commerce, 1992, p. 454.

★ 773 ★

By Education

Earnings of High school Dropouts VS. High School Graduates

Dr. Baldwin of the G.E.D. Testing Service in Washington used the Census Bureau's National Longitudinal Survey of Youth to compare average salaries of high school dropouts, those holding G.E.D. certificates, and high school graduates. For 25-year-olds in 1988, the average annual salary for high school dropouts was $10,379; for those holding the G.E.D. certificate the average salary was $11,777; and for high school graduates the average salary was $15,213. The gap widened for 28-year-olds: thee average salary for dropouts was $12,411, for G.E.D. holders, $12,411, and for high school graduates, $20,195.

Source: "As More Earn Equivalency Diploma, Its Value is Debated," Iver Peterson, *The New York Times*, October 21, 1992, p. B10.

★ 774 ★

By Education

Starting Salaries for College Graduates in Selected Fields

Major	Average Starting Salary, 1992
Advertising	$22,194
Education	23,401
Financial administration	25,382
Geology	28,776
Liberal Arts	21,194
Nursing	29,159
Retailing	20,030

Source: "Estimated Starting Salaries for New College Graduates of 1992," *NEA Today*, 10, April 1992, p. 13.

★ 775 ★

By Education

Starting Salaries for Bachelor's Degree Recipients in Selected Fields of Study

In dollars. Data are average beginning salaries based on offers made by business, industrial, government, and nonprofit and educational employers to graduating students. Data from representative colleges throughout the United States.

Field of Study	Bachelor's			
	Monthly, 1980	Annual		
		1989	1990	1991
Accounting	1,293	25,290	26,391	26,642
Business, general	1,218	22,274	23,529	24,019
Marketing	1,145	22,523	23,543	23,713
Engineering				
Civil	1,554	26,735	28,136	29,658
Chemical	1,801	32,949	35,122	37,492
Computer	(NA)	30,413	31,490	32,280
Electrical	1,690	30,661	31,778	33,190
Mechanical	1,703	30,539	32,064	33,999
Nuclear	1,668	31,281	31,750	32,175
Petroleum	1,987	32,987	35,202	38,882
Engineering tech.	1,585	28,310	29,318	30,098
Chemistry	1,459	26,698	27,494	26,836
Mathematics	1,475	26,789	27,032	27,370
Physics	(NA)	28,296	28,002	29,227
Humanities	1,074	23,010	23,213	23,567

[Continued]

★ 775 ★

Starting Salaries for Bachelor's Degree Recipients in Selected Fields of Study

[Continued]

Field of Study	Bachelor's			
	Monthly, 1980	Annual		
		1989	1990	1991
Social sciences	1,072	20,205	21,627	21,375
Computer science	1,558	28,557	29,804	30,696

Source: U.S. Bureau of the Census. *Statistical Abstract of the United States 1992*. 112th edition. Washington, D.C.: U.S. Department of Commerce, 1992, p. 172. Primary source: College Placement Council, Inc., *A Study of Beginning Offers*, annual. **Remarks**: Source also includes: starting salaries for recipients of master's degree, and doctoral degree. *Note:* NA stands for not available.

★ 776 ★

By Education

Starting Salaries for Master's Degree Recipients in Selected Fields of Study

In dollars. Data are average beginning salaries based on offers made by business, industrial, government, and nonprofit and educational employers to graduating students. Data from representative colleges throughout the United States.

Field of Study	Master's			
	Monthly, 1980	Annual		
		1989	1990	1991
Accounting	1,517	28,874	29,647	30,996
Business, general	1,795	33,903	36,175	35,241
Marketing	1,794	34,462	35,440	43,182
Engineering				
Civil	1,735	30,723	32,336	34,551
Chemical	1,947	36,131	37,862	40,457
Computer	(NA)	37,183	35,748	38,348
Electrical	1,912	36,435	37,526	38,804
Mechanical	1,893	35,260	36,506	38,114
Nuclear	1,831	34,024	36,728	38,086
Petroleum	(NA)	38,475	38,412	43,608
Chemistry	1,688	32,157	32,320	33,575
Mathematics	1,685	31,498	30,069	30,061
Physics	(NA)	29,889	31,480	35,207
Humanities	1,309	25,799	(NA)	(NA)

[Continued]

★ 776 ★

Starting Salaries for Master's Degree Recipients in Selected Fields of Study

[Continued]

Field of Study	Master's			
	Monthly, 1980	Annual 1989	1990	1991
Social sciences	1,298	23,814	(NA)	(NA)
Computer science	1,858	35,823	36,849	37,894

Source: U.S. Bureau of the Census. *Statistical Abstract of the United States 1992.* 112th edition. Washington, D.C.: U.S. Department of Commerce, 1992, p. 172. Primary source: College Placement Council, Inc., *A Study of Beginning Offers*, annual. **Remarks**: Source also includes: starting salaries for recipients of bachelor's degree, and doctoral degree. *Note:* NA stands for not available.

★ 777 ★

By Education

Starting Salaries for Doctoral Degree Recipients in Selected Fields of Study

In dollars. Data are average beginning salaries based on offers made by business, industrial, government, and nonprofit and educational employers to graduating students. Data from representative colleges throughout the United States.

Field of Study	Doctor's			
	Monthly, 1980	Annual 1989	1990	1991
Engineering				
Civil	2,089	37,214	44,481	43,060
Chemical	2,452	47,853	50,524	52,992
Computer	(NA)	53,380	50,526	56,584
Electrical	2,534	48,666	53,147	54,612
Mechanical	2,426	45,893	49,887	46,866
Chemistry	2,261	43,215	45,356	47,911
Mathematics	2,199	37,500	42,775	41,146
Physics	2,303	42,632	41,486	39,913
Computer science	(NA)	50,049	54,788	58,300

Source: U.S. Bureau of the Census. *Statistical Abstract of the United States 1992.* 112th edition. Washington, D.C.: U.S. Department of Commerce, 1992, p. 172. Primary source: College Placement Council, Inc., *A Study of Beginning Offers*, annual. **Remarks**: Source also includes: starting salaries for recipients of bachelor's degree and master's degree. *Note:* NA stands for not available.

★ 778 ★

By Education

College Graduate's Hourly Salary, 1973-1991

Adjusted for inflation.

Year	Average Hourly Salary
1973	$16.45
1981	15.24
1991	14.77

Source: "Young, Smart—and Unemployed," *U.S. News and World Report*, 113, May 25, 1992, p. 20.

By Occupation

★ 779 ★

Annual Salaries of Public School Teachers: an Historical Perspective, 1899-1989

Data in unadjusted dollars.

	1899-1900	1909-10	1919-20	1929-30	1939-40	1949-50	1959-60	1969-70	1979-80	1988-89
Average annual salary	$325	$485	$871	$1,420	$1,441	$3,010	$5,174	$8,840	$16,715	$30,969

Source: The World Almanac and Book of Facts 1992. New York: World Almanac, 1991, p. 213. Primary source: U.S. Department of Education, National Center for Education Statistics. **Remarks**: Source also includes: pupil and teacher enrollment, revenue and expenditures.

★ 780 ★

By Occupation

Salaries of Classroom Teachers by State, 1991-1992

State	Average salary	State	Average salary
U.S. Average	$34,413	Missouri	$28,880
Alabama	26,954	Montana	27,513
Alaska	44,725	Nebraska	27,231
Arizona	31,892	Nevada	33,175
Arkansas	26,569	New Hampshire	32,445
California	41,811	New Jersey	41,381
Colorado	32,926	New Mexico	26,653
Connecticut	47,300	New York	44,200
Delaware	34,548	North Carolina	29,334
District of Columbia	41,256	North Dakota	24,145
Florida	31,119	Ohio	34,359
Georgia	29,680	Oklahoma	25,721
Hawaii	34,528	Oregon	33,656
Idaho	26,759	Pennsylvania	38,540
Illinois	36,623	Rhode Island	36,047
Indiana	33,755	South Carolina	28,209
Iowa	29,196	South Dakota	23,300
Kansas	30,808	Tennessee	28,726
Kentucky	30,880	Texas	29,041
Louisiana	27,037	Utah	26,524
Maine	29,672	Vermont	33,420
Maryland	38,843	Virginia	31,921
Massachusetts	38,066	Washington	34,880
Michigan	40,251	West Virginia	27,298
Minnesota	34,782	Wisconsin	33,873
Mississippi	24,428	Wyoming	29,000

Source: "National Average Salary for Teachers Tops $34,400," *NEA Today*, 10, April 1992.

★ 781 ★

By Occupation

Salaries of Public School Central Office Administrators, 1975-1991

In dollars.

Position	1975	1980	1985	1986	1987	1988	1989	1990	1991
Annual Salary									
Superintendent	30,338	39,344	56,954	60,707	64,580	68,147	71,190	75,425	79,874
Deputy/assoc. superintendent	30,074	37,440	52,877	57,190	60,222	63,872	66,214	69,623	72,428
Assistant superintendent	26,460	33,452	48,003	51,209	53,656	56,894	59,655	62,698	66,553
Administrators for -									
Finance and business	21,850	27,147	40,344	43,200	45,259	47,330	49,933	52,354	55,097
Instructional services	22,608	29,790	43,452	46,110	48,810	50,838	53,716	56,359	59,162
Public relations/information	21,470	24,021	35,287	37,329	38,925	41,305	43,402	44,926	47,938
Staff personnel services	21,470	29,623	44,182	46,269	48,627	51,421	53,972	56,344	59,271
Subject area supervisors	18,601	23,974	34,422	36,797	38,763	41,086	43,555	45,929	48,366

Source: U.S. Bureau of the Census. *Statistical Abstract of the United States 1992*. 112th edition. Washington, D.C.: U.S. Department of Commerce, 1992, p. 152. Primary source: Educational Research Service, Arlington, VA, *National Survey of Salaries and Wages in Public Schools*, annual, vols. 2 and 3. **Remarks**: Source also includes: salaries for school building administrators, auxiliary professional personnel, teachers, secretarial personnel and support personnel.

★ 782 ★

By Occupation

Salaries of Public School Building Administrators, 1975-1991

In dollars.

	1975	1980	1985	1986	1987	1988	1989	1990	1991
Principals									
Elementary	19,061	25,165	36,452	39,024	41,536	43,664	45,909	48,431	51,453
Junior high/middle	21,136	27,625	39,650	42,365	44,861	47,078	49,427	52,163	55,083
Senior high	22,894	29,207	42,094	44,986	47,896	50,512	52,987	55,722	59,106
Assistant principals									
Elementary	15,968	20,708	30,496	32,895	34,347	36,364	38,360	40,916	43,548
Junior high/middle	17,868	23,507	33,793	36,094	37,958	40,093	42,292	44,570	46,981
Senior high	18,939	24,816	35,491	37,616	39,758	41,839	44,002	46,486	49,009

Source: U.S. Bureau of the Census. *Statistical Abstract of the United States 1992*. 112th edition. Washington, D.C.: U.S. Department of Commerce, 1992, p. 152. Primary source: Educational Research Service, Arlington, VA, *National Survey of Salaries and Wages in Public Schools*, annual, vols. 2 and 3. **Remarks**: Source also includes: salaries for central office administrators, teachers, auxiliary professional personnel, secretarial personnel and support personnel.

★ 783 ★

By Occupation

Salaries of Public School Auxiliary Professional Personnel, 1975-1991

In dollars.

Auxillary professional personnel	1975	1980	1985	1986	1987	1988	1989	1990	1991
Counselors	14,479	18,847	27,593	29,388	31,132	32,896	34,244	35,979	38,024
Librarians	12,546	16,764	24,981	26,668	28,390	30,046	31,645	33,469	35,417
School nurses	10,673	13,788	19,944	21,339	22,219	23,692	24,804	26,090	27,713

Source: U.S. Bureau of the Census. *Statistical Abstract of the United States 1992*. 112th edition. Washington, D.C.: U.S. Department of Commerce, 1992, p. 152. Primary source: Educational Research Service, Arlington, VA, *National Survey of Salaries and Wages in Public Schools*, annual, vols. 2 and 3. **Remarks**: Source also includes: salaries for central office administrators, school building administrators, teachers, secretarial personnel and support personnel.

★ 784 ★

By Occupation

Salaries of Public School Secretarial Personnel, 1975-1991

In dollars.

Secretarial/clerical personnel	1975	1980	1985	1986	1987	1988	1989	1990	1991
Central office:									
Secretaries/stenographers	7,318	10,331	15,343	16,383	17,182	18,220	19,045	20,238	21,303
Accounting/payroll clerks	7,588	10,479	15,421	16,604	17,273	18,229	19,143	20,088	21,202
Clerk-typists	6,089	8,359	12,481	13,208	13,729	14,651	15,192	16,125	16,859
School building level:									
Secretaries/stenographers	6,046	8,348	12,504	13,233	13,947	14,749	15,364	16,184	16,953
Library clerks	5,052	6,778	9,911	10,412	10,641	11,234	11,751	12,151	12,696

Source: U.S. Bureau of the Census. *Statistical Abstract of the United States 1992*. 112th edition. Washington, D.C.: U.S. Department of Commerce, 1992, p. 152. Primary source: Educational Research Service, Arlington, VA, *National Survey of Salaries and Wages in Public Schools*, annual, vols. 2 and 3. **Remarks**: Source also includes: salaries for central office administrators, school building administrators, teachers, auxiliary professional personnel, and support personnel.

★ 785 ★

By Occupation

Salaries of Public School Support Personnel, 1975-1991

In dollars.

Hourly Wage Rate	1975	1980	1985	1986	1987	1988	1989	1990	1991
Other support personnel:									
Teacher aides:									
Instructional	2.91	4.06	5.89	6.20	6.43	6.72	7.05	7.43	7.77
Noninstructional	2.81	3.89	5.60	5.91	6.14	6.45	6.69	7.08	7.43
Custodians	3.54	4.88	6.90	7.28	7.51	7.82	8.19	8.54	9.05
Cafeteria workers	2.61	3.78	5.42	5.76	5.92	6.23	6.56	6.77	7.19
Bus drivers	3.75	5.21	7.27	7.72	8.06	8.31	8.78	9.21	9.52

Source: U.S. Bureau of the Census. *Statistical Abstract of the United States 1992.* 112th edition. Washington, D.C.: U.S. Department of Commerce, 1992, p. 152. Primary source: Educational Research Service, Arlington, VA, *National Survey of Salaries and Wages in Public Schools*, annual, vols. 2 and 3. **Remarks**: Source also includes: salaries for central office administrators, school building administrators, teachers, auxiliary professional personnel, and secretarial personnel.

★ 786 ★

By Occupation

Annual Median Salaries of School Custodial and Maintenance Staff, 1991-1992

	Median salary	
	1991	1992
Custodial	$17,800	$18,660
Maintenance	22,596	24,000

Source: "A Quick Look at Some M&O Results," *AS&U* (American School and University), 64, April 1992, p. 36.

★ 787 ★

By Occupation

Salary of Preschool Teachers and Family Day-Care Providers

According to the results of a nationwide survey done by *Working Mother* magazine of 1,733 working mothers, a preschool teacher has an average salary of $11,500 per year, and a family day-care provider has an hourly salary of $4.04.

Source: "Mothers Frequently Change Caregivers," *Christian Science Monitor*, December 8, 1992, p. 9.

★ 788 ★

By Occupation

Salaries for College Administrative Personnel, 1991

Position	Average salary	
	2-year College	4-year College
President	$80,677	$96,703
Chief Academic Officer	59,964	76,805
Chief Business Officer	58,726	70,134
Chief Development Officer	46,618	62,667
Chief Facilities Officer	42,159	49,769
Chief Purchasing Officer	37,414	41,290
Director Data Processing	44,582	51,204

Source: "Tough Times Force Tough Measures," *AS&U*, (American School & University), 64, January 1992, p. 37.

★ 789 ★

By Occupation

Earnings in Division I Athletic Programs

Position	Average Salary	School Benefits	Outside Income	Average Total Earnings
Athletic director	$80,047	$7,380	$16,890	$86,206
Head of women's programs	48,319	3,359	3,873	49,921
Head coach of men's basketball	69,390	19,955	38,686	110,180
Head coach of women's basketball	39,838	5,003	6,651	44,996
Coach (man) of women's teams	32,231			
Coach (woman) of women's teams	31,651			

Source: "Opportunities Aren't Equal in Colleges," *The New York Times*, September 25, 1992, p. B9.

★ 790 ★

By Occupation

Salaries of College and University Faculty at Campuses with Doctoral Programs, 1992-1993

Doctoral Institutions	All	Public	Private	Church-related
Professor	$66,780	$63,250	$80,280	$70,770
Associate professor	47,220	45,840	53,860	50,640
Assistant professor	40,110	38,880	45,510	42,090
Instructor	28,240	27,170	35,590	33,560
Lecturer	33,200	32,890	35,180	29,560
No rank	34,490	33,150	39,420	29,440
All	52,450	50,070	63,240	52,810

Source: "Faculty Salaries Increased 2.5% in 1992-93, AAUP Survey Finds," *The Chronicle of Higher Education*, 40, April 14, 1993, p. A22. Primary source: American Association of University Professors. **Remarks**: Source also includes: faculty salaries for public, private and church-related colleges and universities at campuses with master's programs, and at four-year and two-year institutions.

★ 791 ★

By Occupation

Salaries of College and University Faculty at Campuses with Master's Programs, 1992-1993

	All	Public	Private	Church-related
Professor	$54,760	$54,240	$57,060	$55,970
Associate professor	43,680	43,430	44,020	44,800
Assistant professor	36,160	36,160	35,360	37,090
Instructor	27,590	27,600	26,310	29,380
Lecturer	27,790	27,300	29,860	36,650
No rank	33,970	34,980	32,130	32,420
All	43,950	43,790	44,140	44,800

Source: "Faculty Salaries Increased 2.5% in 1992-93, AAUP Survey Finds," *The Chronicle of Higher Education*, 40, April 14, 1993, p. A22. Primary source: American Association of University Professors. **Remarks**: Source also includes: faculty salaries for public, private and church-related colleges and universities at campuses with doctoral programs, and at four-year and two-year institutions.

★ 792 ★

By Occupation

Salaries of College Faculty at Four-Year Institutions, 1992-1993

4-year institutions	All	Public	Private	Church-related
Professor	$48,390	$48,600	$54,620	$43,210
Associate professor	38,900	40,180	41,710	36,220
Assistant professor	32,420	33,430	34,400	30,500
Instructor	26,230	26,440	27,480	25,520
Lecturer	29,250	26,980	37,340	24,480
No rank	35,870	40,100	37,040	28,500
All	38,430	38,860	42,420	36,160

Source: "Faculty Salaries Increased 2.5% in 1992-93, AAUP Survey Finds," *The Chronicle of Higher Education*, 40, April 14, 1993, p. A22. Primary source: American Association of University Professors. **Remarks**: Source also includes: faculty salaries for public, private and church-related colleges and universities at campuses with doctoral programs and master's programs, and at two-year institutions.

★ 793 ★

By Occupation

Salaries of College Faculty at Two-Year Institutions, 1992-1993

2-year institutions with academic ranks	All	Public	Private	Church-related
Professor	$47,310	$47,820	$36,710	$32,010
Associate professor	39,300	39,760	31,800	29,290
Assistant professor	33,800	34,300	27,100	24,850
Instructor	28,460	28,850	21,950	21,550
Lecturer	25,280	25,440	18,500	20,050
No rank	33,950	34,460	32,430	-
All	37,800	38,320	29,000	27,080

Source: "Faculty Salaries Increased 2.5% in 1992-93, AAUP Survey Finds," *The Chronicle of Higher Education*, 40, April 14, 1993, p. A22. Primary source: American Association of University Professors. **Remarks**: Source also includes: faculty salaries for public, private and church-related colleges and universities at campuses with doctoral programs and master's programs, and at four-year institutions.

★ 794 ★

By Occupation

Salaries of College and University Faculty by Sex, 1992-1993

Rank	Average Salary					
	Public		Private		Church-related	
	Men	Women	Men	Women	Men	Women
Professor	$59,240	$52,900	$70,180	$59,970	$52,430	$46,720
Associate Professor	44,810	41,840	48,070	44,410	41,940	38,650
Assistant Professor	38,110	35,310	40,300	36,390	34,600	32,710
Instructor	28,670	27,010	29,340	27,750	27,350	26,200

Source: "Faculty Salaries Increased 2.5% in 1992-93, AAUP Survey Finds," *The Chronicle of Higher Education*, 40, April 14, 1993, p. A22. Primary source: American Association of University Professors. **Remarks**: Source also includes: faculty salaries for public, private and church-related colleges and universities at campuses with doctoral programs and master's programs, and at four-year and two-year institutions.

★ 795 ★

By Occupation

Schools That Pay the Highest Professors' Salaries, 1993

School	Average Professor's Salary
Harvard University	$92,200
California Institute of Technology	90,900
Princeton University	86,500
Yale University	86,100
Massachusetts Institute of Technology	83,300
University of Chicago	82,000
Camden College	82,000
Columbia University	82,000
New York University	81,200

Source: Money College Guide 1993, p. 98. **Remarks**: Source also includes: schools with the lowest professors' salaries.

★ 796 ★

By Occupation

Schools That Pay the Lowest Professors' Salaries, 1993

School	Average Professor's Salary
Tougaloo College	$23,700
Bethel College (TN)	27,500
MacMurray College	27,900
Union College (KY)	28,500
Concordia College (MI)	28,800
Tabor College	29,000
Ottawa University	29,200
Lambuth University	29,800
Dana College	30,000
Union College (NE)	30,700

Source: Money College Guide 1993, p. 98. **Remarks**: Source also includes: schools with the largest professors' salaries.

★ 797 ★

By Occupation

Median Salaries for Library Personnel by Sex, 1992

Position	Median salary	
	Men	Women
Manager	$46,150	$42,000
Section head	40,000	38,300
Librarian	35,000	34,434

Source: "Women Gaining in SLA's 11th Biennial Salary Survey," *Library Journal*, February 15, 1993, pp. 110-111. Primary source: Special Libraries Association's Biennial Salary Survey 1993. **Remarks**: Source also includes: median salaries for library personnel by geographic area and age.

★ 798 ★

By Occupation

Median Salaries for Library Personnel by Geographic Area of the United States, 1992

Area	Median salary		
	Manager	Section Head	Librarian
East South Central	$35,300	$34,500	$35,500
Mountain	36,500	35,000	30,000
West South Central	38,000	32,680	30,160
West North Central	39,113	34,600	29,100
East North Central	40,000	37,017	33,000

[Continued]

★ 798 ★

Median Salaries for Library Personnel by Geographic Area of the United States, 1992

[Continued]

Area	Median salary		
	Manager	Section Head	Librarian
South Atlantic	41,141	38,738	34,580
New England	43,150	39,476	33,600
Pacific	46,275	39,000	36,628
Middle Atlantic	47,250	42,000	37,500

Source: "Women Gaining in SLA's 11th Biennial Salary Survey," *Library Journal*, February 15, 1993, pp. 110-111. Primary source: Special Libraries Association's Biennial Salary Survey 1993. **Remarks**: Source also includes: median salaries for library personnel by sex and age.

★ 799 ★

By Occupation

Median Salaries for Library Personnel by Age, 1992

Age	Median Salary
20-29	$29,380
30-39	37,500
40-49	41,451
50-59	41,000
60 & over	42,000

Source: "Women Gaining in SLA's 11th Biennial Salary Survey," *Library Journal*, February 15, 1993, pp. 110-111. Primary source: Special Libraries Association's Biennial Salary Survey 1993. **Remarks**: Source also includes: median salaries for library personnel by sex and geographic area.

★ 800 ★

By Occupation

Starting Salaries for Library Personnel, 1985-1991

Year	Average Beginning Salary	Average Salary Increase
1985	$19,753	$962
1986	20,874	1,121
1987	22,247	1,373
1988	23,491	1,244
1989	24,581	1,090
1990	25,306	725
1991	25,583	277

Source: "Average Salary Index: Starting Library Positions, 1985-1991," *Library Journal*, October 15, 1992, p. 35.

★ 801 ★

By Occupation

Hourly Wages of Nursing Personnel, 1990

Position	Mean Hourly Wage
Head Nurse	
Highest Mean	$19.69
Lowest Mean	15.36
Registered Nurse	
Highest Mean	16.43
Lowest Mean	12.19
Mean Hourly	14.29

Source: U.S. Bureau of the Census. *Statistical Abstract of the United States 1992*. 112th edition. Washington, D.C.: U.S. Department of Commerce, 1992, p. 111. Primary source: American Hospital Association, Chicago, IL, 1990 Human Resource Survey.

★ 802 ★

By Occupation

Physicians' Mean Annual Income, 1982-1989

Specialty	1982	1983	1985	1987	1989
General/Family Practice	$71,400	$66,900	$77,900	$91,500	$95,900
Internal Medicine	86,900	94,600	102,000	121,800	146,500
Surgery	128,600	144,300	155,000	187,900	220,500
Pediatrics	70,500	70,800	76,200	85,300	104,700
Obstetrics/Gynecology	112,300	118,100	124,300	163,200	194,300
All	97,700	104,100	112,200	132,300	155,800

Source: U.S. Bureau of the Census. *Statistical Abstract of the United States 1992*. 112th edition. Washington, D.C.: U.S. Department of Commerce, 1992, p. 110. Primary source: American Hospital Association, Chicago, IL, *Socioeconomic Characteristics of Medical Practice, 1990/1991*.

★ 803 ★

By Occupation

Physicians' Mean Annual Income, 1991

Specialty	Mean Annual Salary
Family Practice	$111,000
Psychiatry	128,000
Internal Medicine	150,000
Pathology	198,000
Anesthesiology	221,000
Obstetrics/Gynecology	222,000
Radiology	230,000
Surgeon	234,000
All	170,000

Source: "Doctors' Income," *The Wall Street Journal*, March 2, 1993, p. A20. Primary source: American Medical Association.

★ 804 ★

By Occupation

Salary of Physicians Employed by Health Maintenance Organizations, 1992

According to a survey by the William M. Mercer Co., the average physician employed by a health maintenance organization or hospital received $139,732 in pay and bonuses in 1992. The average self-employed doctor earned $185,600 after expenses in 1990, according to the American Medical Association.

Source: "Away From Politics," *International Herold Tribune*, February 25, 1993, p. 3.

★ 805 ★

By Occupation

What Americans Think Physicians Earn

According to a nationwide telephone survey by Families U.S.A., a foundation working on health-care issues in Washington, Americans believe that doctors earn far less than they actually do earn. Those surveyed thought that specialists made $100,000 after expenses; in reality the average income for radiologists is $230,000 and for anesthesiologists, $221,000. Those surveyed thought top hospital executives earned $150,000; in reality their average annual income is $235,000 after expenses.

Source: "Public Critical of Pay," Philip J. Hilts, *The New York Times*, April 1, 1993, p. B9.

★ 806 ★

By Occupation

Players' Salaries by Selected Spectator Sport, 1980-1990

Sport	Players' Average Salary				
	1980	1983	1985	1987	1990
Baseball	$144,000	$289,000	$371,000	$412,000	$589,000
Basketball	170,000	249,000	325,000	440,000	817,000
Football	79,000	134,000	194,000	204,000	350,000
Hockey	108,000	130,000	158,000	172,000	254,000

Source: U.S. Bureau of the Census. *Statistical Abstract of the United States 1992*. 112th edition. Washington, D.C.: U.S. Department of Commerce, 1992, p. 239. Primary source: Major League Baseball Players Association, New York, NY; National Basketball Association, New York, NY; National Football League Players Association, New York, NY; National Hockey League Players Association, Toronto, Canada.

★ 807 ★

By Occupation

Baseball Players' Salary, 1969-1991

Year	Average Salary	Series Share
1969	$24,909	$18,338
1991	891,188	119,579

Source: "Fatter Paychecks," *USA Today*, October 20, 1992, p. C1.

★ 808 ★

By Occupation

Signing Bonus for Baseball Players, 1990-1992

For baseball players picked in the first round of draft, the average bonus was $243,000 in 1990, $355,000 in 1991, and $477,000 in 1992.

Source: "Big Bonus for Being First," *USA Today*, September 28, 1992, p. C1. Primary source: Baseball America.

★ 809 ★

By Occupation

Starting Salaries for Football Players by Position Played, 1990

Position	Average Starting Salary
Punter	$233,000
Placekicker	252,000
Tight End	345,000
Offensive Lineman	387,000
Defensive Back	395,000
Defensive Lineman	444,000
Linebacker	494,000
Wide Receiver	494,000
Running Back	667,000
Quarterback	1,250,000
All	460,000

Source: Krantz, Les. *The Best and Worst of Everything.* New York: Prentice Hall General Reference, 1991, p. 30. Primary source: Inside Sports, 1991.

★ 810 ★

By Occupation

Salaries of the Highest Paid Baseball Teams

Team	Average Salary
Oakland	$806,554
Boston	777,683
N.Y. Mets	758,575
N.Y. Yankees	725,872
California	695,070
Kansas City	692,973
Los Angeles	685,780
Milwaukee	678,581
San Francisco	666,927
St. Louis	636,794

Source: Krantz, Les. *The Best and Worst of Everything.* New York: Prentice Hall General Reference, 1991, p. 12. Primary source: Major League Baseball Players Association.

★ 811 ★

By Occupation

Salaries of the Lowest Paid Baseball Teams

Team	Average Salary
Baltimore	$279,326
Seattle	388,649
Atlanta	414,443
White Sox	422,199
Philadelphia	461,484
Texas	481,290
Minnesota	495,270
Cleveland	508,756
Cubs	518,050
Pittsburgh	592,390

Source: Krantz, Les. *The Best and Worst of Everything.* New York: Prentice Hall General Reference, 1991, p. 12. Primary source: Major League Baseball Players Association.

★ 812 ★

By Occupation

Annual Total Compensation in Selected Industries, 1980-1990

In dollars.

Industry	Annual Total Compensation				
	1980	1985	1988	1989	1990
Domestic industries	18,815	25,266	28,736	29,702	31,101
Agriculture, forestry, and fisheries	9,836	12,833	14,086	15,249	17,104
Mining	28,181	38,367	42,794	44,443	46,514
Construction	22,033	27,581	30,466	31,465	32,836
Manufacturing	22,023	29,999	33,765	34,937	36,503
Transportation	25,029	30,895	33,004	33,717	35,197
Communication	28,341	39,980	44,197	45,905	47,393
Electric, gas, and sanitary services	27,358	39,429	44,005	46,343	47,822
Wholesale trade	21,598	28,853	34,048	35,138	36,857
Retail trade	12,490	15,625	17,443	17,886	18,462
Finance, insurance, and real estate	18,968	28,014	34,443	35,327	36,679
Services	15,236	21,395	25,309	26,271	27,791
Government and government enterprises	19,681	27,821	32,044	33,548	35,334

Source: U.S. Bureau of the Census. *Statistical Abstract of the United States 1992.* 112th edition. Washington, D.C.: U.S. Department of Commerce, 1992, p. 410. Primary source: U.S. Bureau of Economic Analysis, *The National Income and Product Accounts of the United States,* Vol. 2, 1958-1988, and *Survey of Current Business,* July issues. **Remarks**: Source also includes: annual wages and salaries, 1980-1990, for the industries listed above. *Notes:* Wage and salary payments include executives' compensation, bonuses, tips, and payments-in-kind; total compensation includes, in addition to wages and salaries, employer contributions for social insurance, employer contributions to private pension and welfare funds, director fees, jury and witness fees, etc.

★ 813 ★

By Occupation

Hourly Earnings in Selected Industries, 1970-1990

Private Industry Group	Current Dollars						Constant (1982) Dollars					
	1970	1980	1985	1989	1990	1991	1970	1980	1985	1989	1990	1991
Average hourly earnings	3.23	6.66	8.57	9.66	10.02	10.34	8.03	7.78	7.77	7.64	7.53	7.46
Manufacturing	3.35	7.27	9.54	10.48	10.83	11.18	8.33	8.49	8.65	8.28	8.14	8.07
Mining	3.85	9.17	11.98	13.26	13.69	14.21	9.58	10.71	10.86	10.48	10.29	10.25
Construction	5.24	9.94	12.32	13.54	13.78	14.01	13.03	11.61	11.17	10.70	10.35	10.11
Transportation, public utilities	3.85	8.87	11.40	12.60	12.96	13.23	9.58	10.36	10.34	9.96	9.74	9.55
Wholesale trade	3.43	6.95	9.15	10.39	10.79	11.16	8.53	8.12	8.30	8.21	8.11	8.05
Retail trade	2.44	4.88	5.94	6.53	6.76	7.00	6.07	5.70	5.39	5.16	5.08	5.05
Finance, insurance, real estate	3.07	5.79	7.94	9.53	9.97	10.42	7.64	6.76	7.20	7.53	7.49	7.52
Services	2.81	5.85	7.90	9.38	9.83	10.24	6.99	6.83	7.16	7.42	7.39	7.39

Source: U.S. Bureau of the Census. *Statistical Abstract of the United States 1992*. 112th edition. Washington, D.C.: U.S. Department of Commerce, 1992, p. 410. Primary source: U.S. Bureau of Labor Statistics, Bulletin 2370, supplement to *Employment and Earnings*, July 1991, and *Employment and Earnings*, monthly.

★ 814 ★

By Occupation

Weekly Earnings in Selected Industries, 1970-1990

	Current Dollars						Constant (1982) Dollars					
	1970	1980	1985	1989	1990	1991	1970	1980	1985	1989	1990	1991
Average weekly earnings	120	235	299	334	346	355	298	275	271	264	260	256
Manufacturing	133	289	386	430	442	455	332	337	350	340	332	328
Mining	164	397	520	570	604	631	409	464	471	451	454	455
Construction	195	368	464	513	526	534	486	430	421	406	395	385
Transportation, public utilities	156	351	450	490	504	511	388	410	408	387	379	368
Wholesale trade	137	267	351	395	411	425	341	312	318	312	309	307
Retail trade	82	147	175	189	195	200	205	172	158	149	146	144
Finance, insurance, real estate	113	210	289	341	357	373	281	245	262	270	268	269
Services	97	191	257	306	320	333	240	223	233	242	241	240

Source: U.S. Bureau of the Census. *Statistical Abstract of the United States 1992*. 112th edition. Washington, D.C.: U.S. Department of Commerce, 1992, p. 410. Primary source: U.S. Bureau of Labor Statistics, Bulletin 2370, supplement to *Employment and Earnings*, July 1991, and *Employment and Earnings*, monthly.

★ 815 ★

By Occupation

Hourly Wages in Timber-Based Industries, 1970-1991

	Average hourly earnings: (dollars)					
	1970	1980	1985	1989	1990	1991
Lumber and wood products	2.97	6.57	8.25	8.84	9.09	9.28
Logging	---	8.64	10.92	11.13	11.23	11.15
Sawmills and planning mills	2.84	6.70	8.52	9.03	9.23	9.42
Millwork, plywood, and structural members	---	6.44	8.09	8.73	9.05	9.30
Paper and allied products	3.44	7.84	10.83	11.96	12.30	12.70
Furniture and fixtures	2.77	5.49	7.17	8.25	8.52	8.77

Source: U.S. Bureau of the Census. *Statistical Abstract of the United States 1992*. 112th edition. Washington, D.C.: U.S. Department of Commerce, 1992, p. 673. Primary source: U.S. Bureau of Labor Statistics, Bulletin 2370, supplement to *Employment and Earnings*, July 1991, and *Employment and Earnings*, monthly. **Remarks**: Also included in source: number of employees in the above-named industries.

★ 816 ★

By Occupation

Starting Salaries in Local Police Departments, 1990

Population served	Average base starting salary		
	Entry-level officer	Sergeant	Chief of police
All sizes	$18,910	$25,420	$30,240
1,000,000 or more	26,500	40,420	85,320
500,000 to 999,999	25,110	33,900	67,300
250,000 to 499,999	25,370	35,120	62,770
100,000 to 249,999	24,960	34,330	59,190
50,000 to 99,999	23,300	32,140	51,600
25,000 to 49,999	23,150	31,200	46,400
10,000 to 24,999	21,630	28,950	40,930
2,500 to 9,999	18,710	24,110	29,960
Under 2,500	18,870	19,960	20,610

Source: U.S. Department of Justice. Bureau of Justice Statistics. *Sourcebook of Criminal Justice Statistics 1991*. Washington, D.C.: U.S. Department of Justice, 1992, p. 44. Primary source: U.S. Department of Justice, Bureau of Justice Statistics, *State and Local Police Departments, 1990*, Bulletin NCJ-133284 (Washington, DC: U.S. Department of Justice, February 1992) p. 6, Table 14.

★817★

By Occupation

Starting Salaries in Sheriffs' Departments, 1990

Population served	Average base starting salary		
	Entry-level deputy	Sergeant	Sheriff
All sizes	$17,420	$21,870	$33,530
1,000,000 or more	26,180	35,530	80,350
500,000 to 999,999	22,950	32,100	61,490
250,000 to 499,999	20,130	26,860	53,710
100,000 to 249,999	19,530	24,830	44,900
50,000 to 99,999	18,300	23,100	38,460
25,000 to 49,999	16,950	21,510	33,050
10,000 to 24,999	16,940	20,840	30,320
Under 10,000	15,860	18,560	24,530

Source: U.S. Department of Justice. Bureau of Justice Statistics. *Sourcebook of Criminal Justice Statistics 1991*. Washington, D.C.: U.S. Department of Justice, 1992, p. 44. Primary source: U.S. Department of Justice, Bureau of Justice Statistics, *Sheriffs' Departments, 1990*, Bulletin NCJ-133283 (Washington, DC: U.S. Department of Justice, February 1992) p. 7, Table 14.

★818★

By Occupation

Salaries of City Chiefs of Police, 1991

	Salary levels	
	Mean	Median
	$42,414	$39,648
Region		
Northeast	46,665	45,562
North Central	39,256	37,008
South	36,254	33,212
West	54,725	50,728
City type		
Central	60,076	58,164
Suburban	46,450	44,696
Independent	32,985	31,354
Form of government		
Mayor-council	38,006	35,002
Council-manager	46,861	43,514
Commission	41,858	40,376
Town meeting	46,095	46,640
Representative town meeting	47,306	45,294

Source: U.S. Department of Justice. Bureau of Justice Statistics. *Sourcebook of Criminal Justice Statistics 1991*. Washington, D.C.: U.S. Department of Justice, 1992, p. 59.

★ 819 ★

By Occupation

Median Weekly Earnings for Managerial and Professional Occupations by Sex, 1990-1991

	Median Weekly Earnings			
	1990		1991	
	Men	Women	Men	Women
Executive, administrative, and managerial	$722	$484	$737	$495
Professional specialty	712	525	744	548
All	717	507	741	519

Source: The World Almanac and Book of Facts 1992. New York: World Almanac, 1991, p. 175. Primary source: U.S. Dept. of Labor, Bureau of Labor Statistics. **Remarks**: Also included in source: median weekly earnings in technical, sales, and administrative support; service occupations; precision production, craft, and repair; operators, fabricators, and laborers; and farming, forestry, and fishing.

★ 820 ★

By Occupation

Median Weekly Earnings for Technical, Sales, and Administrative Support Personnel by Sex, 1990-1991

	Median Weekly Earnings			
	1990		1991	
	Men	Women	Men	Women
Technicians and related support	$561	$426	$564	$442
Sales occupations	506	284	499	311
Administrative support, including clerical	438	332	456	349
All	494	329	496	351

Source: The World Almanac and Book of Facts 1992. New York: World Almanac, 1991, p. 175. Primary source: U.S. Dept. of Labor, Bureau of Labor Statistics. **Remarks**: Also included in source: median weekly earnings in managerial and professional specialty; service occupations; precision production, craft, and repair; operators, fabricators, and laborers; and farming, forestry, and fishing.

★ 821 ★

By Occupation

Median Weekly Earnings for Service Occupations by Sex, 1990-1991

	Median Weekly Earnings			
	1990		1991	
	Men	Women	Men	Women
Private household	-	$161	-	$164
Protective services	$454	369	$494	436
Service, other	273	231	279	244
All	317	231	320	243

Source: The World Almanac and Book of Facts 1992. New York: World Almanac, 1991, p. 175. Primary source: U.S. Dept. of Labor, Bureau of Labor Statistics. **Remarks**: Also included in source: median weekly earnings in managerial and professional specialty; technical, sales, and administrative support; precision production, craft, and repair; operators, fabricators, and laborers; and farming, forestry, and fishing.

★ 822 ★

By Occupation

Median Weekly Earnings for Precision Production, Craft, and Repair Occupations by Sex, 1990-1991

	Median Weekly Earnings			
	1990		1991	
	Men	Women	Men	Women
Mechanics & repairers	$477	$476	$472	$541
Construction trades	486	-	478	-
Other precision, production craft, and repair	503	295	517	320
All	487	319	488	354

Source: The World Almanac and Book of Facts 1992. New York: World Almanac, 1991, p. 175. Primary source: U.S. Dept. of Labor, Bureau of Labor Statistics. **Remarks**: Also included in source: median weekly earnings in managerial and professional specialty; technical, sales, and administrative support; service occupations; operators, fabricators, and laborers; and farming, forestry, and fishing.

★ 823 ★

By Occupation

Median Weekly Earnings for Operators, Fabricators, and Laborers by Sex, 1990-1991

	Median Weekly Earnings			
	1990		1991	
	Men	Women	Men	Women
Machine operators, assemblers, inspectors	$391	$261	$403	$272
Transportation & material moving	426	296	421	320
Handlers, equipment cleaners, helpers, laborers	306	249	316	269
All	379	262	391	275

Source: The World Almanac and Book of Facts 1992. New York: World Almanac, 1991, p. 175. Primary source: U.S. Dept. of Labor, Bureau of Labor Statistics. **Remarks**: Also included in source: median weekly earnings in managerial and professional specialty; technical, sales, and administrative support; service occupations; precision production, craft, and repair; and farming, forestry, and fishing.

★ 824 ★

By Occupation

Median Weekly Earnings for Farming, Forestry, and Fishing by Sex, 1990-1991

	Median Weekly Earnings			
	1990		1991	
	Men	Women	Men	Women
Framing, forestry, fishing	$262	$214	$269	$220

Source: The World Almanac and Book of Facts 1992. New York: World Almanac, 1991, p. 175. Primary source: U.S. Dept. of Labor, Bureau of Labor Statistics. **Remarks**: Also included in source: median weekly earnings in managerial and professional specialty; technical, sales, and administrative support; service occupations; precision production, craft, and repair; operators, fabricators,and laborers.

★ 825 ★

By Occupation

Salary of Household Domestics by Sex

According to an analysis of U.S. Census data by the *Los Angeles Times*, men who clean houses earn twice as much as the average cleaning woman, whose annual salary is $10,251 in 1989. Child care was one of the few occupations in which women earned more than men, but not by much. Women's average salary of $411 was just $7.90 a week more than men earned.

Source: Los Angeles Public Library/State of California, *Scan/Info*, 5, January 1993, p. 23. Primary source: *Los Angeles Times*, December 29, 1992, p. A1.

★ 826 ★

By Occupation

Occupations in Which Women are the Majority, but Earn Less Than Men

Occupation	Average Salary in 1989	
	Men	Women
Child Care	$22,413	$17,428
Domestics	19,015	10,287
Health/lab technicians	31,682	25,783
Nurses	42,878	35,580
Orderlies/health aides	22,164	17,605
Secretaries	27,540	21,837
Social workers	30,834	26,250
Teachers	38,724	28,465

Source: Los Angeles Public Library/State of California, *Scan/Info*, 5, January 1993, p. 24. Primary source: *Los Angeles Times*, December 29, 1992, p. A1. Analysis of U.S. Census data by *Los Angeles Times*.

★ 827 ★

By Occupation

High-Paying Outdoor Careers

Career	Average Annual Income
Major League Baseball Player	$424,896
National Football League Player	209,090
Major League Umpire	52,560
Geologist	41,420
Oceanographer	38,695
Line Installer	31,174
Surveyor	30,210
Electrician	29,474
Conservationist	28,196
Jockey	27,490

Source: Krantz, Les. *The Best and Worst of Everything.* New York: Prentice Hall General Reference, 1991, pp. 285-286. Primary source: *Jobs Rated Almanac.* **Remarks**: Source also includes: High-paying professional careers.

★ 828 ★

By Occupation

Low Paying Jobs

Job	Average Weekly Salary
Child Care workers	$127
Food counter worker	167
Cleaners and servants	185
Kitchen workers	191
Food preparation workers	195
Waiter's assistants	199
Cashiers	202
Waiters & waitresses	203
Textile and sewing machine operators	205

Source: Krantz, Les. *The Best and Worst of Everything*. New York: Prentice Hall General Reference, 1991, p. 284. Primary source: *Occupational Outlook Quarterly*, Fall 1990. **Remarks**: Source also includes: high paying jobs.

★ 829 ★

By Occupation

Tobacco Farming in the United States

Tobacco is the United States' most profitable crop partly due to a government price-support program. The Department of Agriculture reported that in 1991 farmers earned a gross average income of $3,862 for every acre of tobacco planted; but only $691 for peanuts, $380 for cotton, $262 for feed corn, and $101 for wheat.

Source: "Tobacco Country Quaking Over Cigarette Tax Increase," Ronald Smothers, *The New York Times*, March 22, 1993, p. A1.

★ 830 ★

By Occupation

Afghan Opium Farmers' Earnings

In a country where the average per capita income is only $70, farmers in Afghanistan can earn an average of $200 per year from an opium harvest. This is ten times what he could earn from wheat. An Afghan farmer receives an average of $22 per pound for raw opium, but by the time it reaches the United States, it sells for $40,000 to $100,000 per pound.

Source: "After War, a Deadly Harvest," Klaus Reisinger, *U.S. News & World Report*, 113, August 17, 1992.

★ 831 ★

By Occupation

Wages of Migrant Workers

Western Michigan is one of the leading employers of migrant workers in the United States. According to Michigan government officials, a migrant family of four earned an average of $7,200 in 1991. This is a little over one-half of the $13,924 a family needs to escape poverty.

Source: "Tide of Migrant Labor Tells of a Law's Failure," Peter Kilborn, *The New York Times*, November 4, 1992, p. A24.

★ 832 ★

By Occupation

Expert Witness Fees

Specialty	Average Fee		
	For Research	For Deposition	To Testify
Accountant	$175	$200	$200
Antique Appraiser	800	-	1,600
Actuary	1,500	-	2,300
Computer Scientist	125	150	125
Economist	150	190	200
Environmental toxicologist	150	200	200
Mechanical engineer	135	165	165

Source: "Opinions with a Price," Anne Kates Smith, *U.S. News & World Report*, 113, July 20, 1992, p. 64. Primary source: National Forensic Center, "Guide to Experts' Fees 1992-1993."

★ 833 ★

By Occupation

Jobs with a Future in the 1990s and Beyond

Occupation	Average Salary	
	1991	2000
Auto mechanic	$26,460	$36,000
Carpenter	23,152	34,650
Chef	13,230	26,000
Cook	40,000	60,000
Corporate personnel trainer	30,870	50,000
Cosmetologist	13,325	18,000
Environmental scientist	39,060	68,200
Flight engineer	46,305	65,000
Hotel manager	55,252	80,000
Human resources manager	81,585	122,160
Insurance claims examiner	22,176	32,260
Insurance salesman	54,000	86,400
Interior designer	31,972	47,000

[Continued]

★ 833 ★

Jobs with a Future in the 1990s and Beyond

[Continued]

Occupation	Average Salary	
	1991	2000
Labor relations specialist	40,091	61,810
Photographer	25,357	35,882
Pilot	92,610	125,000
Public relations specialist	31,972	50,000
Radio/TV news reporter (starting)	28,972	44,000
Restaurant manager	40,792	65,000
Travel agent	21,110	27,000

Source: Kleiman, Carol. *The 100 Best Jobs for the 1990s & Beyond.* Chicago: Dearborn Financial Publishing, Inc., 1992. **Remarks:** Also included in source: Detailed job descriptions for 100 jobs that have a future in the 1990s and beyond.

★ 834 ★

By Occupation

Salary Ranges in "Up and Coming" Professions

Profession	Average Salary		
	Entry	Midlevel	Top
Environmental Accountant	$30,000-50,000	$50,000-80,000	$125,000
Network Administrator	19,000	36,000	71,000
Outplacement Consultant	50,000	85,000	125,000
Special-Education Teacher	21,500	32,800	50,000
Civil Engineer	36,000	57,000	105,000
Toxicologist	40,000-50,000	50,000-65,000	70,000-100,000
Investment Professional	20,000-30,000	40,000-60,000	100,000
Restaurant Site Selector	30,000	50,000	80,000-95,000
Nurse Practitioner	33,000	42,000	50,000
Training Manager	50,000	64,000	78,000
Actuary	25,000-30,000	45,000-65,000	66,000-100,000
Intellectual Property Lawyer	70,000-90,000	100,000-150,000	225,000-425,000
Chief Information Officer	33,000-57,000	58,000-73,000	74,000-300,000
Family Physician	90,000	103,000	120,000
Member Services Director	18,000-20,000	32,000-38,000	45,000-60,000
Technical Administrative Assistant	16,000-20,000	26,000-32,000	50,000
Electronic Publishing Specialists	28,000-34,000	40,000-60,000	75,000-
Merchandise Manager	30,000-40,000	45,000-80,000	100,000-150,000
Computation Chemist	40,000	50,000-60,000	100,000
Wireless Specialist	45,000	60,000	100,000

Source: "Hot Tracks in 20 Professions," *U.S. News & World Report*, 113, October 26, 1992, pp. 104-110.

★ 835 ★

By Occupation

1992 Median Salaries in Hotel/Motel Administration

Position	Median Salary
General Manager	$55,000
Director, Sales & Marketing	40,000
Controller	37,000
Human Resources Manager	37,000
Executive Housekeeper	35,000

Source: "1993 $," *Working Woman* January 1993, p. 41. Primary source: Roth Young Personnel Service of Minneapolis, Inc., Executive Recruiters.

★ 836 ★

By Occupation

1992 Median Salaries in Computing

Position	Median Salary
Sales Manager	$71,000
MIS Director	68,000
Programming Development Manager	61,000
Systems Development Manager	58,000
Database Mgt. Analyst*	40,000
Systems Software Programmer*	39,000
Software Engineer*	38,000

Source: "1993 $," *Working Woman* January 1993, p. 41. Primary source: *1992 Computer Salary Survey and Career Planning Guide*, Source Service. *Note:* * With 2-3 yrs. experience.

★ 837 ★

By Occupation

1992 Average Salaries in Advertising

Position	Average Salary
Chief Executive Officer	$135,000
Account Manager	50,000
Copywriter	45,000
Art Director	40,500
Account Executive	38,500
Media Planner/Buyer	29,000

[Continued]

★ 837 ★

1992 Average Salaries in Advertising

[Continued]

Position	Average Salary
Marketing Manager	52,000
Advertising Manager	48,000

Source: "1993 $," *Working Woman* January 1993, p. 40. Primary source: Ogilvy & Mather and Jerry Fields Associates.

★ 838 ★

By Occupation

1992 Average Salaries in Banking

Position	Average Salary
Head of Lending	$190,000-200,000
Commercial Lender (3+ yrs.)	52,000-77,000
Commercial Real Estate Mortgage Lender	48,000-68,000
Commercial Lender	39,000-51,000
Operations Officer	36,000-46,000
Corporate Trust Officer	36,000-46,000
Branch Manager	34,000-42,000
Residential Real Estate Mortgage Lender	31,000-41,000

Source: "1993 $," *Working Woman* January 1993, p. 40. Primary source: *Robert Half and Accountemps Salary Guide 1992.*

★ 839 ★

By Occupation

1992 Average Salaries in Insurance

Position	Average Salary
Managing Actuary	$88,000
Manager, Auditing	62,237
Account Manager	56,685
Senior Auditor	40,842
Supervisor, Life Claims	39,242
Supervisor, Group Health Claims	33,566
Auditor	33,311

Source: "1993 $," *Working Woman* January 1993, p. 42. Primary source: 1992 LOMA (Life Office Management Association) Managerial, Professional and Technical Compensation Survey.

★ 840 ★

By Occupation

1992 Average Compensation in Management Consulting

Position	Average Compensation (includes salary + bonus)
Compensation & Benefits	$169,233
Strategic Planning	146,976
Production Management	113,750
Multiple Specialties Firm	106,571
MIS Consulting	104,000
Health Care	102,014
Human Resources	86,176
Marketing	81,000

Source: "1993 $," *Working Woman* January 1993, p. 42. Primary source: ACME Survey of US Key Management Information.

★ 841 ★

By Occupation

1992 Salaries in Newspaper Publishing

Position	Average salary
Publisher	$116,396
Editor	71,396
Financial Executive	63,427
Operations Executive	62,248
Advertising Executive	61,096
Managing Editor	54,953
Circulation Executive	53,005
Reporter (1-4 yrs.)	30,816

Source: "1993 $," *Working Woman* January 1993, p. 42. Primary source: Newspaper Association of America.

★ 842 ★

By Occupation

1992 Median Salaries in Public Libraries

Position	Median Salary	
	Medium-sized	Large
Director	$43,000	$64,914
Assistant Director	36,340	51,551
Branch	32,323	40,284
Cataloger	28,343	32,571

[Continued]

★ 842 ★

1992 Median Salaries in Public Libraries

[Continued]

Position	Median Salary	
	Medium-sized	Large
Children's	27,851	34,091
Reference	27,839	32,197
Entry-Level	22,980	25,680

Source: "1993 $," *Working Woman* January 1993, p. 42. ALA (American Library Association) Survey of Librarian Salaries, 1992.

★ 843 ★

By Occupation

Security Guards

The average security guard, working for one of 10,000 security companies in the United States, earns $5 to $7 per hour. Time spent in weapons training averaged a scant eight hours. The number of these underpaid and undertrained guards has grown since 1980 to almost twice the size of the public law-enforcement community. In 1971, a report by Rand Corp. stated that the average security guard was an aging white male, who was underpaid, undertrained, undersupervised, and underscreened. According to Hallcrest Systems of McLean, VA., this situation is the same today, except that he is younger, and black or Hispanic.

Source: "Thugs in Uniform," Richard Behar, *Time*, 140, March 9, 1992., pp. 44-47.

★ 844 ★

By Occupation

Railroad Employees' Vs. Truckers' Annual Income

The average unionized railroad employee earns $56,000 per year, including fringe benefits. This is more than any other industry earns, except investment banking, stock brokerage, and oil. For train crews, usually an engineer, a conductor, and one or two brakemen, a workday is a 108-mile trip. Since the average train covers 175 miles in an eight-hour period, railroad officials are forced to pay a 70% premium for the extra miles. For every dollar in wages, railroads pay 40 cents in fringe benefits. On the other hand, truckers earn an average of $30,000 per year, and receive only 15 cents in fringe benefits for every dollar in wages.

Source: "Competition: Comeback Ahead for Railroads," *Fortune*, 123, June 17, 1991, p. 108.

By Race/Ethnicity

★ 845 ★

Household Income of Native Americans Living on Reservations

According to figures from the United States government, Indians living on reservations have an average household income of $20,025. The government spends $3,500 per person on programs for these Indians. Compare this to the $500 per capita income of the Mapuche Indians in Peru (government help to Indian groups is almost nonexistent in Latin America).

Source: "Latin American Indians: Old Ills, New Politics," Nathaniel C. Nash, *The New York Times*, August 24, 1992, p. A6. **Remarks**: Source also includes: short summaries of selected Indian groups in Latin America.

★ 846 ★

By Race/Ethnicity

Annual Income of Black Families in the Suburbs, 1990

According to the 1990 Census, Black families living in the suburbs have an average yearly income of $32,000, which is 55% higher than that of Black families living in the central cities.

Source: "The Most Blacks," *American Demographics*, 14, September 1992, p. 32. Primary source: 1990 Census. **Remarks**: Source also includes: number of Black residents living in 40 suburban areas of the United States and the average number living in these 40 areas.

★ 847 ★

By Race/Ethnicity

Median Household Income by Race/Ethnicity, 1970-1989

Households as of March of following year. Based on Current Population Survey.

Year	Median income in current dollars (dol.)				Median income in constant (1989) dollars (dol.)			
	All households[1]	White	Black	Hispanic[2]	All households[1]	White	Black	Hispanic[2]
1970	8,734	9,097	5,537	(NA)	27,913	29,073	17,696	(NA)
1975	11,800	12,340	7,408	8,865	27,197	28,442	17,074	20,432
1976	12,686	13,289	7,902	9,569	27,646	28,960	17,221	20,853
1977	13,572	14,272	8,422	10,647	27,771	29,203	17,233	21,786
1978	15,064	15,660	9,411	11,803	28,649	29,783	17,898	22,447
1979	16,461	17,259	10,133	13,042	28,115	29,478	17,307	22,276
1980	17,710	18,684	10,764	13,651	26,651	28,117	16,198	20,543
1981	19,074	20,153	11,309	15,300	26,020	27,491	15,427	20,871
1982	20,171	21,117	11,968	15,178	25,919	27,135	15,379	19,503
1983[3]	21,018	22,035	12,473	15,794	26,167	27,433	15,529	19,663
1984	22,415	23,647	13,471	16,992	26,751	28,222	16,077	20,279
1985	23,618	24,908	14,819	17,465	27,218	28,704	17,078	20,127

[Continued]

★ 847 ★

Median Household Income by Race/Ethnicity, 1970-1989

[Continued]

Year	Median income in current dollars (dol.)				Median income in constant (1989) dollars (dol.)			
	All households[1]	White	Black	Hispanic[2]	All households[1]	White	Black	Hispanic[2]
1986	24,897	26,175	15,080	18,352	28,168	29,614	17,061	20,763
1987[4]	26,061	27,458	15,672	19,336	28,447	29,972	17,107	21,106
1988	27,225	28,781	16,407	20,359	28,537	30,168	17,198	21,340
1989	28,906	30,406	18,083	21,921	28,906	30,406	18,083	21,921

Source: "Money Income of Households—Median Household Income in Current and Constant (1989) Dollars, by Race and Hispanic Origin: 1970 to 1989," *Statistical Abstract of the United States*, 1991, p. 449. Primary source: U.S. Bureau of the Census, *Current Population Reports*, series P-60, No. 168, and unpublished data. *Notes:* NA stands for not available. 1. Includes other races not shown separately. 2. Hispanic persons may be of any race. 3. Beginning 1983, data based on revised Hispanic population controls; data not directly comparable to prior years. 4. Beginning 1987, based on revised processing procedures; data not directly comparable with prior years.

★ 848 ★

By Race/Ethnicity

Median Household Income by Race and Age

	Median income	
	Black	White
Age of householder		
Under 65	$21,011	$35,646
15-24	9,816	19,662
25-34	18,339	31,859
35-44	26,011	40,423
45-54	26,910	44,098
55-64	19,226	34,249
65 and over	9,902	17,539
65-74	11,974	21,089
75+	7,831	13,714

Source: "Percent of Households and Median Income of Households by Selected Characteristics and Race," *The State of Black America 1992*, 1992, p. 85. Primary source: U.S. Department of Commerce, Bureau of the Census, *Money Income of Households, Families, and Persons in the United States: 1990*, September 1991, Table 1. **Remarks**: Source also includes: median household income by race/ethnicity and size of household.

★ 849 ★

By Race/Ethnicity

Median Household Income by Race and Size

	Median income	
	Black	White
Number of persons in household		
One	$10,156	$15,981
Two	20,122	32,561
Three	21,474	38,930
Four	25,683	43,363
Five	24,342	40,715
Six	26,742	40,420
Seven or more	22,361	40,822

Source: "Percent of Households and Median Income of Households by Selected Characteristics and Race," *The State of Black America 1992*, 1992, p. 85. Primary source: U.S. Department of Commerce, Bureau of the Census, *Money Income of Households, Families, and Persons in the United States: 1990*, September 1991, Table 1. **Remarks**: Source also includes: median household income by race/ethnicity and age of household.

★ 850 ★

By Race/Ethnicity

Median Household Income by Race and Number of Earners, 1990

	Median income	
	Black	White
All households	$18,676	$31,231
Number of earners		
No earners	5,870	12,395
One earner	17,040	25,801
Two earners or more	36,404	45,705
Two earners	33,657	42,498
Three earners	42,897	54,264
Four earners or more	60,323	66,876

Source: "Percent of Households and Median Income of Households by Selected Characteristics and Race," *The State of Black America 1992*, 1992, p. 85. Primary source: U.S. Department of Commerce, Bureau of the Census, *Money Income of Households, Families, and Persons in the United States: 1990*, September 1991, Table 1.

★ 851 ★

By Race/Ethnicity

Median Household Income by Race and Type of Residence, 1990

	Median income	
	Black	White
Type of residence		
Nonfarm residence	$18,734	$31,216
Inside metro areas	20,121	33,460
Inside metro areas-large	21,086	35,837
Inside central cities	18,156	29,630
Outside central cities	28,444	39,670
Inside metro areas-small	17,562	30,043
Inside central cities	16,402	26,845
Outside central cities	21,517	31,881
Outside metro areas	13,119	24,887

Source: "Percent of Households and Median Income of Households by Selected Characteristics and Race," *The State of Black America 1992*, 1992, p. 85. Primary source: U.S. Department of Commerce, Bureau of the Census, *Money Income of Households, Families, and Persons in the United States: 1990*, September 1991, Table 1. **Remarks**: Source also includes: household income by race and type of household.

★ 852 ★

By Race/Ethnicity

Median Household Income by Race and Type of Household, 1990

	Median income	
	Black	White
Type of household		
Family households	$21,899	$37,219
Married couple family	33,893	40,433
Single-male headed	24,048	32,869
Single-female headed	12,537	20,867
Nonfamily households	11,789	18,449
Male householder nonfarm	15,451	23,778
Female householder nonfarm	8,661	14,629

Source: "Percent of Households and Median Income of Households by Selected Characteristics and Race," *The State of Black America 1992*, 1992, p. 85. Primary source: U.S. Department of Commerce, Bureau of the Census, *Money Income of Households, Families, and Persons in the United States: 1990*, September 1991, Table 1. **Remarks**: Source also includes: household income by race and type of residence.

★ 853 ★

By Race/Ethnicity

Worker's per Capita Annual Income, 1970-1990

1990 dollars.

	Per capita income	
	Black	White
1990	$9,017	$15,265
1989	9,220	15,701
1988	9,138	15,353
1987	8,796	15,121
1986	8,594	14,730
1982	7,260	12,903
1980	7,620	13,059
1978	8,087	13,625
1974	7,636	12,399
1972	7,470	12,407
1970	6,296	11,298

Source: "Per Capita Income, Aggregate Income, and Income Gap Selected Years," *The State of Black America 1992*, 1992, p. 75. Primary source: U.S. Department of Commerce, Bureau of the Census, *Money Income of Households, Families, and Persons in the U.S.: 1990*, September 1991, Series P-60, No. 174, Table B-8.

★ 854 ★

By Race/Ethnicity

Worker's per Capita Weekly Income, 1979-1990

	Black	White
1990	$329	$427
1989	336	431
1988	331	415
1987	317	404
1986	318	404
1985	308	394
1984	306	391
1983	307	383
1982	304	377
1981	307	381
1980	298	381
1979	328	406

Source: "Median Weekly Earnings of Full-Time Wage and Salary Workers, by Race and Sex, 1979-1990," *The State of Black America 1992*, 1992, p. 111. Primary source: Bureau of Labor Statistics, *Handbook of Labor Statistics*, June 1985, p. 94; *Employment and Earnings*, January 1986-1991.

★ 855 ★

By Race/Ethnicity

Median Family Income by Race/Ethnicity, 1990

	1990 Median income (dols.)	1989 median income (in 1990 dols.)
Families		
All families	35,353	36,062
Race and Hispanic origin of householder		
White	36,915	37,919
Black	21,423	21,301
Hispanic origin[1]	23,431	24,713
Type of family		
All races		
Married-couple families	39,895	40,630
Female householder, no husband present	16,932	17,330
White		
Married-couple families	40,331	41,326
Female householder, no husband present	19,528	19,970
Black		
Married-couple families	33,784	32,306
Female householder, no husband present	12,125	12,258
Hispanic origin[1]		
Married-couple families	27,996	28,862
Female householder, no husband present	11,914	12,380

Source: "Comparison of Income Summary Measures Between 1990 and 1989, by Selected Characteristics," *Money Income of Households, Families, and Persons in the United States: 1990*, 1991, p. 3. Primary source: U.S. Department of Commerce, Economics and Statistics Administration, Bureau of the Census, Current Population Reports: Consumer Income, Series P-60, No. 174, 1991. **Remarks**: Source also includes: median household income by race and per capita income by race. *Note:* 1. Persons of Hispanic origin may be of any race.

★ 856 ★

By Race/Ethnicity

Per Capita Income for Hispanics in Selected Cities, 1989

Metropolitan area	Per capita income for Hispanic suburban residents	Per capita income for Hispanic central-city residents
Newark, NJ	$13,673	$8,821
Washington, DC-MD-VA	13,365	12,038
Oakland, CA	12,636	9,059
Miami-Hialeah, FL	12,441	8,910
New York, NY	12,084	8,430
Philadelphia, PA-NJ	11,984	6,066
San Francisco, CA	11,719	11,400

[Continued]

★ 856 ★

Per Capita Income for Hispanics in Selected Cities, 1989

[Continued]

Metropolitan area	Per capita income for Hispanic suburban residents	Per capita income for Hispanic central-city residents
Boston, MA	11,127	7,602
Chicago, IL	10,817	7,464
Denver, CO	10,730	7,778
Sacramento, CA	10,195	9,691
Las Vegas, NV	9,730	8,879
Orlando, FL	9,442	9,528
Dallas, TX	9,012	7,214
Los Angeles-Long Beach, CA	8,832	7,241
San Diego, CA	8,786	8,159
Houston, TX	8,603	7,011
San Antonio, TX	8,496	7,032
Phoenix, AZ	7,311	7,721
El Paso, TX	4,429	6,538

Source: "Suburban Hispanic Incomes," *American Demographics*, 15, April 1993, p. 35. **Remarks**: Source also includes: per capita income for Hispanic suburban residents, Hispanic central-city residents and Hispanic metro residents for 38 cities in the United States.

By Religious Affiliation

★ 857 ★

Median Household Income by Religious Affiliation, 1989

Religious Affiliation	Median Household Income
Jehovah's Witness	$20,900
Seventh Day Adventist	22,700
Muslim	24,700
Methodist	25,100
Mormon	25,700
Christian Scientist	25,800
Lutheran	25,900
Church of Christ	26,600
Roman Catholic	27,700
Disciples of Christ	27,800

[Continued]

★ 857 ★

Median Household Income by Religious Affiliation, 1989

[Continued]

Religious Affiliation	Median Household Income
Hindu	27,800
Buddhist	28,500
Presbyterian	29,000
Congregational/UCC	30,400
Eastern Orthodox	31,500
Episcopalian	33,000
Agnostic	33,300
Unitarian	34,800
Jewish	36,700
All	28,900

Source: "The Richest Congregations," *American Demographics*, 13, December 1991, p. 13. Primary source: Barry A. Kosmin, based on the National Survey of Religious Identification.

By Sex

★ 858 ★

Housework

According to a study by Cornell University, household labor done by married women has an average value of $5.50 per hour after taxes, $7.64 before taxes, or $10,000 annually. However, for married men, housework averages $9.60 per hour after taxes, but only $6,600 per year. Although men earn a higher rate, they do only 27% of the work women do.

Source: "Housework of Married Women Worth $5.50 an Hour After Taxes," *Christian Science Monitor*, January 13, 1993, p. 7. **Remarks**: This analysis used 1981 time-use data and 1988 dollar values of time per hour for married men and women aged 18 to 65.

★859★

By Sex

Median Income per Capita by Sex and Age, 1970-1989

Age	Female			Male		
	1970	1980	1989	1970	1980	1989
Total with income	$5,440	$11,591	$19,612	$9,184	$19,173	$28,511
15-19 yrs.	3,783	6,779	-	3,950	7,753	-
20-24 yrs.	4,928	9,407	13,653	6,655	12,109	15,501
25-34 yrs.	5,923	12,190	19,706	9,126	17,724	24,991
35-44 yrs.	5,531	12,239	21,498	10,258	21,777	32,370
45-54 yrs.	5,588	12,116	20,905	9,931	22,323	35,356
55-64 yrs.	5,468	11,931	19,895	9,071	21,053	34,505
65 and over	4,884	12,342	21,505	6,754	17,307	34,110

Source: The Universal Almanac 1992. Kansas City, Mo: Andrews and McMeel, 1991, p. 231. Primary source: U.S. Bureau of the Census, *Current Population Reports*.

★860★

By Sex

Jobs Narrowing the Gender-Wage Gap, 1991

For every dollar earned by men, here is what was earned by women. The national average was 70 cents in 1991.

Occupation	1983	1991	Change
Bank Tellers	$.80	$.92	+$.12
Bill Collectors	.58	.74	+.16
Computer Scientists	.75	.86	+.11
Designers	.53	.69	+.16
Personnel Trainers	.66	.80	+.14
Police Protection	.70	.84	+.14
Public Relations Specialists	.74	.84	+.10
Purchasing Agents	.63	.74	+.11
Secretaries	.77	.97	+.20

Source: "The Gender-Wage Gap," *Working Woman*, January 1993, p. 43. **Remarks**: Source also includes: occupations in which the gender-wage gap widened.

★ 861 ★

By Sex

Jobs Widening the Gender-Wage Gap, 1991

For every dollar earned by men, here is what was earned by women. The national average was 70 cents.

Occupation	1983	1991	Change
Financial managers	$.64	$.59	-$.05
Doctors	.82	.54	-.28
Lawyers	.89	.75	-.14

Source: "The Gender-Wage Gap," *Working Woman*, January 1993, p. 43. **Remarks**: Source also includes: occupations in which the gender-wage gap narrowed.

★ 862 ★

By Sex

Women's Median Wage as a Percentage of Men's, by Selected Occupations, 1992

Profession	Women's Wage as a % of Men's
Data-entry keyers	95.0%
Secretaries	91.6
Pharmacists	90.1
Engineers	85.6
Computer programmers	84.1
Lawyers	78.0
Doctors	72.2
Managers of marketing, advertising, pub. rel.	68.5
Machinists, assemblers, inspectors	67.7
Financial managers	62.4

Source: "Across-the-Board Dollar Disparity," *The Wall Street Journal*, June 9, 1993, p. B1. Primary source: U.S. Bureau of Labor Statistics.

By Size of Company

★ 863 ★

Wages and Benefits by Size of Company, 1991

Benefits include life/health insurance, pension, vacation, holidays, sick pay, unemployment taxes, and social security.

Size of company	Average Hourly Wage	Average Hourly Benefits
500 or more workers	$14.26	$6.32
100-499 workers	10.32	3.99
1-99 workers	10.00	3.36

Source: "Tall Order for Small Business," James E. Ellis, *Business Week*, 23, April 19,1993, p. 115.

Disposable Income

★ 864 ★

Per Capita Disposable Income, 1991

In early 1991, per capita disposable income was $11,295, rising later in the year to $11,400.

Source: U.S. Department of Commerce. International Trade Administration. *U.S. Industrial Outlook '92*. Washington, D.C.: U.S. Department of Commerce, 1992, p. 38-1.

Executives

★ 865 ★

Annual Compensation for CEOs in the Top 30 Companies in Selected Countries

Country	Average Compensation (in U.S. $)
France	$800,000
Germany	800,000
Great Britain	1,100,000
Japan	525,000
United States	3,200,000

Source: "Modest Paychecks for Japan's Global Heavyweights," *Business Week*, 22, January 27, 1992, p. 31. *Note:* average compensation includes salary, bonus, and long-term incentives.

★ 866 ★

Executives

Annual Compensation for Executives of Organizations with $250 million in Annual Sales, in Selected Countries, 1990

Country	Average Compensation (in U.S. $)
Argentina	$252,000
Australia	233,000
Canada	389,000
Germany	377,000
Great Britain	308,000
Korea	138,000
Japan	308,000
Singapore	211,000
Sweden	233,000
United States	633,000

Source: "Executive Pay in 10 Countries," *Management Review*, 80, May 1992, p. 13. *Note:* average compensation includes salary, bonus, and long-term incentives.

★ 867 ★

Executives

Annual Compensation for CEOs of Blue-chip Companies

According to a *Business Week* survey, CEOs of blue-chip companies receive an average of $2.5 million per year in total compensation, which includes salary, bonus, and stock. The results of a *Forbes's* magazine survey of executives of 800 companies showed that $1 million was the average compensation.

Source: "Executive Pay Up 26%, Tops Workers 104 Times," John R. Oravec, *AFL-CIO News*, 37, May 25, 1992, p. 10.
Remarks: Source also includes: pay of a CEO vs. a factory worker.

★ 868 ★

Executives

Pay of CEOs Vs. Factory Workers

According to Rudy Oswald, an AFL-CIO economist, in 1991 CEOs' compensation increased $10,000 a week, while the average factory worker received a pay raise of $13 per week (3%). In 1980, the average pay of a CEO was 42 times higher than for a factory worker, by 1990 the CEO made 85 times as much, and in 1991 it was 104 times greater than the factory worker's.

Source: "Executive Pay Up 26%, Tops Workers 104 Times," John R. Orvec, *AFL-CIO News*, 37, May 25, 1992, p. 10.

★ 869 ★

Executives

Pay of CEOs, 1960-1992

Year	Annual Average Pay			
	CEO	Worker	Teacher	Engineer
1960	$190,383	$4,665	$4,995	$9,828
1970	548,787	6,933	8,626	14,695
1980	624,996	15,008	15,970	28,486
1992	3,842,247	24,411	34,098	58,240

Source: "The Widening Gap Between CEO Pay and What Others Make," *Business Week*, 23, April 26, 1993, p. 56. Primary source: Bureau of Labor Statistics, National Education Association, and *Business Week*.

★ 870 ★

Executives

Salaries of the Directors of Charitable Foundations

According to a survey by the *Chronicle of Philanthropy*, the median salary of chief executives of the 100 largest foundations in the United States was $155,000 in 1991.

Source: "Top Salaries at Foundations Reported," *The New York Times*, September 8, 1992, p. D14.

★ 871 ★

Executives

Annual Compensation for Legal Executives

According to a survey of more than 1,000 organizations by William Mercer Inc., a human resources management consulting firm, the average annual compensation, or salary plus bonus, for legal executives of top American companies is $200,000.

Source: "Top Corporate Lawyers Earn Big Bucks," *The Christian Science Monitor*, August 28, 1991, p. 9.

★ 872 ★

Executives

Salaries of Hospital CEOs

Chief executive officers of hospitals have an average salary of $200,000. This represents 123 to 142 percent rise from 1981 to 1991 for CEOs at church, state, municipal and private nonprofit hospitals.

Source: "Hospital Executives Pay Rose Sharply in Decade," Felicity Barringer, *The New York Times*, September 30, 1992, p. A14. Primary source: a private medical association survey.

★ 873 ★

Executives

Pay of Male Executives Vs. Women Executives

According to a survey of 439 top-ranking women at large industrial and service companies in the United States, the average total compensation of women executives rose from $92,000 in 1982 to $187,000 in 1992. Even though women worked the same hours as men, an average of 56 hours per week, their pay was only two-thirds that of men's whose average compensation was $289,000 (based on a 1989 comparable Korn/Ferry survey).

Source: "Executive Women Make Major Gains in Pay and Status," Sue Shellenbarger, *The Wall Street Journal*, June 30, 1992, p. A3.

Government Employees

★874★

Earnings of State Government Employees by State, 1990

State	Average October Earnings 1990 (dol.)	State	Average October Earnings 1990 (dol.)	State	Average October Earnings 1990 (dol.)
U.S.	2,472	Kentucky	2,141	North Dakota	2,057
Alabama	2,196	Louisiana	2,047	Ohio	2,510
Alaska	3,543	Maine	2,352	Oklahoma	1,975
Arizona	2,334	Maryland	2,609	Oregon	2,302
Arkansas	1,922	Massachusetts	2,541	Pennsylvania	2,437
California	3,209	Michigan	2,858	Rhode Island	2,586
Colorado	2,765	Minnesota	2,936	South Carolina	1,956
Connecticut	3,018	Mississippi	1,824	South Dakota	1,979
Delaware	2,245	Missouri	1,965	Tennessee	2,055
Florida	2,095	Montana	2,072	Texas	2,192
Georgia	2,037	Nebraska	2,075	Utah	2,000
Hawaii	2,317	Nevada	2,502	Vermont	2,302
Idaho	2,100	New Hampshire	2,352	Virginia	2,267
Illinois	2,520	New Jersey	2,859	Washington	2,459
Indiana	2,496	New Mexico	2,100	West Virginia	1,919
Iowa	2,936	New York	2,997	Wisconsin	2,503
Kansas	2,077	North Carolina	2,372	Wyoming	2,045

Source: U.S. Bureau of the Census. *Statistical Abstract of the United States 1992*. 112th edition. Washington, D.C.: U.S. Department of Commerce, 1992, p. 308. Primary source: U.S. Bureau of the Census, *Public Employment*, series GE, No. 1, annual. **Remarks**: Source also includes: earnings for local government employees, total full-time equivalent employment, and full-time equivalent employment per 10,000 population, all by state, 1990.

★ 875 ★

Government Employees

Earnings of Local Government Employees by State, 1990

State	Average October Earnings (dol.) local 1990	State	Average October Earnings (dol.) local 1990	State	Average October Earnings (dol.) local 1990
U.S.	2,364	Kentucky	1,823	North Dakota	2,138
Alabama	1,749	Louisiana	1,713	Ohio	2,236
Alaska	3,491	Maine	1,978	Oklahoma	1,761
Arizona	2,540	Maryland	2,776	Oregon	2,322
Arkansas	1,545	Massachusetts	2,554	Pennsylvania	2,403
California	3,073	Michigan	2,646	Rhode Island	2,656
Colorado	2,292	Minnesota	2,552	South Carolina	1,848
Connecticut	2,854	Mississippi	1,543	South Dakota	1,733
Delaware	2,458	Missouri	2,052	Tennessee	1,883
District of Columbia	3,024	Montana	1,959	Texas	1,952
Florida	2,247	Nebraska	2,089	Utah	2,092
Georgia	1,872	Nevada	2,574	Vermont	2,090
Hawaii	2,536	New Hampshire	2,215	Virginia	2,248
Idaho	1,772	New Jersey	2,698	Washington	2,515
Illinois	2,463	New Mexico	1,783	West Virginia	1,862
Indiana	2,036	New York	2,795	Wisconsin	2,372
Iowa	2,024	North Carolina	2,065	Wyoming	2,110
Kansas	1,979				

Source: U.S. Bureau of the Census. *Statistical Abstract of the United States 1992.* 112th edition. Washington, D.C.: U.S. Department of Commerce, 1992, p. 308. Primary source: U.S. Bureau of the Census, *Public Employment*, series GE, No. 1, annual. **Remarks**: Source also includes: earnings for state government employees, total full-time equivalent employment, and full-time equivalent employment per 10,000 population, all by state, 1990.

★ 876 ★

Government Employees

Earnings of City Government Employees by Year, 1970-1990

Year	Average Earnings In October, Full-Time Employees (dollars)		Year	Average Earnings In October, Full-Time Employees (dollars)	
	Education	Other		Education	Other
1970	838	681			
1975	1,130	972	1983	1,962	1,739
1976	1,207	1,035	1984	2,033	1,831
1977	1,280	1,087	1985	2,117	1,953
1978	1,341	1,148	1986	2,186	2,044
1979	1,414	1,231	1987	2,406	2,163
1980	1,501	1,338	1988	2,566	2,220

[Continued]

★ 876 ★

Earnings of City Government Employees by Year, 1970-1990

[Continued]

Year	Average Earnings In October, Full-Time Employees (dollars)		Year	Average Earnings In October, Full-Time Employees (dollars)	
	Education	Other		Education	Other
1981	1,686	1,500	1989	2,669	2,343
1982	1,791	1,625	1990	2,683	2,449

Source: U.S. Bureau of the Census. *Statistical Abstract of the United States 1992*. 112th edition. Washington, D.C.: U.S. Department of Commerce, 1992, p. 309. Primary source: U.S. Bureau of the Census, *Census of Governments, 1972, 1977, 1982, and 1987*, vol. 3, no. 2, *Compendium of Public Employment and City Employment*, series, GE, No. 2, annual. **Remarks**: Also included in source: total number of employees, amount of the October payroll, average annual percent change and full-time equivalent employment, all for 1970-1990.

★ 877 ★

Government Employees

Earnings of City Government Employees in Selected U.S. Cities, 1990

Cities	Average Earnings in October, Full-Time Employees (dol.) 1990	Cities	Average Earnings in October, Full-Time Employees (dol.) 1990	Cities	Average Earnings in October, Full-Time Employees (dol.) 1990
New York, NY	2,783	Seattle, WA	3,274	Pittsburgh, PA	2,240
Los Angeles, CA	3,488	El Paso, TX	1,973	Sacramento, CA	3,021
Chicago, IL	3,002	Cleveland, OH	2,521	Minneapolis, MN	2,815
Houston, TX	2,061	New Orleans, LA	1,623	Tulsa, OK	2,555
Philadelphia, PA	2,843	Nashville-Davidson, TN	2,510	Honolulu, HI	2,600
San Diego, CA	3,019	Denver, CO	2,649	Cincinnati, OH	2,634
Detroit, MI	2,663	Austin, TX	2,186	Miami, FL	3,771
Dallas, TX	1,945	Fort Worth, TX	2,171	Fresno, CA	2,944
Phoenix, AZ	2,876	Oklahoma City, OK	2,430	Omaha, NE	3,051
San Antonio, TX	2,227	Portland, OR	3,305	Toledo, OH	2,847
San Jose, CA	3,453	Kansas City, MO	2,390	Buffalo, NY	2,596
Baltimore, MD	2,540	Long Beach, CA	3,413	Wichita, KS	2,083
Indianapolis, IN	1,910	Tucson, AZ	2,528	Santa Ana, CA	4,144
San Francisco, CA	3,648	St. Louis, MO	2,363	Mesa, AZ	2,898
Jacksonville, FL	2,582	Charlotte, NC	2,399	Colorado Springs, CO	2,618
Columbus, OH	2,416	Atlanta, GA	2,286	Tampa, FL	2,505
Milwaukee, WI	2,431	Virginia Beach, VA	2,232	Newark, NJ	1,698
Memphis, TN	2,287	Alburquerque, NM	2,040	St. Paul, MN	3,265
Washington, DC	2,930	Baton Rouge, LA	2,235	Louisville, KY	2,180
Boston, MA	2,391	Oakland, CA	3,948	Anaheim, CA	3,728

Source: U.S. Bureau of the Census. *Statistical Abstract of the United States 1992*. 112th edition. Washington, D.C.: U.S. Department of Commerce, 1992, p. 310. Primary source: U.S. Bureau of the Census, *City Employment*, series GE, No. 2, annual. **Remarks**: Also included in source: total number of employees, amount of the October payroll, and full-time equivalent employment, all in 63 U.S. cities.

Median Income

★ 878 ★

World-Wide Blue-Collar Salary

In 1990, the world-wide average salary for blue-collar workers was $26,565. In 1989, the average salary was $26,342 for men and $21,615 for women.

Source: The Universal Almanac 1992. Kansas City, MO: Andrews and McMeel, 1991, p. 116.

★ 879 ★

Median Income

Median Household Income for Americans

According to the Census Bureau, the median household income, after adjusting for inflation, was $30,126 in 1992, a decline of $1,077 from 1991.

Source: "Ranks of U.S. Poor Reach 35.7 Million, the Most Since '64," Robert Pear, *The New York Times*, September 4, 1992, p. A1.

★ 880 ★

Median Income

Median Family Income

Family unit	Median Income
Married Couples with Children	$41,260
Single Father with Children	25,211
Single Mother with Children	13,092

Source: "Median Income of Families with Children," *USA Today*, May 6, 1993, p. D1. Primary source: U.S. Bureau of the Census, 1990 data.

★ 881 ★

Median Income

Median Income of the Aged, 1962-1988

Unit	Median Income	
	1962	1988
Married couples	$11,047	$20,305
Nonmarried persons	4,412	7,928

Source: U.S. Department of Health & Human Resources. *Fast Facts & Figures About Social Security*. Washington D.C.: U.S. Department of Health & Human Resources, 1990, p. 4.

Savings

★ 882 ★

Savings as a Percentage of After-Tax Income in Selected Countries, 1993

Country	Average Savings of After-Tax Income
Canada	11.3%
France	12.2%
Germany	14.8%
Great Britain	12.3%
Japan	18.1%
United States	4.1%

Source: "Savings by Nation," *The Wall Street Journal*, April 5, 1993, p. A1. Primary source: ISI Group Inc.

★ 883 ★

Savings

Saving for Retirement

The results of a Gallup Poll conducted by Phoenix Mutual Life Insurance showed that the average "baby boomer" has saved only $35,000 for retirement: $25,000 in a retirement plan and 10,000 in other personal savings.

Source: "How You Can Afford to Retire," Jack Egan, *U.S. News & World Report*, 113, May 25, 1992, pp. 67-70.

Union Membership

★ 884 ★

Weekly Earnings for Union Members in General

According to 1990 data compiled by the Bureau of Labor Statistics, membership in trade unions was worth $119 more in the average worker's weekly paycheck. Full-time unionized wage and salary workers' median weekly earnings were $509, vs. $390 for nonunion workers. The monetary advantage of union membership is worth more than $6,100 a year. This data does not include employer-paid benefits, which in 1989 averaged $6.95 per hour for union workers, compared to $3.80 an hour for nonunion workers.

Source: "Economy: Union Pay Well Ahead of Nonunion," John R. Oravec, *AFL-CIO News*, 36, February 18, 1991, p. 10. **Remarks**: Source also includes: union pay vs. nonunion pay by sex and race/ethnicity.

★ 885 ★

Union Membership

Weekly Earnings for Union Vs. Nonunion Workers by Sex and Race/Ethnicity

According to 1990 data compiled by the Bureau of Labor Statistics, the average weekly pay for unionized female workers was $122 more than for non-union females. Male union workers earned an average of $85 more than non-union males. Among black workers, the union advantage meant $138 more per week; for Hispanic union workers it was $137 more a week; and for white union members it was $119 more each week.

Source: "Economy: Union Pay Well Ahead of Nonunion," John R. Oravec, *AFL-CIO News*, 36, February 18, 1991, p. 10. **Remarks**: Source also includes: union pay vs. nonunion pay in general terms.

Chapter 12
POLLUTION AND RECYCLING

Automobile Pollution

★ 886 ★

Carbon Dioxide Emitted Over an Automobile's Lifetime

Auto with Gas Mileage of -	Average Amount of Carbon Dioxide Over a Car's Lifetime
18.0 mpg	58.0 tons
27.5 mpg	38.0
45.0 mpg	23.0
60.0 mpg	17.0

Source: Sierra, July/August 1990, p. 23. Primary source: Environmental Action Foundation.

★ 887 ★

Automobile Pollution

Carbon Dioxide Emitted in One Year of Driving

Auto with Gas Mileage of -	Average Amount of Carbon Dioxide in One Year of Driving (15,000 miles)
20.0 mpg	14,250 pounds
27.5 mpg	10,355
35.0 mpg	8,151

Source: "Fuel Facts," *Consumer Reports*, 56, April 1991, p. 210.

★ 888 ★

Automobile Pollution

Carbon Dioxide Emitted by Gasoline

For every gallon of gasoline burned, 19 pounds of carbon dioxide are released.

Source: "A Traffic Jam of Potential Polluters," *The New York Times*, November 24, 1992, p. C1. **Remarks**: Source also includes: average number of hours wasted by being stuck in traffic jams every day; annual cost of traffic accidents to the country.

Commercial Waste and Pollution

★ 889 ★

Annual Production of Toxic Material

Each year in the United States, on average:

Toxic chemicals produced	350 billion pounds
Hazardous substances are shipped	8 billion pounds
New pesticides are introduced	2,200 products
New cosmetic products introduced	500 products
Chemically based products produced by the cosmetic industry	20,000 products

Source: Pohanish, Richard. *Hazardous Substances Resource Guide*. Detroit, MI: Gale Research Inc., 1993, p. xv.

★ 890 ★

Commercial Waste and Pollution

Plastic Packaging

An average of 250 million cubic feet of molded or loose plastic "peanuts" are used for packaging in the United States every year, an amount large enough to fill about ten 85-story skyscrapers.

Source: "Tis the Season to Recycle," *Consumer Reports*, 57, December 1992, p. 745.

★ 891 ★

Commercial Waste and Pollution

McDonald Restaurant Waste

Based on a week-long waste audit of two McDonald restaurants in Sycamore, IL and Denver, CO in November 1990, the average McDonald's restaurant serves about 2,000 customers and produces 238 pounds of waste per day.

Source: The Wall Street Journal, April 17, 1991, p. B1. Primary source: McDonald's and the Environmental Defense Fund.

★ 892 ★

Commercial Waste and Pollution

Air Pollution from Beef Production

For every pound of meat it yields, each head of beef cattle emits about one-third pound of methane. Add that to the carbon released from fuels burned in animal farming, and every pound of beef produced has the same greenhouse-warming effect as a 25-mile drive in an average American automobile.

Source: "Fat of the Land: We Can't Keep Eating the Way We Do," Alan Thein Durning, *USA Today* (magazine), November 1992, p. 27.

★ 893 ★

Commercial Waste and Pollution

Hidden Environmental Costs

According to Tellus Institute, the hidden environmental costs, including air pollution.

Material	Product	Cost/Ton	Cost/ Package
Aluminum			
Recycled	12 oz. can	$313	$.60
Virgin	12 oz. can	1,933	3.72
Glass	1/2 gal.	55-85	9.60-12.16
Recycled and virgin			
HDPE	1/2 gal.	344	2.56
Paperboard	1/2 gal.	330	3.84
PET	1/2 gal.	854	11.52
PVC	container for 100 nails/ screws	5,053	7.44

Source: "Environmental Costs of Packaging," *Garbage*, 4, December/January 1993, p. 26.

★ 894 ★

Commercial Waste and Pollution

Waste Generated by Newspapers

An average newspaper subscription, such as *The San Francisco Chronicle*, received every day, produces 550 pounds of wastepaper per subscription per year; an average *New York Times* Sunday edition produces 8 million pounds of waste paper.

Source: Audubon magazine, 92, March 1990, p. 4.

★ 895 ★

Commercial Waste and Pollution

Discarded Computers

An average of 10 million old mainframes, personal computers and work stations are discarded by American businesses and individuals each year. According to a Carnegie Mellon University study, if computers continue to be discarded at this rate, by the year 2005, 150 million computer carcasses will be crammed into American landfills. The landfill space needed would average an acre of land dug to a depth of three and a half miles, or room to stack 15 Empire State Buildings end to end.

Source: "Recycling Answer Sought for Computer Junk," *The New York Times*, April 14, 1993, p. A1. **Remarks**: Source also includes: a method of cutting electricity consumption of personal computers and the resultant savings in energy.

★ 896 ★

Commercial Waste and Pollution

Paper Waste in Offices

Every year the average office worker generates 180 pounds of high-grade paper. Every year the average 100-person office generates 378,000 sheets of copier paper, or enough paper, if stacked, to equal seven stories. Every year Americans throw away enough office and writing paper to build a 12 foot wall reaching from Los Angeles to New York City.

Source: Kimball, Debi. *Recycling in America*. Denver, CO: ABC-CLIO, Inc., 1992, p. 50.

Hazardous Substances

★ 897 ★

Radon Testing

The average cost of testing a home's radon level is $10 and the mitigation cost is $500 to $2,000. According to the Environmental Protection Agency, the average cost to test a school's radon level is $1,000 and the average cost to fix this problem is just a few thousand dollars.

Source: "Statements on Introduced Bills and Joint Sessions," Frank R. Lautenber, *Congressional Record*, April 9, 1991, p. 4244.

★ 898 ★

Hazardous Substances

Deleading a House

Since lead-based paint was not banned from residential use until 1978, three-fourths of the houses built before 1980 have the potential to contain lead-based paint, which if eaten or swallowed, can accumulate in the blood and cause brain damage. The average cost to delead an entire house is $5,000-$7,700.

Source: "Your Home: High Cost of Getting Lead Out of the House," Sherry Harowitz, *Kiplinger's Personal Finance Magazine*, 45, August 1991, p. 72.

Household Waste and Pollution

★ 899 ★

Solid Waste Generated per Family

According to one study, Americans produce about 230 million tons of refuse a year—5.1 pounds per person a day or about 1,900 pounds per year. Another survey reported that a typical suburban family of three generated 40 pounds of garbage weekly with the components of the content given in percentages below.

Type of solid waste	Percent
Aluminum	1%
Polystyrene meat trays, cups, egg cartons, and packing	3%
Disposable diapers	3%
Wood, textiles, old clothing	5%
Metal cans and nails	5%
Plastic soda bottles and bags	5%
Glass	8%

[Continued]

★ 899 ★

Solid Waste Generated per Family

[Continued]

Type of solid waste	Percent
Miscellaneous	10%
Food	11%
Paper	21%
Yard waste and grass clippings	23%

Source: Good Housekeeping, 209, September 1989, p. 272.

★ 900 ★

Household Waste and Pollution

Waste Generated per Capita by State, 1990

State	Average Waste per Capita (tons)	State	Average Waste per Capita (tons)	State	Average Waste per Capita (tons)
Alabama	1.073	Louisiana	1.000	Ohio	1.287
Alaska	0.900	Maine	0.750	Oklahoma	0.818
Arizona	0.911	Maryland	1.600	Oregon	0.870
Arkansas	0.750	Massachusetts	1.118	Pennsylvania	0.760
California	1.580	Michigan	1.271	Rhode Island	1.000
Colorado	0.606	Minnesota	0.952	South Carolina	1.147
Connecticut	0.878	Mississippi	0.692	South Dakota	1.071
Delaware	0.923	Missouri	1.000	Tennessee	0.812
Florida	1.300	Montana	0.750	Texas	1.059
Georgia	0.709	Nebraska	0.733	Utah	0.647
Hawaii	0.909	Nevada	1.000	Vermont	0.600
Idaho	0.750	New Hampshire	1.000	Virginia	1.520
Illinois	1.293	New Jersey	1.250	Washington	1.150
Indiana	0.818	New Mexico	0.660	West Virginia	1.315
Iowa	0.821	New York	1.230	Wisconsin	0.750
Kansas	0.640	North Carolina	0.937	Wyoming	1.100
Kentucky	1.243	North Dakota	0.692		

Source: The 1992 Information Please Environmental Almanac. Boston, MA: Houghton Mifflin Company, 192, p. 192. Primary source: Biocycle, March, 1990.

★ 901 ★

Household Waste and Pollution

Waste Generated by Americans

Americans produce an average of 148.1 million tons of garbage a year, or an average of 3.43 pounds of garbage per person per day, up from 2.5 pounds in 1960. If this rate goes unchecked, the average American would produce over 90,000 pounds of garbage in his/her lifetime.

Source: Rukeyser, Louis. *Louis Rukeyser's Business Almanac.* New York: Simon & Shuster, 1991, p. 28.

★ 902 ★

Household Waste and Pollution

Waste Generated by Type of Material, 1960-1988

Type of material	Daily Average Waste in Pounds per Person			
	1960	1970	1980	1988
Paper	0.91	1.19	1.32	1.60
Glass	0.20	0.34	0.36	0.28
Metals	0.32	0.38	0.35	0.34
Plastics	0.01	0.08	0.19	0.32
Rubber & leather	0.06	0.09	0.10	0.10
Textiles	0.05	0.05	0.06	0.09
Wood	0.09	0.11	0.12	0.14
Other	0.00	0.02	0.07	0.07
Total nonfood product waste	1.65	2.26	2.57	2.94
Food wastes	0.37	0.34	0.32	0.29
Yard wastes	0.61	0.62	0.66	0.70
Miscellaneous inorganic waste	0.04	0.05	0.05	0.06
Total	2.66	3.27	3.61	4.00

Source: The 1992 Information Please Environmental Almanac. Boston, MA: Houghton Mifflin Company, 1992, p. 108. Primary source: U.S. Environmental Protection Agency, 1990. **Remarks**: Details may not add up to totals because of rounding.

★ 903 ★

Household Waste and Pollution

Waste Generated per Person in Selected Cities Worldwide, 1988

City	Average Daily Waste per Capita
Los Angeles	6.4 pounds
Philadelphia	5.8
Chicago	5.0
New York	4.0
Tokyo	3.0
Paris	2.4
Toronto	2.4
Hamburg	1.9
Rome	1.5

Source: The 1992 Information Please Environmental Almanac. Boston, MA: Houghton Mifflin Company, 1992, p. 108. Primary source: National Solid Wastes Management Association.

★ 904 ★

Household Waste and Pollution

Yard Waste

An average of three tons of grass clippings are generated every year by mowing a half-acre of lawn. This is enough to fill 465 one-bushel bags. Twenty to fifty percent of the landfills of the United States are made up of yard waste.

Source: "The Green Way to a Green Yard," *Consumer Reports*, 56, June 1991, p. 407.

★ 905 ★

Household Waste and Pollution

Hazardous Household Waste in General

Fifteen pounds of hazardous waste are generated in the average home each year. Waste per participant brought to household hazardous waste collection sites averages 116 pounds. This includes first-time collections to which people bring years of accumulated wastes. One participant may represent more than one household.

Source: Garbage, 2, March/April 1990, p. 16. Primary source: Dana Duxbury and Associates.

★906★

Household Waste and Pollution

Hazardous Household Waste by Type of Waste

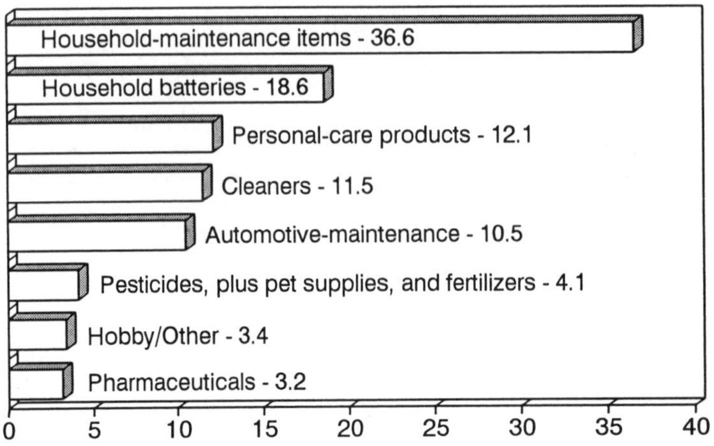

Hazardous household waste by type, shown as percent of total.

Type of waste	Percent of all waste
Household-maintenance items[1]	36.6
Household batteries	18.6
Personal-care products[2]	12.1
Cleaners	11.5
Automotive-maintenance products[3]	10.5
Pesticides, plus pet supplies, and fertilizers	4.1
Hobby/other[4]	3.4
Pharmaceuticals	3.2

Source: Garbage, March/April, 1990, p. 16 from The Garbage Project, University of Arizona. *Notes:* 1. Mostly paint, plus thinners, adhesives. 2. Includes nail polish and remover, hair spray, deodorant. 3. Mostly oil. 4. Includes pool chemicals, lighter fluid, art supplies.

★907★

Household Waste and Pollution

Gifts Wrapped per Household

The number of Christmas gifts wrapped by the average household is 30.

Source: Garbage, 2, November/December 1990, p. 63. Primary source: Hallmark Cards, Inc.; A.C. Neilsen; and National Christmas Association.

★ 908 ★

Household Waste and Pollution

Waste Generated per Year by Type of Waste

Americans throw away enough aluminum cans to produce 6,000 DC-10s; enough plastic and paper cups and plates to give the world six picnics; and an average of 180 million razors every year. In addition, the British dispose of 2.5 billion diapers, and the Japanese dump 30 million single-roll cameras each year.

Source: "How Much is Enough," Alan Duening, *Technology Review*, 94, May-June 1991, p. 60.

★ 909 ★

Household Waste and Pollution

Paper Products per Person per Year

In the United States, every person uses an average of 600 pounds of paper products each year, 70% of which ends up in landfills. An average of 40% of waste in landfills is paper.

Source: "Paper Recycling: Fact from Fiction," *Popular Science*, 239, October 1991, pp. 13-14.

★ 910 ★

Household Waste and Pollution

Used Automobile Oil

An average of 200 to 400 million gallons of waste oil are generated annually by car owners who change their own oil, or enough to equal 36 *Exxon Valdez* spills per year. Since only 10-14 percent is disposed of properly, most of the waste oil ends up in the ground, in streams, and in sewers. One gallon of used oil can make a million gallons of fresh water undrinkable, and a pint of oil can produce a slick that covers an acre of water.

Source: "Trickle Down," *Consumer Reports*, 56, April 1991, pp. 210-211.

★911★

Household Waste and Pollution

Air Pollution Generated by Selected Machines

Gasoline-Burning Machines Used for One Hour	Average Number of Miles a Car is Driven to Equal Same Pollution
Outboard motor	800 miles
Farm tractor	500 miles
Chain saw	200 miles
Leaf blower	100 miles
Lawn mower	50 miles

Source: "Cars are not the Only Polluters," *U.S. News & World Report*, 113, August 17, 1992, p. 13.

★912★

Household Waste and Pollution

Carbon Dioxide Emitted from Household Appliances

Appliance	Carbon Dioxide Added to Atmosphere[1]	
	Time Measurement	Pounds
Color television	per hour	.64
Steam iron	per hour	.85
Vacuum cleaner	per hour	1.70
Air conditioner, room	per hour	4.00
Toaster oven	per hour	12.80
Ceiling fan	per day	4.00
Refrigerator, frostless	per day	12.80
Waterbed heater	per day	24.00
Waterbed heater with thermostat	per day	12.80
Clothes dryer	per load	10.00
Dishwasher	per load	2.60
Toaster	per use	.12
Microwave oven	per 5-minute use	.25
Coffeemaker	per brew	.50

Source: National Wildlife, Feb-March 1990, p. 53. *Notes:* 1. At room temperature and sea level, every pound of carbon dioxide occupies 8.75 cubic feet, about half the size of a refrigerator.

★913★

Household Waste and Pollution

Carbon Dioxide Emitted from the Transportation Sector per Capita

The amount of carbon dioxide emitted from the transportation sector every year averages 9.2 tons per person, or 2.3 billion tons total.

Source: "Global Warming and Other Environmental Consequences of Energy Strategies," John M. DeCicco. U.S. Senate. committee on Environment and Public Works. Hearing, March 20, April 26, 1991. Washington, D.C., Government Printing Office, 1991.

★914★

Household Waste and Pollution

Waste Generated by Type of Material

Type of Material	Average Waste in Pounds per Person
Glass	85 per year
Steel cans	6 per month
Plastic	190 per year
Plastic containers	2 per month

Source: Kimball, Debi. *Recycling in America*. Denver, CO: ABC-CLIO, Inc., 1992, pp. 38-42.

★915★

Household Waste and Pollution

Environmental Costs of Cloth Vs. Disposable Diapers

	Average Weekly Usage of Diapers	
	Cloth	Disposable
Solid waste	0.24 pounds	22.18 pounds
Water pollution	0.117 pounds	0.012 pounds
Air pollution	0.860 pounds	0.093 pounds
Energy consumption	78,890 Btu's	23,290 Btu's
Water consumption	144.0 gallons	23.6 gallons
Raw materials	3.6 pounds	25.3 pounds

Source: "Choosing Diapers," *The New York Times*, October 23, 1992, p. B2. Primary source: Arthur D. Little, Inc.

Recycling

★916★

Recycled Newspaper

To save a 35-40 foot tree, a stack of newspapers, averaging four feet in thickness, must be recycled.

Source: Heloise Hints for a Healthy Planet. New York: Perigee, 1990, p. 67.

★917★

Recycling

State and Local Government Expenditures for Solid Waste Management and Sewerage, by State, 1990

State	Total Expenditures[1] Per capita (dol.)	State	Total Expenditures[1] Per capita (dol.)	State	Total Expenditures[1] Per capita (dol.)
United States	114	Kentucky	66	North Dakota	50
Alabama	57	Louisiana	100	Ohio	105
Alaska	168	Maine	115	Oklahoma	72
Arizona	108	Maryland	130	Oregon	87
Arkansas	56	Massachusetts	145	Pennsylvania	111
California	121	Michigan	109	Rhode Island	125
Colorado	88	Minnesota	112	South Carolina	77
Connecticut	167	Mississippi	64	South Dakota	45
Delaware	146	Missouri	66	Tennessee	84
District of Columbia	280	Montana	52	Texas	95
Florida	125	Nebraska	53	Utah	84
Georgia	86	Nevada	72	Vermont	101
Hawaii	165	New Hampshire	132	Virginia	122
Idaho	78	New Jersey	212	Washington	166
Illinois	97	New Mexico	97	West Virginia	58
Indiana	78	New York	188	Wisconsin	149
Iowa	93	North Carolina	94	Wyoming	152
Kansas	48				

Source: U.S. Bureau of the Census. *Statistical Abstract of the United States 1992.* 112th edition. Washington, D.C.: U.S. Department of Commerce, 1992, p. 215. Primary source: U.S. Bureau of the Census, *Governmental Finances*, 1989-1990, series GF, No. 5. **Remarks**: Source also includes: total expenditures for solid waste management and sewerage, as well as separate figures for each, all by state, 1990. *Note:* 1. Includes capital expenditures.

★918★

Recycling

Recovery and Disposal of Municipal Solid Waste, 1960-1988

Item and material	1960	1965	1970	1975	1980	1985	1986	1987	1988
Waste generated									
Per person per day (lb.)	2.66	3.00	3.27	3.26	3.61	3.71	3.80	3.92	4.00
Materials recovered									
Per person per day (lb.)	0.18	0.19	0.23	0.25	0.35	0.38	0.42	0.45	0.52
Combustion for energy recovery									
Per person per day (lb.)	(NA)	0.01	0.02	0.02	0.06	0.17	0.22	0.36	0.59
Combustion without energy recovery									
Per person per day (lb.)	0.82	0.75	0.66	0.45	0.27	0.10	0.07	0.05	0.02
Landfilled, other disposal									
Per person per day (lb.)	1.67	2.05	2.37	2.54	2.93	3.06	3.09	3.06	2.87

Source: U.S. Bureau of the Census. *Statistical Abstract of the United States 1992*. 112th edition. Washington, D.C.: U.S. Department of Commerce, 1992, p. 216. Primary source: Franklin Associates, Ltd., Prairie Village, KS, *Characterization of Municipal Solid Waste in the United States: 1990 update*. Prepared for the U.S. Environmental Protection Agency. **Remarks**: Source also includes: percentage of the total waste generated by type of material.

★919★

Recycling

Cost of Curbside Collection of Recyclable Materials

According to the National Solid Wastes Management Association, a curbside pickup program costs cities $70-$100 per ton (often more than landfilling), of which an average of $25 per ton is used to pay reprocessing facilities which do not break even by reselling the materials alone.

Source: "Business Aim to Buy More Recycled Goods," *The Christian Science Monitor*, September 17, 1992, p. 9.

★920★

Recycling

Recycling Costs to Commercial Waste Haulers

According to Waste Management, a commercial waste hauler, the cost of collecting and sorting an average ton of recyclables is $175, which is only worth about $44 afterwards.

Source: "Two Major Garbage Rivals Find Their Profits Trashed," Jeff Bailey, *The Wall Street Journal*, March 4, 1993, p. B4. **Remarks**: Source also includes: price per ton of recyclable materials, 1988-1992; waste material collected as a percentage of total recovered and net cost to recover as a percentage of total, by type of material.

★921★

Recycling

Reusable Vs. Disposable Hospital Gowns

According to Medline, a hospital supply company, the per use cost of disposable hospital gowns is $4.00, compared to $3.00 per use for reusable gowns. This includes sterilization, cleaning, and a pro-rata share of the initial $80.00 price for each of its 80 lifetime uses.

But Baxter, another hospital supply company, argues that the average disposable gown costs $3.10, compared with $3.60 for reusable gowns, including all associated cleaning and repackaging costs.

Source: "Hospitals Are Returning to Reusable Surgical Supplies," George Anders, *The Wall Street Journal*.

★922★

Recycling

Natural Resources Saved by Recycling

Every day office workers generate an average of .51 pounds of recyclable paper. One ton of recycled wastepaper would save an average of 7,000 gallons of water, 3.3 cubic yards of landfill space, three barrels of oil, 17 trees, and 4,000 kilowatt-hours of electricity, or energy to power the average home for six months.

Source: "The 'Greening' of Law Offices," Jonathan E. Rinde, *Trial*, 27, May 1991, p. 119.

★923★

Recycling

Glass Recycling

A recycled glass container can save the amount of energy needed to light a 100-watt bulb for four hours. One ton of glass that is recycled saves the energy equivalent of nine gallons of fuel oil. Thirty percent of the average glass soft drink bottle is made up of recycled glass.

Source: Kimball, Debi. *Recycling in America*. Denver, CO: ABC-CLIO, Inc., 1992, p. 39.

★924★

Recycling

Steel Recycling

An average of 9,000 steel cans are recovered for recycling every minute of every day, from the 100 million used every day by Americans. Recycling these cans can save 74% of the energy needed to produce them from virgin materials.

Source: Kimball, Debi. *Recycling in America*. Denver, CO: ABC-CLIO, Inc., 1992, p. 38.

Chapter 13
TRAVEL, ENTERTAINMENT, AND RECREATION

Entertainment

★ 925 ★

Cost of a National League Football Game

According to the newsletter, *Team Marketing Report*, in 1992, the average cost of four to attend a National League football game, including entrance fees, concessions, two caps, two programs, and parking was $163.70.

Source: "Average Ticket Prices Up, But 12 Teams Hold Line," Michael Hiestand, *USA Today*, September 2, 1992, p. 3C.

★ 926 ★

Entertainment

Cost of a Major League Baseball Game

According to the newsletter, *Team Marketing Report*, in 1993, the average cost of four to attend a major league baseball game, was $90.81, a 4.7% rise in price, compared with 10.4% rise in 1992.

Source: "A Special Background Report on Trends in Industry and Finance," *The Wall Street Journal*, April 1, 1993, p. A1.

★ 927 ★

Entertainment

Disney World and Euro Disneyland

An average London-Orlando package deal to Disney World for plane, hotel and park entry is $115 per day; but $200 per day for this same kind of package deal for the British to travel to Euro Disney. Even without the lodging or travel expenses, a family of four spends an average of $280 per day at Euro Disney, according to a Parisian newspaper.

Source: "The Mouse Isn't Roaring," *Business Week*, 22, August 24, 1992, p. 38.

★928★

Entertainment

Movie Videos

In 1992, in a population where 75% of all households had a VCR machine (compared with only 5.7% of households in 1982), the average number of movie videos rented was 38. Forty percent of a movie's revenues came from these video rentals, whereas, only 25% came from theater showings.

Source: "The Balance Sheet, Please...," *U.S. News & World Report*, 114, April 5, 1993, p. 16.

★929★

Entertainment

Movie Attendance and the VCR

According to the Motion Picture Association of America, movie-goers saw an average of 6.7 movies in 1991, down from 7.1 in 1990.

Source: "VCR Wars," *American Demographics*, 14, August 1992, p. 6. **Remarks**: Source also includes: number of videos sold per video outlet store.

★930★

Entertainment

Movie Attendance in Selected Countries

Country	Year	Average Number of Movies per Person
Angola	1979	0.9
Brazil	1977	1.9
Canada	1980	4.2
Chile	1981	1.3
Egypt	1978	1.2
France	1981	3.5
Germany, West (former)	1981	2.3
Great Britain	1981	1.5
India	1981	6.8
Indonesia	1980	1.0
Japan	1981	1.3
Kuwait	1981	3.5
Netherlands	1981	2.0
Nigeria	1979	0.3
Poland	1981	2.8
Sweden	1981	2.7
United States	1981	4.6
USSR (former)	1981	15.9
Zambia	1976	0.3

Source: Brandreth, Gyles. *Your Vital Statistics*. New York: Citadel, 1986, p. 34.

★ 931 ★

Entertainment

Winners of *Jeopardy!*

The average winner on *Jeopardy!* earns $11,500 per show. But every year out of the tens of thousands who inquire about becoming a contestant, an average of 20,000 take the qualifying exam, from which about 450 people actually make it onto the show.

Source: "Going for the Grand Prize," Anthony Cook, *Money*, 21, September 1992, p. 131.

★ 932 ★

Entertainment

Annual Spending on Entertainment, 1984-1989

Other equipment includes pets, toys, playground, sports, exercise, and photographic equipment.

Year	Average Amount Spent on Entertainment			
	Fees & Admissions	TV, Radio, Sound Equip.	Other Equip.	Reading
1984	$313	$322	$420	$132
1985	320	371	479	141
1986	308	371	470	140
1987	323	379	491	142
1988	353	416	560	150
1989	377	429	618	157
ALL	371	454	597	153

Source: U.S. Bureau of the Census. *Statistical Abstract of the United States 1992*. 112th edition. Washington, D.C.: U.S. Department of Commerce, 1992, p. 235. Primary source: U.S. Bureau of Labor Statistics, *Consumer Expenditure Survey*, annual. **Remarks**: Source also includes: average annual expenditures for entertainment and reading by age, by area of country, and by income, 1990.

★ 933 ★

Entertainment

Annual Spending on Entertainment by Age, 1990

Other equipment includes pets, toys, playground, sports, exercise, and photographic equipment.

Year	Average Amount Spent on Entertainment			
	Fees & Admissions	TV, Radio, Sound Equip.	Other Equip.	Reading
Under 25 yrs.	$218	$344	$272	$75
25-34 yrs.	342	503	628	134
35-44 yrs.	527	583	726	188
45-54 yrs.	438	592	937	184
55-64 yrs.	365	390	752	163

[Continued]

★ 933 ★

Annual Spending on Entertainment by Age, 1990

[Continued]

Year	Average Amount Spent on Entertainment			
	Fees & Admissions	TV, Radio, Sound Equip.	Other Equip.	Reading
65-74 yrs.	329	307	277	156
75 yrs. +	153	161	109	112

Source: U.S. Bureau of the Census. *Statistical Abstract of the United States 1992.* 112th edition. Washington, D.C.: U.S. Department of Commerce, 1992, p. 235. Primary source: U.S. Bureau of Labor Statistics, *Consumer Expenditure Survey*, annual. **Remarks**: Source also includes: average annual expenditures for entertainment and reading by year, 1984-1989 and by area of country, and income, 1990.

★ 934 ★

Entertainment

Annual Spending on Entertainment by Region of the United States, 1990

Other equipment includes pets, toys, playground, sports, exercise, and photographic equipment.

Region	Average Amount Spent on Entertainment			
	Fees & Admissions	TV, Radio, Sound Equip.	Other Equip.	Reading
Northeast	$398	$429	$505	$171
Midwest	347	433	576	160
South	306	439	521	131
West	482	528	842	164

Source: U.S. Bureau of the Census. *Statistical Abstract of the United States 1992.* 112th edition. Washington, D.C.: U.S. Department of Commerce, 1992, p. 235. Primary source: U.S. Bureau of Labor Statistics, *Consumer Expenditure Survey*, annual. **Remarks**: Source also includes: average annual expenditures for entertainment and reading by year, 1984-1989, and by age and income, 1990.

★ 935 ★

Entertainment

Annual Spending on Entertainment by Income, 1990

Other equipment includes pets, toys, playground, sports, exercise, and photographic equipment.

Income	Average Amount Spent on Entertainment			
	Fees & Admissions	TV, Radio, Sound Equip.	Other Equip.	Reading
Under $5000	$121	$216	$154	$59
$5,000-$9,999	87	204	288	64
$10,000-$14,999	123	278	211	90
$15,000-$19,999	191	319	312	107
$20,000-$29,999	301	404	422	142

[Continued]

★935★

Annual Spending on Entertainment by Income, 1990

[Continued]

Income	Average Amount Spent on Entertainment			
	Fees & Admissions	TV, Radio, Sound Equip.	Other Equip.	Reading
$30,000-$39,999	400	527	576	182
$40,000-$49,999	536	639	1,066	202
$50,000 +	902	843	1,391	298

Source: U.S. Bureau of the Census. *Statistical Abstract of the United States 1992.* 112th edition. Washington, D.C.: U.S. Department of Commerce, 1992, p. 235. Primary source: U.S. Bureau of Labor Statistics, *Consumer Expenditure Survey*, annual. **Remarks**: Source also includes: average annual expenditures for entertainment and reading by year, 1984-1989, and by age and region of the country, 1990.

★936★

Entertainment

Watching, Listening, and Reading per Capita

Activity	Average Time Spent Daily per Capita
Watching television	3 hrs., 48 min.
Listening to radio	3 hrs., 21 min.
Reading newspapers	34 minutes
Reading magazines	20 minutes

Source: "Top Cities for Watching, Listening, and Reading," *Fortune*, 126, May 18, 1992, p. 18. Primary source: Young & Rubicom. **Remarks**: Source also includes: cities spending the most time per capita, watching television, listening to the radio, and reading newspapers and magazines.

★937★

Entertainment

Cities Watching, Listening, and Reading the Most

According to Young & Rubicam, an ad agency:

Greenwood-Greenville, Mississippi watches the most television per capita: 4 hours, 39 minutes per day.

Miami-Fort Lauderdale, Florida, listens to the most radio per capita: 3 hours, 43 minutes per day.

Charlottesville, Virginia reads newspapers the most per capita: 53 minutes per day.

[Continued]

★937★

Cities Watching, Listening, and Reading the Most

[Continued]

Boston, Massachusetts reads magazines the most per capita: 29 minutes per day.

Youngstown, Ohio spends the most time per capita on all forms of media: 8 hours, 50 minutes.

Source: "Top Cities for Watching, Listening, and Reading," *Fortune*, 126, May 18, 1992, p. 18. **Remarks**: Source also includes: average number of hours Americans spend watching television, listening to the radio, and reading newspapers and magazines.

★938★

Entertainment

Television Viewing per Week in Selected Cities

Cities	Average Time per Week	Cities	Average Time per Week
Hartford/New Haven	49 hrs., 15 min.	Chicago	50 hrs., 30 min.
Los Angeles	49 hrs., 15 min.	Sacramento/Stockton	50 hrs., 30 min.
Memphis	49 hrs., 15 min.	Philadelphia	53 hrs., 00 min.
New Orleans	49 hrs., 15 min.	Atlanta	54 hrs., 15 min.
New York City	49 hrs., 15 min.	Houston	55 hrs., 30 min.
Seattle-Tacoma	49 hrs., 15 min.	Dallas/Fort Worth	56 hrs., 45 min.
Washington, DC	49 hrs., 15 min.	Detroit	56 hrs., 45 min.
Boston	50 hrs., 30 min.		

Source: Krantz, Les. *The Best and Worst of Everything.* New York: Prentice Hall General Reference, 1991, p. 101. Primary source: Nielsen Media Research.

★939★

Entertainment

Television Viewing by Sex and Age

Age and Sex	Average Time per Week
55 and older	
Women	38 hrs., 53 min.
Men	37 hrs., 32 min.
35-54	
Women	32 hrs., 14 min.
Men	27 hrs., 01 min.
18-34	
Women	28 hrs., 43 min.
Men	25 hrs., 44 min.

[Continued]

★ 939 ★

Television Viewing by Sex and Age

[Continued]

Age and Sex	Average Time per Week
Teens	
Female	21 hrs., 18 min.
Male	22 hrs., 36 min.
Children	
6-11	23 hrs., 17 min.
2-5	25 hrs., 43 min.

Source: Krantz, Les. *The Best and Worst of Everything.* New York: Prentice Hall General Reference, 1991, p. 45. Primary source: Nielsen Media Research.

★ 940 ★

Entertainment

Television Viewing by Time of Day

Age & Sex	Average Time per Week				
	Mon-Fri 10am-4:30pm	Mon-Fri 4:30-7:30	Mon-Fri 11:30pm-1am	Mon-Sat 8-11pm	Sat 7am-1pm
Women, 18+	6h, 30m	4h, 55m	1h,13m	9h,46m	46m
12-17	2h, 56m	3h, 41m	47m	7h, 07m	49m
18-24	5h, 10m	3h, 18m	1h, 11m	6h, 43m	35m
25-54	5h, 34m	4h, 10m	1h, 25m	9h, 22m	43m
55+	8h, 49m	6h, 59m	1h, 29m	11h, 49m	52m
Men, 18+	4h, 11m	4h, 06m	1h, 25m	9h, 02m	47m
12-17	2h, 28m	3h, 32m	47m	7h, 16m	59m
18-24	3h, 33m	2h, 42m	1h, 13m	5h, 25m	32m
25-54	3h, 23m	3h, 27m	1h, 26m	8h, 48m	47m
55+	6h, 32m	6h, 26m	1h, 28m	11h, 37m	58m
Children					
2-5	6h, 47m	4h, 40m	28m	5h, 31m	1h, 35m
6-11	2h, 34m	4h, 04m	23m	5h, 53m	1h, 26m

Source: The World Almanac and Book of Facts 1992. New York: World Almanac, 1991, p. 317. Primary source: Nielsen Media Research, February 1991.

★ 941 ★

Entertainment

Television Time per Household

Year	Average Time per Household per Day
1971	6 hrs., 2 min.
1989	7 hrs., 1 min.

Source: The Universal Almanac 1992. Kansas City, MO: Andrews and McMeel, 1991, p. 251. Primary source: Television Bureau of Advertising, 1991.

★ 942 ★

Entertainment

Violence in Children's Television Programs

According to a report to the National Cable Television Association by Annenberg School of Communications at the University of Pennsylvania, there are an average of 5.2 violent acts per program and 17.3 per hour in children's programming created for cable; on broadcast television there are an average of 7.8 violent acts per program and 32 per hour.

Source: "Kids Shows Most Violent, Research Finds," Diane Duston, *Detroit Free Press*, January 28, 1993, p. 3A.

★ 943 ★

Entertainment

Children's' Viewing of Murder on Television

The average child in the United States sees 26,000 murders on television by his/her 18th birthday.

Source: Guinness Book of Records 1992. New York: Bantam Book, 1992, p. 510.

★ 944 ★

Entertainment

Children's Viewing of Commercials on Television

Each week the average child in the United States views 350 commercials on television.

Source: "Troubling TV Ads," Marc Silver, *U.S. News & World Report*, 114, February 1, 1993, pp. 65-67.

★ 945 ★

Entertainment

Commercial Time on Television

In an average hour of prime time television, there are nine minutes of commercials, or about 15% of each hour.

Source: "Barrage of Ads in Super Bowl Blurs Messages," Kevin Goldman, *The Wall Street Journal*, February 3, 1993. **Remarks**: Source also includes: a list of commercials and length of each for one hour of Super Bowl XXVII.

★ 946 ★

Entertainment

High School Graduates and Television Viewing

The average American teenager has viewed 22,000 hours of television (which includes 15.8 examples of sexual language or imagery per hour of prime time television), by his/her high school graduation.

Source: "Long Island Lolita," *U.S. News & World Report*, 114, January 4, 1993, p. 104.

★ 947 ★

Entertainment

Television Viewing by Type of Program

Type of Program	Average Number of Homes Viewing
Movies	10,300,000
Comedy series	9,600,000
News/reality shows	9,300,000
Drama series	8,400,000

Source: "Love Those Movies on TV," *USA Today*, June 1, 1993, p. D1. Primary source: Satellite ORBIT analysis of Nielsen ratings for ABC, CBS, Fox and NBC in 1992-1993 season.

★ 948 ★

Entertainment

Cable Television Rates, 1955-1990

Year	Average Monthly Rate	Average Number of Subscribers (1,000)
1955	$5.00	250
1960	5.00	750
1965	5.00	1,500
1970	5.50	5,100
1975	6.50	9,800
1980	7.85	17,500
1985	10.24	35,430
1987	13.27	41,200
1989	15.97	47,390
1990	17.58	50,455

Source: U.S. Bureau of the Census. *Statistical Abstract of the United States 1992*. 112th edition. Washington, D.C.: U.S. Department of Commerce, 1992, p. 556. Primary source: Paul Kagan Associates, Inc., Carmel, CA, *The Cable TV Financial Databook*, annual; *The Kagan Census of Cable*. **Remarks**: Source also includes: total revenues for cable television, 1955-1990; Pay TV monthly rates and percent of homes with pay TV, 1955-1990.

Recreation

★ 949 ★

Toys in Tokyo

According to a study conducted by Product Science Research Institute, children in Tokyo own an average of 411 toys apiece totaling $1,530 per child. All related toys were counted individually, e.g. each car in a collection was counted separately.

Source: "Toy Boxes Full in Tokyo," *Detroit Free Press*, January 28, 1993, p. 1E.

★950★

Recreation

Rail Trails in the United States

Rails to Trails Conservancy, a Washington group that's leading the movement to turn abandoned railroads into biking and hiking paths, estimates that there are 514 of these trails totaling 6,384 miles. The average length of each is 12.4 miles. Here is a breakdown by state.

State	Total Miles	Number of Trails
West Virginia	191.5	10
Missouri	209.7	4
New York	298.7	24
Illinois	323.9	29
Washington	433.1	30
Minnesota	486.5	22
Iowa	544.4	37
Pennsylvania	545.7	45
Michigan	710.4	48
Wisconsin	751.8	39

Source: "Rail-Trail Mileage in Top 10 States," *The Christian Science Monitor*, October 22, 1992, p. 8.

★951★

Recreation

Turning Abandoned Railroads into Biking and Hiking Paths

Rails to Trails Conservancy, a Washington group that's leading the movement to turn abandoned railroads into biking and hiking paths, estimates that an average of 3,000 miles of railroads are abandoned each year in the United States. Some 5,000 miles of these have been turned into biking and hiking trails.

Source: "Rail-to-Trail Movement Reaches a Milestone," *The Wall Street Journal*, September 21, 1992, p. B1.

★ 952 ★

Recreation

Hikers on the Appalachian Trail

The Appalachian Trail:

An average of 175 hikers, out of 1,900 who attempt it, travel the full Appalachian Trail (2,143 miles) each year.

The average time needed to accomplish this feat is 175 days, but the fastest time anyone has claimed to finish the trip is 52 days, 9 hours, and 41 minutes.

An average of 600 calories per hour are burned from climbing the hills, and an average of 23 pounds are lost by a "thru-hiker."

An average of 5 million steps are needed to travel from Maine to Georgia.

An average pair of leather hiking boots lasts 1,200 to 1,600 miles.

Source: "Database," *U.S. News & World Report*, 113, March 23, 1992, p. 14.

★ 953 ★

Recreation

Vacationers per Week

In any given week, an average of 2.3 million people are on paid vacation. During the summer months on paid vacation per week: June, 3.3 million; July, 5.9 million; and August 5.7 million.

Source: "Vacation Time is Here," *USA Today*, May 24, 1993, p. 1A. Primary source: Bureau of Labor Statistics.

★ 954 ★

Recreation

The Game of Chess

In the average game of chess, there are 10^{120} possible moves, according to one estimate. A computer would need 10^{101} years to consider all of the moves and find the best response.

Source: "Playing to Win," Elisabeth Geake, *New Scientist*, 136, September 19, 1992, p. 24.

★955★

Recreation

Fishing and Hunting as a Recreation

One in four adult Americans fish for sport (46.6 million total), 84% of whom fish in freshwater, and 26% in salt water. They spent an average of $604 per fisherman, or a total of $28.2 billion, in 1985. By contrast, an average of one in ten adult Americans hunt for sport (16.7 million). They spent an average of $830. The majority of fishermen are anglers and live in urban areas, whereas the majority of hunters live in rural areas.

Source: Council on Environmental Quality. *Environmental Trends*. Washington, DC: Council on Environmental Quality, 1989, pp. 106-107.

Travel (General)

★956★

Hotel Room Rates Per Night

The average hotel room rate was $69 per night in 1992, up from $54 in 1987.

Source: "Lower Cost Is No Boon," *USA Today*, September 8, 1992, p. 8E. Primary source: Smith Travel Research. **Remarks**: Source also includes: average daily car-rental rates, average daily food expenses, number of business trips taken, number of business travelers, average miles flown per trip, and average cents per mile for unrestricted coach airfare, all 1987/1988- 1991/1992.

★957★

Travel (General)

Hotel Room Occupancy Rate, 1989-1992

68% is considered the minimum for sustained profitability.

Year	Average Occupancy Rate of Hotel Rooms
1989	62.6%
1991	60.2
1992	61.7

Source: "Hotel Occupancy Rate Rises, New Figures Show," Edwin McDowell, *The New York Times*, February 9, 1993, p. D4. Primary source: Smith Travel Research.

★ 958 ★

Travel (General)

Travel Expenses by Region of the United States, 1993

Region	Average Cost for Hotel, Car, and Food per Day
Northwest	$193
West	178
Midwest	169
South	165

Source: "Travel Costs Creep Up," *USA Today*, April 28, 1993, p. 1B. Primary source: Corporate Travel.

★ 959 ★

Travel (General)

Duty-Free Purchases Made by Japanese and Kuwaiti Travelers, 1990

The high spenders in 1990, in terms of duty-free purchases, were Japanese travelers who spent an average of $442 (U.S. dollars), and Kuwaiti travelers who spent an average of $600 on duty-free purchases while forced to live outside their own country because of the Persian Gulf War.

Source: "Market Changes in Duty Free," *Airline, Ship & Catering Onboard Services*, 23, October-November 1991, p. 1.

★ 960 ★

Travel (General)

Spending by Japanese Tourists

In 1988, excluding airfare, Japanese tourists spend an average of $2,218 abroad per person, or a total of $18 billion. A survey by Japan Travel Bureau (JTB) reported that Japanese newlyweds traveling overseas on their honeymoon, spend an average of $750 per day and stayed an average of one week.

Source: "Hot Dog! Here Come the Tourists," *Business Tokyo*, February 1990, p. 42.

★ 961 ★

Travel (General)

Pleasure Trips in One Household, 1985-1990

Based on monthly telephone survey of 1,500 U.S. adults.

Household characteristic	1980	1985	1989	1990
Average household members on trip	2.2	2.1	2.0	2.1
Average nights per trip	---	5.6	4.4	4.4
Average miles per trip in U.S.	---	1,010	950	867

Source: U.S. Bureau of the Census. *Statistical Abstract of the United States 1992.* 112th edition. Washington, D.C.: U.S. Department of Commerce, 1992, p. 245. Primary source: U.S. Travel Center, Washington, D.C., *National Travel Survey*, annual. **Remarks**: Source also includes: business trips in one household, 1985-1990.

★ 962 ★

Travel (General)

Daily Cost of Meals and Lodging in the Most Expensive United States' Cities

City	Average Daily Cost of Meals and Lodging	City	Average Daily Cost of Meals and Lodging
New York City	$297	San Francisco	$194
Washington, D.C.	248	Atlantic City	191
Honolulu	238	Los Angeles	190
Chicago	215	Anchorage	167
Boston	197	Philadelphia	161

Source: "Big Apple Takes Big Bite Out of Travel Budget," *Parade* magazine, January 24, 1993, p. 9. Primary source: Runzheimer International, a Wisconsin-based consulting firm.

★ 963 ★

Travel (General)

Daily Cost of Meals and Lodging in Selected European Cities

City	Average Daily Cost of Meals and Lodging	
	1990	1992
Berlin	$201	$255
London	302	356
Madrid	232	238
Milan	287	319
Paris	332	314
Rome	271	278

Source: "Many Americans Stick Close to Home," Brett Pulley, *The Wall Street Journal*, July 31, 1992, p. B1. Primary source: Runzheimer International, a Wisconsin-based consulting firm.

★964★

Travel (General)

Foreign Visitors in the United States

Every Minute:
An average of 41 foreign visitors arrive in the United States.
Visitors spend an average of $22,000.
Visitors generate an average of $291 in local taxes, $851 in state taxes, and $1,392 in federal taxes.

One job is created for every 55 foreign tourists arriving in the United States.

Source: Rukeyser, Louis. *Louis Rukeyser's Business Almanac.* New York: Simon & Shuster, 1991, p. 552.

★965★

Travel (General)

Cost at Youth Hostels per Person per Night

The International Youth Hostel Federation, with 342,000 beds in 4,900 facilities in 60 countries, has an average rate of $10-15 per person per night.

Source: "Hostel System Offers Inexpensive Nightly Rates," Lucy Izon, *Los Angeles Times,* October 6, 1991, p. L20.

Travel and Traffic Congestion

★966★

Travel in London and Other Major Cities, 1900 and 1988

In 1900, the average speed of horse-drawn carriages was 8 miles per hour; in 1988 automobiles could travel no faster crossing central London. Other cities in the same predicament: in Lagos, Nigeria, traffic traveled 8 mph; in Copenhagen, 9 mph; in New York and Brisbane, Australia, 9.9 mph; in Paris, 10.5 mph; and in Stockholm, 11.2 mph.

Source: How in the World. Pleasantville, NY: The Reader's Digest Association, 1990, p. 50.

★967★

Travel and Traffic Congestion

Travel on Los Angeles Freeways, 1970 and 1992

In 1972, automobiles traveled along Los Angeles freeways at an average speed of 60 miles per hour; in 1982, they crawled along at 17 miles per hour.

Source: "Driving Into the 21st Century," Samia El-Badry and Peter K. Nance, *American Demographics*, 14, September 1992, p. 46. **Remarks**: Source also includes: average number of miles driven by men and women, 1983 and 1990; average cost of traffic congestion per metropolitan area and per vehicle, 1988.

★968★

Travel and Traffic Congestion

Vehicles on Hong-Kong Roads

According to a report by the International Road Federation in Geneva, the city with the highest density of vehicles on its roads is Hong Kong: 261 vehicles per 0.6 miles.

Source: "Travel Update," *International Herald Tribune*, November 19, 1992, p. 2.

★969★

Travel and Traffic Congestion

America's Busiest Highways

City/Route	Average Two-Way Traffic per Day
George Washington Bridge, New York City/I-95	249,300
Dan Ryan/Kennedy Expressway, Chicago/I-90/94	248,000
Santa Monica/San Bernardino, Freeway, Los Angeles/I-10	243,000
Seattle/I-5	224,000
Oakland Bay Bridge, San Francisco/I-5	223,000
Dallas/I-35	217,700
New Orleans/I-10	187,600
Houston/I-610	169,000
Miami/I-95	169,000
San Diego/I-5	161,000

Source: Krantz, Les. *The Best and Worst of Everything*. New York: Prentice Hall General Reference, 1991, p. 273. Primary source: Federal Highway Administration.

★ 970 ★

Travel and Traffic Congestion

Cost of Traffic Congestion per Vehicle in Selected Cities, 1988

Cost of congestion measured by delays, fuel costs, and insurance costs.

City	Average Cost per Vehicle	City	Average Cost per Vehicle
Boston	$830	Miami	770
Chicago	470	New York	1,030
Dallas	600	Philadelphia	570
Detroit	520	San Francisco-Oakland	780
Houston	660	Seattle-Everett	680
Los Angeles	880	Washington	1,050

Source: "Driving Into the 21st Century," Samia El-Badry and Peter K. Nance, *American Demographics*, 14, September 1992, p. 46. Primary source: Texas Transportation Institute at Texas A&M University. **Remarks**: Source also includes: average number of miles driven by men and women, 1983 and 1990; average speed on Los Angeles freeways, 1970 and 1990; average cost of traffic congestion per metropolitan area, 1988.

★ 971 ★

Travel and Traffic Congestion

Cost of Traffic Congestion per Metropolitan Area of Selected Cities, 1988

Cost of congestion measured by delays, fuel costs, and insurance costs.

City	Average Cost per Metro-politan Area (in millions)	City	Average Cost per Metro-politan Area (in millions)
Boston	$1,280	Miami	$1,040
Chicago	1,880	New York	6,040
Dallas	960	Philadelphia	1,550
Detroit	1,510	San Francisco-Oakland	2,340
Houston	1,470	Seattle-Everett	800
Los Angeles	6,880	Washington	1,730

Source: "Driving Into the 21st Century," Samia El-Badry and Peter K. Nance, *American Demographics*, 14, September 1992, p. 46. Primary source: Texas Transportation Institute at Texas A&M University. **Remarks**: Source also includes: average number of miles driven by men and women, 1983 and 1990; average speed on Los Angeles freeways, 1970 and 1990; average cost of traffic congestion per vehicle in selected metropolitan areas, 1988.

Travel by Airplane

★ 972 ★

Number of Miles per Airplane Trip

The average number of miles per airplane trip was 862 in 1991, down from 1,200 in 1987, 1,100 in 1988, and 1,000 in 1990.

Source: "Lower Cost Is No Boon," *USA Today*, September 8, 1992, p. 8E. Primary source: Smith Travel Research. **Remarks**: Source also includes: average hotel room rates, average daily car-rental rates, average daily food expenses, number of business trips taken, number of business travelers, and average cents per mile for unrestricted coach airfare, all 1987/1988-1991/1992.

★ 973 ★

Travel by Airplane

Per Mile Cost for Unrestricted Coach Airfare

The average cost of unrestricted coach airfare was 23 cents per mile in 1992, down from 26 cents in 1987 and 34 cents in 1991.

Source: "Lower Cost Is No Boon," *USA Today*, September 8, 1992, p. 8E. Primary source: Smith Travel Research. **Remarks**: Source also includes: average hotel room rates, average daily car-rental rates, average daily food expenses, number of business trips taken, number of business travelers, and average miles flown per trip, all 1987/1988-1991/1992.

Travel by Automobile

★ 974 ★

Daily Car-Rental Rates

The average daily car-rental rate was $56 in 1992, down from $65 in 1989.

Source: "Lower Cost Is No Boon," *USA Today*, September 8, 1992, p. 8E. Primary source: Smith Travel Research. **Remarks**: Source also includes: average hotel room rates, average daily food expenses, number of business trips taken, number of business travelers, average miles flown per trip, and average cents per mile for unrestricted coach airfare, all 1987/1988-1991/1992.

★975★

Travel by Automobile

Number of Automobile Trips Taken in a Lifetime

The average American takes approximately 50,000 automobile trips in his/her lifetime.

Source: Krantz, Les. *What the Odds Are: A to Z Odds on Everything You Hoped or Feared Could Happen.* New York: Prentice Hall General Reference, 1992, p. 27. **Remarks**: Source also includes: states with the most deaths resulting from automobile accidents and states with the largest death rate per mile driven.

★976★

Travel by Automobile

States With the Most Deaths Caused by Automobile Accidents

State	Average Number of Deaths per Year
California	5,223
Texas	3,568
Florida	2,875
New York	2,224

Source: Krantz, Les. *What the Odds Are: A to Z Odds on Everything You Hoped or Feared Could Happen.* New York: Prentice Hall General Reference, 1992, p. 27. Primary source: National Highway Safety Board. **Remarks**: Source also includes: number of trips the average American takes in his/her lifetime, and states with the most deaths per miles driven.

★977★

Travel by Automobile

States With the Most Deaths per Automobile Miles Driven

State	Average Number of Deaths per Million Vehicle Miles
Arizona	4.5
Mississippi	4.0
New Mexico	4.0
South Carolina	3.7

Source: Krantz, Les. *What the Odds Are: A to Z Odds on Everything You Hoped or Feared Could Happen.* New York: Prentice Hall General Reference, 1992, p. 27. Primary source: National Highway Safety Board. **Remarks**: Source also includes: number of trips the average American takes in his/her lifetime and states with the most deaths caused by automobile accidents.

★978★

Travel by Automobile

Cars and Trucks Added to the Roads Each Day

An average of 7,700 additional cars and trucks are crowding the roads of the United States every day.

Source: "Can We Prevent a Grim Future for American Transportation?" Dick Swett, member of the Public Works and Transportation Committee, *USA Today* magazine, March 1992, p. 14.

★979★

Travel by Automobile

Automobile Miles Driven by Men and Women, 1983 and 1990

Sex/Age	Average Number of Miles per Year	
	1983	1990
Men		
All	14,000	16,600
Aged 20-34	15,800	18,300
Aged 35-54	17,785	18,900
Women		
All	6,400	9,500
Aged 20-34	7,100	11,200
Aged 35-54	5,985	10,500

Source: "Driving Into the 21st Century," Samia El-Badry and Peter K. Nance, *American Demographics*, 14, September 1992, p. 46. **Remarks**: Source also includes: average speed on Los Angeles freeways, 1970 and 1990; average cost of traffic congestion per metropolitan area and per vehicle, 1988.

★980★

Travel by Automobile

Miles Driven by Rental Cars

An average rental car is driven 2,000-3,000 miles each month.

Source: "Rental Car Firms Jack Up Their Prices," James S. Hirsch, *The Wall Street Journal*, November 4, 1992, p. B1. **Remarks**: Source also includes: actual rental prices charged by the four largest rental-car companies in various United States' cities.

★ 981 ★

Travel by Automobile

Family Automobile Travel

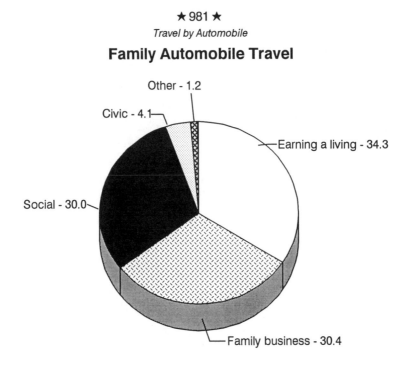

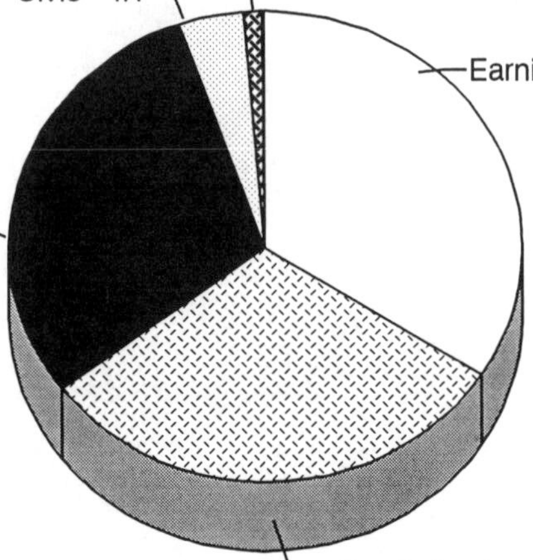

Purpose	Percent
Earning a living	34.3
Family business	30.4
Social	30.0
Civic	4.1
Other	1.2

Source: Consumer's Research, January 1989, p. 21, from U.S. Department of Transportation, Federal Highway Administration 1983 to 1984 National Transportation Study.

★ 982 ★

Travel by Automobile

Automobile Trips per Day by Sex

According to a Federal Highway Administration study, in 1990, female drivers made more but shorter trips per day than men: an average of 3.13 for women, compared with 3.04 for men.

Source: "Odds and Ends," *The Wall Street Journal*, October 19, 1992, p. B1.

★ 983 ★

Travel by Automobile

Automobile Usage in Japan and the United States

Automobiles are used less in Japan than in the United States: an average of 4,400 miles per year in Japan, vs. 9,500 miles in the United States. They are also younger: averaging 4.8 years in Japan vs. 7.6 years in the United States.

Source: "Business: Parts Makers Go Spare," *Economist*, 320, July 6, 1991, p. 68.

★ 984 ★

Travel by Automobile

Persons per Automobile in Selected Countries, 1989

Country	Average Number of People per Automobile	Country	Average Number of People per Automobile
China	822	Brazil	13.0
India	408	Argentina	7.9
Sudan	230	Poland	7.9
Yemen	314	Saudi Arabia	7.3
Zimbabwe	174	Greece	6.1
Philippines	161	Israel	5.9
Ecuador	150	Hungary	4.6
Honduras	148	Kuwait	4.4
Egypt	127	Japan	3.8
Nigeria	86	Spain	3.5
Nicaragua	75	Great Britain	2.6
Thailand	67	France	2.4
Libya	44	Sweden	2.3
Turkey	39	Australia	2.2
Colombia	35	Canada	2.1
USSR (former)	23	Luxembourg	2.0
Mexico	14	United States	1.7

Source: MVMA Motor Vehicle Facts and Figures '91. Detroit, MI: Motor Vehicle Manufacturing Association, 1991, pp. 36-37.

★985★

Travel by Automobile

Vehicle Miles Traveled by Households with Vehicles, 1988

Year	Household with Vehicles	
	Average Number of Miles Traveled per Household	Vehicles Miles Traveled per Vehicle
1983	16,830	9,399
1985	17,402	9,855
1988	18,595	10,246

Source: MVMA Motor Vehicle Facts and Figures '91. Detroit, MI: Motor Vehicle Manufacturing Association, 1991, p. 44. Primary source: U.S. Department of Energy, Energy Information Administration, *Household Vehicles Energy Consumption, 1988*. **Remarks**: Source also includes: number of households, number of households with vehicles, number of vehicles, and vehicle miles traveled for 1983-1988.

★986★

Travel by Automobile

Household Ownership of Automobiles by Region, 1988

Region	Average per Household	
	Number of Vehicles	Vehicle Miles Traveled
Northwest	1.7	17,997
Midwest	1.8	18,518
South	1.8	18,859
West	1.9	18,783
Metropolitan area	1.8	18,586
Central city	1.6	15,669
Outside central city	1.9	20,385
Nonmetropolitan	1.8	18,627

Source: MVMA Motor Vehicle Facts and Figures '91. Detroit, MI: Motor Vehicle Manufacturing Association, 1991, p. 45. Primary source: U.S. Department of Energy, Energy Information Administration, *Household Vehicles Energy Consumption, 1988*. **Remarks**: Source also includes: for 1988: number of households, number of vehicles per household, vehicle miles traveled, fuel consumption, all by household size, and by family income.

★ 987 ★

Travel by Automobile

Household Ownership of Automobiles by Income, 1988

Income	Average per Household	
	Number of Vehicles	Vehicle Miles Traveled
Less than $10,000	1.3	10,932
$10,000-$14,999	1.4	12,978
$15,000-$19,999	1.6	15,837
$20,000-$24,999	1.7	17,813
$25,000-$34,999	1.8	18,617
$35,999-$49,999	2.2	24,170
$50,000-$74,999	2.3	25,555
$75,000-	2.4	27,428

Source: MVMA Motor Vehicle Facts and Figures '91. Detroit, MI: Motor Vehicle Manufacturing Association, 1991, p. 45. Primary source: U.S. Department of Energy, Energy Information Administration, *Household Vehicles Energy Consumption, 1988.* **Remarks**: Source also includes: for 1988: number of households, number of vehicles per household, vehicle miles traveled, fuel consumption, all by region, and household size.

★ 988 ★

Travel by Automobile

Vehicle Miles Traveled per Vehicle, 1940-1989

Year	Average Annual Miles Traveled per Automobile	Year	Average Annual Miles Traveled per Automobile
1940	9,080	1980	9,141
1950	9,020	1985	9,560
1960	9,446	1987	9,878
1970	10,272	1988	10,121
1975	9,690	1989	10,382

Source: MVMA Motor Vehicle Facts and Figures '91. Detroit, MI: Motor Vehicle Manufacturing Association, 1991, p. 53. Primary source: U.S. Department of Transportation, Federal Highway Administration, *Highwat Facts*, annual. **Remarks**: Source also includes: average annual miles traveled by buses, motorcycles, trucks; total miles traveled by passenger vehicles and trucks; fuel consumed; and average fuel consumption per vehicle, all 1940-1990.

Travel on Public Transit Systems

★ 989 ★

Busiest Public Transit Systems in the World

City	Average Number of Passengers per Year (in millions)	City	Average Number of Passengers per Year (in millions)
Moscow	2,426	New York City	1,006
Tokyo	1,694	Buenos Aires	930
Bombay	1,407	Osaka	835
Paris	1,156	Leningrad	763
Mexico City	1,117	Calcutta	600

Source: Krantz, Les. *The Best and Worst of Everything.* New York: Prentice Hall General Reference, 1991, p. 107. Primary source: International Union of Public Transport, Brussels.

Miscellaneous

★ 990 ★

Pit Stops on the NASCAR's Auto Racing Circuit

In a 500-mile race on the NASCAR circuit, the average stock car makes three pit stops to make adjustments to car's handling, to refuel, and change tires. The average pit stop in the 1950's took four minutes; in the 1960's, about one minute; in 1990, less than 22 seconds.

Source: "Gentlemen, Start Your Production Lines," Joseph Blackburn, *Christian Science Monitor*, August 30, 1991, p. 19.

Chapter 14
VITAL STATISTICS AND LIFESPANS

Age of Population

★ 991 ★

Median Age of the United States' Population by Race, 1980-1990

Year	All	White	Black	Hispanic	Other
1980	30.0	30.9	24.9	23.2	26.8
1990	32.8	34.0	28.1	29.3	26.3

Source: "The Legacy of the 1980s," Judith Waldrop and Thomas Exter, *American Demographics*, 14, March 1991, p. 32-38. **Remarks**: Also included in source: the percent of the population by race and Hispanic origin.

★ 992 ★

Age of Population

Median Age of the United States' Population, 1820-2030

Year	Age
1820	16.7
1950	30.0
1970	28.0
1980	30.0
1990	33.0
2010	39.0
2030	41.8

Source: The Universal Almanac 1992. Kansas City, MO: Andrews and McMeel, 1991, p. 208. Primary source: U.S. Bureau of the Census.

★ 993 ★

Age of Population

Median Age of the Population in Selected Countries

Country	1990	2010
Nigeria	16.3	18.1
Mexico	20.0	26.5
Brazil	22.9	29.2
China	25.4	33.9
Korea	25.7	34.4
U.S.	32.9	37.4
Great Britain	35.7	40.0
Japan	37.2	42.2
Italy	36.2	42.4

Source: "Whose Population is Aging Fastest," *Fortune*, 126, September 21, 1992, p. 16.

★ 994 ★

Age of Population

Aging Population in Europe, Japan, and the United States

Because of rising life expectancy and declining birth rates, the population in many countries is aging. In 1992, fewer than one in five people in the 12 countries of the European Community were over 60 years old; by the year 2020 more than one in four people will be over 60. Also, the percent of the population over age 75 is expected to double to 12% by then. To be country specific, in Great Britain, there were 4.3 people working for each person over 65 in 1980; this ratio is expected to change to 3 to 1 by the year 2040. But in the United States and Japan the change will be more dramatic. By 2040, the segment of the population over 65 years in the United States is expected to increase around 75%; in Japan a 149% increase is expected. Because of this increase in the aged population, proposals to raise the age of retirement and pension eligibility are being studied in various countries.

Source: "Anxiety in a Graying Europe," William E. Schmidt, *International Herald Tribune*, July 14, 1993, p. 5.

★ 995 ★

Age of Population

Cities in the United States With the Oldest Population

City	Median Age
Denver, CO	38.7
New Orleans, LA	39.9
Atlanta, GA	40.5
Dallas, TX	40.9
Los Angeles, CA	41.1
Philadelphia, PA	42.1
New York, NY	42.2
Detroit, MI	43.1
Chicago, IL	45.7
Sarasota, FL	51.1

Source: Krantz, Les. *The Best and Worst of Everything.* New York: Prentice Hall General Reference, 1991, p. 66. **Remarks**: Also included in source: United States' cities with the youngest populations.

★ 996 ★

Age of Population

Cities in the United States With the Youngest Population

City	Median Age
Champaign-Urbana, IL	26.6
Killeen-Temple, TX	26.5
State College, PA	26.4
Brownsville-Harlingen, TX	25.7
Salt Lake City-Ogden, UT	25.5
McAllen-Edinburg-Mission, TX	25.3
Bryan-College Station, TX	25.1
Laredo, TX	24.8
Jacksonville, NC	24.3
Provo-Orem, UT	21.5

Source: Krantz, Les. *The Best and Worst of Everything.* New York: Prentice Hall General Reference, 1991, p. 66. **Remarks**: Also included in source: United States' cities with the oldest populations.

★ 997 ★

Age of Population

Median Age of Once-Married Mothers, 1920-1954

Median age at	All mothers born 1920-54	1920-24	1925-29	1930-34	1935-39	1940-44	1945-49	1950-54
First marriage	20.9	21.5	21.0	20.5	20.3	20.9	21.2	21.0
Birth of first child	22.9	23.9	23.2	22.4	21.9	22.6	23.2	23.4
Birth of last child	29.2	32.1	31.3	30.2	28.9	28.6	28.6	28.0

Source: Arthur J. Norton and Louisa F. Miller, "The Family Life Cycle: 1985," *Work and Family Patterns of American Women*, Current Population Reports Special Studies Series P-23, Table B, No. 165, 1990, p. 3 (Washington, DC, 1990). **Remarks**: Also included in source: the total once-married, currently married mothers; average number of children per woman; and the number of years between age at first marriage and first birth and first birth and last birth.

★ 998 ★

Age of Population

Median Age of Twice-Married Mothers, 1920-1954

Median age at--	All mothers born 1920-54	1920-24	1925-29	1930-34	1935-39	1940-44	1945-49	1950-54
First marriage	19.0	19.1	18.9	18.6	18.8	19.2	19.3	19.0
Birth of first child	20.8	21.9	21.7	20.4	20.4	20.6	20.8	21.0
Birth of last child	27.6	28.8	29.9	28.7	27.5	26.9	26.8	26.8
Separation before divorce	26.1	27.8	27.1	28.9	28.9	27.6	25.7	23.9
Divorce	27.3	28.9	29.2	30.5	30.2	28.8	26.7	25.0
Remarriage	30.9	35.4	34.5	35.1	34.9	33.3	30.1	28.1

Source: Arthur J. Norton and Louuisa F. Miller, "The Family Life Cycle: 1985," *Work and Family Patterns of American Women*, Current Population Reports Special Studies Series P-23, No. 165, 1990, p. 4. (Washington DC, 1990). **Remarks**: Also included in source: figures for the twice-married, currently married mothers; average number of children per woman; and the number of years between age at first marriage and first birth and first birth and last birth. *Notes:* Excludes separated women and women whose first marriage ended in widowhood.

★ 999 ★

Age of Population

Median Age of Divorced Mothers, 1920-1954

Median age at--	All mothers born 1920-54	1920-24	1925-29	1930-34	1935-39	1940-44	1945-49	1950-54
First marriage	20.4	21.9	21.5	21.2	20.2	20.1	20.0	19.8
Birth of first child	22.0	23.7	23.2	22.8	22.1	21.5	21.7	20.8
Birth of last child	27.2	31.0	30.4	29.1	28.5	27.0	26.3	24.0
Separation before divorce	32.7	44.0	43.7	40.2	37.5	34.4	31.2	27.3
Divorce	34.2	46.3	46.3	41.5	39.5	36.1	32.5	28.7

Source: Arthur J. Norton and Louisa F. Miller, "The Family Life Cycle: 1985," *Work and Family Patterns of American Women*, Current Population Reports, Special Studies Series P-23, No. 165, 1990, p. 4 (Washington, DC, 1990). **Remarks**: Also included in source: the total once-married, currently divorced mothers; average number of children per woman; and the number of years between age at first marriage and first birth and first birth and last birth.

★ 1000 ★

Age of Population

Age of First-time Home Buyers Vs. Repeat Home Buyers, 1976-1989

Year	Average Age	
	First-time Buyer	Repeat Buyer
1976	28.1	35.9
1980	28.3	36.4
1984	29.1	37.8
1985	28.4	38.4
1986	30.9	39.5
1987	29.6	39.1
1988	30.3	38.9
1989	29.6	39.4

Source: U.S. Bureau of the Census. *Statistical Abstract of the United States 1992*. 112th edition. Washington, D.C.: U.S. Department of Commerce, 1992, p. 724. Primary source: Chicago Title Insurance Company, Chicago, IL, *The Guarantor*, bimonthly. **Remarks**: Source also includes: median purchase price; average monthly payment; and down payment of first-time and repeat buyers, all for 1976-1990.

★ 1001 ★

Age of Population

Age of First-time Home Buyers Vs. Repeat Home Buyers, 1990-1992

Year	Average Age	
	First-time Buyer	Repeat Buyer
1990	30.5	39.1
1991	30.7	39.8
1992	31.0	40.8

Source: Chicago Title and Trust Family of Title Insurers, Chicago, IL, *Who's Buying Homes in America*, 1993, p. 3. **Remarks**: Source also includes: median price of homes, average price of homes, average monthly payment, monthly payment as a percentage of income, average number of houses looked at, average number of months buyers looked before deciding, all for 1990-1992.

Size of Population

★ 1002 ★

Population per Square Mile by State, 1991

State	Population per sq mile of land area 1991	State	Population per sq mile of land area 1991	State	Population per sq mile of land area 1991
U.S.	71.3	Kentucky	93.5	North Dakota	9.2
Alabama	80.6	Louisiana	97.6	Ohio	267.1
Alaska	1.0	Maine	40.0	Oklahoma	46.2
Arizona	33.0	Maryland	497.2	Oregon	30.4
Arkansas	45.5	Massachusetts	765.0	Pennsylvania	266.9
California	194.8	Michigan	164.9	Rhode Island	961.1
Colorado	32.6	Minnesota	55.7	South Carolina	118.2
Connecticut	679.2	Mississippi	55.3	South Dakota	9.3
Delaware	347.9	Missouri	74.9	Tennessee	120.2
District of Columbia	9,742.9	Montana	5.6	Texas	66.2
Florida	245.9	Nebraska	20.7	Utah	21.5
Georgia	114.3	Nevada	11.7	Vermont	61.3
Hawaii	176.7	New Hampshire	123.2	Virginia	158.7
Idaho	12.6	New Jersey	1,046.0	Washington	75.4
Illinois	207.6	New Mexico	12.8	West Virginia	74.8
Indiana	156.4	New York	382.4	Wisconsin	91.2
Iowa	50.0	North Carolina	138.3	Wyoming	4.7
Kansas	30.5				

Source: U.S. Bureau of the Census. *Statistical Abstract of the United States 1992*. 112th edition. Washington, D.C.: U.S. Department of Commerce, 1992, p. 23. Primary source: U.S. Bureau of the Census. **Remarks**: Also included in source: total population of the 50 states for 1970-1991, rank, and the percent change for 1970-1991.

★ 1003 ★

Size of Population

Population per Square Mile in Regions Around the World, 1990

Region	Population per square mile
Africa	58.1
Asia	302.5
Australia	6.1
Caribbean	374.2
Central America	120.4
Europe	264.5
North America	37.4
Oceania	31.6
South America	44.2

[Continued]

★ 1003 ★

Population per Square Mile in Regions Around the World, 1990

[Continued]

Region	Population per square mile
USSR (former)	33.7
World total	82.5

Source: The Universal Almanac 1992. Kansas City, MO: Andrews and McMeel, 1991, p. 293. Primary source: Population Reference Bureau, Inc., *World Population Data Sheet* (1991) and Central Intelligence Agency, *World Factbook 1991* (1990). **Remarks**: Also included in source: total area in square miles and kilometers, its percentage of world population, the total population, and population per square kilometer.

★ 1004 ★

Size of Population

Population per Square Mile in Selected Countries, 1991

Country	Population per sq. mile 1991	Country	Population per sq. mile 1991	Country	Population per sq. mile 1991
World total	107	Burundi	589	Ethiopia	125
Afghanistan	66	Byelarus	126	Fiji	105
Albania	315	Cambodia	105	Finland	42
Algeria	28	Cameroon	63	France	269
Andorra	305	Canada	8	Gabon	11
Angola	18	Cape Verde	248	Gambia, The	227
Antigua and Barbuda	376	Central African Republic	12	Georgia	201
Argentina	31	Chad	11	Germany	588
Armenia	287	Chile	46	West Germany (former)	667
Aruba	860	China: Mainland	320	Ghana	176
Australia	6	Taiwan	1,659	Greece	199
Austria	240	Colombia	84	Grenada	640
Azerbaijan	213	Comoros	569	Guatemala	221
Bahamas, The	65	Congo	18	Guinea	79
Bahrain	2,247	Costa Rica	159	Guinea-Bissau	95
Bangladesh	2,255	Cote d'Ivoire	106	Guyana	10
Barbados	1,534	Cuba	251	Haiti	591
Belgium	850	Cyprus	199	Honduras	115
Belize	26	Czechoslovakia	325	Hungary	296
Benin	113	Denmark	314	Iceland	7
Bhutan	88	Djibouti	41	India	757
Bolivia	17	Dominica	298	Indonesia	274
Botswana	6	Dominican Republic	395	Iran	93
Brazil	48	Ecuador	101	Iraq	117
Brunei	195	Egypt	142	Ireland	131
Bulgaria	209	El Salvador	677	Israel	581
Burkina	89	Equatorial Guinea	35	Italy	509
Burma	166	Estonia	96	Jamaica	595

[Continued]

★ 1004 ★

Population per Square Mile in Selected Countries, 1991

[Continued]

Country	Population per sq. mile 1991	Country	Population per sq. mile 1991	Country	Population per sq. mile 1991
Japan	814	New Zealand	32	Spain	204
Jordan	97	Nicaragua	81	Sri Lanka	697
Kazakhstan	16	Niger	17	Sudan	30
Kenya	115	Nigeria	348	Surinam	6
Kiribati	257	North Korea	469	Swaziland	129
Kuwait	320	Norway	36	Sweden	54
Kyrgyzstan	57	Oman	19	Switzerland	442
Laos	46	Pakistan	391	Syria	182
Latvia	107	Panama	84	Tajikistan	96
Lebanon	857	Papua New Guinea	22	Tanzania	79
Lesotho	154	Paraguay	31	Thailand	288
Liberia	73	Peru	45	Togo	181
Libya	6	Philippines	571	Tonga	369
Liechtenstein	458	Poland	322	Trinidad and Tobago	649
Lithuania	147	Portugal	294	Tunisia	138
Luxembourg	389	Qatar	122	Turkey	197
Madagascar	54	Romania	263	Turkmenistan	19
Malawi	260	Russia	22	Tuvalu	928
Malaysia	142	Rwanda	820	Uganda	242
Maldives	1,953	Saint Kitts and Nevis	290	Ukraine	220
Mali	18	Saint Lucia	649	United Arab Emirates	74
Malta	2,876	Saint Vincent and the Grenadines	873	United Kingdom	617
Mauritania	5			United States	71
Mauritius	1,514	San Marino	1,004	Uruguay	47
Mexico	121	Sao Tome and Principe	347	Uzbekistan	118
Moldova	331	Saudi Arabia	22	Vanuatu	30
Monaco	38,477	Senegal	107	Venezuela	59
Mongolia	4	Seychelles	392	Vietnam	538
Morocco	152	Sierra Leone	155	Western Samoa	173
Mozambique	50	Singapore	11,441	Yemen	49
Namibia	5	Solomon Islands	33	Zaire	43
Nauru	1,151	Somalia	28	Zambia	30
Nepal	371	South Africa	86	Zimbabwe	72
Netherlands	1,146	South Korea	1,138		

Source: U.S. Bureau of the Census. *Statistical Abstract of the United States 1992*. 112th edition. Washington, D.C.: U.S. Department of Commerce, 1992, pp. 820-822. Primary source: U.S. Bureau of the Census, *World Population Profile: 1991*. **Remarks**: Also included in source: the population of each country, its rank, and its area per square mile.

★ 1005 ★

Size of Population

Population per Square Mile in Selected Cities Worldwide, 1991

City & country	Population per square mile 1991	City & country	Population per square mile 1991
Tokyo-Yokohama, Japan	25,018	Chicago, United States	8,568
Mexico City, Mexico	40,037	Bangkok, Thailand	58,379
Sao Paulo, Brazil	41,466	Bogota, Colombia	74,851
Seoul, South Korea	49,101	Madras, India	51,270
New York, United States	11,480	Beijing, China	38,156
Osaka-Kobe-Kyoto, Japan	28,025	Hong Kong, Hong Kong	247,501
Bombay, India	127,461	Santiago, Chile	42,018
Calcutta, India	56,927	Pusan, South Korea	92,735
Rio de Janeiro, Brazil	44,952	Tianjin, China	98,990
Buenos Aires, Argentina	21,790	Bangalore, India	96,041
Moscow, Soviet Union (former)	27,562	Nagoya, Japan	15,606
Manila, Philippines	54,024	Milan, Italy	13,806
Los Angeles, United States	9,126	Leningrad, Soviet Union (former)	33,614
Cairo, Egypt	97,106	Madrid, Spain	68,385
Jakarta, Indonesia	130,026	Dhaka, Bangladesh	138,108
Tehran, Iran	87,312	Lahore, Pakistan	76,779
London, United Kingdom	10,429	Shenyang, China	109,974
Delhi, India	63,612	Barcelona, Spain	48,584
Paris, France	20,185	Baghdad, Iraq	41,843
Karachi, Pakistan	42,179	Manchester, United Kingdom	11,287
Lagos, Nigeria	142,821	Philadelphia, United States	8,499
Essen, Germany	10,585	San Francisco, United States	9,315
Shanghai, China	88,924	Belo Horizonte, Brazil	48,249
Lima, Peru	56,794	Kinshasa, Zaire	65,732
Taipei, Taiwan	48,517		
Istanbul, Turkey	40,476		

Source: U.S. Bureau of the Census. *Statistical Abstract of the United States 1992.* 112th edition. Washington, D.C.: U.S. Department of Commerce, 1992, pp. 825-826. Primary source: U.S. Bureau of the Census, *World Population Profile: 1991.* **Remarks**: Also included in source: population of each city, its rank, and its area per square mile.

★ 1006 ★

Size of Population

Population per Square Mile in Selected United States' Cities

Cities	Population per square mile
Detroit, MI	983
Milwaukee, WI	985
Washington, DC	1,007
Miami-Hialeah, FL	1,010
San Jose, CA	1,172
Cleveland, OH	1,208

[Continued]

★ 1006 ★

Population per Square Mile in Selected United States' Cities

[Continued]

Cities	Population per square mile
Bridgeport, CT	1,323
Philadelphia, PA	1,390
Honolulu, HI	1,404
Oakland, CA	1,449
Trenton, NJ	1,452
Newark, NJ	1,495
Boston, MA	1,555
San Francisco, CA	1,589
Los Angeles-Long Beach, CA	2,209
Bergen-Passaic, NJ	3,047
Anaheim-Santa Ana, CA	3,107
Chicago, IL	3,227
New York, NY	7,511
Jersey City, NJ	11,976

Source: "Shoulder to Shoulder: Metros with the Highest Population Densities," *Sales & Marketing Management*, 144, August 19, 1991, p. A-20.

★ 1007 ★

Size of Population

Increase in the United States' Population Due to Immigration

The Census Bureau estimates that legal and illegal immigration combined will increase the United States' population an average of 800,000 every year for the next 6 decades. This group of immigrants and their offspring will account for 21 percent of the population by the year 2050.

Source: Los Angeles Public Library/State of California, *Scan/Info*, 5, December 1992, pp. 16-18. Primary source: *New York Times*, December 4, 1992, p. A1, A4.

★ 1008 ★

Size of Population

Male-Female Ratio of the United States' Population

According to the Census Bureau, the male-female ratio in the United States is an average of 95 men to 100 women.

Source: "The Men Are Here, and Here to Stay," *The New York Times*, December 28, 1992, p. A12. **Remarks**: Also included in source: a break-down of the male-female ratio across the United States, including the lowest ratio: 78 men to 100 women in South Boston, VA; and the highest ratio of 212 men to every 100 women in West Feliciana Parish, LA (the Louisiana State Penitentiary located there accounts for the imbalance).

Size of Household

★ 1009 ★

Household Size by Metropolitan Areas in the United States, 1990

Metropolitan area	Persons per household
New York-Northern New Jersey-Long Island, NY-NJ-CT	2.67
Los Angeles-Anaheim-Riverside, CA	2.91
Chicago-Gary-Lake County, IL-IN-WI	2.72
San Francisco-Oakland-San Jose, CA	2.61
Philadelphia-Wilmington-Trenton, PA-NJ-DE-MD	2.66
Detroit-Ann Arbor, MI	2.67
Boston-Lawrence-Salem, MA-NH	2.61
Washington, DC-MD-VA	2.62
Dallas-Fort Worth, TX	2.64
Houston-Galveston-Brazoria, TX	2.75
Miami-Fort Lauderdale, FL	2.58
Atlanta, GA	2.64
Cleveland-Akron-Lorain, OH	2.56
Seattle-Tacoma, WA	2.49
San Diego, CA	2.69
Minneapolis-St. Paul, MN-WI	2.58
St. Louis, MO-IL	2.59
Baltimore, MD	2.64
Pittsburgh-Beaver Valley, PA	2.46
Phoenix, AZ	2.59
Tampa-St. Petersburg-Clearwater, FL	2.32
Denver-Boulder, CO	2.46
Cincinnati-Hamilton, OH-KY-IN	2.61
Milwaukee-Racine, WI	2.61
Kansas City, MO-KS	2.55
Sacramento, CA	2.60
Portland-Vancouver, OR-WA	2.52
Norfolk-Virginia Beach-Newport News, VA	2.69
Columbus, OH	2.54
San Antonio, TX	2.82

Source: U.S. Bureau of the Census. *Statistical Abstract of the United States 1992*. 112th edition. Washington, D.C.: U.S. Department of Commerce, 1992, p. 50. Primary source: U.S. Bureau of the Census, unpublished data from *1990 Census of Population and Housing Summary Tape File 1C*. **Remarks**: Also included in source: total number of households and the percent change, 1980-1990; total number of families; total number of married couples, with and without children; number of one-parent families; and number of one-person households, for 60 cities in the United States.

★ 1010 ★

Size of Household

Household Size by Type of Household and Presence of Children, 1991

Type of household and presence of children	Persons per household 1991
Total households	2.63
Family households	3.23
With own children under 18	3.95
Without own children under 18	2.54
Married couple family	3.26
With own children under 18	4.17
Without own children under 18	2.46
Male householder, no spouse present	3.11
With own children under 18	3.49
Without own children under 18	2.85
Female householder, no spouse present	3.12
With own children under 18	3.24
Without own children under 18	2.92
Nonfamily households	1.22
Living alone	1.00
Male householder	1.32
Living alone	1.00
Female householder	1.15
Living alone	1.00

Source: U.S. Bureau of the Census. *Statistical Abstract of the United States 1992*. 112th edition. Washington, D.C.: U.S. Department of Commerce, 1992, p. 51. Primary source: U.S. Bureau of the Census, *Current Population Reports*, series P-20, No. 458, earlier reports, and unpublished data. **Remarks**: Also included in source: total number of households, 1980-1991 and the percent change, 1980-1991; the percent distribution, and number of persons in households, 1991 and the percent distribution.

★ 1011 ★

Size of Household

Household Size, 1960-1991

Year	Average Size of Household
1960	3.33
1970	3.14
1975	2.94
1980	2.76
1985	2.69
1989	2.62
1990	2.63
1991	2.63

Source: U.S. Bureau of the Census. *Statistical Abstract of the United States 1992*. 112th edition. Washington, D.C.: U.S. Department of Commerce, 1992, p. 46. Primary source: U.S. Bureau of the Census, *Current Population Reports*, series P-20, No. 458. **Remarks**: Also included in source: number of households, families, subfamilies, married couple families, and unrelated individuals, 1960-1991.

★ 1012 ★

Size of Household

Family Size, 1960-1991

Year	Average Size of Family
1960	3.67
1970	3.58
1975	3.42
1980	3.29
1985	3.23
1989	3.16
1990	3.17
1991	3.18

Source: U.S. Bureau of the Census. *Statistical Abstract of the United States 1992*. 112th edition. Washington, D.C.: U.S. Department of Commerce, 1992, p. 46. Primary source: U.S. Bureau of the Census, *Current Population Reports*, series P-20, No. 458. **Remarks**: Also included in source: number of households, families, subfamilies, married couple families, and unrelated individuals, 1960-1991.

★ 1013 ★

Size of Household

Household Size by Age and Race

Age of householder	Average Number of Persons per household		
	Total	Black	Hispanic
15-24 years	2.38	2.67	3.13
25-29 years	2.64	2.90	3.52
30-34 years	3.03	2.93	3.68
35-44 years	3.28	3.35	3.98
45-54 years	2.96	3.06	3.63
55-64 years	2.35	2.72	2.98
65-74 years	1.89	2.26	2.29
75 years-	1.56	1.93	1.96

Source: U.S. Bureau of the Census. *Statistical Abstract of the United States 1992*. 112th edition. Washington, D.C.: U.S. Department of Commerce, 1992, p. 48. Primary source: U.S. Bureau of the Census, *Current Population Reports*, series P-20, No. 458 and series P-60, No. 174.. **Remarks**: Also included in source: number of households by race, Hispanic origin, age, size and marital status.

★ 1014 ★

Size of Household

Household Size by Race and Region of the United States

Region	Average Number of Persons per household		
	Total	Black	Hispanic
Northeast	2.63	2.83	3.18
Midwest	2.58	2.82	3.43
South	2.63	2.96	3.34
West	2.71	2.53	3.62

Source: U.S. Bureau of the Census. *Statistical Abstract of the United States 1992*. 112th edition. Washington, D.C.: U.S. Department of Commerce, 1992, p. 48. Primary source: U.S. Bureau of the Census, *Current Population Reports*, series P-20, No. 458 and series P-60, No. 174. **Remarks**: Also included in source: number of households by race, Hispanic origin, age, size and marital status.

★ 1015 ★

Size of Household

Household Size by State, 1990

As of April 1.

State	Persons per household 1990	State	Persons per household 1990	State	Persons per household 1990
U.S.	2.63	Kentucky	2.60	North Dakota	2.55
Alabama	2.62	Louisiana	2.74	Ohio	2.59
Alaska	2.80	Maine	2.56	Oklahoma	2.53
Arizona	2.62	Maryland	2.67	Oregon	2.52
Arkansas	2.57	Massachusetts	2.58	Pennsylvania	2.57
California	2.79	Michigan	2.66	Rhode Island	2.55
Colorado	2.51	Minnesota	2.58	South Carolina	2.68
Connecticut	2.59	Mississippi	2.75	South Dakota	2.59
Delaware	2.61	Missouri	2.54	Tennessee	2.56
District of Columbia	2.26	Montana	2.53	Texas	2.73
Florida	2.46	Nebraska	2.54	Utah	3.15
Georgia	2.66	Nevada	2.53	Vermont	2.57
Hawaii	3.01	New Hampshire	2.62	Virginia	2.61
Idaho	2.73	New Jersey	2.70	Washington	2.53
Illinois	2.65	New Mexico	2.74	West Virginia	2.55
Indiana	2.61	New York	2.63	Wisconsin	2.61
Iowa	2.52	North Carolina	2.54	Wyoming	2.63
Kansas	2.53				

Source: U.S. Bureau of the Census. *Statistical Abstract of the United States 1992.* 112th edition. Washington, D.C.: U.S. Department of Commerce, 1992, p. 49. Primary source: U.S. Bureau of the Census, *Census of Population: 1970; 1980 Census of Population*, vol. 1, chapter B; and unpublished data from *1990 Census of Population and Housing Summary Tape File 1C*. **Remarks**: Also included in source: number of households, 1970-1990 and the percent change; number of families; and number of nonfamily households.

★ 1016 ★

Size of Household

Household Size in Selected Regions Worldwide, 1970-1985

Region	Number of persons per household			
	1970	1975	1980	1985
Developed countries	3.36	3.18	3.03	2.90
Northern America	3.23	3.04	2.88	2.73
Japan	3.90	3.60	3.39	3.22
Oceania	3.65	3.52	3.40	3.25
Northern Europe	3.00	2.87	2.75	2.67
Western Europe	2.98	2.86	2.73	2.64
Southern Europe	3.67	3.53	3.40	3.28
Eastern Europe	3.10	2.97	2.84	2.73
USSR (former)	3.57	3.32	3.14	2.98
Developing countries	5.07	5.00	4.89	4.76
Eastern Africa	4.99	4.97	4.95	4.93

[Continued]

★ 1016 ★

Household Size in Selected Regions Worldwide, 1970-1985

[Continued]

Region	Number of persons per household			
	1970	1975	1980	1985
Northern Africa	5.44	5.41	5.31	5.15
Middle Africa	4.58	4.60	4.62	4.66
Southern Africa	5.00	5.05	5.09	5.09
Western Africa	4.90	4.93	4.97	5.00
Caribbean	4.67	4.64	4.56	4.44
Middle America	5.01	4.99	4.90	4.80
Temperate South America	4.02	3.85	3.69	3.53
Tropical South America	5.19	5.10	4.96	4.81
China	4.58	4.42	4.21	4.01
East Asia[1]	5.63	5.33	5.02	4.64
South Asia	5.46	5.44	5.39	5.28

Source: 1989 Report on the World Social Situation, 1989 (New York: United Nations, 1989), p. 2. Primary source: United Nations, Department of International Economic and Social Affairs, "Estimates and projections of households" (provisional report, 22 January 1979), table 5. *Note:* 1. Excluding China and Japan.

★ 1017 ★

Size of Household

Size of American Households, 1790, 1900, 1940, 1950, 1992

Year	Average Number of People per Household
1790	5.8
1900	4.8
1940	3.7
1950	3.3
1990	2.6

Source: "Myths of the 1950s," Diane Crispell, *American Demographics*, 14, August 1992, pp. 38-43. **Remarks**: Also included in source: median age at first marriage, birth rates, suburban population, and crime rates.

★ 1018 ★

Size of Household

Children per Household, 1991

In 1991, the average number of children per American family was .96, dropping from two children for a typical family in 1977. Among America's 24 million two-parent families with children, there are an average of 1.88 "own" children (householder's biological sons and daughters, stepchildren, and adopted children) living in each household. Of the seven million families headed by a single mother, 1.74 is the average number of children per family; and the one million families headed by a single dad, have an average of 1.5 children. Along racial lines in households with children, Hispanics have an average of 2.18 children; blacks, an average of 1.90; and whites, 1.87.

Source: "Fractional Families," *American Demographics*, 14, December 1992, p. 6.

Birth

★ 1019 ★

Number of Children per Woman in Regions Worldwide, 1982

Region	Average no. of children per woman	Region	Average no. of children per woman
World	3.8	Asia	3.9
Developed countries	2.0	South-West Asia	5.8
Developing countries	4.4	Middle South Asia	5.5
		South-East Asia	4.7
Africa	6.4	East Asia	2.3
Northern Africa	6.2		
Western Africa	6.8	Europe	2.0
Eastern Africa	6.6	Northern Europe	1.8
Middle Africa	6.0	Western Europe	1.6
Southern Africa	5.2	Eastern Europe	2.3
		Southern Europe	2.3
North America	1.8		
		USSR (former)	2.4
Latin America	4.5		
Middle Latin America	5.3	Oceania	2.8
Caribbean	3.8		
Tropical South Latin America	4.6		
Temperate South Latin America	2.9		

Source: "Global Health-for-all and Other Indicators Relating to Women, Health and Development," *Women, Health and Development*, Annex, p. 40 (1985). Primary source: UNESCO,; WHO estimates; Population Reference Bureau and United Nations Population Division. **Remarks**: Also included in source: percentage of literate adults, percentage of births attended by trained personnel, percentage of infants with low birth weight, infant mortality rate, percentage enrolled in schools, and percentage of women 15-19 married.

★ 1020 ★

Birth

Median Birth Weight, 1970-1989

Represents registered births. Excludes births to nonresidents of the United States.

Year	Median Birth Weight		
	Total	White	Black
1970	7 lb.-4 oz.	7 lb.-5 oz.	6 lb.-14 oz.
1975	7 lb.-5 oz.	7 lb.-7 oz.	6 lb.-15 oz.
1980	7 lb.-7 oz.	7 lb.-8 oz.	7 lb.-0 oz.
1981	7 lb.-7 oz.	7 lb.-8 oz.	7 lb.-0 oz.
1982	7 lb.-7 oz.	7 lb.-8 oz.	7 lb.-0 oz.
1983	7 lb.-7 oz.	7 lb.-8 oz.	7 lb.-0 oz.
1984	7 lb.-7 oz.	7 lb.-9 oz.	7 lb.-0 oz.
1985	7 lb.-7 oz.	7 lb.-9 oz.	7 lb.-0 oz.
1986	7 lb.-7 oz.	7 lb.-9 oz.	7 lb.-0 oz.
1987	7 lb.-7 oz.	7 lb.-9 oz.	7 lb.-0 oz.
1988	7 lb.-7 oz.	7 lb.-9 oz.	7 lb.-0 oz.
1989	7 lb.-7 oz.	7 lb.-8 oz.	6 lb.-15 oz.

Source: U.S. Bureau of the Census. *Statistical Abstract of the United States 1992*. 112th edition. Washington, D.C.: U.S. Department of Commerce, 1992, p. 68. Primary source: U.S. National Center for Health Statistics, *Vital Statistics of the United States*, annual; *Monthly Vital Statistics Report*, and unpublished data. **Remarks**: Also included in source: number of births attended and percent of births with low birth weight.

★ 1021 ★

Birth

Population Increase

Worldwide, an average of nine babies are born and three people die every two seconds. The population grows an average of 93 million people per year, 7.7 million per month, 1.8 million per week, 254,000 per day, 10,600 per hour, 3 people each second. The annual average growth will be 94 million by the year 2000 and 98 million by 2020, at which time 98% of the increase will be in developing countries.

Source: The Universal Almanac 1992. Kansas City, MO: Andrews and McMeel, 1991, p. 316. Primary source: U.S. Bureau of the Census, *Worldwide Population Profile: 1989* (1989).

★ 1022 ★

Birth

Birth Rates: Tuesday and September

According to the National Center for Health Statistics (NCHS), Tuesday is the biggest day for births. In 1988, an average of 11,700 babies were born on a Tuesday—9% higher than the daily average for that year. And September is the month for high birth rates. In 1990, there were an average of 12,100 babies born every day in September, compared with a daily average of 11,400 for the entire year.

Source: "Seasons: The Birthday Boost," Judith Waldrop, *American Demographics*, 13, September 1991, p. 4.

★ 1023 ★

Birth

Babies Born in the United States, 1946-1992

During the years 1946 to 1964, an average of four million babies were born each year, or 76 million total; between 1965 and 1976, an average of 3.4 million babies were born, or a total of 41 million babies; since 1977, an average of 3.7 million babies were born each year, or 56 million.

Source: "How to Talk to Young Adults," Susan Mitchell, *American Demographics*, 15, April 1993, p. 50.

Marriage and Divorce

★ 1024 ★

Median Age at Marriage in Selected Countries

Country	Year	Median age at marriage	
		Female	Male
Australia	1981	23.0	25.0
Canada	1981	22.0	25.0
China	1982	22.2	24.5
France	1982	23.8	26.2
Germany, Federal Republic	1982	24.0	28.0
Hungary	1981	20.0	24.0
India	1981	17.9	23.3
Israel	1980	21.8	25.3

[Continued]

★ 1024 ★

Median Age at Marriage in Selected Countries

[Continued]

Country	Year	Median age at marriage	
		Female	Male
Japan	1980	25.0	28.0
United States	1982	23.0	25.0

Source: Selected from "Median Age at Marriage and Median Age of Mother at First Birth, Selected Countries; Selected Years, 1980 to 1985," *Children's Well- Being: An International Comparison*, A Report of the Select Committee on Children, Youth, and Families, 101st Cong., 2d sess., March 1990, p. 120. Primary source: U.S. Bureau of the Census, Center for International Research, International Data Base; and United Nations, 1988, *Demographic Yearbook 1986*, table 26. Also in source: Median age of mother at first birth. The median age at marriage is the age at which 50% of the population is married. Median age at marriage calculated at the U.S. Bureau of the Census based on reported distribution of persons by marital status.

★ 1025 ★

Marriage and Divorce

Median Age at First Marriage, 1890-1990

Year	Men	Women
1890	26.1	22.0
1900	25.9	21.9
1910	25.1	21.6
1920	24.6	21.2
1930	24.3	21.3
1940	24.3	21.5
1950	22.8	20.3
1955	22.6	20.2
1960	22.8	20.3
1965	22.8	20.6
1970	23.2	20.8
1975	23.5	21.1
1980	24.7	22.0
1985	25.5	23.3
1989	26.2	23.8
1990	26.1	23.9

Source: The World Almanac and Book of Facts 1992. New York: World Almanac, 1991, p. 943. Primary source: Bureau of the Census.

★ 1026 ★

Marriage and Divorce

Median Age of Women at Marriage, 1970-1988

Year	First marriage	Remarriage	
		Divorced	Widowed
1970	20.6	30.1	51.2
1975	20.8	30.2	52.4
1977	21.1	30.2	53.1
1978	21.4	30.5	52.6
1979	21.6	30.8	55.2
1980	21.8	31.0	53.6
1981	22.0	31.2	53.6
1982	22.3	31.6	54.1
1983	22.5	32.0	54.0
1984	22.8	32.5	54.2
1985	23.0	32.8	54.6
1986	23.3	33.1	53.3
1987	23.6	33.3	53.9
1988	23.7	33.6	53.9

Source: U.S. Bureau of the Census. *Statistical Abstract of the United States 1992.* 112th edition. Washington, D.C.: U.S. Department of Commerce, 1992, p. 91. Primary source: U.S. National Center for Health Statistics, *Vital Statistics of the United States*, annual; *Monthly Vital Statistics Report*, and unpublished data. **Remarks**: Also included in source: median age of men at marriage and marriage rates for men and women.

★ 1027 ★

Marriage and Divorce

Median Age of Men at Marriage, 1970-1988

Year	First marriage	Remarriage	
		Divorced	Widowed
1970	22.5	34.5	58.7
1975	22.7	33.6	59.4
1977	23.0	33.6	60.1
1978	23.2	33.8	59.7
1979	23.4	33.9	61.7
1980	23.6	34.0	61.2
1981	23.9	34.4	61.0
1982	24.1	34.9	61.7
1983	24.4	35.3	62.0
1984	24.6	35.9	62.4
1985	24.8	36.1	62.7
1986	25.1	36.6	62.9
1987	25.3	36.7	62.8
1988	25.5	37.0	63.0

Source: U.S. Bureau of the Census. *Statistical Abstract of the United States 1992.* 112th edition. Washington, D.C.: U.S. Department of Commerce, 1992, p. 91. Primary source: U.S. National Center for Health Statistics, *Vital Statistics of the United States*, annual; *Monthly Vital Statistics Report*, and unpublished data. **Remarks**: Also included in source: median age of women at marriage, marriage rates for men and women.

★1028★

Marriage and Divorce

Median Duration of Marriage, 1970-1988

Year	Median duration of marriage (years)
1970	6.7
1975	6.5
1980	6.8
1981	7.0
1982	7.0
1983	7.0
1984	6.9
1985	6.8
1986	6.9
1987	7.0
1988	7.1

Source: U.S. Bureau of the Census. *Statistical Abstract of the United States 1992.* 112th edition. Washington, D.C.: U.S. Department of Commerce, 1992, p. 92. Primary source: U.S. National Center for Health Statistics, *Vital Statistics of the United States*, annual; *Monthly Vital Statistics Report*, and unpublished data. **Remarks**: Also included in source: median age at divorce and the number of children involved per divorce decree.

★1029★

Marriage and Divorce

Median Age at Divorce by Sex, 1970-1988

Year	Median age at divorce (years)	
	Men	Women
1970	32.9	29.8
1975	32.2	29.5
1980	32.7	30.3
1981	33.1	30.6
1982	33.6	31.1
1983	34.0	31.5
1984	34.3	31.7
1985	34.4	31.9
1986	34.6	32.1
1987	34.9	32.5
1988	35.1	32.6

Source: U.S. Bureau of the Census. *Statistical Abstract of the United States 1992.* 112th edition. Washington, D.C.: U.S. Department of Commerce, 1992, p. 92. Primary source: U.S. National Center for Health Statistics, *Vital Statistics of the United States*, annual; *Monthly Vital Statistics Report*, and unpublished data. **Remarks**: Also included in source: median duration of marriage and the number of children per divorce decree.

★ 1030 ★

Marriage and Divorce

Number of Children per Divorce Decree, 1970-1988

Year	Average Number of Children per Decree
1970	1.22
1975	1.08
1980	0.98
1981	0.97
1982	0.94
1983	0.94
1984	0.92
1985	0.92
1986	0.90
1987	0.89
1988	0.89

Source: U.S. Bureau of the Census. *Statistical Abstract of the United States 1992*. 112th edition. Washington, D.C.: U.S. Department of Commerce, 1992, p. 92. Primary source: U.S. National Center for Health Statistics, *Vital Statistics of the United States*, annual; *Monthly Vital Statistics Report*, and unpublished data. **Remarks**: Also included in source: median duration of marriage and the median age at divorce.

Accidental Death

★ 1031 ★

The Cost of Accidental Death

Cost by Type of Accident	Equivalent to:
Motor vehicle $96.1 billion	$11,500 rebate on each new car sold, 1991 93 cents of every dollar spent on gas & oil, 1991
Work $63.3 billion	46 cents of every dollar of corporate dividends, 1991 20 cents of every dollar of pre-tax corporate profits, 1991
Home $22.0 billion	$26,200 rebate on each new home built in 1991 14 cents of every dollar of property tax paid, 1991
Public	$20,300 bonus for each police officer and fireman

[Continued]

★ 1031 ★

The Cost of Accidental Death

[Continued]

$13.9 billion	$1.5 million grant to each public library in U.S.
ALL	39 cents of every dollar paid in federal income tax, 1991
$177.2 billion	20 cents of every dollar spent on food in the U.S., 1991

Source: National Safety Council. *Accident Facts 1992.* Itasca, IL: National Safety Council, Safety & Health News Center, 1992, p. 25.

★ 1032 ★

Accidental Death

Civilian Casualties During War

In World War I: an average of 137,000 civilians died each month, or seven million by Armistice Day, 1918. In World War II: an average of 388,000 civilians died every month from 1939-1945.

Source: "War, Wealth, and a New Era in History," Alvin Toffler and Heidi Toffler, *World Monitor*, 4, May 1991, p. 49.

★ 1033 ★

Accidental Death

Deaths from Lawn Mowers

According to the Consumer Product Safety Commission, each year in the United States there are an average of 75 deaths due to lawn mower accidents, most often involving persons over the age of 65 or under the age of five. Every year there are around 55,000 injuries involving lawn mowers.

Source: Wellness Letter. University of California at Berkeley, School of Public Health, vol. 9, issue 10, July 1993, p. 1.

★ 1034 ★

Accidental Death

Babies Who Die Before Age One in Selected Countries, 1991

Country	Income per Capita	Average Number of Deaths per 1,000 births
Brazil	$2,920	55
Mexico	2,870	30
Russia	3,220	23
China	370	22
United States	22,560	9

Source: "Infant Mortality: How Brazil Compares," *The New York Times*, May 14, 1993, p. A8. Primary source: UNICEF; World Bank Atlas.

★ 1035 ★

Accidental Death

Deaths from Hair Dryers

Since January 1991, hair dryers have been made with a safety plug that automatically shuts off if they fall into water. But plugged-in dryers made before that time are lethal if they land in water. An average of 17 people a year, most of them children under 10, died in this fashion before hair dryers became safer.

Source: "In the Stores," *Consumer Reports*, 57, August 1992, p. 532.

★ 1036 ★

Accidental Death

Accidental Deaths by Type of Accident, 1991

Type of Accident	One Every -	Average Number of Deaths per		
		Hour	Day	Week
Motor vehicle				
Deaths	5 minutes	5	119	840
Injuries	4 seconds	180	4,400	30,800
Work				
Deaths	53 minutes	1	190	190
Injuries	19 seconds	190	32,700	32,700
Workers-off-the-job				
Deaths	16 minutes	4	640	640
Injuries	13 seconds	290	48,100	48,100
Home				
Deaths	26 minutes	2	390	390
Injuries	10 seconds	350	59,600	59,600
Public nonmotor vehicle				
Deaths	29 minutes	2	350	350

[Continued]

★ 1036 ★

Accidental Deaths by Type of Accident, 1991

[Continued]

Type of Accident	One Every -	Average Number of Deaths per		
		Hour	Day	Week
Injuries	14 seconds	260	44,200	44,200
ALL				
Deaths	5 minutes	10	1,690	1,690
Injuries	4 seconds	980	165,400	165,400

Source: National Safety Council. *Accident Facts 1992.* Itasca, IL: National Safety Council, Safety & Health News Center, 1992, p. 25.

Suicide

★ 1037 ★

Suicide in the Military

In the combined armed services, an average of 250 active members commit suicide every year. The 1991 rate for the combined services was 12.6 per 100,000; the army's was 14.3 per 100,000. The rate for males between the ages of 18 and 34 (a majority of military personnel fall within this group) was 20 suicides per 100,000 in 1991.

Source: "18% Rise in Suicides in the Army is Found Between 1987 and 1991," *New York Times,* September 8, 1992, p. A14.

★ 1038 ★

Suicide

Deaths Due to Suicide

Each year an average of 30,000 people commit suicide, or 1 person every 17.1 minutes.

Source: National Center for Health Statistics. *Health United States 1991.* Washington, D.C.: U.S. Department of Health and Human Services, 1992.

Life Expectancy

★ 1039 ★

Life Expectancy in Selected Countries, 1991-2000

Country	Expectation of Life at Birth (years)		Country	Expectation of Life at Birth (years)		Country	Expectation of Life at Birth (years)	
	1991	2000		1991	2000		1991	2000
United States	75.7	77.0	Ghana	54.6	57.5	Panama	74.0	76.5
Afghanistan	43.5	47.9	Greece	77.7	79.3	Papua New Guinea	55.4	59.1
Albania	75.1	77.5	Guatemala	63.2	66.9	Paraguay	69.7	71.4
Algeria	66.7	69.6	Guinea	42.8	47.0	Peru	64.3	68.1
Angola	44.3	48.9	Haiti	53.6	56.1	Philippines	64.6	66.9
Argentina	70.9	72.3	Honduras	66.0	70.0	Poland	72.9	76.1
Australia	77.0	78.7	Hungary	71.6	75.5	Portugal	74.7	77.4
Austria	77.3	78.9	India	57.2	61.4	Romania	71.9	75.6
Bangladesh	53.0	57.5	Indonesia	61.0	66.6	Rwanda	52.5	56.8
Belgium	77.1	78.8	Iran	64.5	69.1	Saudi Arabia	65.9	70.6
Benin	50.5	54.7	Iraq	67.0	71.0	Senegal	55.1	60.0
Bolivia	61.5	65.5	Ireland	75.5	77.9	Sierra Leone	44.8	49.7
Brazil	65.2	67.5	Israel	77.0	78.5	Singapore	74.8	76.9
Bulgaria	72.7	76.0	Italy	78.1	79.6	Somalia	55.9	60.9
Burkina	52.2	56.4	Jamaica	73.6	75.9	South Africa	64.2	67.0
Burma	54.9	58.0	Japan	79.2	80.8	South Korea	69.7	72.3
Burundi	52.4	56.4	Jordan	71.2	73.2	Soviet Union (former)	69.8	73.2
Cambodia	49.3	54.6	Kenya	61.5	64.6	Spain	78.3	79.7
Cameroon	51.0	55.2	Kuwait	73.6	76.7	Sri Lanka	71.1	73.0
Canada	77.5	79.2	Laos	50.2	54.7	Sudan	53.0	56.8
Central African Republic	47.1	51.5	Lebanon	68.4	71.4	Sweden	77.8	79.3
Chad	39.8	43.5	Liberia	56.4	60.4	Switzerland	79.1	80.1
Chile	73.4	75.1	Libya	68.1	72.5	Syria	69.4	73.2
China: Mainland	70.0	72.4	Madagascar	52.6	56.7	Tanzania	52.0	56.1
Taiwan	74.6	76.3	Malawi	49.2	52.9	Thailand	68.5	72.1
Colombia	71.0	74.1	Malaysia	68.1	70.9	Togo	55.6	59.7
Congo	54.2	58.3	Mali	46.1	50.3	Tunisia	71.9	74.8
Costa Rica	76.8	78.9	Mexico	72.2	75.4	Turkey	69.8	74.5
Cote d'Ivoire	54.3	58.3	Mongolia	65.1	68.4	Uganda	51.0	55.0
Cuba	75.6	76.4	Morocco	64.6	69.8	United Arab Emirates	70.9	73.0
Czechoslovakia	72.9	76.3	Mozambique	47.4	51.5	United Kingdom	76.5	78.5
Denmark	75.9	78.0	Nepal	50.6	55.2	Uruguay	72.6	72.9
Dominican Republic	67.2	70.2	Netherlands	77.8	79.3	Venezuela	74.2	77.1
Ecuador	66.2	69.0	New Zealand	75.5	78.0	Vietnam	64.7	68.8
Egypt	60.8	65.0	Nicaragua	62.5	66.8	Yemen	49.9	54.6
El Salvador	65.5	69.6	Niger	51.0	55.5	Yugoslavia (former)	73.0	76.3
Ethiopia	51.3	55.4	Nigeria	48.9	52.7	Zaire	53.9	57.9
Finland	75.8	78.1	North Korea	69.0	71.5	Zambia	56.4	60.4
France	77.8	79.3	Norway	77.1	78.8	Zimbabwe	61.7	65.7
Germany	75.8	77.9	Pakistan	56.6	59.6			

Source: U.S. Bureau of the Census. *Statitsical Abstract of the United States 1992*. 112th edition. Washington, D.C.: U.S. Department of Commerce, 1992, pp. 824-825. Primary source: U.S. Bureau of the Census, *World Population Profiles: 1991*. **Remarks**: Also included in source: crude birth rate, crude death rate, infant mortality rate, and total fertility rate.

★ 1040 ★

Life Expectancy

Life Expectancy Worldwide, 1955-2000

In 1950-1955, life expectancy worldwide was 47.4 years, rising to 64.5 years by 1995-2000. The lowest life expectancy is 39.4 years in Ethiopia and Sierre Leone for men, 42 for females in Afghanistan; the highest is 81.9 years for women in Japan, 75.8 years for men in Japan.

Source: Guinness Book of Records 1992. New York: Bantam Book, 1992, p. 197.

★ 1041 ★

Life Expectancy

Life Expectancy by Race, Sex, and Age: 1989

Age in 1989 (years)	Expectation of Life in Years				
	Total	White		Black	
		Male	Female	Male	Female
At birth	75.3	72.7	79.2	64.8	73.5
1	75.0	72.3	78.8	65.2	73.8
5	71.1	68.5	74.9	61.4	70.0
10	66.2	63.6	70.0	56.5	65.1
15	61.3	58.6	65.1	51.6	60.2
20	56.6	54.0	60.2	47.1	55.3
25	51.9	49.4	55.3	42.7	50.6
30	47.2	44.7	50.5	38.5	45.9
35	42.5	40.1	45.7	34.3	41.3
40	37.9	35.6	40.9	30.4	36.8
45	33.4	31.1	36.1	26.6	32.4
50	28.9	26.7	31.5	23.0	28.2
55	24.7	22.5	27.1	19.6	24.1
60	20.8	18.7	22.9	16.4	20.4
65	17.2	15.2	19.0	13.6	17.0
75	10.9	9.4	11.9	8.8	11.0
85	6.2	5.3	6.5	5.6	6.7

Source: U.S. Bureau of the Census. *Statistical Abstract of the United States 1992.* 112th edition. Washington, D.C.: U.S. Department of Commerce, 1992, p. 77. Primary source: U.S. National Center for Health Statistics, *Vital Statistics of the United States*, annual; and unpublished sources. **Remarks**: Also included in source: expected deaths per 1,000 alive at specified age by race and sex.

★ 1042 ★

Life Expectancy

Life Expectancy at Birth, 1970-2010

Year	Total			White			Black and other			Black		
	Total	Male	Female	Total	Male	Female	Total	Male	Female	Total	Male	Female
1970	70.8	67.1	74.7	71.7	68.0	75.6	65.3	61.3	69.4	64.1	60.0	68.3
1975	72.6	68.8	76.6	73.4	69.5	77.3	68.0	63.7	72.4	66.8	62.4	71.3
1976	72.9	69.1	76.8	73.6	69.9	77.5	68.4	64.2	72.7	67.2	62.9	71.6
1977	73.3	69.5	77.2	74.0	70.2	77.9	68.9	64.7	73.2	67.7	63.4	72.0
1978	73.5	69.6	77.3	74.1	70.4	78.0	69.3	65.0	73.5	68.1	63.7	72.4
1979	73.9	70.0	77.8	74.6	70.8	78.4	69.8	65.4	74.1	68.5	64.0	72.9
1980	73.7	70.0	77.4	74.4	70.7	78.1	69.5	65.3	73.6	68.1	63.8	72.5
1981	74.2	70.4	77.8	74.8	71.1	78.4	70.3	66.1	74.4	68.9	64.5	73.2
1982	74.5	70.9	78.1	75.1	71.5	78.7	71.0	66.8	75.0	69.4	65.1	73.7
1983	74.6	71.0	78.1	75.2	71.7	78.7	71.1	67.2	74.9	69.6	65.4	73.6
1984	74.7	71.2	78.2	75.3	71.8	78.7	71.3	67.4	75.0	69.7	65.6	73.7
1985	74.7	71.2	78.2	75.3	71.9	78.7	71.2	67.2	75.0	69.5	65.3	73.5
1986	74.8	71.3	78.3	75.4	72.0	78.8	71.2	67.2	75.1	69.4	65.2	73.5
1987	75.0	71.5	78.4	75.6	72.2	78.9	71.3	67.3	75.2	69.4	65.2	73.6
1988	74.9	71.5	78.3	75.6	72.3	78.9	71.2	67.1	75.1	69.2	64.9	73.4
1989	75.3	71.8	78.6	76.0	72.7	79.2	71.2	67.1	75.2	69.2	64.8	73.5
1990, prel.	75.4	72.0	78.8	76.0	72.6	79.3	72.4	68.4	76.3	70.3	66.0	74.5
Projections												
1995	76.3	72.8	79.7	76.8	73.4	80.2	(NA)	(NA)	(NA)	72.4	68.8	76.0
2000	77.0	73.5	80.4	77.5	74.0	80.9	(NA)	(NA)	(NA)	73.5	69.9	77.1
2005	77.6	74.2	81.0	78.1	74.6	81.5	(NA)	(NA)	(NA)	74.6	71.0	78.1
2010	77.9	74.4	81.3	78.3	74.9	81.7	(NA)	(NA)	(NA)	75.0	71.4	78.5

Source: U.S. Bureau of the Census. *Statistical Abstract of the United States 1992*. 112th edition. Washington, D.C.: U.S. Department of Commerce, 1992, p. 76. Primary source: U.S. National Center for Health Statistics, *Vital Statistics of the United States*, annual; *Monthly Vital Statistics Report*; and unpublished sources.

★ 1043 ★

Life Expectancy

Life Expectancy by Year, 1900-1987

Specified age and year	All races			White			Black		
	Both sexes	Male	Female	Both sexes	Male	Female	Both sexes	Male	Female
	Remaining life expectancy in years								
1900[1,2]	47.3	46.3	48.3	47.6	46.6	48.7	33.0[3]	32.5[3]	33.5[3]
1950[2]	68.2	65.6	71.1	69.1	66.5	72.2	60.7	58.9	62.7
1960[2]	69.7	66.6	73.1	70.6	67.4	74.1	63.2	60.7	65.9
1970	70.9	67.1	74.8	71.7	68.0	75.6	64.1	60.0	68.3
1975	72.6	68.8	76.6	73.4	69.5	77.3	66.8	62.4	71.3

[Continued]

★ 1043 ★

Life Expectancy by Year, 1900-1987

[Continued]

Specified age and year	All races			White			Black		
	Both sexes	Male	Female	Both sexes	Male	Female	Both sexes	Male	Female
1980	73.7	70.0	77.4	74.4	70.7	78.1	68.1	63.8	72.5
1986	74.8	71.3	78.3	75.4	72.0	78.8	69.4	65.2	73.5

Source: "Life Expectancy at Birth and at 65 Years of Age, according to Race and Sex: United States, Selected Years 1900-1987," *Health United States - 1988*, March 1989, p. 53. Primary source: *Vital Statistics Rates in the United States, 1940-1960*; *Vital Statistics of the United States, 1970*; *Monthly Vital Statistics Report, 1987-1988*. National Center for Health Statistics, U.S. Government Printing Office, Washington, DC. *Notes:* 1. Death registration area only. The death registration area increased from 10 States and the District of Columbia in 1900 to the coterminous United States in 1933. 2. Includes deaths of nonresidents of the United States. 3. Figure is given for all other population.

★ 1044 ★

Life Expectancy

Life Expectancy at Age 65 Years in 1900-1987

Specified age and year	All races			White			Black		
	Both sexes	Male	Female	Both sexes	Male	Female	Both sexes	Male	Female
	Remaining life expectancy in years								
At 65 years									
1900[1,2]	11.9	11.5	12.2	-	11.5	12.2	-	10.4	11.4
1950[2]	13.9	12.8	15.0	-	12.8	15.1	13.9	12.9	14.9
1960[2]	14.3	12.8	15.8	14.4	12.9	15.9	13.9	12.7	15.1
1970	15.2	13.1	17.0	15.2	13.1	17.1	14.2	12.5	15.7
1975	16.1	13.8	18.1	16.1	13.8	18.2	15.0	13.1	16.7
1980	16.4	14.1	18.3	16.5	14.2	18.4	15.1	13.0	16.8
1986	16.8	14.7	18.6	16.9	14.8	18.7	15.4	13.4	17.0

Source: "Life Expectancy at Birth and at 65 Years of Age, according to Race and Sex: United States, Selected Years 1900-1987," *Health United States - 1988*, March 1989, p. 53. Primary source: *Vital Statistics Rates in the United States, 1940-1960*; *Vital Statistics of the United States, 1970*; *Monthly Vital Statistics Report, 1987-1988*. National Center for Health Statistics, U.S. Government Printing Office, Washington, DC. *Notes:* 1. Death registration area only. The death registration area increased from 10 States and the District of Columbia in 1900 to the coterminous United States in 1933. 2. Includes deaths of nonresidents of the United States.

★ 1045 ★

Life Expectancy

Life Expectancy of Humans in Selected Time Periods

Time Period	Average Lifespan (years)
Neanderthal man	29
Cro-Magnon man	32
Man in the Copper Age	36
Man in the Bronze Age	38

[Continued]

★1045★

Life Expectancy of Humans in Selected Time Periods

[Continued]

Time Period	Average Lifespan (years)
Greek and Roman man	36
Fifth-century man (England)	30
Fourteenth-century man (England)	38
Seventeenth century man (England)	51
Eighteenth century man (Europe)	45

Source: Kendig, Frank, and Richard Hutton. *Life-Spans*. New York: Holt, Rinehart and Winston, 1979, p. 8.

★1046★

Life Expectancy

Life Expectancy of Right-Handers Vs. Left-Handers

According to a study done by psychologists Diane Halpern and Stanley Coren, published in the *New England Journal of Medicine*, right-handers live longer than left-handers: an average of 75 years for those who are right-handed, compared to 66 for left-handed people. Right-handed men lived an average of ten years longer than left-handed men; while right-handed women lived an average of five years longer than left-handed women. Those who are left-handed are six times more likely to die in accidents; four times more likely to do so while driving.

Source: "Lifestyle: Health: The Right Stuff for a Longer Life," *Newsweek*, 117, April 15, 1991, p. 58.

★1047★

Life Expectancy

Life Expectancy and Good Health for Men

According to a study published in the journal *Circulation*, April 1991: the average man whose blood pressure is under control will add one year to his life; getting cholesterol under 200 increases his life expectancy by eight months; eliminating smoking adds ten months; and weight control adds seven months. For men who have one of these risk factors the benefit is greater. A 35-year-old man who reduces cholesterol from 250 to 200 adds one year to his life expectancy; reducing weight by 30% to the ideal level, adds one more year; and eliminating death from heart disease would extend his life expectancy by 3.1 years.

Source: Los Angels Public Library/State of California, *Scan/Info*, 3, May 1991, p. 19. Primary source: *San Francisco Chronicle*, April 26, 1991, p. A30. **Remarks**: Also included in source: Life expectancy as it relates to good health in women.

★ 1048 ★

Life Expectancy

Life Expectancy and Good Health for Women

According to a study published in the journal *Circulation*, April 1991: the average woman whose blood pressure is under control will add five months to her life; getting cholesterol under 200 increases her life expectancy by ten months; eliminating smoking adds eight months; and weight control adds five months. Eliminating death from heart disease extends the average life expectancy by 3.3 years.

Source: Los Angeles Public Library/State of California, *Scan/Info*, 3, May 1991, p. 19. Primary source: *San Francisco Chronicle*, April 26, 1991, p. A30. **Remarks**: Also included in source: Life expectancy as it relates to good health in men.

★ 1049 ★

Life Expectancy

Life Expectancy and Retirement

Based on average life expectancy, a man retiring at age 55 can expect to spend 22 years in retirement; a woman, 27 years.

Source: "The Early Retirement Crunch," Mary Rowland, *The New York Times*, June 30, 1991, p. 46, 64F.

★ 1050 ★

Life Expectancy

Life Expectancy of Dogs and Cats

The average life expectancy for cats, unaltered and living in a household, is 13-15 years for males, 15-17 years for females. Neutering adds an average of one to two more years. For dogs, 12 years is the average lifespan.

Source: Guinness Book of Records 1992. New York: Bantam Book, 1992, pp. 70, 74.

★ 1051 ★

Life Expectancy

Life Expectancy of Animals in Captivity

Animal	Average Life Expectancy (years)	Animal	Average Life Expectancy (years)
Bear (Black)	18	Horse	20
Beaver	5	Kangaroo	7
Buffalo	15	Leopard	12
Camel	12	Lion	15
Cow	15	Monkey	15
Deer	8	Mouse	12
Elephant	40	Rabbit	5
Fox	7	Rhinoceros	15
Giraffe	10	Sheep	12
Goat	8	Squirrel	10
Gorilla	20	Tiger	16
Hippopotamus	25	Zebra	15

Source: The World Almanac and Book of Facts 1992. New York: World Almanac, 1991, p. 659.

★ 1052 ★

Life Expectancy

Life Expectancy of Selected Species of Birds

Type	Average Life Expectancy (years)	Type	Average Life Expectancy (years)
Cardinals	30	Penguins	34
Geese	25	Sparrows	20
Hawks	15	Storks	35
Mockingbirds	10	Swans	70
Ostriches	25	Woodpeckers	10
Parrots	50		

Source: Kendig, Frank, and Richard Hutton. *Life-Spans*. New York: Holt, Rinehart and Winston, 1979, pp. 52-58.

★ 1053 ★

Life Expectancy

Life Expectancy of Selected Species of Fish

Type	Average Life Expectancy (years)	Type	Average Life Expectancy (years)
Carp	20 to 25	Perch	5 to 6
Cod	15	Pike	10
Dolphins	3 to 4	Salmon	10
Goldfish	7	Walleyes	6 to 7
Herrings	10	White Bass	4 to 5

Source: Kendig, Frank, and Richard Hutton. *Life-Spans*. New York: Holt, Rinehart and Winston, 1979, pp. 61-67.

★ 1054 ★

Life Expectancy

Life Expectancy of Selected Species of Insects

Type	Average Life Expectancy	Type	Average Life Expectancy
Bees		Flies	19 to 30 days
queen	6 years	Grasshoppers	5 months
worker	6 months	Mosquitoes	2 months
drones	8 weeks	Spiders	4 to 7 years
Cockroaches	40 days	Termites	
Crickets	9 to 14 weeks	queen	50 years
Earthworms	5 to 10 years	workers	20 years
Fireflies	few weeks		

Source: Kendig, Frank, and Richard Hutton. *Life-Spans*. New York: Holt, Rinehart and Winston, 1979, pp. 71-76.

★ 1055 ★

Life Expectancy

Life Expectancy of Selected Species of Trees

Type	Average Life Expectancy (years)	Type	Average Life Expectancy (years)
Souther poplar	100	Red spruce	275
Red maple	110	California cedar	350
Dogwood	115	Eastern hemlock	350
Slash Pine	200	Red fir	350
Pacific yew	200	Sugar pine	400
American elm	235	White oak	450
White fir	260	Douglas fir	750
Sugar maple	275	Redwood	1,000
American beech	275	Sequoia	2,500
White ash	275	Bristlecone pine	3,000+

Source: Kendig, Frank, and Richard Hutton. *Life-Spans*. New York: Holt, Rinehart and Winston, 1979, pp. 107-108.

★ 1056 ★

Life Expectancy

Lifespans of Airlines

According to Avmark, an appraisal and consulting company that keeps track of the world's airlines, the average age of the major carriers of United States jetliners is 12.67 years; worldwide, the fleet of more than 8,000 airliners is 13.1 years.

	Average Age of Fleet
American Airlines	
Delta	9.48
Piedmont	10.34
American	10.84
Continental	12.12
United	14.87
TWA	15.29
Northwest	15.47
Foreign Airlines	
Singapore	4.50
Lufthansa	7.70
All Nippon	8.16
Japan Air	9.14
Air France	9.59
Alitalia	9.94
British Air	10.02

[Continued]

★ 1056 ★

Lifespans of Airlines

[Continued]

	Average Age of Fleet
Malaysian	10.27
Air Canada	13.00

Source: Bluman, Allen. *Elementary Statistics*. Dubuque, IA: William C. Brown Publishers, 1992, p. 346. Primary source: Associated Press. *Note:* Figures are based on the age as of July 1, 1988.

★ 1057 ★

Life Expectancy

Lifespan of an Automobile 1941-1990

Year	Mean (years)	Median (years)	Year	Mean (years)	Median (years)
1941	5.5	4.9	1970	5.6	4.9
1946	9.0	8.8	1974	5.7	5.2
1948	8.8	8.0	1976	6.2	5.5
1950	7.8	6.2	1979	6.4	5.9
1954	6.2	4.8	1981	6.9	6.0
1956	5.6	5.1	1983	7.4	6.5
1960	5.9	5.4	1985	7.6	6.9
1964	6.0	5.3	1989	7.6	6.5
1966	5.7	4.8	1990	7.8	6.5

Source: MVMA Motor Vehicle Facts and Figures '91. Detroit, MI: Motor Vehicle Manufacturing Association, 1991, p. 26. *Notes:* Mean: the sum of the products of units multiplied by age, divided by the total units; median: a value in an ordered set of values below and above which there are an equal number of values.

★ 1058 ★

Life Expectancy

Lifespans of Household Appliances

Appliance	Average Lifespan (years)	Low	High
Dishwasher	10	5	14
Air-conditioner	11	5	15
Microwave	11	5	14
Gas water heater	11	8	14
Gas dryer	13	12	14
Automatic washer	13	12	14
Electric dryer	14	12	16
Electric water heater	14	10	18
Freezer	16	10	22
Electric furnace	16	8	20
Electric stove	17	10	30
Refrigerator	17	10	20
Gas furnace	19	5	40
Gas stove	19	10	30
Oil furnace	20	7.5	40

Source: "Appliances Life Spans," *New York Times*, December 1, 1990, p. 50. Primary source: *Appliance Magazine*, Dana Chase Publications, 1990.

★ 1059 ★

Life Expectancy

Lifespan of a Light Bulb

Type	Average Lifespan (hours)
Fluorescent (compact)	10,000
Halogen	3,000
Incandescent	
75-watt long-life	3,000
75-watt	750
lower wattage	1,000

Source: "Bright Ideas in Light Bulbs," *Consumer Reports*, 57, October 1992, pp. 664-670.

★ 1060 ★

Life Expectancy

Lifespan of United States' Currency

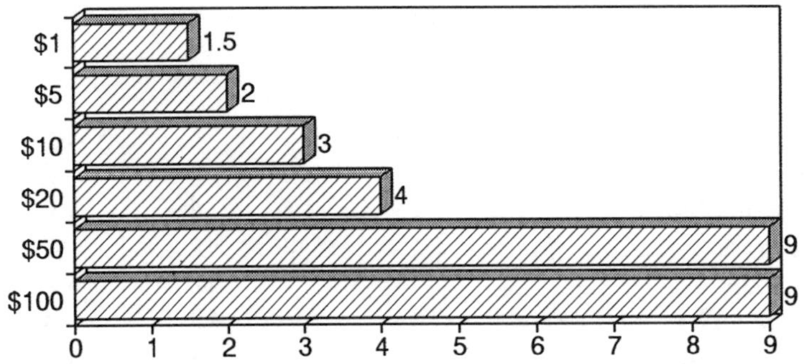

Denomination of Currency	Years
$1	1.5
$5	2
$10	3
$20	4
$50	9
$100	9

Source: Los Angeles Public Library/State of California, *Scan/Info*, 4, January 1992, p. 7. Primary source: *San Francisco Chronicle*, October 14, 1991, p. D4. The U.S. Treasury Department.

Miscellaneous

★ 1061 ★

Metropolitan Areas with the Highest Percentage of Households Speaking a Foreign Language

Metropolitan Area	Percentage
Miami	41.6%
Los Angeles	38.3%
New York	28.4%
San Francisco	26.8%
San Diego	25.0%
National metro average	15.8%

Source: "Demographics," *U.S. News & World Report*, 113, August 10, 1992, p. 15. Primary source: U.S. Census, 1990.

Chapter 15
WEATHER AND NATURE

Avalanches

★ 1062 ★

Frequency of Avalanches in Selected Parts of the World

The Austrian Ministry of Agriculture and Forestry estimates an average of 2,700 avalanches fall regularly into inhabited areas in Austria. Swiss researchers estimate that tens of thousands fall in that country. In remote mountain areas of America, thousands of avalanches fall every winter.

Source: Tufty, Barbara. *1001 Questions About Earthquakes, Avalanches, Floods and Other Natural Disasters.* New York: Dover Publications, Inc., 1978, p. 147.

★ 1063 ★

Avalanches

Pressure Exerted By an Avalanche

An average large avalanche on an open slope has been measured to exert 4,500 pounds per square foot.

Source: Tufty, Barbara. *1001 Questions About Earthquakes, Avalanches, Floods and Other Natural Disasters.* New York: Dover Publications, Inc., 1978, p. 147.

★ 1064 ★

Avalanches

Speed of an Avalanche

Dry loose-snow avalanches travel an average of 20-30 miles per hour and wet loose-snow avalanches, an average of 22 miles per hour. Avalanches made of sluggish, damp snow are the slowest, traveling 10 to 15 miles an hour and can be outdistanced by a good skier. The avalanches of the airborne powder kind, are the fastest, traveling an average speed of 100 miles per hour.

Source: Tufty, Barbara. *1001 Questions About Earthquakes, Avalanches, Floods and Other Natural Disasters.* New York: Dover Publications, Inc., 1978, pp. 145- 146.

★ 1065 ★

Avalanches

Destructive Power of an Avalanche

An average of 150 people are killed by avalanches every year throughout the world.

Source: Tufty, Barbara. *1001 Questions About Earthquakes, Avalanches, Floods and Other Natural Disasters*. New York: Dover Publications, Inc., 1978, p. 258.

Earthquakes

★ 1066 ★

Most Active Earthquake Region in the United States

Central Nevada is the most active earthquake region in the United States, having been shaken by an average of 5,000 earthquakes per year since 1769, the date when records were first kept.

Source: Tufty, Barbara. *1001 Questions About Earthquakes, Avalanches, Floods and Other Natural Disasters*. New York: Dover Publications, Inc., 1978, p. 35.

★ 1067 ★

Earthquakes

Frequency of Earthquakes

An average of two earthquakes shake the earth every minute, or about one million earthquakes per year. An average of 20 major earthquakes, measuring 7 to 7.9 on the Richter Scale, break out every year. Great earthquakes, measuring 8 or above, occur an average of one every two to three years.

Source: Tufty, Barbara. *1001 Questions About Earthquakes, Avalanches, Floods and Other Natural Disasters*. New York: Dover Publications, Inc., 1978, p. 18.

★ 1068 ★

Earthquakes

Destructive Power of an Earthquake

An average of 15,000 people are killed every year from earthquakes and the fires, floods, and waves that result from them.

Source: Tufty, Barbara. *1001 Questions About Earthquakes, Avalanches, Floods and Other Natural Disasters*. New York: Dover Publications, Inc., 1978, p. 1.

Fire

★ 1069 ★

Heat of a Fire

The temperature of 12 gallons of water can be raised 86 degrees by a fire from one pound of average wood. A heavy brush fire of 40 acres can give off enough energy to equal an atomic bomb.

Source: Tufty, Barbara. *1001 Questions About Earthquakes, Avalanches, Floods and Other Natural Disasters*. New York: Dover Publications, Inc., 1978, p. 299.

★ 1070 ★

Fire

Speed of a Fire

A fire backing into an oxygen rich wind, burns an average of 120-180 feet of grass and pine needles per hour. On the other hand, crown fires that race through the treetops burn 15-20 miles per hour, as fast as a man can run.

Source: Tufty, Barbara. *1001 Questions About Earthquakes, Avalanches, Floods and Other Natural Disasters*. New York: Dover Publications, Inc., 1978, p. 298.

★ 1071 ★

Fire

Height of Flames

A high-intensity fire throws out flames an average of 50-150 feet. But it may have a convection column as high as 25,000 feet or more.

Source: Tufty, Barbara. *1001 Questions About Earthquakes, Avalanches, Floods and Other Natural Disasters*. New York: Dover Publications, Inc., 1978, p. 298.

★ 1072 ★

Fire

Frequency of Forest Fires

According to Forest Service reports, in the 1960s an average of 364 fires occurred each day, or 100,000 fires per year, and burned an average of 7 million acres a year.

Source: Tufty, Barbara. *1001 Questions About Earthquakes, Avalanches, Floods and Other Natural Disasters*. New York: Dover Publications, Inc., 1978, p. 299.

Floods

★ 1073 ★

Greatest Flood on the Potomac River (1936)

The maximum daily average flow of water was 426,000 cubic feet per second, for a total of 275.5 billion gallons of water, during the greatest flood on record for the Potomac River, 1936. The average daily flow was enough water to meet the needs of the Washington metropolitan area for four years.

Source: Tufty, Barbara. *1001 Questions About Earthquakes, Avalanches, Floods and Other Natural Disasters.* New York: Dover Publications, Inc., 1978, p. 200.

★ 1074 ★

Floods

Death and Destruction From Flooding

An average of 75,000 Americans are driven from their homes by floods every year; an average of 80 persons are killed annually by floods.

Source: Bair, Frank E. *The Weather Almanac.* 6th ed. Detroit, MI: Gale Research Inc., 1992, p. 127.

★ 1075 ★

Floods

Frequency of Floods

On an average, once every two years a river overflows its banks and invades low places; on average, every 100 years a major flood occurs on a river.

Source: Tufty, Barbara. *1001 Questions About Earthquakes, Avalanches, Floods and Other Natural Disasters.* New York: Dover Publications, Inc., 1978, p. 201.

Humidity

★1076★

Least Humid Cities in the United States

For period of record through 1990.

City	Average Relative Humidity (P.M.)
Phoenix, AZ	23.0
El Paso, TX	27.0
Albuquerque, NM	29.0
Reno, NV	32.0
Denver, CO	40.0
Boise, ID	43.0
Salt Lake City, UT	43.0
Cheyenne, WY	44.0
Great Fall, MT	45.0
Sacramento, CA	46.0

Source: U.S. Bureau of the Census. *Statistical Abstract of the United States, 1992*. 112th ed. Washington, D.C.: U.S. Department of Commerce, p. 227. Primary source: U.S. National Oceanic and Atmospheric Administration, *Comparative Climatic Data*, annual. **Remarks**: Average relative humidity for 69 U.S. cities is listed.

★1077★

Humidity

Most Humid Cities in the United States

For period of record through 1990.

City	Average Relative Humidity (P.M.)
Cincinnati, OH	81
Juneau, AK	73
Sault Ste. Marie, MI	67
Los Angeles, CA	64
Milwaukee, WI	64
New Oreleans, LA	63
Duluth, MN	63
Buffalo, NY	63
San Diego, CA	62
Indianapolis, IN	62

[Continued]

★ 1077 ★

Most Humid Cities in the United States

[Continued]

City	Average Relative Humidity (P.M.)
Cleveland, OH	62
Seattle-Tacoma, WA	62

Source: U.S. Bureau of the Census. *Statistical Abstract of the United States, 1992*. 112th ed. Washington, D.C. U.S. Department of Commerce, p. 227. Primary source: U.S. National Oceanic and Atmospheric Administration, *Comparative Climatic Data*, annual. **Remarks**: Average relative humidity for 69 U.S. cities is listed.

Hurricanes

★ 1078 ★

Frequency of Hurricanes

In an average year, more than one hundred disturbances with hurricane potential are observed in the Atlantic, Gulf, and Caribbean; but on average only 10 of these reach the tropical storm stage and only an average of six mature into hurricanes. On average, two of these hurricanes strike the United States, where they are apt to kill from 50 to 100 people, somewhere between Texas and Maine, and cause hundreds of millions of dollars in property damage. In a worse-than-average year, the same storms cause several hundred deaths, and property damage totaling billions of dollars.

Source: Bair, Frank E. *The Weather Almanac*. 6th ed. Detroit, MI: Gale Research Inc., 1992, p. 46.

★ 1079 ★

Hurricanes

Energy of a Hurricane

The average heat energy released by a hurricane in one day can be the equivalent of energy released by fusion of four hundred, 20-megaton hydrogen bombs. This released energy, if converted to electricity, could supply the electrical needs of the entire United States for about six months.

Source: Bair, Frank E. *The Weather Almanac*. 6th ed. Detroit, MI: Gale Research, Inc., 1992, p. 51.

★ 1080 ★

Hurricanes

Cost of Preparing for a Hurricane

When a hurricane is forecast to move inland on a projected path, the coastal area placed under warning is about 300 miles in length. The average cost of preparations, whether a hurricane strikes or not, is more than $50 million for the Gulf Coast. This estimate covers the cost of boarding up homes, closing down businesses and manufacturing plants, evacuating oil rigs, etc. It does not include economic losses due to disruption of commerce activities such as sales, tourists canceling reservations, etc.

Source: Bair, Frank E. *The Weather Almanac.* 6th ed. Detroit, MI: Gale Research Inc., 1992, p. 59.

★ 1081 ★

Hurricanes

Duration of an Atlantic Ocean Hurricane

The average life of an Atlantic Ocean hurricane is 9 days. Those that occur in August last the longest, having an average life of 12 days.

Source: Tufty, Barbara. *1001 Questions Answered About Hurricanes, Tornadoes and Other Natural Air Disasters.* New York: Dover Publications, Inc., 1987, p. 32.

★ 1082 ★

Hurricanes

Waves and Wind Speed of a Hurricane

The waves travel faster than the whole hurricane system: an average speed of 30 to 50 miles an hour for the waves, but only 12 miles an hour for the forward movement of the system itself. However, the winds that generate these waves travel at an average speed of 100 miles an hour.

Source: Tufty, Barbara. *1001 Questions Answered About Hurricanes, Tornadoes and Other Natural Air Disasters.* New York: Dover Publications, Inc., 1987, p. 34.

★ 1083 ★

Hurricanes

The Wind Circulation of a Hurricane

Hurricane winds are divided into three regions. The outer portion of hurricane winds blow with an average force of 40 miles an hour, extending out from a diameter of 350-400 miles to within 20-60 miles of the center. The second region is of winds blowing with an average force of 74 miles an hour, having a diameter of 100 miles to within 15 miles from the eye. The third region is of maximum winds, having an average speed of 120-150 miles an hour. The winds immediately surrounding the eye of the storm blow in a circular band, which is an averdage of 5-30 miles wide.

Source: Tufty, Barbara. *1001 Questions Answered About Hurricanes, Tornadoes and Other Natural Air Disasters*. New York: Dover Publications, Inc., 1987, p. 22.

★ 1084 ★

Hurricanes

Duration of a Hurricane Over Any Given Spot

Hurricanes of average size take about 10 hours to pass a given spot. The first part takes an average of 4 hours; the eye, or quiet time, is an average of 20-30 minutes; and the other half lasts another 4-5 hours until the winds calm down and the rains cease.

Source: Tufty, Barbara. *1001 Questions Answered About Hurricanes, Tornadoes and Other Natural Air Disasters*. New York: Dover Publications, Inc., 1987, p. 24.

★ 1085 ★

Hurricanes

Number of Hurricanes and Typhoons Worldwide

Each year an average of 60 full-fledged hurricanes and typhoons rise out of the tropical seas.

Source: Tufty, Barbara. *1001 Questions Answered About Hurricanes, Tornadoes and Other Natural Air Disasters*. New York: Dover Publications, Inc., 1987, p. 25.

★ 1086 ★

Hurricanes

Speed of a Hurricane

A hurricane travels slowly in its early stages, with an average speed of only 10-15 miles per hour. As it matures, its average speed increases to 20-30 miles an hour. Then it picks up more speed—traveling an average of 60 miles per hour. An average hurricane travels about 3,000 miles before it dies out, which breaks down into about 300-400 miles per day.

Source: Tufty, Barbara. *1001 Questions Answered About Hurricanes, Tornadoes and Other Natural Air Disasters*. New York: Dover Publications, Inc., 1987, p. 23.

★ 1087 ★

Hurricanes

Eye of a Hurricane

The eye of a hurricane, or the quiet, innermost part, is an average of 14 to 20 miles across. In the Pacific, the eyes are larger, as much as 50 miles in diameter.

Source: Tufty, Barbara. *1001 Questions Answered About Hurricanes, Tornadoes and Other Natural Air Disasters*. New York: Dover Publications, Inc., 1987, p. 18.

Ice

★ 1088 ★

Size of hailstones

The average hailstone is about one-quarter inch in diameter. Many are reported much larger: golf ball size; oval shaped, 1.5 to 2 inches long; or even as large as a baseball, 3-4 inches in diameter.

Source: Tufty, Barbara. *1001 Questions Answered About Hurricanes, Tornadoes and Other Natural Air Disasters*. New York: Dover Publications, Inc., 1987, p. 151.

★ 1089 ★

Ice

Thickness of Ice in Antarctica

The average thickness of the ice that covers Antarctica is 7,100 feet. However, at its thickest point, it is 15,700 feet. This is ten times taller than the Sears Tower in Chicago, the world's tallest building.

Source: World Book Encyclopedia. 22 vol., Chicago, IL: World Book, 1990, vol. 1, pp. 530, 532.

★ 1090 ★

Ice

Formation of Icebergs

An average of 7,500 large icebergs break off the thick glaciers along the Greenland coast per year, according to estimates of U.S. Coast Guard researchers.

Source: Tufty, Barbara. *1001 Questions Answered About Hurricanes, Tornadoes and Other Natural Air Disasters.* New York: Dover Publications, Inc., 1987, p. 302.

★ 1091 ★

Ice

Location Where the Most Icebergs Are Formed

An average of 5,400 icebergs are formed per year from twelve large glaciers in the Northeast and Disko Bays along the western coast of Greenland.

Source: Tufty, Barbara. *1001 Questions Answered About Hurricanes, Tornadoes and Other Natural Air Disasters.* New York: Dover Publications, Inc., 1987, p. 303.

★ 1092 ★

Ice

TNT and Gasoline Needed to Break Up an Iceberg

An average size iceberg would need an estimated two thousand tons of TNT to break it up; or 2 million gallons of gasoline to melt it down.

Source: Tufty, Barbara. *1001 Questions Answered About Hurricanes, Tornadoes and Other Natural Air Disasters.* New York: Dover Publications, Inc., 1987, p. 305.

Lightning

★ 1093 ★

The Energy of a Lightning Stroke

An average lightning stroke has 30,000 amperes of electricity. A single discharge can carry from 10 to 100 million volts. However, lightning has been measured to carry as many as 345,000 amperes. This is enough electricity to light 600,000 60-watt light bulbs for as long as the duration of the flash.

Source: Tufty, Barbara. *1001 Questions Answered About Hurricanes, Tornadoes and Other Natural Air Disasters.* New York: Dover Publications, Inc., 1987, p. 126.

★ 1094 ★

Lightning

Frequency of Lightning

It is estimated that lightning strikes the earth an average of 100 times each second.

Source: Bair, Frank E. *The Weather Almanac.* 6th ed. Detroit, MI: Gale Research, Inc., 1992, p. 120.

★ 1095 ★

Lightning

The Destructive Power of Lightning

The average death toll for lightning is greater than for tornadoes or hurricanes. According to data assembled by the National Center for Health Statistics, lightning kills an average of 150 Americans per year and injures an average of 250. Property loss—fire and other damage to structures, aircraft damage, livestock deaths and injuries, forest fires, disruption of electromagnetic transmissions and other effects—is estimated to be an average of $100 million annually.

Source: Bair, Frank E. *The Weather Almanac.* 6th ed. Detroit, MI: Gale Research, Inc., 1992, p. 120.

Oceans

★ 1096 ★

Height of Atlantic Ocean Waves

Average Atlantic Ocean waves were 30 percent higher in the mid 1980s than during the 1960s, according to Sheldon Bacon, a marine physicist at the Institute of Oceanographic Sciences in Wormley. Measurements showed average wave heights of 9.2 feet in the early 1960s and 12.5 in the late 1980s in the mid-Atlantic; and average waves of 5.9 feet in 1962 compared with 7.9 feet in the 1980s closer to shore, off the coast of England.

Source: Waite, Teresa L., "Atlantic Waves Higher, Independent of Winds," *New York Times*, December 15, 1992, p. C14.

★ 1097 ★

Oceans

Depth of the Pacific Ocean

The average depth of the Pacific Ocean is 13,740 feet. It is the world's largest ocean, measuring 64.2 million miles in area, and accounting for 45.9 percent of the world's oceans.

Source: Guinness Book of Records 1992. New York: Bantam Book, 1992, p. 18.

Planets

★ 1098 ★

Distance of the Planets From the Sun

Planets	Average Distance From the Sun (in miles)
Mercury	35,983,000
Venus	67,237,700
Earth	92,955,900
Mars	141,634,800
Jupiter	483,612,200
Saturn	888,184,000
Uranus	1,782,000,000
Neptune	2,794,000,000
Pluto	3,666,000,000

Source: The Guinness Book of Answers. 8th ed. Enfield, Eng.: Guinness Publishing, 1991, p. 14.

Precipitation

★ 1099 ★

Speed of a raindrop

The speed of a raindrop varies with drop size and wind speed. A typical raindrop falls an average of 600 feet per minute or 7 miles per hour.

Source: Forrester, Frank H. *1001 Questions Answered About the Weather.* New York: Grosset & Dunlap, 1957, p. 53.

★ 1100 ★

Precipitation

Rainiest Cities in the United States

In inches. Based on 30-year period, 1961 through 1990.

City	Average annual rainfall
Mobile, AL	63.96
New Orleans, LA	61.88
Miami, FL	55.91
Jackson, MS	55.37
Juneau, AK	54.31
Memphis, TN	52.10
Jacksonville, FL	51.32
Little Rock, AR	50.86
Atlanta, GA	50.77
Columbia, SC	49.91
Nashville, TN	47.30

Source: U.S. Bureau of the Census. *Statistical Abstract of the United States 1992*. 112th ed. Washington, D.C.: U.S. Department of Commerce, p. 224. Primary source: U.S. National Oceanic and Atmospheric Administration, *Climatography of the United States*, No. 81.

★ 1101 ★

Precipitation

Driest Cities in the United States

In inches. Based on 30-year period, 1961 through 1990.

City	Average annual rainfall
Reno, NV	7.53
Phoenix, AZ	7.66
El Paso, TX	8.81
Albuquerque, NM	8.88
San Diego, CA	9.90
Boise, ID	12.11
Cheyenne, WY	14.40
Great Falls, MT	15.21
Denver, CO	15.40
Bismark, ND	15.47

Source: U.S. Bureau of the Census. *Statistical Abstract of the United States 1992*. 112th ed. Washington, D.C.: U.S. Department of Commerce, p. 224. Primary source: U.S. National Oceanic and Atmospheric Administration, *Climatography of the United States*, No. 81.

★1102★

Precipitation

Greatest Annual Rainfall in the World

Tutunendo, Colombia has the greatest annual rainfall in the world, with an average of 463.4 inches per year.

Source: Guinness Book of Records 1992. New York: Bantam Book, 1992, p. 44.

★1103★

Precipitation

Greatest Annual Snowfall in the United States

Blue Canyon, California had the greatest average annual snowfall in the United States in 1989, with 240.8 inches.

Source: Guinness Book of Records 1992. New York: Bantam Book, 1992, p. 44.

★1104★

Precipitation

Precipitation Falling on the United States

On the continental United States an average of 4,400 billion gallons of rain, snow, and other precipitation fall daily, or an average of 30 inches of precipitation per year.

Source: Tufty, Barbara. *1001 Questions About Earthquakes, Avalanches, Floods and Other Natural Disasters.* New York: Dover Publications, Inc., 1978, pp. 264- 265.

★1105★

Precipitation

Ground Storage and Evaporation of Water in the United States

An average of 1,100 billion gallons, or 8.5 inches of water, sink into the soil annually to become ground storage. An average of 21.5 inches is evaporated or absorbed by plants every year.

Source: Tufty, Barbara. *1001 Questions About Earthquakes, Avalanches, Floods and Other Natural Disasters.* New York: Dover Publications, Inc., 1978, pp. 264- 265.

★ 1106 ★

Precipitation

Lowest Annual Rainfall in the World

An annual average of 0.02 inches of rain falls at Arica, on the northern desert of Chile. Many years may pass with no rain at all.

Source: Tufty, Barbara. *1001 Questions About Earthquakes, Avalanches, Floods and Other Natural Disasters*. New York: Dover Publications, Inc., 1978, p. 254.

★ 1107 ★

Precipitation

Water in the Atmosphere

On a daily average, 40,000 billion gallons of water exist in the atmosphere over the United States, either as clouds or water vapor.

Source: Williams, Jack. *The Weather Book*. New York: Vintage Books, 1992, p. 86.

★ 1108 ★

Precipitation

Cities of the United States With the Most Precipitation

Days with precipitation of .01 inch or more. For period of record through 1990.

City	Average number of days
Juneau, AK	220
Buffalo, NY	169
Sault Ste. Marie, MI	166
Seattle-Tacoma, WA	156
Cleveland, OH	154
Pittsburgh, PA	154
Burlington, VT	154
Portland, OR	152
Charleston, WV	151

Source: U.S. Bureau of the Census. *Statistical Abstract of the United States 1992*. 112th ed. Washington, D.C.: U.S. Department of Commerce, p. 225. Primary source: U.S. National Oceanic and Atmospheric Administration, *Comparative Climatic Data*, annual.

★1109★

Precipitation

Cities of the United States With the Least Amount of Precipitation

Days with precipitation of .01 inch or more. For period of record through 1990.

City	Average number of days
Los Angeles, CA	35
Phoenix, AZ	36
San Diego, CA	42
El Paso, TX	48
Reno, NV	51
Sacramento, CA	57
Albuquerque, NM	61
San Francisco, CA	62
Dallas-Fort Worth, TX	78
Oklahoma City, OK	82

Source: U.S. Bureau of the Census. *Statistical Abstract of the United States 1992*. 112th ed. Washington, D.C.: U.S. Department of Commerce, p. 225. Primary source: U.S. National Oceanic and Atmospheric Administration *Comparative Climatic Data*, annual.

★1110★

Precipitation

Cities of the United States With the Most Snow

In inches. For period of record through 1990.

City	Average annual snowfall
Sault Ste. Marie	116.8
Juneau, AK	99.4
Buffalo, NY	91.3
Duluth, MN	76.9
Burlington, VT	76.8
Portland, ME	70.4
Albany, NY	64.3
Concord, NH	63.6
Denver, CO	59.8
Great Falls, MT	59.3
Salt Lake City, UT	57.8

Source: U.S. Bureau of the Census. *Statistical Abstract of the United States 1992*. 112th ed. Washington, D.C.: U.S. Department of Commerce, p. 226. Primary source: U.S. National Oceanic and Atmospheric Administration, *Comparative Climatic Data*, annual.

★1111★

Precipitation

Water Content of Snow

An average of 10 inches of snow is equal to 1 inch of water. Heavy, wet snow has a high water content: 4 to 5 inches may contain 1 inch of water. A dry, powdery snow might have 15 inches of snow equal to 1 inch of water. Or an inch of wet snow averages 5,300 gallons of water; an inch of powdery snow averages 1,300 gallons.

Source: Mineral Information Service, a Publication of the California Division of Mines and Geology, 23:189 (September 1970).

★1112★

Precipitation

Size of Thunderclouds

The average size of thunderclouds is 5 to 10 miles in diameter and they tower 25,000 to 40,000 feet high. In about ten minutes, a cloud expands from a mile in diameter at a height of 15,000 feet.

Source: Tufty, Barbara. *1001 Questions Answered About Hurricanes, Tornadoes and Other Natural Air Disasters*. New York: Dover Publications Inc., 1987, p. 113.

★1113★

Precipitation

Duration of a Thunderstorm

The average length of a thunderstorm is twenty-five minutes. It drops an average of half a million tons of water, about .75 of an inch of rain over a 9-square-mile area.

Source: Field, Frank. *Doctor Frank Field's Weather Book*. New York: Putnam, 1981, p. 196.

★1114★

Precipitation

Occurrence of Thunderstorms

An average of 2,000 thunderstorms are forming, growing or dying somewhere on earth every minute. An average of 16 million such storms are formed per year.

Source: Tufty, Barbara. *1001 Questions Answered About Hurricanes, Tornadoes and Other Natural Air Disasters*. New York: Dover Publications, Inc., 1987, p. 114.

★1115★

Precipitation

Energy of Thunderstorms

An average thunderstorm, 25,000-40,000 feet high, releases more energy per minute than a 120-kiloton nuclear bomb, according to some scientists.

Source: Tufty, Barbara. *1001 Questions Answered About Hurricanes, Tornadoes and Other Natural Air Disasters.* New York: Dover Publications, Inc., 1987, p. 114.

★1116★

Precipitation

Thunderstorms Per Year Worldwide

The greatest number of thunderstorms occur at Kampala, Uganda, which has an average of 242 thunderstorm days per year. The island of Java, in Indonesia, has an average of 223 thunderstorm days each year. The doldrum belt along the equator averages only 75 to 150 thunderstorm days annually.

Source: Tufty, Barbara. *1001 Questions Answered About Hurricanes, Tornadoes and Other Natural Air Disasters.* New York: Dover Publications, Inc., 1987, p. 115.

★1117★

Precipitation

Thunderstorms Per Year in the United States

The Central Plains states have an average of only 45 thunderstorm days per year, while the Rocky Mountain states average 50 to 70 days per year, and the southeastern states, the most stormy area outside the tropics, 70 to 90 days.

Source: Tufty, Barbara. *1001 Questions Answered About Hurricanes, Tornadoes and Other Natural Air Disasters.* New York: Dover Publications, Inc., 1987, p. 115.

Sunshine

★1118★

Sunniest Cities in the United States

For period of record through 1990.

City	Average Percentage of Possible Sunshine
Phoenix, AZ	86.0
El Paso, TX	83.0
Reno, NV	79.0
Sacramento, CA	78.0
Albuquerque, NM	76.0
Los Angeles, CA	73.0
Miami, FL	73.0
Denver, CO	70.0
Honolulu, HI	69.0
San Diego, CA	68.0
Oklahoma City, OK	68.0

Source: U.S. Bureau of the Census. *Statistical Abstract of the United States 1992*. 112th ed. Washington, D.C.: U.S. Department of Commerce, p. 227. Primary source: U.S. National Oceanic and Atmospheric Administration, *Comparative Climatic Data*, annual. **Remarks**: Average percentage of possible sunshine for 69 U.S. cities is listed.

★1119★

Sunshine

Cloudiest Cities in the United States

For period of record through 1990.

City	Average Percentage of Possible Sunshine
Juneau, AK	30.0
Charleston, WV	40.0
Pittsburgh, PA	46.0
Seattle-Tacoma, WA	46.0
Sault Ste. Marie, MI	47.0
Portland, OR	48.0
Buffalo, NY	49.0
Burlington, VT	49.0

[Continued]

★1119★

Cloudiest Cities in the United States

[Continued]

City	Average Percentage of Possible Sunshine
Cleveland, OH	49.0
Columbus, OH	49.0

Source: U.S. Bureau of the Census. *Statistical Abstract of the United States 1992*. 112th ed. Washington, D.C.: U.S. Department of Commerce, p. 227. Primary source: U.S. National Oceanic and Atmospheric Administration, *Comparative Climatic Data*, annual. **Remarks**: Average percentage of possible sunshine for 69 U.S. cities is listed.

Temperature

★1120★

Lowest Annual Temperature in the United States

The lowest average annual temperature recorded in the United States is 9.6 degrees Fahrenheit at Barrow, Alaska. It also has the coolest summers, with an average temperature of 36.7 degrees Fahrenheit. The lowest average winter temperature is minus 15.6 degrees Fahrenheit at Barter Island on the Arctic coast of Alaska.

Source: Bair, Frank E. *The Weather Almanac*. 6th ed. Detroit, MI: Gale Research Inc., 1992, p. 358.

★1121★

Temperature

Lowest Annual Mean Temperature in the United States

The lowest annual mean temperature in the 48 contiguous states is at Mt. Washington, New Hampshire (elevation 6,262), with 27.0 degrees Fahrenheit, and the lowest mean summer temperature of 47.2 degrees Fahrenheit.

Source: Bair, Frank E. *The Weather Almanac*. 6th ed. Detroit, MI: Gale Research Inc., 1992, p. 358.

★ 1122 ★

Temperature

Highest Annual Temperature in the World

The highest average annual temperature in the world is 88 degrees Fahrenheit at Lugh, Somalia, East Africa.

Source: Bair, Frank E. *The Weather Almanac*. 6th ed. Detroit, MI: Gale Research Inc., 1992, p. 357.

★ 1123 ★

Temperature

Highest Annual Temperature in the United States

In the United States the highest annual average temperature is Key West, Florida with 77.7 degrees Fahrenheit; the highest summer average is in Death Valley, California, with 98.2 degrees Fahrenheit; and the highest winter average is in Key West, Florida with 70.2 degrees Fahrenheit.

Source: Bair, Frank E. *The Weather Almanac*. 6th ed. Detroit, MI: Gale Research Inc., 1992, p. 357.

★ 1124 ★

Temperature

Global Surface Temperature

Today the average global temperature is about 60 degrees Fahrenheit, 5 to 9 degrees warmer than in the last ice age. Based on an analysis of how the earth's climate responded to changes in atmospheric heat-trapping influences in the distant past, the warming trend points towards an average increase of 4-6 degrees Fahrenheit if the atmospheric carbon dioxide doubles from its present level.

Source: Stevens, William, "Estimates of Warming Gain More Precision and Warn of Disaster," *The New York Times*, December 15, 1992, p. C9.

★ 1125 ★

Temperature

Temperatures in Selected Cities in the United States

Temperatures are in Farenheit degrees.

State	Station	Annual avg.	State	Station	Annual avg.
Alabama	Mobile	67.5	Montana	Great Falls	44.8
Alaska	Juneau	40.6	Nebraska	Omaha	50.6
Arizona	Phoenix	72.6	Nevada	Reno	50.8
Arkansas	Little Rock	61.8	New Hampshire	Concord	45.1
California	Los Angeles	63.0	New Jersey	Atlantic City	53.0
Colorado	Denver	50.3	New Mexico	Albuquerque	56.2
Connecticut	Hartford	49.9	New York	Albany	47.4
Delaware	Wilmington	54.2		New York	54.7
District of Columbia	Washington	58.0	North Carolina	Raleigh	59.3
Florida	Miami	75.9	North Dakota	Bismarck	41.6
Georgia	Atlanta	61.3	Ohio	Cleveland	49.6
Hawaii	Honolulu	77.2	Oklahoma	Oklahoma City	60.0
Idaho	Boise	50.9	Oregon	Portland	53.6
Illinois	Chicago	49.0	Pennsylvania	Philadelphia	54.3
Indiana	Indianapolis	52.3	Rhode Island	Providence	50.4
Iowa	Des Moines	49.9	South Carolina	Columbia	63.1
Kansas	Wichita	56.2	South Dakota	Sioux Falls	45.5
Kentucky	Louisville	56.1	Tennessee	Memphis	62.3
Louisiana	New Orleans	68.1	Texas	Dallas -	
Maine	Portland	45.4		Ft. Worth	65.4
Maryland	Baltimore	55.1	Utah	Salt Lake City	52.0
Massachusetts	Boston	51.3	Vermont	Burlington	44.6
Michigan	Detroit	48.6	Virginia	Richmond	57.7
Minnesota	Minneapolis-		Washington	Seattle-Tacoma	52.0
	St. Paul	44.9	West Virginia	Charleston	55.0
Mississippi	Jackson	64.2	Wisconsin	Milwaukee	46.1
Missouri	St. Louis	56.1	Wyoming	Cheyenne	45.6

Source: U.S. Bureau of the Census. *Statistical Abstract of the United States 1992*. 112th ed. Washington, D.C.: U.S. Department of Commerce, p. 219. Primary source: U.S. National Oceanic and Atmospheric Administration, *Climatography of the United States*, No. 81.

★ 1126 ★

Temperature

The Ten Warmest Years Since 1880

Year	Average temperature (Deg. F)
1990	59.81
1981, 1988	59.64
1987	59.56
1980, 1983	59.51
1989	59.45

[Continued]

★1126★

The Ten Warmest Years Since 1880

[Continued]

Year	Average temperature (Deg. F)
1973	59.31
1977, 1986	59.30

Source: The New York Times, January 10, 1991, p. A1. Primary source: NASA/Goddard Institute for Space Studies.

★1127★

Temperature

Average Monthly Temperatures in Selected Cities Worldwide

Temperatures are in Farenheit degrees.

	J	F	M	A	M	J	J	A	S	O	N	D
In Europe												
Amsterdam	36	36	41	46	54	59	63	63	57	52	43	39
Athens	48	50	54	59	68	77	81	79	73	64	57	52
Berlin	48	34	39	46	55	63	64	63	57	46	39	34
Dublin	41	41	43	46	52	57	59	59	55	50	45	39
Geneva	32	30	41	48	55	63	64	64	57	48	41	36
London	41	43	45	50	55	61	64	64	61	55	48	43
Madrid	41	45	50	55	61	70	75	75	68	59	48	43
Moscow	16	16	25	39	54	63	64	63	52	39	27	18
Paris	37	39	45	50	57	63	66	64	61	52	45	39
Prague	34	36	37	46	55	61	66	63	57	46	37	34
Rome	46	48	52	57	63	72	75	75	70	63	55	48
Stockholm	27	27	30	39	50	59	64	63	54	45	37	32
Vienna	30	32	39	48	57	63	66	66	59	50	39	32
Other Continents												
Bombay, India	75	75	77	81	88	84	82	82	81	82	81	79
Darwin, Australia	82	82	84	84	82	79	77	79	82	84	86	84
Douala, Cameroon	75	77	75	75	75	73	72	72	73	73	72	75
Jeddah, Saudi Arabia	73	77	81	84	86	90	91	88	86	82	81	77
New York	32	32	41	52	61	72	77	75	68	59	46	36
Montreal, Canada	14	16	27	43	55	64	70	68	59	48	36	19
Peking, China	23	25	39	59	81	88	88	86	79	68	50	23

Source: The Guinness Book of Answers. 8th ed. Enfield, Eng.: Guinness Publishing, 1991, p. 99.
Remarks: The source also gives centigrade temperatures.

★1128★

Temperature

Annual Temperature in the United States, 1895-1990

From 1895 to 1990, the average annual temperature in the United States was 52.5 degrees Fahrenheit. There were no upward or downward trends during these years. However, the beginning of the century was cool; the 1920s to the 1950s were warm, the 1960s and 70s were cool, and the 1980s were warm. The highest average temperature was 54.67 degrees Fahrenheit in 1937; the lowest was 50.72 degrees Fahrenheit in 1917.

Source: Williams, Jack. *The Weather Book.* New York: Vintage Books, 1992, p. 186.

Tornadoes, Cyclones, Tsunamis

★1129★

Size of a Tornado Path

On the average, tornado paths are only a quarter of a mile wide and seldom more than 16 miles long.

Source: Bair, Frank E. *The Weather Almanac.* 6th ed. Detroit, MI: Gale Research, Inc., 1992, p. 84.

★1130★

Tornadoes, Cyclones, Tsunamis

Forward Speed of a Tornado

The forward speed of a tornado averages 40 miles per hour.

Source: Bair, Frank E. *The Weather Almanac.* 6th ed. Detroit, MI: Gale Research, Inc., 1992, p. 84.

★1131★

Tornadoes, Cyclones, Tsunamis

The Weakest Tornadoes

The weakest tornadoes are an average of 1.1 miles long, 46 yards wide, with winds of 73 miles per hour. The area affected is .06 square miles.

Source: Council on Environmental Quality. *Environmental Trends.* Washington, DC: Council on Environmental Quality, 1989, p. 144.

★ 1132 ★

Tornadoes, Cyclones, Tsunamis

The Most Violent Tornadoes

The most violent tornadoes are an average of 34.2 miles long, 616 yards wide, with winds of 261 miles per hour. The area of devastation is 11.88 square miles.

Source: Council on Environmental Quality. *Environmental Trends*. Washington DC: Council on Environmental Quality, 1989, p. 144.

★ 1133 ★

Tornadoes, Cyclones, Tsunamis

Destructive Power of Tornadoes

Tornadoes killed an average of 102 people each year during the period, 1953-1985.

Source: Tufty, Barbara. *1001 Questions Answered About Hurricanes, Tornadoes and Other Natural Air Disasters*. New York: Dover Publications, Inc., 1987, p. 355.

★1134★

Tornadoes, Cyclones, Tsunamis

Frequency of Tornadoes in the United States

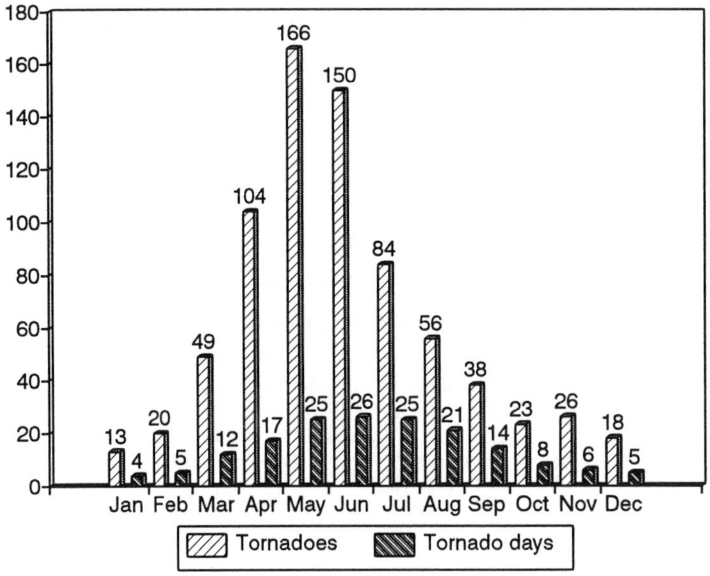

During the period 1953-1969, an average of 642 tornadoes per year occurred in the United States, about half of them during three months—April, May, and June. For the same period, the annual average number of tornado days—days on which one or more tornadoes were reported—was 159. Average annual frequency by states for this period ranges from 103 tornadoes in Texas to less than three in most of the northeastern and far western states.

Month	Tornadoes	Tornado days
January	13	4
February	20	5
March	49	12
April	104	17
May	166	25
June	150	26
July	84	25
August	56	21
September	38	14
October	23	8
November	26	6
December	18	5

Source: Bair, Frank E. *The Weather Almanac.* 6th ed. Detroit, MI: Gale Research, Inc., 1992, pp. 85, 112.

★ 1135 ★

Tornadoes, Cyclones, Tsunamis

Cyclones

A cyclone travels an average of 600 miles per day and lasts an average of one week. A cyclone, with winds averaging 60 miles per hour, often spreads out more than 1,000 miles in diameter.

Source: Tufty, Barbara. *1001 Questions Answered About Hurricanes, Tornadoes and Other Natural Air Disasters.* New York: Dover Publications, Inc., 1987, p. 60.

Wind

★ 1136 ★

Windiest Cities in the United States

For period of record through 1990.

City	Average Wind Speed (m.p.h)
Cheyenne, WY	13.0
Great Falls, MT	12.8
Boston, MA	12.5
Oklahoma City, OK	12.4
Wichita, KS	12.3
Buffalo, NY	12.0
Milwaukee, WI	11.6
Honolulu, HI	11.4
Duluth, MN	11.1
Sioux Falls, SD	11.1

Source: U.S. Bureau of the Census. *Statistical Abstract of the United States 1992*. 112th ed. Washington, D.C.: U.S. Department of Commerce, p. 227. Primary source: U.S. National Oceanic and Atmospheric Administration, *Comparative Climatic Data*, annual. **Remarks**: In 1990 Chicago ranked 21st in the list of the 68 windy cities with an average wind speed of 10.3 miles per hour. Average wind speed for 58 other U.S. cities is also listed.

★ 1137 ★

Wind

Highest Surface Wind

The highest surface wind ever recorded was on Mount Washington, New Hampshire at an elevation of 6288 feet. On April 12, 1934 its wind was 231 miles per hour and its average wind speed was 35 miles per hour.

Source: Williams, Jack. *The Weather Book*. New York: Vintage Books., 1992, p. 43.

Plants

★1138★

Daily Plant Growth by Selected Plants

Plant	Average Daily Growth
Bamboo	35.4 inches
Seaweed	17.7
Bermuda grass	5.9
Spruce growing on polar tree line	0.0003
Lichens	0.0001

Source: Diagram Group. *Comparisons.* New York: St. Martin's Press, 1980, p. 186.

Animals

★1139★

Gestation Period of Selected Animals

Animal	Average Gestation Period (days)	Animal	Average Gestation Period (days)
Bear (Black)	219	Hippopotamus	238
Beaver	122	Horse	330
Buffalo	278	Kangaroo	42
Bactrian Camel	406	Leopard	98
Cat	63	Lion	100
Cow	284	Monkey	164
Deer	201	Mouse	21
Dog	61	Rabbit	31
Elephant	645	Rhinoceros	450
Fox	52	Sheep	154
Giraffe	425	Squirrel	44
Goat	151	Tiger	105
Gorilla	257	Zebra	365

Source: The World Almanac and Book of Facts 1992. New York: World Almanac, 1991, p. 659.

★ 1140 ★

Animals

Body Temperatures of Selected Animals, Fish, Birds, and Reptiles

Type	Average Body Temperature (Fahrenheit degrees)	Type	Average Body Temperature (Fahrenheit degrees)
Jay Bird	109.4	Gull	93.2
Sparrow	105.8	Three-toed sloth	91.7
Owl	104.4	Python	83.5
Goat	103.8	Crocodile	78.1
Rabbit	101.3	Bullfrog	76.5
Seal	99.9	Goldfish	74.3
Polar bear	99.1	Anteater	73.9
Man	98.6	Garter snake	70.9
African Elephant	97.5	Perch	60.7
Blue whale	95.9	Salamander	58.6

Source: Diagram Group. *Comparisons*. New York: St. Martin's Press, 1980, p. 154.

★ 1141 ★

Animals

Weight of Selected Animals and Birds

Type	Average Weight	Type	Average Weight
Hummingbird	0.07 oz.	Coyote	75 lb.
Mouse	0.79 oz.	Porpoise	103 lb.
Shrew	3.00 oz.	Alligator	150 lb.
Hamster	4.20 oz.	Llama	375 lb.
Rat	16.05 oz.	Polar bear	715 lb.
Monkey	6 lb.	Horse	950 lb.
Chicken	7 lb.	Cow	1800 lb.
Fox	14 lb.	Walrus	1.6 tons
Raccoon	21 lb.	Elephant	7 tons
Porcupine	27.5 lb.	Blue whale	153.5 tons

Source: Diagram Group. *Comparisons*. New York: St. Martin's Press, 1980, p. 119.

★ 1142 ★

Animals

Heartbeat of Selected Animals

Type	Average Number of Heart Beats per Minute
Elephant	20
Ox	25
Frog	30
In freezing temp.	6-8
Human	60-80
Rabbit	200
Hedgehog	250
In hibernation	3
Mouse	500

Source: Asimov, Isaac. *The Human Body.* New rev. ed. New York: A Mentor Book, 1992, p. 167.

★ 1143 ★

Animals

Breath-Holding Capability of Animals

Animal	Average time in minutes
Human	1
Polar bear	1.5
Pearl diver (human)	2.5
Sea otter	5
Platypus	10
Muskrat	12
Hippopotamus	15
Sea cow	16
Beaver	20
Porpoise	15
Seal	15 to 28
Greenland whale	60
Sperm whale	90
Battlenose whale	120

Source: Taber, Robert W. *1001 Questions Answered About the Oceans and Oceanography.* New York: Dodd, Mead, 1972, p. 15.

★ 1144 ★

Animals

Quills of a Porcupine

The average porcupine has about 30,000 quills.

Source: Costello, David F. *The World of the Porcupine*. Philadelphia, PA: Lippincott, 1966, p. 13.

★ 1145 ★

Animals

Weight of Pigs

The adult Chinese potbellied pig weighs an average 70 to 150 pounds, while adult farm pigs weigh an average of 1,200 to 1,500 pounds. Chinese potbellies are an ancient breed that stand only 16 inches tall. They are intelligent, easily house-broken, and reportedly make affectionate pets.

Source: Mull, Kayla. *Pot-Bellied Pet Pigs*. Orange, CA: All Publishing, 1990, pp. 6-7.

★ 1146 ★

Animals

Weight of Whales

Whale	Average Weight in Tons
Sperm	35
Blue	84
Finback	50
Humpback	33
Right	50
Sei	17
Gray	20
Bowhead	50
Bryde's	17
Minke	10

Source: Duxbury, Alyn C., and Alison Duxbury. *An Introduction to the World's Oceans*. Reading, MA: Addison-Wesley Publishing Co, Inc., 1984, pp. 438- 439.

★1147★

Animals

Spider's Web

The average orb-weaver spider takes 30 to 60 minutes to completely spin its web. The web must be replaced every few days because it loses its stickiness, and thus, its ability to entrap food.

Source: The Illustrated Encyclopedia of Wildlife. 15 vols. Lakeville, CT: Grey Castle Press, 1991, vo. 9, pp. 2264-2265.

Miscellaneous

★1148★

Particles Falling to the Ground

An average of 20 million tons of particles of varying size fall to the earth each day, some no bigger than a grain of sand.

Source: Begley, Sharon, "The Science of Doom," *Newsweek*, 120:56 (November 23, 1992).

★1149★

Miscellaneous

Water Speeds of Selected Oceans, Currents, Rivers, Bores

Name	Average Speed
Antarctic Circumpolar Current	0.9 mph
Mississippi River	2.0
Amazon River	1.5-3.0
Gulf Stream	4.6-5.8
Pentland Firth	12.4
Ganges Bore	16.9
Saltstraumen Current	18
Lava Falls, Colorado River	30

Source: Diagram Group. *Comparisons.* New York: St. Martin's Press, 1980, pp. 186- 187.

SUBJECT INDEX

The *Subject Index* holds references to more than 1,400 subjects, concepts, countries, states, cities, organizations, and the like. Each entry is followed by one or more page numbers marked p. or pp. After the page references, table references are provided within brackets. The phrase *p. 340 [*339*]* means that the item appears on page 340 in Table 339.

Numbers following p. or pp. are page references. Numbers in [ ] are table references.

Numbers following p. or pp. are page references. Numbers in [ ] are table references.

Numbers following p. or pp. are page references. Numbers in [ ] are table references.

Numbers following p. or pp. are page references. Numbers in [ ] are table references.

Numbers following p. or pp. are page references. Numbers in [ ] are table references.

Numbers following p. or pp. are page references. Numbers in [ ] are table references.

Subject Index

Numbers following p. or pp. are page references. Numbers in [ ] are table references.

Numbers following p. or pp. are page references. Numbers in [ ] are table references.

Subject Index

Numbers following p. or pp. are page references. Numbers in [ ] are table references.

Geauga County, OH, p. 232 [★498★]
Gender and other topics *See:* topic, or subhead under Women or Men
General Motors Corp., pp. 29-30 [★68-69★]
Geneva, p. 350 [★755★]
— average monthly temperature, p. 515 [★1127★]
Georgia (country), p. 461 [★1004★]
Georgia (state), pp. 39, 55, 64, 87, 92, 121-122, 126-127, 136, 141, 147-150, 153-154, 168-172, 176, 178, 183, 229-230, 248-256, 267-273, 283, 294, 300, 305, 307, 319-321, 350-352, 355-356, 365, 408-409, 419, 426, 460, 469 [★90★, ★128★, ★150★, ★206★, ★217★, ★287★, ★289★, ★294-295★, ★310★, ★317★, ★329-332★, ★338★, ★342★, ★368-372★, ★378★, ★380★, ★394★, ★493-494★, ★530-532★, ★534★, ★536-540★, ★571-574★, ★576-578★, ★595★, ★625-626★, ★642★, ★649★, ★652★, ★682★, ★684★, ★687★, ★756-759★, ★763-764★, ★780★, ★874-875★, ★900★, ★917★, ★1002★, ★1004★, ★1015★]
Germany, pp. 3-4, 30, 38, 49, 62, 69, 71, 74, 152, 155, 166, 213, 262, 282, 315, 357-358, 405, 412, 461, 473, 481 [★4-5★, ★7★, ★71★, ★87★, ★118★, ★146★, ★160★, ★165★, ★171★, ★336★, ★344★, ★364★, ★460★, ★556★, ★594★, ★673★, ★765★, ★767-768★, ★865-866★, ★882★, ★1004★, ★1039★]
— Democratic Republic (former), p. 1 [★1★]
— East (former) | 82, 85 | 193, 202
— Federal Republic (former), pp. 1, 172-174, 473 [★1★, ★373-375★, ★1024★]
— West (former), pp. 82, 85, 175, 304, 356, 430, 461 [★193★, ★202★, ★376★, ★647★, ★765★, ★930★, ★1004★]
Gestation periods, p. 520 [★1139★]
Ghana, pp. 358, 461, 481 [★768★, ★1004★, ★1039★]
Gift expenditures, pp. 209-210 [★451-453★]
Gifts, pp. 108, 422 [★263★, ★907★]
Glaciers *See:* Icebergs
Glass recycling, p. 428 [★923★]
Global warming, p. 513 [★1124★]
GNP per capita by country, pp. 358-359 [★768-769★]
Government aid programs, pp. 62, 222, 257-258 [★145★, ★479★, ★542-543★]
See also: Federal aid programs; Income assistance benefits; Welfare
— health and medical care expenditure, p. 324 [★692★]
— health and medical care expenditures, p. 283 [★596★]
— immigrants, p. 258 [★545★]
Government documents, p. 32 [★76★]
Government salaries *See:* City government salaries; State and local government salaries
Grain consumption *See:* Baked goods consumption; Breakfast cereal consumption; Cereal consumption
Grand Forks, ND, p. 54 [★127★]
Grand Island, NE, p. 198 [★423★]
Grand Rapids, MI, p. 221 [★477★]
Great Britain, pp. 3, 38, 49, 62, 69, 71, 116, 152, 155, 179,

Numbers following p. or pp. are page references. Numbers in [ ] are table references.

Great Britain continued:
184, 213, 356-357, 405, 412, 423, 430, 451, 456 [★4★, ★87★, ★118★, ★146★, ★160★, ★165★, ★278★, ★336★, ★344★, ★383★, ★395-396★, ★460★, ★765★, ★767★, ★865-866★, ★882★, ★908★, ★930★, ★984★, ★993-994★]
See also: Britain; United Kingdom; names of individual countries
Great Falls, MT, pp. 505, 508, 519 [★1101★, ★1110★, ★1136★]
Greece, pp. 62, 82, 85, 282, 356, 358, 451, 461, 481 [★146★, ★193★, ★202★, ★594★, ★765★, ★768★, ★984★, ★1004★, ★1039★]
Greene County, MS, p. 233 [★499★]
Greenland
— icebergs, pp. 502 [★1090-1091★]
Greenville, NC, p. 224 [★484★]
Greenville-Spartanburg, SC, p. 225 [★486★]
Greeting cards, pp. 32, 101 [★78★, ★243-244★]
Grenada, p. 461 [★1004★]
Grocery stores *See:* Supermarkets; Retail sales
Grooming, pp. 327-328 [★695-697★]
Grooming aids
— expenditures, p. 221 [★477★]
Grooming time (men), p. 12 [★22★]
Gross National Product *See:* GNP
Growth rate (plants), p. 520 [★1138★]
Grundy County, TN, p. 233 [★499★]
Guatemala, pp. 358, 461, 481 [★768★, ★1004★, ★1039★]
Guinea, pp. 358, 461, 481 [★768★, ★1004★, ★1039★]
Guinea-Bissau, pp. 358, 461 [★768★, ★1004★]
Guns (handguns) in crime, p. 116 [★278★]
Guyana, pp. 358, 461 [★768★, ★1004★]
Hailstones, p. 501 [★1088★]
Haines Borrough, AK, p. 354 [★762★]
Hair, pp. 336, 343 [★720-723★, ★741★]
Hair dryer safety, p. 479 [★1035★]
Haiti, pp. 358, 461, 481 [★768★, ★1004★, ★1039★]
Hamburg, p. 421 [★903★]
Hammond, IN, p. 109 [★264★]
Handbags, p. 111 [★269★]
Handicapped *See:* Nursing homes
Handicapped students (costs of educating), p. 180 [★388★]
Handwriting, p. 43 [★103★]
Hanover County, VA, p. 233 [★498★]
Hartford, CT, pp. 221, 261 [★477★, ★554★]
Hartford/New Haven, CT, p. 434 [★938★]
Harvard University, p. 372 [★795★]
Hawaii, pp. 39, 55, 87, 92, 122, 136, 141, 147-150, 153, 168-172, 176, 178, 183, 229-230, 248-256, 267-273, 283, 294-295, 300, 305, 307, 319-321, 350-352, 355-356, 365, 408-409, 419, 426, 460, 469 [★90★, ★128★, ★206★, ★217★, ★289★, ★310★, ★317★, ★329-332★, ★338★, ★368-372★, ★378★, ★380★, ★394★, ★493-494★, ★530-532★, ★534★, ★536-540★, ★571-574★, ★576-578★, ★595★, ★625-626★, ★642★, ★649★, ★652★, ★682★, ★684★, ★687★, ★756-759★, ★763-764★, ★780★, ★874-875★, ★900★, ★917★, ★1002★,

Subject Index

Numbers following p. or pp. are page references. Numbers in [ ] are table references.

Numbers following p. or pp. are page references. Numbers in [ ] are table references.

Numbers following p. or pp. are page references. Numbers in [ ] are table references.

Numbers following p. or pp. are page references. Numbers in [ ] are table references.

Subject Index

Numbers following p. or pp. are page references. Numbers in [ ] are table references.

Numbers following p. or pp. are page references. Numbers in [ ] are table references.

Subject Index

Numbers following p. or pp. are page references. Numbers in [ ] are table references.

Numbers following p. or pp. are page references. Numbers in [ ] are table references.

Numbers following p. or pp. are page references. Numbers in [ ] are table references.

Numbers following p. or pp. are page references. Numbers in [ ] are table references.

Numbers following p. or pp. are page references. Numbers in [ ] are table references.

Numbers following p. or pp. are page references. Numbers in [ ] are table references.

Numbers following p. or pp. are page references. Numbers in [ ] are table references.

Numbers following p. or pp. are page references. Numbers in [ ] are table references.

Numbers following p. or pp. are page references. Numbers in [ ] are table references.

Numbers following p. or pp. are page references. Numbers in [ ] are table references.

Subject Index

Numbers following p. or pp. are page references. Numbers in [ ] are table references.

Numbers following p. or pp. are page references. Numbers in [ ] are table references.

Numbers following p. or pp. are page references. Numbers in [ ] are table references.

SELECTIVE NUMERICAL VALUES LOCATOR

The Numeric Index holds more than 5,300 references to values shown in the tables of *Gale Book of Averages*. The arrangement of items is by major topic (e.g., *Area*), by subtopic (e.g., *Acres*), and then by numerical value in ascending order. Each item is followed by a page reference marked by 'p.' and a table reference in brackets. The phrase *p. 96 [*229*]* refers to page 96 and Table 229. The word "average" is implied where otherwise not stated.

Area

Acres

0.42, cropland available per capita, 2025 (projected), p. 96 [★229★]
0.69, cropland available per capita, 1990, p. 96 [★229★]
3, average Japanese farm size, p. 40 [★91★]
90, pizza consumed daily, p. 78 [★184★]
199, average farm size, 1860, p. 39 [★89★]
462, average farm size, 1987, p. 40 [★91★]
5.4 million, annual average farmland lost to farming, 1950-1982, p. 39 [★89★]
7 million, damaged by forest fires annually in the United States, p. 495 [★1072★]

Square feet

17, skin of average woman, p. 343 [★740★]
20, skin of average man, p. 343 [★740★]
50.9, area per jail inmate, 1988, p. 123 [★291★]
54.3, area per jail inmate, 1983, p. 123 [★291★]
400-450, paint coverage per gallon of paint (recommended), p. 111 [★267★]
650, average area of Japanese home, p. 100 [★239★]
650, paint coverage per gallon of paint (average roller), p. 111 [★267★]
900, average area of U.S. apartment, 1986, p. 100 [★238★]
1,900, average area of U.S. single-family home, p. 100 [★240★]

Square inches

48, average skin area that men shave, p. 327 [★696★]
412, average skin area that women shave, p. 327 [★696★]

Square miles

0.06, affected by weakest tornadoes, p. 516 [★1131★]
9, thunderstorms, p. 509 [★1113★]
11.88, affected by strongest tornadoes, p. 517 [★1132★]
64.2 million, Pacific Ocean, p. 504 [★1097★]

Distance

Miles

10.73, average one-way commute, 1985, p. 60 [★140★]

Energy

Amperes

30,000, average lightning stroke, p. 502 [★1093★]

BTU

198, energy used per capita in New York, 1989, p. 87 [★206★]
206, energy used per capita in Rhode Island, 1989, p. 87 [★206★]
226, energy used per capita in New Hampshire, 1989, p. 87 [★206★]
231, energy used per capita in Vermont, 1989, p. 87 [★206★]
232, energy used per capita in Massachusetts, 1989, p. 87 [★206★]
236, energy used per capita in Florida, 1989, p. 87 [★206★]
237, energy used per capita in Connecticut, 1989, p. 87 [★206★]
245, energy used per capita in California, 1989, p. 87 [★206★]
258, energy used per capita in Arizona, 1989, p. 87 [★206★]
269, energy used per capita in Hawaii, 1989, p. 87 [★206★]
269, energy used per capita in Maryland, 1989, p. 87 [★206★]
274, energy used per capita in Colorado, 1989, p. 87 [★206★]
279, energy used per capita in Maine, 1989, p. 87 [★206★]
290, energy used per capita in District of Columbia, 1989, p. 87 [★206★]
290, energy used per capita in Wisconsin, 1989, p. 87 [★206★]
294, energy used per capita in Missouri, 1989, p. 87 [★206★]
295, energy used per capita in North Carolina, 1989, p. 87 [★206★]
297, energy used per capita in South Dakota, 1989, p. 87 [★206★]
298, energy used per capita in Michigan, 1989, p. 87 [★206★]
298, energy used per capita in Pennsylvania, 1989, p. 87 [★206★]
302, energy used per capita in New Jersey, 1989, p. 87 [★206★]
302, energy used per capita in Virginia, 1989, p. 87 [★206★]
303, energy used per capita in Illinois, 1989, p. 87 [★206★]
307, energy used per capita in Minnesota, 1989, p. 87 [★206★]
315, energy used per capita in Georgia, 1989, p. 87 [★206★]
318, energy used per capita in Utah, 1989, p. 87 [★206★]
323, energy used per capita in Oregon, 1989, p. 87 [★206★]
326, energy used per capita in Iowa, 1989, p. 87 [★206★]
327, energy used per capita in Nebraska, 1989, p. 87 [★206★]
327.6, energy used per capita in United States, 1989, p. 87 [★206★]
331, energy used per capita in South Carolina, 1989, p. 87 [★206★]
343, energy used per capita in Arkansas, 1989, p. 87 [★206★]
345, energy used per capita in Nevada, 1989, p. 87 [★206★]
346, energy used per capita in Delaware, 1989, p. 87 [★206★]
354, energy used per capita in Ohio, 1989, p. 87 [★206★]

Energy - continued

BTU

357, energy used per capita in Tennessee, 1989, p. 87 [★206★]
366, energy used per capita in New Mexico, 1989, p. 87 [★206★]
367, energy used per capita in Idaho, 1989, p. 87 [★206★]
377, energy used per capita in Mississippi, 1989, p. 87 [★206★]
396, energy used per capita in Kentucky, 1989, p. 87 [★206★]
396, energy used per capita in Washington, 1989, p. 87 [★206★]
399, energy used per capita in Alabama, 1989, p. 87 [★206★]
401, energy used per capita in Oklahoma, 1989, p. 87 [★206★]
409, energy used per capita in Kansas, 1989, p. 87 [★206★]
430, energy used per capita in West Virginia, 1989, p. 87 [★206★]
434, energy used per capita in Montana, 1989, p. 87 [★206★]
446, energy used per capita in Indiana, 1989, p. 87 [★206★]
483, energy used per capita in North Dakota, 1989, p. 87 [★206★]
570, energy used per capita in Texas, 1989, p. 87 [★206★]
803, energy used per capita in Wyoming, 1989, p. 87 [★206★]
804, energy used per capita in Louisiana, 1989, p. 87 [★206★]
1,075, energy used per capita in Alaska, 1989, p. 87 [★206★]
64, million energy used annually for five or more unit bldg. homes, 1987, p. 89 [★210★]
71, million energy used annually in house built 1980 or later, 1987, p. 88 [★209★]
76, million energy used annually for mobile home homes, 1987, p. 89 [★210★]
86, million energy used annually in house built 1975-1979, 1987, p. 88 [★209★]
93, million energy used annually for two to four-unit bldg. homes, 1987, p. 89 [★210★]
95, million energy used annually in house built 1970-1974, 1987, p. 88 [★209★]
99, million energy used annually for single-family attached homes, 1987, p. 89 [★210★]
100, million energy used annually in house built 1960-1969, 1987, p. 88 [★209★]
104, million energy used annually in house built 1940-1949, 1987, p. 88 [★209★]
105 million, annual household energy usage, 1985, p. 88 [★207★]
110, million energy used annually in house built 1950-1959, 1987, p. 88 [★209★]
115, million energy used annually for single-family detached homes, 1987, p. 89 [★210★]
120, million energy used annually in house built 1939 or earlier, 1987, p. 88 [★209★]
126 million, annual household energy usage, 1980, p. 88 [★207★]
868 million, average industrial energy consumption per employee, 1988, p. 36 [★84★]

Calories

6-12, expended in a kiss, p. 97 [★233★]
88, burned by a secretary in 1 hour, p. 345 [★745★]
105, burned by an executive in 1 hour, p. 345 [★745★]
600, average burned per hour hiking Appalachian Trail hills, p. 440 [★952★]
2200, daily recommended intake, women, p. 97 [★231★]

Kilotons

120, released in 1 minute by a thunderstorm, p. 510 [★1115★]

Energy - continued

Kilowatt hours

1, energy used by electric clothes iron, p. 90 [★212★]
5-50, energy used annually by electric mower, p. 86 [★203★]
10-150, energy used annually by ceiling fan, p. 86 [★203★]
17-50, energy used annually by clock, p. 86 [★203★]
20-300, energy used annually by coffee maker, p. 86 [★203★]
25-400, energy used annually by computer, p. 86 [★203★]
75-1,000, energy used annually by color television, p. 86 [★203★]
103, energy used annually by clothes washer, p. 86 [★203★]
165, energy used annually by dishwasher, p. 86 [★203★]
200-1,000, energy used annually by aquarium/terrarium, p. 86 [★203★]
200-1,000, energy used annually by dehumidifier, p. 86 [★203★]
200-400, energy used annually by bottled water dispenser, p. 86 [★203★]
650, energy used annually by electric range/oven, p. 86 [★203★]
993, energy used annually by electric clothes dryer, p. 86 [★203★]
4,000, electricity saved by recycling 1 ton of wastepaper, p. 428 [★922★]
4,000, electricity used to power average home for six months, p. 428 [★922★]
7,380, energy used annually per household, 1971, p. 84 [★201★]
8,825, energy used annually per household, 1981, p. 84 [★201★]
9,738, energy used annually per household, 1991, p. 84 [★201★]

Megatons

8,000, released by hurricane in one day, p. 498 [★1079★]

Miles per gallon

5.54, fuel consumed per bus, 1970, p. 67 [★155★]
5.75, fuel consumed per bus, 1975, p. 67 [★155★]
5.95, fuel consumed per bus, 1980, p. 67 [★155★]
5.84, fuel consumed per bus, 1985, p. 67 [★155★]
5.36, fuel consumed per bus, 1990, p. 67 [★155★]
7.85, fuel consumed per truck, 1970, p. 67 [★155★]
8.99, fuel consumed per truck, 1975, p. 67 [★155★]
9.54, fuel consumed per truck, 1980, p. 67 [★155★]
9.79, fuel consumed per truck, 1985, p. 67 [★155★]
10.62, fuel consumed per truck, 1990, p. 67 [★155★]
13.52, fuel consumed per car, 1970, p. 67 [★155★]
13.52, fuel consumed per car, 1975, p. 67 [★155★]
15.46, fuel consumed per car, 1980, p. 67 [★155★]
18.20, fuel consumed per car, 1985, p. 67 [★155★]
19.9, fuel consumed per vehicle, 1978, p. 67 [★156★]
21, fuel consumed per car, 1990, p. 67 [★155★]
23.5, fuel consumed per vehicle, 1980, p. 67 [★156★]
27, fuel consumed per vehicle, 1985, p. 67 [★156★]
27.8, fuel consumed per vehicle, 1990,1,, p. 67 [★156★]

Volts

10-100 million, lightning stroke, p. 502 [★1093★]

Length

Feet

3, blood vessels in square centimeter of skin, p. 343 [★741★]
5 ft. 3 3/4 in., height of average woman, p. 337 [★725★]
5 ft. 9 in., height of average man, p. 337 [★725★]
5.9, east-Atlantic Ocean wave heights in 1962, p. 503 [★1096★]
7.9, east-Atlantic Ocean wave heights in 1980s, p. 503 [★1096★]
9.2, mid-Atlantic Ocean wave heights in early 1960s, p. 503 [★1096★]

Numbers following p. are page references. Numbers in [ ] are table references.

Length - continued

Feet

12, nerves in square centimeter of skin, p. 343 [★741★]
12.5, mid-Atlantic Ocean wave heights in late 1980s, p. 503 [★1096★]
50-150, high-intensity fire average flame height, p. 495 [★1071★]
7,100, average ice cover thickness in Antarctica, p. 501 [★1089★]
13,740, depth of Pacific Ocean, p. 504 [★1097★]
15,000, cloud height, p. 509 [★1112★]
15,700, maximum ice cover thickness in Antarctica, p. 501 [★1089★]
25,000, high-intensity fire maximum convection column height, p. 495 [★1071★]
25,000-40,000 thundercloud height, p. 509 [★1112★]

Inches

0.02, lowest annual rainfall, p. 507 [★1106★]
0.25, hailstone diameter, p. 501 [★1088★]
0.75, rain dropped by a thunderstorm, p. 509 [★1113★]
1.5-2, hailstone diameter, p. 501 [★1088★]
4-5, wet snow contains 1 inch of water, p. 509 [★1111★]
7.53, annual average rainfall in Reno, NV, p. 505 [★1101★]
7.66, annual average rainfall in Phoenix, AZ, p. 505 [★1101★]
8.5, water absorbed annually by soil in U.S., p. 506 [★1105★]
8.81, annual average rainfall in El Paso, TX, p. 505 [★1101★]
8.88, annual average rainfall in Albuquerque, NM, p. 505 [★1101★]
9.90, annual average rainfall in San Diego, CA, p. 505 [★1101★]
12.11, annual average rainfall in Boise, ID, p. 505 [★1101★]
14.40, annual average rainfall in Cheyenne, WY, p. 505 [★1101★]
15.21, annual average rainfall in Great Falls, MT, p. 505 [★1101★]
15.40, annual average rainfall in Denver, CO, p. 505 [★1101★]
15.47, annual average rainfall in Bismark, ND, p. 505 [★1101★]
16, height of a Chinese potbellied pig, p. 523 [★1145★]
21.5, water released annually by soil in U.S., p. 506 [★1105★]
24, spine of average woman, p. 333 [★713★]
28, spine of average man, p. 333 [★713★]
30, precipitation annually in continental U.S., p. 506 [★1104★]
47.30, annual average rainfall in Nashville, TN, p. 505 [★1100★]
49.91, annual average rainfall in Columbia, SC, p. 505 [★1100★]
50.77, annual average rainfall in Atlanta, GA, p. 505 [★1100★]
50.86, annual average rainfall in Little Rock, AR, p. 505 [★1100★]
51.32, annual average rainfall in Jacksonville, FL, p. 505 [★1100★]
52.10, annual average rainfall in Memphis, TN, p. 505 [★1100★]
54.31, annual average rainfall in Juneau, AK, p. 505 [★1100★]
55.37, annual average rainfall in Jackson, MS, p. 505 [★1100★]
55.91, annual average rainfall in Miami, FL, p. 505 [★1100★]
57.8, annual snowfall in Salt Lake City, UT, p. 508 [★1110★]
59.3, annual snowfall in Great Falls, MT, p. 508 [★1110★]
59.8, annual snowfall in Denver, CO, p. 508 [★1110★]
61.88, annual average rainfall in New Orleans, LA, p. 505 [★1100★]
63.6, annual snowfall in Concord, NH, p. 508 [★1110★]
63.96, annual average rainfall in Mobile, AL, p. 505 [★1100★]
64.3, annual snowfall in Albany, NY, p. 508 [★1110★]
70.4, annual snowfall in Portland, ME, p. 508 [★1110★]
76.8, annual snowfall in Burlington, VT, p. 508 [★1110★]
76.9, annual snowfall in Duluth, MN, p. 508 [★1110★]
91.3, annual snowfall in Buffalo, NY, p. 508 [★1110★]
99.4, annual snowfall in Juneau, AK, p. 508 [★1110★]
116.8, annual snowfall in Sault Ste. Marie, MI, p. 508 [★1110★]
240.8, greatest annual snowfall in the U.S., 1989, p. 506 [★1103★]

Length - continued

Inches

463.4, greatest annual rainfall, p. 506 [★1102★]

Miles

0.25, tornado path width, p. 516 [★1129★]
1.1, length of weakest tornadoes, p. 516 [★1131★]
12.4, average length of paths converted by Rails to Trails, p. 439 [★950★]
14-20, diameter of hurricane eye, p. 501 [★1087★]
16, tornado path length, p. 516 [★1129★]
34.2, length of strongest tornadoes, p. 517 [★1132★]
50, diameter of Pacific hurricane eye, p. 501 [★1087★]
258.98, average train trip per passenger, 1987, p. 53 [★125★]
300-400, distance covered by a hurricane in a day, p. 501 [★1086★]
350-400, hurricane diameter, p. 500 [★1083★]
600, distance traveled by a cyclone in a day, p. 519 [★1135★]
862, average distance per airplane trip, 1991, p. 447 [★972★]
1,000, cyclone diameter, p. 519 [★1135★]
1,020, average distance per business trip, 1990, p. 59 [★139★]
2,000-3,000, monthly distance an average rental car is driven, p. 449 [★980★]
3,000, average length of railroad tracks abandoned each year, p. 439 [★951★]
3,000, distance covered by a hurricane, p. 501 [★1086★]
4,400, average distance automobile is driven in Japan per year, p. 451 [★983★]
9,080, average annual miles travelled per automobile, 1940, p. 453 [★988★]
9,164, automobile miles driven per capita in Montana, p. 64 [★150★]
9,246, automobile miles driven per capita in Oklahoma, p. 64 [★150★]
9,277, automobile miles driven per capita in Maine, p. 64 [★150★]
9,417, automobile miles driven per capita in Alabama, p. 64 [★150★]
9,491, automobile miles driven per capita in South Dakota, p. 64 [★150★]
9,500, average automobile miles driven by women, 1990, p. 449 [★979★]
9,500, average distance automobile is driven per year, p. 451 [★983★]
9,638, automobile miles driven per capita in Virginia, p. 64 [★150★]
9,657, automobile miles driven per capita in Vermont, p. 64 [★150★]
9,950, automobile miles driven per capita in New Mexico, p. 64 [★150★]
10,083, automobile miles driven per capita in Georgia, p. 64 [★150★]
10,165, automobile miles driven per capita in Delaware, p. 64 [★150★]
10,246, average number of automobile miles travelled per vehicle, 1988, p. 452 [★985★]
10,382, average annual miles travelled per automobile, 1988, p. 453 [★988★]
14,000, average automobile miles driven by men, 1983, p. 449 [★979★]
18,595, average number of automobile miles travelled per household, 1988, p. 452 [★985★]

Numbers following p. are page references. Numbers in [ ] are table references.

Length - continued

Miles

23,499, average miles of passenger train road operated, 1987, p. 53 [★125★]
35,983,000, Mercury distance from the sun, p. 504 [★1098★]
67,237,700, Venus distance from the sun, p. 504 [★1098★]
92,955,900, Earth distance from the sun, p. 504 [★1098★]
141,634,800, Mars distance from the sun, p. 504 [★1098★]
483,612,200, Jupiter distance from the sun, p. 504 [★1098★]
888,184,000, Saturn distance from the sun, p. 504 [★1098★]
1,782,000,000, Uranus distance from the sun, p. 504 [★1098★]
2,794,000,000, Neptune distance from the sun, p. 504 [★1098★]
3,666,000,000, Pluto distance from the sun, p. 504 [★1098★]

Yards

46, width of weakest tornadoes, p. 516 [★1131★]
616, width of strongest tornadoes, p. 517 [★1132★]

Mass

Milligrams

40, caffeine in a 6-oz. cup of tea, p. 80 [★188★]
45, caffeine in a 12-oz. can of cola, p. 80 [★188★]
227, daily caffeine consumption, p. 80 [★188★]

Money

Cents

1.19, average weekly supermarket profit per sales dollar, 1990, p. 55 [★129★]
3, cost per pound of weight lost for 12-week TOPS program, p. 242 [★518★]
7 to 9, cost per cloth diaper, p. 200 [★429★]
8, average per coupon handling fees paid to retailers, p. 56 [★131★]
8.93, water costs per 100 gallons in Chicago (lake-fed), p. 95 [★224★]
10.3-11.9, water costs per 100 gallons in Miami (aquifer-fed), p. 95 [★224★]
10.575, average revenue per train passenger-mile, 1987, p. 53 [★125★]
12.6, water costs per 100 gallons in New York City, p. 95 [★224★]
13 to 17, cost per diaper service diaper, p. 200 [★429★]
13.7, water costs per 100 gallons in Tucson, p. 95 [★224★]
14.6, water costs per 100 gallons in Santa Barbara, p. 95 [★224★]
18.31, cost per mile of owning and operating an automobile, 1975, p. 189 [★403★]
23, average cost per mile of unrestricted coach airfare, 1992, p. 447 [★973★]
23.3, operating cost per mile of compact automobile, p. 190 [★406★]
24, average movie admission price, 1940, p. 47 [★113★]
24, average savings (medical, etc.) by sedentary person walking or running a mile, p. 331 [★707★]
25, cost per disposable diaper, p. 200 [★429★]
30.6, operating cost per mile of large automobile, p. 190 [★406★]
39.2, operating cost per mile of van, p. 190 [★406★]
41, cost per mile of owning and operating an automobile, 1990, p. 189 [★403★]
53, average movie admission price, 1950, p. 47 [★113★]
68, bank cost to process a check, p. 244 [★524★]
69, average movie admission price, 1960, p. 47 [★113★]

Money - continued

Cents

70, earned by women per dollar earned by men, 1991, p. 402 [★860★]

Dollars

$0.24, health care providers' administrative and billing costs, p. 287 [★605★]
$0.80, cost of incandescent light bulb, p. 205 [★439★]
$0.84, school library average expenditure per student on computer software, 1984-1985, p. 157 [★348★]
$1, Canadian physician cost of paperwork per patient, p. 311 [★666★]
$1, cost to test blood for HIV infection, p. 288 [★608★]
$1.02, weekly expenditures for coffee and tea, 1989, p. 206 [★443★]
$1.10, cost per pound of weight lost for 12-week Weight Watchers diet program, p. 242 [★518★]
$1.11, hourly cost of child care by relative, 1990, p. 198 [★425★]
$1.12, average price per gallon of gasoline, 1992, p. 37 [★86★]
$1.17, average price per gallon of gasoline, 1991, p. 37 [★86★]
$1.23, average price per gallon of gasoline, 1970, p. 37 [★86★]
$1.25, average price per gallon of gasoline, 1990, p. 37 [★86★]
$1.33, average price per gallon of gasoline, 1965, p. 37 [★86★]
$1.35, hourly cost of family day care child care, 1990, p. 198 [★425★]
$1.39, average price per gallon of gasoline, 1975, p. 37 [★86★]
$1.45, average price per gallon of gasoline, 1960, p. 37 [★86★]
$1.49, school library average expenditure on periodicals per student, 1984-1985, p. 157 [★348★]
$1.54, average price per gallon of gasoline, 1985, p. 37 [★86★]
$1.67, hourly cost of day care center child care, 1990, p. 198 [★425★]
$1.79, average price per gallon of gasoline in Canada, 1992, p. 38 [★87★]
$1.80, school library average expenditure on A-V materials per student, 1984-1985, p. 157 [★348★]
$1.85, average hourly wage in Mexico, p. 357 [★766★]
$2, federal mass transit funds to Indiana, 1990, p. 255 [★539★]
$2-3, annual health budget per person in developing countries, p. 288 [★608★]
$2.05, average hourly benefits in Portugal, p. 356 [★765★]
$2.10, average price per gallon of gasoline, 1980, p. 37 [★86★]
$2.30, hourly cost of in-home care child care, 1990, p. 198 [★425★]
$2.52, weekly expenditures for soft drinks, 1989, p. 206 [★443★]
$2.6, federal mass transit funds to North Carolina, 1990, p. 255 [★539★]
$2.6, federal mass transit funds to South Dakota, 1990, p. 255 [★539★]
$2.63, average hourly benefits in Greece, p. 356 [★765★]
$2.64, average hourly wage in Brazil, p. 357 [★766★]
$2.65, magazine newsstand price, 1990, p. 241 [★517★]
$2.68, price per volume of trade paperback Juveniles books, 1991, p. 165 [★362★]
$2.70, average hourly wage in Portugal, p. 356 [★765★]
$2.7, federal mass transit funds to Kansas, 1990, p. 255 [★539★]
$2.83, public school expenditure per capita, 1899-1900, p. 178 [★381★]
$3, federal research on arthritis per person, p. 289 [★611★]
$3, Los Angeles average annual cost of traffic congestion per automobile, p. 44 [★104★]

Numbers following p. are page references. Numbers in [ ] are table references.

Money - continued

Dollars

$3, West Virginia nursing home expenditure per capita, 1966, p. 321 [★687★]
$3.20, average hourly wage in Hong Kong, p. 357 [★766★]
$3.21, average price per gallon of gasoline in Great Britain, 1992, p. 38 [★87★]
$3.41, school library average expenditure per student on computer hardware, 1984-1985, p. 157 [★348★]
$3.45, median annual cost per square foot for school maintenance and operations, 1991, p. 180 [★387★]
$3.56, price per volume of mass market paperback juveniles books, 1991, p. 164 [★361★]
$3.59, average price per gallon of gasoline in Germany, 1992, p. 38 [★87★]
$3.60, average price per gallon of gasoline in France, 1992, p. 38 [★87★]
$3.68, price per volume of trade paperback religion books, 1991, p. 165 [★362★]
$3.71, average price per gallon of gasoline in Japan, 1992, p. 38 [★87★]
$3.75, bank fee for monthly checking account fee, 1991, p. 243 [★523★]
$3.80, added hourly value of employer-paid benefits, nonunion members, 1989, p. 413 [★884★]
$4, Mississippi nursing home expenditure per capita, 1966, p. 321 [★687★]
$4, public funds expenditure per capita on health care, 1929, p. 283 [★596★]
$4, public funds expenditure per capita on health care, 1935, p. 283 [★596★]
$4.04, family day-care provider's average hourly salary, p. 368 [★787★]
$4.05, price per volume of mass market paperback fiction books, 1991, p. 164 [★361★]
$4.08, average hourly wage in Greece, p. 356 [★765★]
$4.08, cost per pound of weight lost for 12-week Diet Center diet program, p. 242 [★518★]
$4.12, average hourly benefits in Great Britain, p. 356 [★765★]
$4.18, average price per gallon of gasoline in Sweden, 1992, p. 38 [★87★]
$4.20, price per volume of trade paperback fiction books, 1991, p. 165 [★362★]
$4.22, average hourly benefits in United States, p. 356 [★765★]
$4.25, average hourly benefits in Japan, p. 356 [★765★]
$4.57, price per volume of mass market paperback total books, 1991, p. 164 [★361★]
$4.71, average price per gallon of gasoline in Italy, 1992, p. 38 [★87★]
$4.71, price per volume of trade paperback poetry, drama books, 1991, p. 165 [★362★]
$4.71, public school expenditure per capita, 1909-1910, p. 178 [★381★]
$4.87, price per volume of trade paperback sports, recreation books, 1991, p. 165 [★362★]
$4.91, price per volume of trade paperback Biography books, 1991, p. 165 [★362★]
$5 to 10, purchase price of gerbil, 1991, p. 211 [★456★]
$5 to 10, purchase price of hamster, 1991, p. 211 [★456★]

Money - continued

Dollars

$5 to 10, purchase price of turtles, 1991, p. 211 [★456★]
$5, average monthly cable television rates, 1955, p. 438 [★948★]
$5, fecal-occult blood test, 1990, p. 291 [★617★]
$5, German health insurance payment for doctor visit, p. 315 [★673★]
$5, New Mexico nursing home expenditure per capita, 1966, p. 321 [★687★]
$5-7, hourly wage of average security guard, p. 393 [★843★]
$5.18, price per volume of trade paperback literature books, 1991, p. 165 [★362★]
$5.21, price per volume of trade paperback travel books, 1991, p. 165 [★362★]
$5.29, price per volume of mass market paperback sports, recreation books, 1991, p. 164 [★361★]
$5.36, price per volume of mass market paperback language books, 1991, p. 164 [★361★]
$5.46, average food expenditure per airline passenger, 1990, p. 28 [★66★]
$5.57, price per volume of trade paperback philosophy, psychology books, 1991, p. 165 [★362★]
$5.72, price per volume of trade paperback education books, 1991, p. 165 [★362★]
$5.81, price per volume of trade paperback history books, 1991, p. 165 [★362★]
$5.93, price per volume of trade paperback books, 1991, p. 165 [★362★]
$5.98, Arkansas expenditures per capita for public libraries, 1989, p. 153 [★338★]
$6, District of Columbia nursing home expenditure per capita, 1966, p. 321 [★687★]
$6, Hawaii nursing home expenditure per capita, 1966, p. 321 [★687★]
$6, North Carolina nursing home expenditure per capita, 1966, p. 321 [★687★]
$6, public funds expenditure per capita on health care, 1940, p. 283 [★596★]
$6, South Carolina nursing home expenditure per capita, 1966, p. 321 [★687★]
$6, Tennessee nursing home expenditure per capita, 1966, p. 321 [★687★]
$6, Virginia nursing home expenditure per capita, 1966, p. 321 [★687★]
$6, Wyoming nursing home expenditure per capita, 1966, p. 321 [★687★]
$6.03, price per volume of trade paperback sociology, economics books, 1991, p. 165 [★362★]
$6.08, price per volume of mass market paperback poetry, drama books, 1991, p. 164 [★361★]
$6.24, price per volume of mass market paperback history books, 1991, p. 164 [★361★]
$6.24, school library average expenditure per student on books, 1984-1985, p. 157 [★348★]
$6.27, price per volume of trade paperback art books, 1991, p. 165 [★362★]
$6.35, cost per pound of weight lost for 12-week Medifast diet program, p. 242 [★518★]
$6.45, price per volume of mass market paperback literature books, 1991, p. 164 [★361★]

Money - continued

Dollars

$6.58, Mississippi expenditures per capita for public libraries, 1989, p. 153 [★338★]
$6.61, price per volume of mass market paperback law books, 1991, p. 164 [★361★]
$6.80, cost per pound of weight lost for 12-week registered dietitian diet program, p. 242 [★518★]
$6.81, price per volume of mass market paperback philosophy, psychology books, 1991, p. 164 [★361★]
$6.88, price per volume of mass market paperback sociology, economics books, 1991, p. 164 [★361★]
$6.95, added hourly value of employer-paid benefits, trade union members, 1989, p. 413 [★884★]
$6.95, price per volume of mass market paperback religion books, 1991, p. 164 [★361★]
$7, annual energy costs for microwave, p. 201 [★432★]
$7, Nevada nursing home expenditure per capita, 1966, p. 321 [★687★]
$7, physician cost of paperwork per patient, p. 311 [★666★]
$7.04, Kentucky expenditures per capita for public libraries, 1989, p. 153 [★338★]
$7.09, price per volume of trade paperback business books, 1991, p. 165 [★362★]
$7.19, school cafeteria worker's average hourly wage, 1991, p. 368 [★785★]
$7.24, price per volume of mass market paperback biography books, 1991, p. 164 [★361★]
$7.46, West Virginia expenditures per capita for public libraries, 1989, p. 153 [★338★]
$7.56, Montana expenditures per capita for public libraries, 1989, p. 153 [★338★]
$7.61, average hourly benefits in France, p. 356 [★765★]
$7.63, price per volume of trade paperback medicine books, 1991, p. 165 [★362★]
$7.66, weekly allowance for girls 4-12, 1991, p. 347 [★748★]
$7.70, Tennessee expenditures per capita for public libraries, 1989, p. 153 [★338★]
$7.77, instructional teacher aide's average hourly wage, 1991, p. 368 [★785★]
$7.79, price per volume of trade paperback language books, 1991, p. 165 [★362★]
$7.85, price per volume of mass market paperback business books, 1991, p. 164 [★361★]
$7.88, average hourly benefits in Switzerland, p. 356 [★765★]
$7.94, cost per pound of weight lost for 12-week HMR diet program, p. 242 [★518★]
$7.97, price per volume of trade paperback technology books, 1991, p. 165 [★362★]
$8, Alabama nursing home expenditure per capita, 1966, p. 321 [★687★]
$8, Arizona nursing home expenditure per capita, 1966, p. 321 [★687★]
$8, cost of preparing medical insurance check, p. 289 [★610★]
$8, Delaware nursing home expenditure per capita, 1966, p. 321 [★687★]
$8, Georgia nursing home expenditure per capita, 1966, p. 321 [★687★]
$8, Louisiana nursing home expenditure per capita, 1966, p. 321 [★687★]

Money - continued

Dollars

$8, physician's cost per insurance claim submitted, p. 289 [★610★]
$8, property tax per capita in Mesa, AZ, p. 263 [★558★]
$8.01, North Dakota expenditures per capita for public libraries, 1989, p. 153 [★338★]
$8.08, price per volume of mass market paperback education books, 1991, p. 164 [★361★]
$8.13, Alabama expenditures per capita for public libraries, 1989, p. 153 [★338★]
$8.22, Delaware expenditures per capita for public libraries, 1989, p. 153 [★338★]
$8.24, price per volume of mass market paperback medicine books, 1991, p. 164 [★361★]
$8.25, national average electricity rate per kilowatt-hour, p. 205 [★439★]
$8.43, average weekly supermarket sales per square foot, 1990, p. 55 [★129★]
$8.49, average hourly wage in France, p. 356 [★765★]
$8.62, cost per pound of weight lost for 12-week Nutrisystems diet program, p. 242 [★518★]
$8.69, average hourly benefits in Netherlands, p. 356 [★765★]
$8.72, South Carolina expenditures per capita for public libraries, 1989, p. 153 [★338★]
$8.81, price per volume of trade paperback science books, 1991, p. 165 [★362★]
$8.87, weekly allowance for boys 4-12, 1991, p. 347 [★748★]
$8.94, Texas expenditures per capita for public libraries, 1989, p. 153 [★338★]
$9, Kentucky nursing home expenditure per capita, 1966, p. 321 [★687★]
$9, Maryland nursing home expenditure per capita, 1966, p. 321 [★687★]
$9, Utah nursing home expenditure per capita, 1966, p. 321 [★687★]
$9.05, school custodian's average hourly wage, 1991, p. 368 [★785★]
$9.17, price per volume of mass market paperback science books, 1991, p. 164 [★361★]
$9.41, weekly expenditures for meat, 1989, p. 206 [★442★]
$9.49, average hourly wage in Austria, p. 356 [★765★]
$9.52, average hourly wage in Italy, p. 356 [★765★]
$9.52, school bus driver's average hourly wage, 1991, p. 368 [★785★]
$9.58, average hourly benefits in Sweden, p. 356 [★765★]
$9.59, average hourly wage in Great Britain, p. 356 [★765★]
$9.65, Nevada expenditures per capita for public libraries, 1989, p. 153 [★338★]
$9.84, Idaho expenditures per capita for public libraries, 1989, p. 153 [★338★]
$9.91, public school expenditure per capita, 1919-1920, p. 178 [★381★]
$9.94, weekly expenditures for fruits and vegetables, 1989, p. 205 [★441★]
$9.98, cost per pound of weight lost for 12-week Optifast diet program, p. 242 [★518★]
$9.99, average hourly benefits in Italy, p. 356 [★765★]
$10, average cost of testing a home's radon level, p. 418 [★897★]
$10, Michigan nursing home expenditure per capita, 1966, p. 321 [★687★]

Numbers following p. are page references. Numbers in [ ] are table references.

Money - continued

Dollars

$10, New Jersey nursing home expenditure per capita, 1966, p. 321 [★687★]
$10, pap test, 1990, p. 291 [★617★]
$10-15, average daily youth hostel rate, p. 444 [★965★]
$10.06, Georgia expenditures per capita for public libraries, 1989, p. 153 [★338★]
$10.10, average hourly wage in Copenhagen, p. 349 [★755★]
$10.20, average hourly wage in Oslo, p. 349 [★755★]
$10.24, Oklahoma expenditures per capita for public libraries, 1989, p. 153 [★338★]
$10.40, average hourly wage in Los Angeles, p. 349 [★755★]
$10.43, cost per pound of weight lost for 12-week Jenny Craig diet program, p. 242 [★518★]
$10.50, average hourly wage in New York, p. 349 [★755★]
$10.55, North Carolina expenditures per capita for public libraries, 1989, p. 153 [★338★]
$10.66, average hourly wage in Netherlands, p. 356 [★765★]
$10.66, price per volume of trade paperback law books, 1991, p. 165 [★362★]
$10.70, average hourly wage in Helsinki, p. 349 [★755★]
$10.90, average hourly wage in Tokyo, p. 349 [★755★]
$10.93, production workers' average hourly wage, 1990, p. 356 [★764★]
$10.99, Pennsylvania expenditures per capita for public libraries, 1989, p. 153 [★338★]
$11, California nursing home expenditure per capita, 1966, p. 321 [★687★]
$11, Florida nursing home expenditure per capita, 1966, p. 321 [★687★]
$11, Texas nursing home expenditure per capita, 1966, p. 321 [★687★]
$11.22, price per volume of mass market paperback travel books, 1991, p. 164 [★361★]
$11.30, average hourly benefits in Germany, Western, p. 356 [★765★]
$11.34, New Mexico expenditures per capita for public libraries, 1989, p. 153 [★338★]
$11.40, average hourly wage in Luxembourg, p. 349 [★755★]
$11.40, average hourly wage in United States, p. 356 [★765★]
$11.53, Louisiana expenditures per capita for public libraries, 1989, p. 153 [★338★]
$11.56, cost per pound of weight lost for 12-week United Weight Control diet program, p. 242 [★518★]
$11.71, Rhode Island expenditures per capita for public libraries, 1989, p. 153 [★338★]
$11.85, Iowa expenditures per capita for public libraries, 1989, p. 153 [★338★]
$11.93, Florida expenditures per capita for public libraries, 1989, p. 153 [★338★]
$12, annual energy costs for dishwasher, p. 201 [★432★]
$12, Idaho nursing home expenditure per capita, 1966, p. 321 [★687★]
$12, Indiana nursing home expenditure per capita, 1966, p. 321 [★687★]
$12, Iowa state mental health care agency expenditure per capita, 1987, p. 319 [★682★]
$12, Missouri nursing home expenditure per capita, 1966, p. 321 [★687★]

Money - continued

Dollars

$12, Montana nursing home expenditure per capita, 1966, p. 321 [★687★]
$12, nursing home expenditure per capita, 1966, p. 321 [★687★]
$12, Ohio nursing home expenditure per capita, 1966, p. 321 [★687★]
$12, Pennsylvania nursing home expenditure per capita, 1966, p. 321 [★687★]
$12, United States nursing home expenditure per capita, 1966, p. 321 [★687★]
$12.35, bank fee for stop-payment, 1991, p. 243 [★523★]
$12.37, Vermont expenditures per capita for public libraries, 1989, p. 153 [★338★]
$12.40, monthly cost of residential telephone service, 1990, p. 212 [★457★]
$12.50, German health insurance payment for doctor's visit per night, p. 315 [★673★]
$12.66, Maine expenditures per capita for public libraries, 1989, p. 153 [★338★]
$12.72, average hourly wage in Sweden, p. 356 [★765★]
$12.90, average per-customer spending on children's books at college/university bookstores, 1992, p. 162 [★357★]
$13, Arkansas nursing home expenditure per capita, 1966, p. 321 [★687★]
$13, average increase in blue-collar worker's pay, 1991, p. 406 [★868★]
$13, Illinois nursing home expenditure per capita, 1966, p. 321 [★687★]
$13.01, price per volume of hardbound juveniles books, 1990, p. 162 [★359★]
$13.09, average hourly wage in Germany, Western, p. 356 [★765★]
$13.09, South Dakota expenditures per capita for public libraries, 1989, p. 153 [★338★]
$13.16, Nebraska expenditures per capita for public libraries, 1989, p. 153 [★338★]
$13.37, Missouri expenditures per capita for public libraries, 1989, p. 153 [★338★]
$13.40, price per volume of mass market paperback art books, 1991, p. 164 [★361★]
$13.60, average hourly wage in Japan, p. 356 [★765★]
$13.64, Utah expenditures per capita for public libraries, 1989, p. 153 [★338★]
$13.70, average per-customer spending on children's books at general/specialty bookstores, 1992, p. 162 [★357★]
$13.70, average per-customer spending on children's books at national/regional chain bookstores, 1992, p. 162 [★357★]
$13.83, Michigan expenditures per capita for public libraries, 1989, p. 153 [★338★]
$13.87, bank fee for overdrafts, 1991, p. 243 [★523★]
$14, Wisconsin nursing home expenditure per capita, 1966, p. 321 [★687★]
$14.17, bank fee for insufficient funds, 1991, p. 243 [★523★]
$14.46, New Hampshire expenditures per capita for public libraries, 1989, p. 153 [★338★]
$14.50, average hourly wage in Geneva, p. 349 [★755★]
$14.53, Arizona expenditures per capita for public libraries, 1989, p. 153 [★338★]
$14.74, Oregon expenditures per capita for public libraries, 1989, p. 153 [★338★]

Numbers following p. are page references. Numbers in [ ] are table references.

Money - continued

Dollars

$14.77, college graduate's hourly salary, 1991, p. 364 [★778★]
$15 to 25, purchase price of rabbit, 1991, p. 211 [★456★]
$15, average annual advertising expenditures per capita worldwide, 1950, p. 27 [★61★]
$15, Colorado nursing home expenditure per capita, 1966, p. 321 [★687★]
$15, federal mass transit funds to U.S., 1990, p. 255 [★539★]
$15, Maine nursing home expenditure per capita, 1966, p. 321 [★687★]
$15, Rhode Island nursing home expenditure per capita, 1966, p. 321 [★687★]
$15.24, college graduate's hourly salary, 1981, p. 364 [★778★]
$15.51, average hourly wage in Switzerland, p. 356 [★765★]
$15.60, German health insurance payment for electrocardiogram, p. 315 [★673★]
$15.85, Kansas expenditures per capita for public libraries, 1989, p. 153 [★338★]
$15.89, California expenditures per capita for public libraries, 1989, p. 153 [★338★]
$15.92, fee for 3-inch by 5-inch bank safety deposit box, 1992, p. 241 [★515★]
$16, Arizona state mental health care agency expenditure per capita, 1987, p. 319 [★682★]
$16, average hourly wage in Zurich, p. 349 [★755★]
$16, New Hampshire nursing home expenditure per capita, 1966, p. 321 [★687★]
$16, New York nursing home expenditure per capita, 1966, p. 321 [★687★]
$16, public library funding per capita, p. 155 [★343★]
$16, Washington nursing home expenditure per capita, 1966, p. 321 [★687★]
$16.29, Virginia expenditures per capita for public libraries, 1989, p. 153 [★338★]
$16.45, college graduate's hourly salary, 1973, p. 364 [★778★]
$16.53, Wisconsin expenditures per capita for public libraries, 1989, p. 153 [★338★]
$16.60, national debt per capita, 1900, p. 259 [★548★]
$17, Idaho state mental health care agency expenditure per capita, 1987, p. 319 [★682★]
$17, mental health care expenditure per capita, 1969, p. 318 [★681★]
$17, Nebraska nursing home expenditure per capita, 1966, p. 321 [★687★]
$17, Oregon nursing home expenditure per capita, 1966, p. 321 [★687★]
$17.23, Hawaii expenditures per capita for public libraries, 1989, p. 153 [★338★]
$17.58, average monthly cable television rates, 1990, p. 438 [★948★]
$17.80, German health insurance payment for doctor's house call, p. 315 [★673★]
$17.91, public school expenditure per capita, 1939-1940, p. 178 [★381★]
$17.95, Indiana expenditures per capita for public libraries, 1989, p. 153 [★338★]
$18, Kansas nursing home expenditure per capita, 1966, p. 321 [★687★]

Money - continued

Dollars

$18, private funds expenditure per capita on health care, 1935, p. 284 [★597★]
$18, South Dakota nursing home expenditure per capita, 1966, p. 321 [★687★]
$18, Texas state mental health care agency expenditure per capita, 1987, p. 319 [★682★]
$18.01, Colorado expenditures per capita for public libraries, 1989, p. 153 [★338★]
$18.4, average per room selling price of hotels, 1991, p. 63 [★147★]
$18.62, Minnesota expenditures per capita for public libraries, 1989, p. 153 [★338★]
$19, Connecticut nursing home expenditure per capita, 1966, p. 321 [★687★]
$19, North Dakota nursing home expenditure per capita, 1966, p. 321 [★687★]
$19, Oklahoma nursing home expenditure per capita, 1966, p. 321 [★687★]
$19, Utah state mental health care agency expenditure per capita, 1987, p. 319 [★682★]
$19, Vermont nursing home expenditure per capita, 1966, p. 321 [★687★]
$19.03, public school expenditure per capita, 1929-1930, p. 178 [★381★]
$19.18, Massachusetts expenditures per capita for public libraries, 1989, p. 153 [★338★]
$19.25, Illinois expenditures per capita for public libraries, 1989, p. 153 [★338★]
$19.81, Washington expenditures per capita for public libraries, 1989, p. 153 [★338★]
$19.83, price per volume of hardbound fiction books, 1990, p. 162 [★359★]
$19.83, Wyoming expenditures per capita for public libraries, 1989, p. 153 [★338★]
$20, cost of fluorescent light bulb, p. 205 [★439★]
$20, German health insurance payment for diagnosis of pregnancy, p. 315 [★673★]
$21, Nebraska state mental health care agency expenditure per capita, 1987, p. 319 [★682★]
$21, personal funds expenditure per capita on health care, 1935, p. 284 [★598★]
$21, public funds expenditure per capita on health care, 1950, p. 283 [★596★]
$21.16, New Jersey expenditures per capita for public libraries, 1989, p. 153 [★338★]
$22, annual energy costs for color TV, p. 201 [★432★]
$22, Iowa nursing home expenditure per capita, 1966, p. 321 [★687★]
$22, Massachusetts nursing home expenditure per capita, 1966, p. 321 [★687★]
$22, Minnesota nursing home expenditure per capita, 1966, p. 321 [★687★]
$22, Mississippi state mental health care agency expenditure per capita, 1987, p. 319 [★682★]
$22, price per pound of raw opium (Afghanistan), p. 387 [★830★]
$22.10, Connecticut expenditures per capita for public libraries, 1989, p. 153 [★338★]

Money - continued

Dollars

$22.50, German health insurance payment for examination of a baby, p. 315 [★673★]
$23, Arkansas state mental health care agency expenditure per capita, 1987, p. 319 [★682★]
$23, Kentucky state mental health care agency expenditure per capita, 1987, p. 319 [★682★]
$23, private funds expenditure per capita on health care, 1940, p. 284 [★597★]
$23, West Virginia state mental health care agency expenditure per capita, 1987, p. 319 [★682★]
$23.34, Ohio expenditures per capita for public libraries, 1989, p. 153 [★338★]
$23.40, average per-customer spending on children's books at children's bookstores, 1992, p. 162 [★357★]
$23.64, Alaska expenditures per capita for public libraries, 1989, p. 153 [★338★]
$24, New Mexico state mental health care agency expenditure per capita, 1987, p. 319 [★682★]
$24, property tax per capita in Columbus, OH, p. 263 [★558★]
$24, Tennessee state mental health care agency expenditure per capita, 1987, p. 319 [★682★]
$24.16, price per volume of mass market paperback technology books, 1991, p. 164 [★361★]
$24.45, Maryland expenditures per capita for public libraries, 1989, p. 153 [★338★]
$25, administrative costs for a doctor's office visit, p. 289 [★610★]
$25, city government costs per ton to pay reprocessing facilities to take recyclables, p. 427 [★919★]
$25, cost of otocoustic emission test, p. 297 [★636★]
$25, Florida state mental health care agency expenditure per capita, 1987, p. 319 [★682★]
$25, Illinois state mental health care agency expenditure per capita, 1987, p. 319 [★682★]
$25, Louisiana state mental health care agency expenditure per capita, 1987, p. 319 [★682★]
$25, Mexican physician charge for office visit, p. 314 [★672★]
$25, private funds expenditure per capita on health care, 1929, p. 284 [★597★]
$25, property tax per capita in Mobile, AL, p. 263 [★558★]
$25.95, cost of homemade Thanksgiving meal for 10, p. 209 [★450★]
$26, Hawaii state mental health care agency expenditure per capita, 1987, p. 319 [★682★]
$26, nursing home expenditure per capita, 1982, p. 321 [★687★]
$26, personal funds expenditure per capita on health care, 1929, p. 284 [★598★]
$26, personal funds expenditure per capita on health care, 1940, p. 284 [★598★]
$26, public funds expenditure per capita on health care, 1955, p. 283 [★596★]
$27, South Dakota state mental health care agency expenditure per capita, 1987, p. 319 [★682★]
$27.11, annual magazine subscription price, 1990, p. 241 [★517★]
$27.39, average revenue per train passenger, 1987, p. 53 [★125★]
$28, Kansas state mental health care agency expenditure per capita, 1987, p. 319 [★682★]
$28, Montana state mental health care agency expenditure per capita, 1987, p. 319 [★682★]

Money - continued

Dollars

$28, Nevada state mental health care agency expenditure per capita, 1987, p. 319 [★682★]
$28, Oregon state mental health care agency expenditure per capita, 1987, p. 319 [★682★]
$28.13, German health insurance payment for chest X-ray, p. 315 [★673★]
$28.30, weekly expenditures for meals away from home, 1989, p. 207 [★444★]
$29, Alabama state mental health care agency expenditure per capita, 1987, p. 319 [★682★]
$29, monetary loss per rape, 1990, p. 117 [★280★]
$29.46, District of Columbia expenditures per capita for public libraries, 1989, p. 153 [★338★]
$29.48, New York expenditures per capita for public libraries, 1989, p. 153 [★338★]
$29.58, price per volume of hardbound biography books, 1990, p. 162 [★359★]
$30, California state mental health care agency expenditure per capita, 1987, p. 319 [★682★]
$30, Colorado state mental health care agency expenditure per capita, 1987, p. 319 [★682★]
$30, Oklahoma state mental health care agency expenditure per capita, 1987, p. 319 [★682★]
$30, Wyoming state mental health care agency expenditure per capita, 1987, p. 319 [★682★]
$30.41, price per volume of hardbound travel books, 1990, p. 162 [★359★]
$30.52, price per volume of hardbound sports, recreation books, 1990, p. 162 [★359★]
$31, Indiana state mental health care agency expenditure per capita, 1987, p. 319 [★682★]
$31, Missouri state mental health care agency expenditure per capita, 1987, p. 319 [★682★]
$31, Wisconsin state mental health care agency expenditure per capita, 1987, p. 319 [★682★]
$31.31, price per volume of hardbound religion books, 1991, p. 162 [★359★]
$32, Georgia state mental health care agency expenditure per capita, 1987, p. 319 [★682★]
$32, property tax per capita in Toledo, OH, p. 263 [★558★]
$32.12, physician cost of hepatitis B vaccine, p. 316 [★675★]
$32.19, price per volume of hardbound poetry, drama books, 1990, p. 162 [★359★]
$33, Ohio state mental health care agency expenditure per capita, 1987, p. 319 [★682★]
$33.04, cost to operate 60-watt (equivalent) fluorescent light bulb 10,000 hours, p. 205 [★439★]
$34, property tax per capita in Tucson, AZ, p. 263 [★558★]
$34.61, per capita value of property lost to fire, 1990, p. 110 [★266★]
$35, public funds expenditure per capita on health care, 1960, p. 283 [★596★]
$35, Virginia state mental health care agency expenditure per capita, 1987, p. 319 [★682★]
$35.80, price per volume of hardbound literature books, 1990, p. 162 [★359★]
$35.83, cost per mile of owning and operating a subcompact automobile, 1991, p. 191 [★407★]

Numbers following p. are page references. Numbers in [ ] are table references.

Money - continued

Dollars

$36, New Hampshire state mental health care agency expenditure per capita, 1987, p. 319 [★682★]
$36.43, price per volume of hardbound history books, 1990, p. 162 [★359★]
$37, Washington state mental health care agency expenditure per capita, 1987, p. 319 [★682★]
$38, state mental health care agency expenditure per capita, 1987, p. 319 [★682★]
$38, United States state mental health care agency expenditure per capita, 1987, p. 319 [★682★]
$38.72, price per volume of hardbound education books, 1991, p. 162 [★359★]
$39, average Social Security benefits for children of disabled workers 1970, p. 247 [★529★]
$39, public school expenditure per capita, 1949-1950, p. 178 [★381★]
$39.64, physician cost of oral polio vaccine, p. 316 [★675★]
$40, annual local sheriffs' department operating expenses per resident, 1990, p. 139 [★315★]
$40.58, price per volume of hardbound philosophy, psychology books, 1990, p. 162 [★359★]
$41, Delaware state mental health care agency expenditure per capita, 1987, p. 319 [★682★]
$41, North Carolina state mental health care agency expenditure per capita, 1987, p. 319 [★682★]
$41, Rhode Island state mental health care agency expenditure per capita, 1987, p. 319 [★682★]
$42, annual energy costs for range/oven, p. 201 [★432★]
$42, cost of grandparent's birthday gift per child, p. 210 [★453★]
$42, Maine state mental health care agency expenditure per capita, 1987, p. 319 [★682★]
$42, Minnesota state mental health care agency expenditure per capita, 1987, p. 319 [★682★]
$42, North Dakota state mental health care agency expenditure per capita, 1987, p. 319 [★682★]
$42, property tax per capita in Colorado Springs, CO, p. 263 [★558★]
$42.10, price per volume of hardbound sociology, economics books, 1990, p. 162 [★359★]
$42.18, price per volume of hardbound art books, 1991, p. 162 [★359★]
$42.98, price per volume of hardbound language books, 1990, p. 162 [★359★]
$43, New Jersey state mental health care agency expenditure per capita, 1987, p. 319 [★682★]
$43, rise in average Chinese monthly income since 1991 economic reforms, p. 359 [★771★]
$43.64, cost per mile of owning and operating a mid-size automobile, 1991, p. 191 [★407★]
$44, price per ton paid for collected recyclables, p. 427 [★920★]
$44, Vermont state mental health care agency expenditure per capita, 1987, p. 319 [★682★]
$45, average annual expenditures per capita for city and town parks, p. 262 [★555★]
$45, average Social Security benefits for children of retired workers 1970, p. 247 [★529★]
$45, average Social Security benefits for special benefits 1970, p. 247 [★529★]

Money - continued

Dollars

$45, German health insurance payment for doctor's house call at-night, p. 315 [★673★]
$45, property tax per capita in Tulsa, OK, p. 263 [★558★]
$45, South Carolina state mental health care agency expenditure per capita, 1987, p. 319 [★682★]
$45.48, price per volume of hardbound business books, 1991, p. 162 [★359★]
$46, annual energy costs for furnace fan, p. 201 [★432★]
$46, average annual advertising expenditures per capita worldwide, 1989, p. 27 [★61★]
$47, cost of doctor's office visit, p. 292 [★619★]
$48, Connecticut hospital care expenditure per capita, 1966, p. 307 [★652★]
$48, property tax per capita in Riverside, CA, p. 263 [★558★]
$49, Maryland state mental health care agency expenditure per capita, 1987, p. 319 [★682★]
$49, New Mexico nursing home expenditure per capita, 1982, p. 321 [★687★]
$49, property tax per capita in Akron, OH, p. 263 [★558★]
$49, Wyoming nursing home expenditure per capita, 1982, p. 321 [★687★]
$49.85, physician cost of diptheria/tetanus/pertussis vaccine, p. 316 [★675★]
$50, Alaska state mental health care agency expenditure per capita, 1987, p. 319 [★682★]
$50, cost of group psychotherapy session per person, 1991, p. 319 [★683★]
$50, Idaho hospital care expenditure per capita, 1966, p. 307 [★652★]
$50, Pennsylvania state mental health care agency expenditure per capita, 1987, p. 319 [★682★]
$50, public funds expenditure per capita on health care, 1965, p. 283 [★596★]
$50.58, physician cost of measles/mumps/rubella vaccine, p. 316 [★675★]
$51, amalgam dental filling, p. 312 [★668★]
$51, Colorado hospital care expenditure per capita, 1966, p. 307 [★652★]
$51, credit card debt of people with some high school education, p. 214 [★463★]
$52, property tax per capita in Aurora, CO, p. 263 [★558★]
$53, nursing home expenditure per capita, 1982, p. 321 [★687★]
$54, weekly cost of child care, 1988, p. 198 [★424★]
$55, nursing home expenditure per capita, 1982, p. 321 [★687★]
$55.59, cost per mile of owning and operating a full-size automobile, 1991, p. 191 [★407★]
$56, Arkansas hospital care expenditure per capita, 1966, p. 307 [★652★]
$56, average charitable contributions of teen, 1991, p. 195 [★418★]
$56, average daily rental-car cost, 1992, p. 447 [★974★]
$56, Connecticut state mental health care agency expenditure per capita, 1987, p. 319 [★682★]
$56, Georgia hospital care expenditure per capita, 1966, p. 307 [★652★]
$56, lifetime energy costs of average light, p. 201 [★431★]
$56.09, average cost of periodical subscription in Africa, 1992, p. 166 [★364★]

Numbers following p. are page references. Numbers in [ ] are table references.

Money - continued

Dollars

$57, Iowa hospital care expenditure per capita, 1966, p. 307 [★652★]
$57, United States hospital care expenditure per capita, 1982, p. 307 [★652★]
$57.17, average household spending in bookstores, 1987, p. 160 [★354★]
$57.50, cost to operate 60-watt incandescent light bulb 10,000 hours, p. 205 [★439★]
$58, Florida hospital care expenditure per capita, 1966, p. 307 [★652★]
$58, private funds expenditure per capita on health care, 1950, p. 284 [★597★]
$58.20, physician cost of hemophilus influenza vaccine, p. 316 [★675★]
$59, average Social Security benefits for wives and husbands 1970, p. 247 [★529★]
$60, United States hospital care expenditure per capita, 1966, p. 307 [★652★]
$60.75, average cost of recreation periodical subscription, 1992, p. 165 [★363★]
$61, Alabama hospital care expenditure per capita, 1966, p. 307 [★652★]
$61, Michigan state mental health care agency expenditure per capita, 1987, p. 319 [★682★]
$62, Massachusetts state mental health care agency expenditure per capita, 1987, p. 319 [★682★]
$62, West Virginia nursing home expenditure per capita, 1982, p. 321 [★687★]
$62.01, average cost of periodical subscription in South America, 1992, p. 166 [★364★]
$62.08, average cost of language and literature periodical subscription, 1992, p. 165 [★363★]
$63, Alabama hospital care expenditure per capita, 1966, p. 307 [★652★]
$63, Alaska hospital care expenditure per capita, 1966, p. 307 [★652★]
$63, Hawaii hospital care expenditure per capita, 1966, p. 307 [★652★]
$63, Indiana hospital care expenditure per capita, 1966, p. 307 [★652★]
$63, nursing home expenditure per capita, 1982, p. 321 [★687★]
$63, Utah nursing home expenditure per capita, 1982, p. 321 [★687★]
$64, composite dental filling, p. 312 [★668★]
$64.1, federal highway funds to Minnesota, 1990, p. 254 [★538★]
$65, nursing home expenditure per capita, 1982, p. 321 [★687★]
$66, Arizona hospital care expenditure per capita, 1966, p. 307 [★652★]
$66, Florida hospital care expenditure per capita, 1966, p. 307 [★652★]
$67, Delaware hospital care expenditure per capita, 1966, p. 307 [★652★]
$67, District of Columbia hospital care expenditure per capita, 1966, p. 307 [★652★]
$67.14, fee for 10-inch by 10-inch bank safety deposit box, 1992, p. 241 [★515★]
$68, Georgia hospital care expenditure per capita, 1966, p. 307 [★652★]
$69, average hotel room rate, 1992, p. 441 [★956★]
$69, District of Columbia hospital care expenditure per capita, 1966, p. 307 [★652★]
$69, Illinois hospital care expenditure per capita, 1966, p. 307 [★652★]
$69, Iowa hospital care expenditure per capita, 1966, p. 307 [★652★]
$69.10, average cost of history periodical subscription, 1992, p. 165 [★363★]
$69.48, cost of 750 kwh in Wichita, KS, p. 202 [★434★]
$69.65, cost of 750 kwh in Boston, MA, p. 202 [★434★]
$70, Illinois hospital care expenditure per capita, 1966, p. 307 [★652★]
$70, personal funds expenditure per capita on health care, 1950, p. 284 [★598★]
$70-100, city government costs per ton for curbside recyclable pickup programs, p. 427 [★919★]
$70.47, cost of 750 kwh in Miami, FL, p. 202 [★434★]
$71, Idaho hospital care expenditure per capita, 1966, p. 307 [★652★]
$71.68, cost of 750 kwh in Toledo, OH, p. 202 [★434★]
$72, Idaho hospital care expenditure per capita, 1966, p. 307 [★652★]
$72.24, price per volume of hardbound medicine books, 1990, p. 162 [★359★]
$72.66, cost of 750 kwh in Riverside, CA, p. 202 [★434★]
$72.74, cost monthly cellular telephone service, 1991, p. 212 [★459★]
$72.99, cost of 750 kwh in Philadelphia, PA, p. 202 [★434★]
$73, Hawaii hospital care expenditure per capita, 1966, p. 307 [★652★]
$73.98, average cost of law periodical subscription, 1992, p. 165 [★363★]
$73.98, cost of 750 kwh in Pittsburgh, PA, p. 202 [★434★]
$74, Alabama hospital care expenditure per capita, 1966, p. 307 [★652★]
$74, Alaska hospital care expenditure per capita, 1966, p. 307 [★652★]
$74, hospital care cost per day, 1970, p. 305 [★648★]
$74.39, price per volume of hardbound science books, 1990, p. 162 [★359★]
$74.98, cost of 750 kwh in San Diego, CA, p. 202 [★434★]
$75, Connecticut hospital care expenditure per capita, 1966, p. 307 [★652★]
$75, Florida hospital care expenditure per capita, 1966, p. 307 [★652★]
$75, North Carolina nursing home expenditure per capita, 1982, p. 321 [★687★]
$75, private funds expenditure per capita on health care, 1955, p. 284 [★597★]
$75.19, cost of 750 kwh in Newark, NJ, p. 202 [★434★]
$76, Indiana hospital care expenditure per capita, 1966, p. 307 [★652★]
$76, Kansas hospital care expenditure per capita, 1966, p. 307 [★652★]
$76, South Carolina nursing home expenditure per capita, 1982, p. 321 [★687★]

Numbers following p. are page references. Numbers in [ ] are table references.

Money - continued

Dollars

$76, Tennessee nursing home expenditure per capita, 1982, p. 321 [★687★]
$76.80, price per volume of hardbound technology books, 1990, p. 162 [★359★]
$77, annual energy costs for washer/dryer, p. 201 [★432★]
$78, Arizona hospital care expenditure per capita, 1966, p. 307 [★652★]
$78.62, average cost of philosophy and religion periodical subscription, 1992, p. 165 [★363★]
$79, Hawaii hospital care expenditure per capita, 1966, p. 307 [★652★]
$79, nursing home expenditure per capita, 1982, p. 321 [★687★]
$80, hospital care expenditure per capita, 1966, p. 307 [★652★]
$80, United States hospital care expenditure per capita, 1966, p. 307 [★652★]
$81, Delaware hospital care expenditure per capita, 1966, p. 307 [★652★]
$82, Arkansas hospital care expenditure per capita, 1966, p. 307 [★652★]
$82, average Social Security benefits for children of deceased workers 1970, p. 247 [★529★]
$82, cost of grandparent's holiday gift per child, p. 210 [★453★]
$82, Nevada nursing home expenditure per capita, 1982, p. 321 [★687★]
$82.55, average cost of periodical subscription in Canada, 1992, p. 166 [★364★]
$83, credit card debt, high school graduates, p. 214 [★463★]
$83, hospital care cost per day, 1971, p. 306 [★650★]
$83, United States hospital care expenditure per capita, 1966, p. 307 [★652★]
$84, annual energy costs for lighting, p. 201 [★432★]
$84, Arizona hospital care expenditure per capita, 1966, p. 307 [★652★]
$84, nursing home expenditure per capita, 1982, p. 321 [★687★]
$85, added weekly paycheck value for trade union membership for men, 1990, p. 413 [★885★]
$85, Iowa hospital care expenditure per capita, 1966, p. 307 [★652★]
$85, Virginia nursing home expenditure per capita, 1982, p. 321 [★687★]
$85.51, average cost of art and architecture periodical subscription, 1992, p. 165 [★363★]
$86, daily nursing home charges, 1990, p. 223 [★481★]
$86, Georgia hospital care expenditure per capita, 1966, p. 307 [★652★]
$86, nursing home expenditure per capita, 1982, p. 321 [★687★]
$86.37, cost of 750 kwh in New York City, NY, p. 202 [★434★]
$87, average Social Security benefits for widowed mothers 1970, p. 247 [★529★]
$87, public school expenditure per capita, 1959-1960, p. 178 [★381★]
$88, California hospital care expenditure per capita, 1966, p. 307 [★652★]
$88, cost of mammogram, p. 292 [★619★]
$88, Texas nursing home expenditure per capita, 1982, p. 321 [★687★]
$89, Colorado hospital care expenditure per capita, 1966, p. 307 [★652★]

Money - continued

Dollars

$89, Louisiana nursing home expenditure per capita, 1982, p. 321 [★687★]
$89, state government debt per capita in 1985, p. 270 [★575★]
$90, annual expenditures per capita for police protection and correction in West Virginia, 1988, p. 136 [★310★]
$90, annual lottery betting per capita, p. 242 [★520★]
$90, California hospital care expenditure per capita, 1966, p. 307 [★652★]
$90, elderly family out-of-pocket health care costs for hospital, 1991, p. 287 [★606★]
$90, Illinois hospital care expenditure per capita, 1966, p. 307 [★652★]
$90, Mississippi nursing home expenditure per capita, 1982, p. 321 [★687★]
$90, New Hampshire nursing home expenditure per capita, 1982, p. 321 [★687★]
$90.81, average cost for four people to attend a major league baseball game, 1993, p. 429 [★926★]
$91, Connecticut hospital care expenditure per capita, 1966, p. 307 [★652★]
$91, Delaware hospital care expenditure per capita, 1966, p. 307 [★652★]
$91, nursing home expenditure per capita, 1982, p. 321 [★687★]
$92, Montana nursing home expenditure per capita, 1982, p. 321 [★687★]
$93, personal funds expenditure per capita on health care, 1955, p. 284 [★598★]
$93.75, German health insurance payment for doctor's attendance at normal birth, p. 315 [★673★]
$95, mental health care expenditure per capita, 1988, p. 318 [★681★]
$97, New Jersey nursing home expenditure per capita, 1982, p. 321 [★687★]
$99, New York state mental health care agency expenditure per capita, 1987, p. 319 [★682★]
$99.94, average cost of political science periodical subscription, 1992, p. 165 [★363★]
$100, average annual household expenditures for dairy price controls, p. 281 [★591★]
$100, Colorado hospital care expenditure per capita, 1966, p. 307 [★652★]
$100, cost of background check of prospective new hire, p. 40 [★93★]
$100, cost per individual psychotherapy session per person, 1991, p. 319 [★683★]
$100, sigmoidoscopy, 1990, p. 291 [★617★]
$101, annual expenditures per capita for police protection and correction in Mississippi, 1988, p. 136 [★310★]
$101, California hospital care expenditure per capita, 1966, p. 307 [★652★]
$101, farmer's gross average income per acre of wheat, 1991, p. 387 [★829★]
$102, average Social Security benefits for widows and widowers, nondisabled 1970, p. 247 [★529★]
$102, Maryland nursing home expenditure per capita, 1982, p. 321 [★687★]
$103, annual expenditures per capita for police protection and correction in Arkansas, 1988, p. 136 [★310★]

Numbers following p. are page references. Numbers in [ ] are table references.

Money - continued

Dollars

$103, average Social Security benefits for parents, 1970, p. 247 [★529★]
$104, Kentucky nursing home expenditure per capita, 1982, p. 321 [★687★]
$104, nursing home expenditure per capita, 1982, p. 321 [★687★]
$105, cost of pap smear and exam, p. 292 [★619★]
$106, Michigan nursing home expenditure per capita, 1982, p. 321 [★687★]
$108, annual local police department operating expenses per resident, 1990, p. 137 [★312★]
$108, monetary loss per murder, 1990, p. 117 [★279★]
$108, private funds expenditure per capita on health care, 1960, p. 284 [★597★]
$109, annual energy costs for air-conditioning, p. 201 [★432★]
$109, nursing home expenditure per capita, 1982, p. 321 [★687★]
$110, average Social Security benefits for children of disabled workers 1980, p. 247 [★529★]
$110, Indiana hospital care expenditure per capita, 1966, p. 307 [★652★]
$111, cost of physical exam, p. 292 [★619★]
$111, Oklahoma nursing home expenditure per capita, 1982, p. 321 [★687★]
$112, nursing home expenditure per capita, 1982, p. 321 [★687★]
$113, Oregon nursing home expenditure per capita, 1982, p. 321 [★687★]
$113.67, average cost of education periodical subscription, 1992, p. 165 [★363★]
$114, nursing home expenditure per capita, 1982, p. 321 [★687★]
$115, package cost per day from London to Disney World, p. 429 [★927★]
$116, Arkansas hospital care expenditure per capita, 1966, p. 307 [★652★]
$116, average AFDC payment by recipient in 1960, p. 251 [★535★]
$116, Pennsylvania nursing home expenditure per capita, 1982, p. 321 [★687★]
$118, average Social Security benefits for retired workers 1970, p. 247 [★529★]
$119, added weekly paycheck value for trade union membership for Hispanics, 1990, p. 413 [★885★]
$119, added weekly paycheck value of trade union membership, 1990, p. 413 [★884★]
$119.82, average cost of library and information science periodical subscription, 1992, p. 165 [★363★]
$120, mammogram, 1990, p. 291 [★617★]
$121, AFDC benefits by family in Alabama, 1990, p. 251 [★534★]
$122, added weekly paycheck value for trade union membership for women, 1990, p. 413 [★885★]
$123, annual expenditures per capita for police protection and correction in North Dakota, 1988, p. 136 [★310★]
$124, AFDC benefits by family in Mississippi, 1990, p. 251 [★534★]
$124, Kansas government debt per capita, 1990, p. 272 [★577★]
$125.32, average cost of periodical subscription in Australia and New Zealand, 1992, p. 166 [★364★]
$126, personal funds expenditure per capita on health care, 1960, p. 284 [★598★]
$127, annual expenditures per capita for police protection and correction in South Dakota, 1988, p. 136 [★310★]

Money - continued

Dollars

$127, child care worker's average weekly salary, p. 387 [★828★]
$129, nursing home expenditure per capita, 1982, p. 321 [★687★]
$129, public funds expenditure per capita on health care, 1970, p. 283 [★596★]
$130, District of Columbia state mental health care agency expenditure per capita, 1987, p. 319 [★682★]
$131, annual expenditures per capita for police protection and correction in Indiana, 1988, p. 136 [★310★]
$131, average Social Security benefits for disabled workers 1970, p. 247 [★529★]
$133, average AFDC payment by recipient in 1989 (est.), p. 251 [★535★]
$133, average car payment, 1975, p. 192 [★410★]
$135, average annual income of farmers in China, p. 359 [★770★]
$135, college students' monthly discretionary spending, p. 243 [★522★]
$136, average Social Security benefits for children of disabled workers 1983, p. 247 [★529★]
$136.98, average cost of psychology periodical subscription, 1992, p. 165 [★363★]
$137, Pennsylvania state government expenditures per capita for higher education,, p. 150 [★332★]
$137, Washington nursing home expenditure per capita, 1982, p. 321 [★687★]
$138, added weekly paycheck value for trade union membership for Blacks, 1990, p. 413 [★885★]
$139, Missouri nursing home expenditure per capita, 1982, p. 321 [★687★]
$139.44, average cost of periodical subscription in Italy, 1992, p. 166 [★364★]
$140, annual expenditures per capita for police protection and correction in Kentucky, 1988, p. 136 [★310★]
$140, average Social Security benefits for children of retired workers 1980, p. 247 [★529★]
$140, Nebraska nursing home expenditure per capita, 1982, p. 321 [★687★]
$141, annual expenditures per capita for police protection and correction in Nebraska, 1988, p. 136 [★310★]
$142, annual expenditures per capita for police protection and correction in Maine, 1988, p. 136 [★310★]
$143, Ohio nursing home expenditure per capita, 1982, p. 321 [★687★]
$144, annual expenditures per capita for police protection and correction in Iowa, 1988, p. 136 [★310★]
$145, annual expenditures per capita for police protection and correction in Alabama, 1988, p. 136 [★310★]
$145, annual expenditures per capita for police protection and correction in Vermont, 1988, p. 136 [★310★]
$145, average AFDC payment by recipient in 1980, p. 251 [★535★]
$147, annual expenditures per capita for police protection and correction in Idaho, 1988, p. 136 [★310★]
$149, Alaska hospital care expenditure per capita, 1966, p. 307 [★652★]
$149, annual expenditures per capita for police protection and correction in Oklahoma, 1988, p. 136 [★310★]
$149, Vermont nursing home expenditure per capita, 1982, p. 321 [★687★]

Numbers following p. are page references. Numbers in [ ] are table references.

Money - continued

Dollars

$150, cost of family psychotherapy session, 1991, p. 319 [★683★]
$150, Wisconsin nursing home expenditure per capita, 1982, p. 321 [★687★]
$152, Pennsylvania state and local government expenditures per capita for higher education, 1988-1989, p. 149 [★331★]
$153, annual expenditures per capita for police protection and correction in Montana, 1988, p. 136 [★310★]
$153, average AFDC payment by recipient in 1970, p. 251 [★535★]
$153, average annual spending on reading, 1985-1989, p. 431 [★932★]
$154, annual energy costs for water heating, p. 201 [★432★]
$154, North Dakota nursing home expenditure per capita, 1982, p. 321 [★687★]
$154, private funds expenditure per capita on health care, 1965, p. 284 [★597★]
$156, gardener's average gardening expenditures, p. 104 [★251★]
$157, annual expenditures per capita for police protection and correction in Tennessee, 1988, p. 136 [★310★]
$157, average Social Security benefit for disabled workers' children, 1989, p. 246 [★527★]
$157, District of Columbia state and local government expenditures per capita for higher education, 1988-1989, p. 149 [★331★]
$159, annual expenditures per capita for police protection and correction in Kansas, 1988, p. 136 [★310★]
$159, credit card debt for people with some college, p. 214 [★463★]
$161, annual expenditures per capita for police protection and correction in South Carolina, 1988, p. 136 [★310★]
$163, annual expenditures per capita for police protection and correction in Missouri, 1988, p. 136 [★310★]
$163, nursing home expenditure per capita, 1982, p. 321 [★687★]
$163.70, average cost for four people to attend an NFL game, 1992, p. 429 [★925★]
$164, average Social Security benefits for wives and husbands 1980, p. 247 [★529★]
$165, average travel expenses in southern United States, 1993, p. 442 [★958★]
$165, South Dakota nursing home expenditure per capita, 1982, p. 321 [★687★]
$165.36, bank minimum balance required to avoid fee, 1991, p. 243 [★523★]
$166, AFDC benefits by family in Texas, 1990, p. 251 [★534★]
$167, food counter worker's average weekly salary, p. 387 [★828★]
$168, nursing home expenditure per capita, 1982, p. 321 [★687★]
$169, AFDC benefits by family in Louisiana, 1990, p. 251 [★534★]
$169, average travel expenses in Midwest, 1993, p. 442 [★958★]
$170, annual expenditures per capita for police protection and correction in Utah, 1988, p. 136 [★310★]
$171, annual expenditures per capita for police protection and correction in Pennsylvania, 1988, p. 136 [★310★]
$173, average federal food stamp family benefit, p. 261 [★552★]
$175, cost per ton to collect and sort recyclables, p. 427 [★920★]
$175, personal funds expenditure per capita on health care, 1965, p. 284 [★598★]
$175.55, average cost of business and economics periodical subscription, 1992, p. 165 [★363★]

Money - continued

Dollars

$176, average Social Security benefits for children of retired workers 1983, p. 247 [★529★]
$176, Maine nursing home expenditure per capita, 1982, p. 321 [★687★]
$177, annual expenditures per capita for police protection and correction in New Hampshire, 1988, p. 136 [★310★]
$177, annual expenditures per capita for police protection and correction in North Carolina, 1988, p. 136 [★310★]
$177, annual expenditures per capita for police protection and correction in Texas, 1988, p. 136 [★310★]
$178, average travel expenses in the western United States, 1993, p. 442 [★958★]
$179, Missouri state government expenditures per capita for higher education,, p. 150 [★332★]
$180, Illinois state government expenditures per capita for higher education,, p. 150 [★332★]
$182, Florida state government expenditures per capita for higher education,, p. 150 [★332★]
$184, New York nursing home expenditure per capita, 1982, p. 321 [★687★]
$185, cleaners and servants' average weekly salary, p. 387 [★828★]
$186, annual expenditures per capita for police protection and correction in Minnesota, 1988, p. 136 [★310★]
$186, monthly nursing home charge per resident, 1964, p. 322 [★688★]
$187, annual expenditures per capita for police protection and correction in Georgia, 1988, p. 136 [★310★]
$187, Florida state and local government expenditures per capita for higher education, 1988-1989, p. 149 [★331★]
$189.76, average cost for periodical subscription in Asia, 1992, p. 166 [★364★]
$191, kitchen worker's average weekly salary, p. 387 [★828★]
$192, District of Columbia hospital care expenditure per capita, 1966, p. 307 [★652★]
$192, Massachusetts nursing home expenditure per capita, 1982, p. 321 [★687★]
$192, New York state government expenditures per capita for higher education,, p. 150 [★332★]
$192.57, average cost of periodical subscription in France, 1992, p. 166 [★364★]
$193, AFDC benefits by family in Arkansas, 1990, p. 251 [★534★]
$193, AFDC benefits by family in Tennessee, 1990, p. 251 [★534★]
$193, annual expenditures per capita for police protection and correction in Louisiana, 1988, p. 136 [★310★]
$193, average travel expenses in Northwest United States, 1993, p. 442 [★958★]
$194, annual expenditures per capita for police protection and correction in Wisconsin, 1988, p. 136 [★310★]
$194, public school expenditure per capita, 1970, p. 175 [★377★]
$194.34, average cost of United States periodical subscription, 1992, p. 166 [★364★]
$195, food preparation worker's average weekly salary, p. 387 [★828★]
$196, annual expenditures per capita for police protection and correction in Ohio, 1988, p. 136 [★310★]
$196, Connecticut state and local government expenditures per capita for higher education, 1988-1989, p. 149 [★331★]

Numbers following p. are page references. Numbers in [ ] are table references.

Money - continued

Dollars

$198, average annual advertising expenditures per capita, 1950, p. 27 [★61★]
$198, average Social Security benefits for wives and husbands 1970, p. 247 [★529★]
$199, average Social Security benefits for retired worker and wife 1970, p. 247 [★529★]
$199, waiter's assistants average weekly salary, p. 387 [★828★]
$200, Afghan opium farmer's average annual income, p. 387 [★830★]
$200, difference in automobile purchase price, white men and women, p. 193 [★413★]
$200, Massachusetts state government expenditures per capita for higher education,, p. 150 [★332★]
$200, package cost per day from London to Euro Disney, p. 429 [★927★]
$200, public school expenditure per capita, 1969-1970, p. 178 [★381★]
$201, Massachusetts state and local government expenditures per capita for higher education, 1988-1989, p. 149 [★331★]
$202, annual expenditures per capita for police protection and correction in Washington, 1988, p. 136 [★310★]
$202, cashier's average weekly salary, p. 387 [★828★]
$203, annual expenditures per capita for police protection and correction in Rhode Island, 1988, p. 136 [★310★]
$203, Montana state government expenditures per capita for higher education,, p. 150 [★332★]
$203, waiter and waitress' average weekly salary, p. 387 [★828★]
$204, annual expenditures per capita for police protection and correction in Virginia, 1988, p. 136 [★310★]
$204, New Hampshire state and local government expenditures per capita for higher education, 1988-1989, p. 149 [★331★]
$204, South Dakota state and local government expenditures per capita for higher education, 1988-1989, p. 149 [★331★]
$205, textile and sewing machine operator's average weekly salary, p. 387 [★828★]
$206, nursing home expenditure per capita, 1982, p. 321 [★687★]
$207, state government debt in 1970, p. 270 [★575★]
$209, AFDC benefits by family in South Carolina, 1990, p. 251 [★534★]
$210, annual expenditures per capita for police protection and correction in Illinois, 1988, p. 136 [★310★]
$210, cost per household of federal food price support programs, 1991, p. 205 [★440★]
$210, New Jersey state government expenditures per capita for higher education,, p. 150 [★332★]
$211, annual expenditures per capita for police protection and correction in New Mexico, 1988, p. 136 [★310★]
$211, Georgia state government expenditures per capita for higher education,, p. 150 [★332★]
$212, annual expenditures per capita for police protection and correction in Oregon, 1988, p. 136 [★310★]
$212, Montana state and local government expenditures per capita for higher education, 1988-1989, p. 149 [★331★]
$213, cost of Christmas gifts per child, 1992, p. 210 [★452★]
$213, Georgia state and local government expenditures per capita for higher education, 1988-1989, p. 149 [★331★]
$213, Louisiana state government expenditures per capita for higher education,, p. 150 [★332★]

Money - continued

Dollars

$214, city government taxes per capita in San Antonio, TX 1990, p. 279 [★587★]
$214, Rhode Island nursing home expenditure per capita, 1982, p. 321 [★687★]
$215, Louisiana state and local government expenditures per capita for higher education, 1988-1989, p. 149 [★331★]
$215, Mississippi state government expenditures per capita for higher education,, p. 150 [★332★]
$217, Arkansas state and local government expenditures per capita for higher education, 1988-1989, p. 149 [★331★]
$217, Arkansas state government expenditures per capita for higher education,, p. 150 [★332★]
$217, average Social Security benefits for wives and husbands 1983, p. 247 [★529★]
$217, private funds expenditure per capita on health care, 1970, p. 284 [★597★]
$218, Nevada state and local government expenditures per capita for higher education, 1988-1989, p. 149 [★331★]
$218, per capita athletic expenditures on women in 1987, NCAA Div. III, p. 146 [★328★]
$218.88, average cost of geography periodical subscription, 1992, p. 165 [★363★]
$219, AFDC benefits by family in Kentucky, 1990, p. 251 [★534★]
$219, Missouri state and local government expenditures per capita for higher education, 1988-1989, p. 149 [★331★]
$220, Maryland state government expenditures per capita for higher education,, p. 150 [★332★]
$223, cost of hospital outpatient visit, 1990, p. 309 [★657★]
$225, annual household footwear expenditures, 1990, p. 221 [★476★]
$225, average packaging cost per person, p. 63 [★149★]
$225.50, average cost of periodical subscription in Japan, 1992, p. 166 [★364★]
$227, annual expenditures per capita for police protection and correction in Wyoming, 1988, p. 136 [★310★]
$227, West Virginia state and local government expenditures per capita for higher education, 1988-1989, p. 149 [★331★]
$230, annual expenditures per capita for police protection and correction in Connecticut, 1988, p. 136 [★310★]
$230, annual expenditures per capita for police protection and correction in United States, 1988, p. 136 [★310★]
$230.39, physician total cost of childhood immunization vaccines, p. 316 [★675★]
$231, New York state and local government expenditures per capita for higher education, 1988-1989, p. 149 [★331★]
$233, annual expenditures per capita for police protection and correction in Colorado, 1988, p. 136 [★310★]
$233, Illinois state and local government expenditures per capita for higher education, 1988-1989, p. 149 [★331★]
$235, Minnesota nursing home expenditure per capita, 1982, p. 321 [★687★]
$239, annual energy costs for refrigerator/freezer, p. 201 [★432★]
$239, city government taxes per capita in Memphis, TN 1990, p. 279 [★587★]
$240, average Social Security benefits for children of deceased workers 1980, p. 247 [★529★]
$242, AFDC benefits by family in North Carolina, 1990, p. 251 [★534★]

Numbers following p. are page references. Numbers in [ ] are table references.

Money - continued

Dollars

$242, average Social Security benefit for retired workers' children, 1989, p. 246 [★527★]
$243, AFDC benefits by family in West Virginia, 1990, p. 251 [★534★]
$244, Ohio state government expenditures per capita for higher education,, p. 150 [★332★]
$245, annual expenditures per capita for police protection and correction in Massachusetts, 1988, p. 136 [★310★]
$245, New Jersey state and local government expenditures per capita for higher education, 1988-1989, p. 149 [★331★]
$246, average Social Security benefits for widowed mothers 1980, p. 247 [★529★]
$248, annual expenditures per capita for police protection and correction in Hawaii, 1988, p. 136 [★310★]
$248, annual expenditures per capita for police protection and correction in Michigan, 1988, p. 136 [★310★]
$249, state government expenditures per capita for higher education,, p. 150 [★332★]
$250, balance on private-label credit cards, p. 214 [★462★]
$250, Kentucky state and local government expenditures per capita for higher education, 1988-1989, p. 149 [★331★]
$251, Texas state government expenditures per capita for higher education,, p. 150 [★332★]
$252, Idaho state government expenditures per capita for higher education,, p. 150 [★332★]
$252, lifetime energy costs of average color television, p. 201 [★431★]
$253, Rhode Island state and local government expenditures per capita for higher education, 1988-1989, p. 149 [★331★]
$254, average Social Security benefits for wives and husbands 1980, p. 247 [★529★]
$255, Ohio state and local government expenditures per capita for higher education, 1988-1989, p. 149 [★331★]
$258, city government taxes per capita in Milwaukee, WI 1990, p. 279 [★587★]
$258, Oregon state government expenditures per capita for higher education,, p. 150 [★332★]
$260, AFDC benefits by family in Georgia, 1990, p. 251 [★534★]
$260, annual expenditures per capita for police protection and correction in Delaware, 1988, p. 136 [★310★]
$260, Maine state and local government expenditures per capita for higher education, 1988-1989, p. 149 [★331★]
$260, South Carolina state and local government expenditures per capita for higher education, 1988-1989, p. 149 [★331★]
$260, Tennessee state and local government expenditures per capita for higher education, 1988-1989, p. 149 [★331★]
$262, annual expenditures per capita for police protection and correction in Florida, 1988, p. 136 [★310★]
$262, farmer's gross average income per acre of feed corn, 1991, p. 387 [★829★]
$262, Maine state government expenditures per capita for higher education,, p. 150 [★332★]
$264, AFDC benefits by family in Arizona, 1990, p. 251 [★534★]
$264, AFDC benefits by family in New Mexico, 1990, p. 251 [★534★]
$267, AFDC benefits by family in Virginia, 1990, p. 251 [★534★]
$267, annual expenditures per capita for police protection and correction in Maryland, 1988, p. 136 [★310★]

Money - continued

Dollars

$269, AFDC benefits by family in Florida, 1990, p. 251 [★534★]
$270, Idaho state and local government expenditures per capita for higher education, 1988-1989, p. 149 [★331★]
$271, AFDC benefits by family in Indiana, 1990, p. 251 [★534★]
$272, United States state and local government expenditures per capita for higher education, 1988-1989, p. 149 [★331★]
$273, Maryland state and local government expenditures per capita for higher education, 1988-1989, p. 149 [★331★]
$275, AFDC benefits by family in Missouri, 1990, p. 251 [★534★]
$275, AFDC benefits by family in Nevada, 1990, p. 251 [★534★]
$275, annual expenditures per capita for police protection and correction in New Jersey, 1988, p. 136 [★310★]
$276, average Social Security benefits for children of deceased workers 1970, p. 247 [★529★]
$276, average Social Security benefits for parents 1980, p. 247 [★529★]
$276, Colorado state government expenditures per capita for higher education,, p. 150 [★332★]
$278, Kansas state government expenditures per capita for higher education,, p. 150 [★332★]
$279, Mississippi state and local government expenditures per capita for higher education, 1988-1989, p. 149 [★331★]
$280, average cost per day for four people in Euro Disney, p. 429 [★927★]
$280, Wisconsin state government expenditures per capita for higher education,, p. 150 [★332★]
$282, AFDC benefits by family in Idaho, 1990, p. 251 [★534★]
$283, federal income tax average negligence penalty, p. 266 [★568★]
$283, Oklahoma state and local government expenditures per capita for higher education, 1988-1989, p. 149 [★331★]
$286, Arizona state government expenditures per capita for higher education,, p. 150 [★332★]
$286, Colorado state and local government expenditures per capita for higher education, 1988-1989, p. 149 [★331★]
$287, average Social Security benefits for wives and husbands 1983, p. 247 [★529★]
$290, AFDC benefits by family in South Dakota, 1990, p. 251 [★534★]
$291, Minnesota state government expenditures per capita for higher education,, p. 150 [★332★]
$291, Nebraska state government expenditures per capita for higher education,, p. 150 [★332★]
$292, average Social Security benefits for widowed mothers 1970, p. 247 [★529★]
$292, public school expenditure per capita, 1975, p. 175 [★377★]
$293, AFDC benefits by family in Delaware, 1990, p. 251 [★534★]
$293, Texas state and local government expenditures per capita for higher education, 1988-1989, p. 149 [★331★]
$295, Indiana state and local government expenditures per capita for higher education, 1988-1989, p. 149 [★331★]
$296, cost of hospital care per day, Medicare enrollees, 1980, p. 308 [★654★]
$297, Alabama state and local government expenditures per capita for higher education, 1988-1989, p. 149 [★331★]
$298, average Social Security benefits for children of deceased workers 1983, p. 247 [★529★]

Numbers following p. are page references. Numbers in [ ] are table references.

Money - continued

Dollars

$299, AFDC benefits by family in Oklahoma, 1990, p. 251 [*534*]
$300, annual expenditures per capita for police protection and correction in Arizona, 1988, p. 136 [*310*]
$300, California state government expenditures per capita for higher education,, p. 150 [*332*]
$300.35, average cost of general science periodical subscription, 1992, p. 165 [*363*]
$302, personal funds expenditure per capita on health care, 1970, p. 284 [*598*]
$303, Virginia state and local government expenditures per capita for higher education, 1988-1989, p. 149 [*331*]
$303, Virginia state government expenditures per capita for higher education,, p. 150 [*332*]
$304, federal aid per capita in Nevada 1990, p. 269 [*573*]
$308, average car payment, 1992, p. 192 [*410*]
$309, average Social Security benefits for widowed mothers 1983, p. 247 [*529*]
$309, federal aid per capita in Florida 1990, p. 269 [*573*]
$310, annual household expenditures on furniture, p. 221 [*475*]
$311, average Social Security benefits for widows and widowers, nondisabled 1980, p. 247 [*529*]
$311, monthly HMO (health center option) premium per family, 1990, p. 285 [*600*]
$312, property tax per capita in Virginia Beach, VA, p. 262 [*557*]
$315, average annual income of urban Chinese, p. 359 [*770*]
$315, per capita athletic expenditures on men in 1987, NCAA Div. III, p. 146 [*328*]
$316, AFDC benefits by family in Colorado, 1990, p. 251 [*534*]
$316, California state and local government expenditures per capita for higher education, 1988-1989, p. 149 [*331*]
$316, monthly HMO (doctor's office option) premium per family, 1990, p. 285 [*600*]
$316, North Carolina state government expenditures per capita for higher education,, p. 150 [*332*]
$319, monthly health insurance premium (managed-care) per family, 1990, p. 285 [*600*]
$320, cost of mobile telephones, 1992, p. 212 [*458*]
$320, elderly family out-of-pocket health care costs for Medicare, 1991, p. 287 [*606*]
$320, Michigan state government expenditures per capita for higher education,, p. 150 [*332*]
$323, AFDC benefits by family in Ohio, 1990, p. 251 [*534*]
$325, city government taxes per capita in San Diego, CA 1990, p. 279 [*587*]
$325, average teacher's salary, 1899-1900, p. 364 [*779*]
$328, AFDC benefits by family in Kansas, 1990, p. 251 [*534*]
$328, personal funds expenditure per capita on health care, 1971, p. 284 [*598*]
$329, AFDC benefits by family in Illinois, 1990, p. 251 [*534*]
$329, AFDC benefits by family in Nebraska, 1990, p. 251 [*534*]
$329, Black workers' median weekly income, 1990, p. 398 [*854*]
$330, Hawaii state and local government expenditures per capita for higher education, 1988-1989, p. 149 [*331*]
$331, Oregon state and local government expenditures per capita for higher education, 1988-1989, p. 149 [*331*]
$334, Minnesota state and local government expenditures per capita for higher education, 1988-1989, p. 149 [*331*]

Money - continued

Dollars

$334, Washington state and local government expenditures per capita for higher education, 1988-1989, p. 149 [*331*]
$335, Idaho hospital care expenditure per capita, 1982, p. 307 [*652*]
$336, federal aid per capita in New Hampshire 1990, p. 269 [*573*]
$338, North Carolina state and local government expenditures per capita for higher education, 1988-1989, p. 149 [*331*]
$340, cost of cellular telephones, 1992, p. 212 [*458*]
$340, federal aid per capita in Virginia 1990, p. 269 [*573*]
$341, average Social Security benefits for retired workers 1980, p. 247 [*529*]
$341, Kansas state and local government expenditures per capita for higher education, 1988-1989, p. 149 [*331*]
$341, loss per convenience store robbery, 1990, p. 118 [*282*]
$342, Iowa state government expenditures per capita for higher education,, p. 150 [*332*]
$343, AFDC benefits by family in Montana, 1990, p. 251 [*534*]
$343, annual expenditures per capita for police protection and correction in California, 1988, p. 136 [*310*]
$343, average Social Security benefits for widows and widowers, nondisabled 1970, p. 247 [*529*]
$343, city government taxes per capita in Phoenix, AZ 1990, p. 279 [*587*]
$345, Arizona state and local government expenditures per capita for higher education, 1988-1989, p. 149 [*331*]
$347, Nebraska state and local government expenditures per capita for higher education, 1988-1989, p. 149 [*331*]
$350, annual energy costs for space heating, p. 201 [*432*]
$350, average Social Security benefits for parents, 1983, p. 247 [*529*]
$350, Shinto wedding cost per guest, Japan, p. 240 [*512*]
$352, AFDC benefits by family in Wyoming, 1990, p. 251 [*534*]
$352, city government taxes per capita in Jacksonville, FL 1990, p. 279 [*587*]
$352, federal aid per capita in Arizona 1990, p. 269 [*573*]
$352, federal aid per capita in Missouri 1990, p. 269 [*573*]
$353, AFDC benefits by family in Utah, 1990, p. 251 [*534*]
$354, federal aid per capita in Florida, 1990, p. 252 [*536*]
$357, Wyoming state government expenditures per capita for higher education,, p. 150 [*332*]
$361, federal aid per capita in Virginia, 1990, p. 252 [*536*]
$361, gold dental filling, p. 312 [*668*]
$362, personal funds expenditure per capita on health care, 1972, p. 284 [*598*]
$363, average federal income tax delinquency penalty, p. 265 [*564*]
$364, AFDC benefits by family in New Jersey, 1990, p. 251 [*534*]
$365, Michigan state and local government expenditures per capita for higher education, 1988-1989, p. 149 [*331*]
$366, state and local federal funds per capita, 1980, p. 278 [*585*]
$367, annual expenditures per capita for police protection and correction in Nevada, 1988, p. 136 [*310*]
$367, Wisconsin state and local government expenditures per capita for higher education, 1988-1989, p. 149 [*331*]

Numbers following p. are page references. Numbers in [ ] are table references.

Money - continued

Dollars

$368, AFDC benefits by family in Oregon, 1990, p. 251 [*534*]
$368, federal aid per capita in Nevada, 1990, p. 252 [*536*]
$370, New Mexico state government expenditures per capita for higher education,, p. 150 [*332*]
$371, AFDC benefits by family in Maryland, 1990, p. 251 [*534*]
$371, average annual spending on fees and admissions for entertainment, 1984-1989, p. 431 [*932*]
$371, average Social Security benefits for disabled workers 1980, p. 247 [*529*]
$371, property tax per capita in Jersey City, NJ, p. 262 [*557*]
$372, average Social Security benefits for children of deceased workers 1980, p. 247 [*529*]
$374, median rental payment in the United States, 1990, p. 230 [*494*]
$377, AFDC benefits by family in North Dakota, 1990, p. 251 [*534*]
$378, AFDC benefits by family in Iowa, 1990, p. 251 [*534*]
$380, farmer's gross average income per acre of cotton, 1991, p. 387 [*829*]
$380.53, average cost of geology periodical subscription, 1992, p. 165 [*363*]
$381, average Social Security benefit for widowed Black women, 1989, p. 250 [*533*]
$381, average Social Security benefits for widowed mothers 1980, p. 247 [*529*]
$382, federal aid per capita in Texas 1990, p. 269 [*573*]
$383, annual expenditures per capita for police protection and correction in New York, 1988, p. 136 [*310*]
$383, property tax per capita in San Francisco, CA, p. 262 [*557*]
$383, state government expenditure in 1970, p. 270 [*575*]
$384, average AFDC payment by family in 1989 (est.), p. 251 [*535*]
$384, state government revenue in 1970, p. 270 [*575*]
$385, federal aid per capita in New Hampshire, 1990, p. 252 [*536*]
$385, New Mexico state and local government expenditures per capita for higher education, 1988-1989, p. 149 [*331*]
$386, federal aid per capita in Kansas 1990, p. 269 [*573*]
$391, AFDC benefits by family in Pennsylvania, 1990, p. 251 [*534*]
$391, credit card debt, college graduates, p. 214 [*463*]
$391, South Dakota, hospital care cost per day, 1990, p. 305 [*649*]
$392, federal aid per capita in Illinois 1990, p. 269 [*573*]
$394, average Social Security benefits for children of deceased workers 1983, p. 247 [*529*]
$395, Iowa state and local government expenditures per capita for higher education, 1988-1989, p. 149 [*331*]
$396, AFDC benefits by family in U.S., 1990, p. 251 [*534*]
$396, average Social Security benefit for retired Black women, 1989, p. 250 [*533*]
$396, average Social Security benefits for widows and widowers, nondisabled 1983, p. 247 [*529*]
$396, Utah state and local government expenditures per capita for higher education, 1988-1989, p. 149 [*331*]
$397, average Social Security benefits for retired workers 1970, p. 247 [*529*]

Money - continued

Dollars

$397, Colorado hospital care expenditure per capita, 1982, p. 307 [*652*]
$397, Vermont state and local government expenditures per capita for higher education, 1988-1989, p. 149 [*331*]
$398, AFDC benefits by family in District of Columbia, 1990, p. 251 [*534*]
$398, Iowa hospital care expenditure per capita, 1982, p. 307 [*652*]
$399, federal aid per capita in North Carolina 1990, p. 269 [*573*]
$399, Florida hospital care expenditure per capita, 1982, p. 307 [*652*]
$400, household holiday gift expenditures, 1992, p. 209 [*451*]
$400, property tax per capita in Baltimore, MD, p. 262 [*557*]
$400-500, average annual cost per student of homeschooling, p. 179 [*385*]
$401, personal funds expenditure per capita on health care, 1973, p. 284 [*598*]
$404, ceramic filling, p. 312 [*668*]
$404.29, average cost of United Kingdom periodical subscription, 1992, p. 166 [*364*]
$405, Montana, hospital care cost per day, 1990, p. 305 [*649*]
$406, federal aid per capita in Indiana 1990, p. 269 [*573*]
$406, federal aid per capita in Texas, 1990, p. 252 [*536*]
$406, Greece expenditure per capita on health care, 1990, p. 282 [*594*]
$408, average Social Security benefits for widowed mothers 1983, p. 247 [*529*]
$408, city government taxes per capita in Houston, TX 1990, p. 279 [*587*]
$408, elderly family out-of-pocket health care costs for doctor, 1991, p. 287 [*606*]
$411, female child care worker's average weekly salary, p. 385 [*825*]
$412, federal aid per capita in Kansas, 1990, p. 252 [*536*]
$413, federal aid per capita in Colorado 1990, p. 269 [*573*]
$415, AFDC benefits by family in New Hampshire, 1990, p. 251 [*534*]
$416, cost of cancer per capita, 1990, p. 291 [*617*]
$418, city government taxes per capita in San Jose, CA 1990, p. 279 [*587*]
$418, cost of broken arm, p. 292 [*619*]
$419, North Dakota state and local government expenditures per capita for higher education, 1988-1989, p. 149 [*331*]
$420, average charity deduction (federal income tax), 1991, p. 266 [*566*]
$421, average AFDC payment by family in 1980, p. 251 [*535*]
$422, North Dakota state government expenditures per capita for higher education,, p. 150 [*332*]
$424, public school expenditure per capita, 1979-1980, p. 178 [*381*]
$425, AFDC benefits by family in Maine, 1990, p. 251 [*534*]
$425, federal aid per capita in Missouri, 1990, p. 252 [*536*]
$426, Wyoming state and local government expenditures per capita for higher education, 1988-1989, p. 149 [*331*]
$427, federal aid per capita in Oklahoma 1990, p. 269 [*573*]
$427, North Dakota, hospital care cost per day, 1990, p. 305 [*649*]

Numbers following p. are page references. Numbers in [ ] are table references.

Money - continued

Dollars

$427, white workers' median weekly income, 1990, p. 398 [★854★]
$428, Iowa hospital care expenditure per capita, 1982, p. 307 [★652★]
$428, public school expenditure per capita, 1980, p. 175 [★377★]
$431, Connecticut hospital care expenditure per capita, 1982, p. 307 [★652★]
$431, federal aid per capita in Nebraska 1990, p. 269 [★573★]
$433, United States hospital care expenditure per capita, 1982, p. 307 [★652★]
$434, federal aid per capita in Colorado, 1990, p. 252 [★536★]
$434, Idaho hospital care expenditure per capita, 1982, p. 307 [★652★]
$437, federal aid per capita in Indiana, 1990, p. 252 [★536★]
$438, federal aid per capita in Maryland 1990, p. 269 [★573★]
$439, federal aid per capita in Ohio 1990, p. 269 [★573★]
$439, Mississippi, hospital care cost per day, 1990, p. 305 [★649★]
$440, average Social Security benefits for disabled workers 1970, p. 247 [★529★]
$441, average Social Security benefits for retired workers 1983, p. 247 [★529★]
$442, federal aid per capita in Arizona, 1990, p. 252 [★536★]
$442, Japanese travelers' duty-free purchases, 1990, p. 442 [★959★]
$442, loss per gas or service station robbery, 1990, p. 118 [★282★]
$443, Arkansas hospital care expenditure per capita, 1982, p. 307 [★652★]
$443, average AFDC payment by family in 1960, p. 251 [★535★]
$443, federal aid per capita in Pennsylvania 1990, p. 269 [★573★]
$443, Georgia hospital care expenditure per capita, 1982, p. 307 [★652★]
$444, Delaware state and local government expenditures per capita for higher education, 1988-1989, p. 149 [★331★]
$444, federal aid per capita in Georgia 1990, p. 269 [★573★]
$444, federal aid per capita in North Carolina, 1990, p. 252 [★536★]
$445, District of Columbia hospital care expenditure per capita, 1982, p. 307 [★652★]
$445, state and local federal funds per capita, 1985, p. 278 [★585★]
$449, AFDC benefits by family in Washington, 1990, p. 251 [★534★]
$449, average Mississippi widow(er) Social Security benefit, 1990, p. 250 [★532★]
$449, federal aid per capita in Michigan 1990, p. 269 [★573★]
$449, Illinois hospital care expenditure per capita, 1982, p. 307 [★652★]
$450, difference in automobile purchase price, white men and black women, p. 193 [★413★]
$454, average annual spending on audio visual equipment, 1985-1989, p. 431 [★932★]
$456, average Social Security benefits for disabled workers 1983, p. 247 [★529★]
$456, personal funds expenditure per capita on health care, 1974, p. 284 [★598★]
$458, Alaska state and local government expenditures per capita for higher education, 1988-1989, p. 149 [★331★]
$458, Alaska state government expenditures per capita for higher education,, p. 150 [★332★]

Money - continued

Dollars

$458, Hawaii hospital care expenditure per capita, 1982, p. 307 [★652★]
$459, AFDC benefits by family in Wisconsin, 1990, p. 251 [★534★]
$459, city government taxes per capita in Columbus, OH 1990, p. 279 [★587★]
$460, federal aid per capita in Delaware 1990, p. 269 [★573★]
$461, federal aid per capita in Washington 1990, p. 269 [★573★]
$462, federal aid per capita in Illinois, 1990, p. 252 [★536★]
$462, Wyoming, hospital care cost per day, 1990, p. 305 [★649★]
$463, North Carolina government debt per capita, 1990, p. 272 [★577★]
$463, Texas government debt per capita, 1990, p. 272 [★577★]
$463.58, average cost for periodical subscription in Europe, 1992, p. 166 [★364★]
$464, federal aid per capita in Iowa 1990, p. 269 [★573★]
$468, Arizona hospital care expenditure per capita, 1982, p. 307 [★652★]
$468, federal aid per capita in California, 1990, p. 252 [★536★]
$469, average Social Security benefit for retired white women, 1989, p. 250 [★533★]
$470, federal aid per capita in Delaware, 1990, p. 252 [★536★]
$471, federal aid per capita in New Jersey 1990, p. 269 [★573★]
$473, high school prom costs for "him," 1993, p. 241 [★516★]
$473, property tax per capita in Rochester, NY, p. 262 [★557★]
$474, average District of Columbia widow(er) Social Security benefit, 1990, p. 250 [★532★]
$474.94, average cost of biology periodical subscription, 1992, p. 165 [★363★]
$477, average Arkansas widow(er) Social Security benefit, 1990, p. 250 [★532★]
$477, federal aid per capita in United States 1990, p. 269 [★573★]
$477.60, average cost of periodical subscription in Germany, 1992, p. 166 [★364★]
$479, Hawaii hospital care expenditure per capita, 1982, p. 307 [★652★]
$479, monthly nursing home charge per resident, 1973-1974, p. 322 [★689★]
$480, average Alabama widow(er) Social Security benefit, 1990, p. 250 [★532★]
$480, loss per larceny-theft, 1990, p. 119 [★283★]
$480.38, average cost of engineering and technology periodical subscription, 1992, p. 165 [★363★]
$481, city government taxes per capita in Dallas, TX 1990, p. 279 [★587★]
$481, Georgia government debt per capita, 1990, p. 272 [★577★]
$482, average Social Security benefits for widows and widowers, nondisabled 1980, p. 247 [★529★]
$482, average South Carolina widow(er) Social Security benefit, 1990, p. 250 [★532★]
$483, federal aid per capita in Wisconsin 1990, p. 269 [★573★]
$484, federal aid per capita in Georgia, 1990, p. 252 [★536★]
$486, federal aid per capita in Arkansas 1990, p. 269 [★573★]
$486, federal aid per capita in Kentucky 1990, p. 269 [★573★]
$487, average North Carolina widow(er) Social Security benefit, 1990, p. 250 [★532★]
$487, federal aid per capita in Utah, 1990, p. 252 [★536★]
$489, property tax per capita in Richmond, VA, p. 262 [★557★]

Numbers following p. are page references. Numbers in [ ] are table references.

Money - continued

Dollars

$489.45, median annual cost per student for school maintenance and operations, 1991, p. 180 [★386★]
$490, annual household expenditures for women's clothing 1988, p. 220 [★473★]
$490, average production cost of a ton of American steel, p. 50 [★120★]
$490, federal aid per capita in Idaho 1990, p. 269 [★573★]
$490, Nebraska, hospital care cost per day, 1990, p. 305 [★649★]
$490.40, average production cost of a ton of Japanese steel, p. 50 [★120★]
$491, federal aid per capita in Maryland, 1990, p. 252 [★536★]
$492, average Georgia widow(er) Social Security benefit, 1990, p. 250 [★532★]
$492, Georgia hospital care expenditure per capita, 1982, p. 307 [★652★]
$492, property tax per capita in Anchorage, AK, p. 262 [★557★]
$494, federal aid per capita in Nebraska, 1990, p. 252 [★536★]
$495, District of Columbia hospital care expenditure per capita, 1982, p. 307 [★652★]
$495, federal aid per capita in South Carolina 1990, p. 269 [★573★]
$495, Iowa, hospital care cost per day, 1990, p. 305 [★649★]
$496, AFDC benefits by family in Minnesota, 1990, p. 251 [★534★]
$496, average Kentucky widow(er) Social Security benefit, 1990, p. 250 [★532★]
$497, average Tennessee widow(er) Social Security benefit, 1990, p. 250 [★532★]
$497, federal aid per capita in Ohio, 1990, p. 252 [★536★]
$498, Alaska hospital care expenditure per capita, 1982, p. 307 [★652★]
$498, Arizona hospital care expenditure per capita, 1982, p. 307 [★652★]
$498, average annual advertising expenditures per capita, 1989, p. 27 [★61★]
$498, average Social Security benefit for widowed white women, 1989, p. 250 [★533★]
$498, federal aid per capita in Oklahoma, 1990, p. 252 [★536★]
$498, Idaho hospital care expenditure per capita, 1982, p. 307 [★652★]
$500, AFDC benefits by family in Michigan, 1990, p. 251 [★534★]
$500, Mapuche Indians (Peru) per capita income, p. 394 [★845★]
$503, city government taxes per capita in Detroit, MI 1990, p. 279 [★587★]
$506, Hawaii hospital care expenditure per capita, 1982, p. 307 [★652★]
$506.08, average cost of math and computer science periodical subscription, 1992, p. 165 [★363★]
$508, AFDC benefits by family in Rhode Island, 1990, p. 251 [★534★]
$508, federal aid per capita in Tennessee 1990, p. 269 [★573★]
$508, monthly welfare nursing home payment per resident, 1977, p. 324 [★692★]
$509, average Louisiana widow(er) Social Security benefit, 1990, p. 250 [★532★]
$509, federal aid per capita in Alabama 1990, p. 269 [★573★]
$510, AFDC benefits by family in Massachusetts, 1990, p. 251 [★534★]
$511, federal aid per capita in Michigan, 1990, p. 252 [★536★]

Money - continued

Dollars

$512, Indiana hospital care expenditure per capita, 1982, p. 307 [★652★]
$514, federal aid per capita in New Jersey, 1990, p. 252 [★536★]
$514, federal aid per capita in New Mexico 1990, p. 269 [★573★]
$515, average District of Columbia retired worker Social Security benefit, 1990, p. 248 [★530★]
$515, federal aid per capita in Pennsylvania, 1990, p. 252 [★536★]
$516, average Virginia widow(er) Social Security benefit, 1990, p. 250 [★532★]
$517, Alaska hospital care expenditure per capita, 1982, p. 307 [★652★]
$518, AFDC benefits by family in Vermont, 1990, p. 251 [★534★]
$518, average New Mexico widow(er) Social Security benefit, 1990, p. 250 [★532★]
$518, California tuition revenues per student for higher education, 1989, p. 148 [★330★]
$519, city government taxes per capita in Los Angeles, CA 1990, p. 279 [★587★]
$519, federal aid per capita in Minnesota 1990, p. 269 [★573★]
$519, federal aid per capita in Wisconsin, 1990, p. 252 [★536★]
$519, personal funds expenditure per capita on health care, 1975, p. 284 [★598★]
$520, average Mississippi retired worker Social Security benefit, 1990, p. 248 [★530★]
$520, federal aid per capita in Alabama, 1990, p. 252 [★536★]
$522, average family Social Security benefit for nondisabled aged widow or widower, p. 246 [★526★]
$522, Mississippi government debt per capita, 1990, p. 272 [★577★]
$523, average Social Security benefits for widows and widowers, nondisabled 1983, p. 247 [★529★]
$524, average North Dakota widow(er) Social Security benefit, 1990, p. 250 [★532★]
$524, average South Dakota widow(er) Social Security benefit, 1990, p. 250 [★532★]
$525, average District of Columbia disabled worker Social Security benefit, 1990, p. 249 [★531★]
$525, federal aid per capita in West Virginia 1990, p. 269 [★573★]
$526, average Maine widow(er) Social Security benefit, 1990, p. 250 [★532★]
$526, cost of two-year supply of diapers, cloth washed at home, p. 200 [★430★]
$527, average West Virginia widow(er) Social Security benefit, 1990, p. 250 [★532★]
$528, federal aid per capita in Washington, 1990, p. 252 [★536★]
$529, average Oklahoma widow(er) Social Security benefit, 1990, p. 250 [★532★]
$529, average Social Security benefits for retired workers 1980, p. 247 [★529★]
$530, AFDC benefits by family in New York, 1990, p. 251 [★534★]
$530, average Hawaii widow(er) Social Security benefit, 1990, p. 250 [★532★]
$530, Connecticut hospital care expenditure per capita, 1982, p. 307 [★652★]
$532, federal aid per capita in Arkansas, 1990, p. 252 [★536★]
$532, Kansas, hospital care cost per day, 1990, p. 305 [★649★]
$533, average South Dakota disabled worker Social Security benefit, 1990, p. 249 [★531★]

Numbers following p. are page references. Numbers in [ ] are table references.

Money - continued

Dollars

$533, city government taxes per capita in Indianapolis, IN 1990, p. 279 [★587★]
$533, federal aid per capita in Connecticut 1990, p. 269 [★573★]
$534, Arkansas, hospital care cost per day, 1990, p. 305 [★649★]
$534, average Maine disabled worker Social Security benefit, 1990, p. 249 [★531★]
$536, city government taxes per capita in Chicago, IL 1990, p. 279 [★587★]
$536, Iowa hospital care expenditure per capita, 1982, p. 307 [★652★]
$536, Minnesota, hospital care cost per day, 1990, p. 305 [★649★]
$536, state tax per capita in New Hampshire 1990, p. 268 [★572★]
$537, average Alaska widow(er) Social Security benefit, 1990, p. 250 [★532★]
$537, Tennessee government debt per capita, 1990, p. 272 [★577★]
$538, average Mississippi disabled worker Social Security benefit, 1990, p. 249 [★531★]
$538, average Texas widow(er) Social Security benefit, 1990, p. 250 [★532★]
$539, average Arkansas retired worker Social Security benefit, 1990, p. 248 [★530★]
$539, average family Social Security benefit for disabled worker, alone, p. 246 [★526★]
$539, federal aid per capita in California 1990, p. 269 [★573★]
$539, Indiana hospital care expenditure per capita, 1982, p. 307 [★652★]
$540, Colorado hospital care expenditure per capita, 1982, p. 307 [★652★]
$540, federal aid per capita in Hawaii, 1990, p. 252 [★536★]
$540, monthly mortgage payment, 1983, p. 223 [★483★]
$540, state government debt in 1980, p. 270 [★575★]
$541, Alabama hospital care expenditure per capita, 1982, p. 307 [★652★]
$541, federal aid per capita in Minnesota, 1990, p. 252 [★536★]
$542, federal aid per capita in South Carolina, 1990, p. 252 [★536★]
$544, military contracts per capita, 1990, p. 260 [★549★]
$547, Idaho, hospital care cost per day, 1990, p. 305 [★649★]
$547, military contracts per capita in Mississippi 1990, p. 260 [★550★]
$548, average Missouri widow(er) Social Security benefit, 1990, p. 250 [★532★]
$548, military contracts per capita in Rhode Island 1990, p. 260 [★550★]
$549, Alabama hospital care expenditure per capita, 1982, p. 307 [★652★]
$549, federal aid per capita in Massachusetts 1990, p. 269 [★573★]
$550, federal aid per capita in Hawaii 1990, p. 269 [★573★]
$550, state and local federal funds per capita, 1990, p. 278 [★585★]
$551, average North Carolina disabled worker Social Security benefit, 1990, p. 249 [★531★]
$552, Alaska hospital care expenditure per capita, 1982, p. 307 [★652★]
$552, average family Social Security benefit for retired worker, alone, p. 246 [★526★]

Money - continued

Dollars

$552, Delaware hospital care expenditure per capita, 1982, p. 307 [★652★]
$553, average Arkansas disabled worker Social Security benefit, 1990, p. 249 [★531★]
$553, average Kentucky retired worker Social Security benefit, 1990, p. 248 [★530★]
$554, Wisconsin, hospital care cost per day, 1990, p. 305 [★649★]
$555, average Alabama retired worker Social Security benefit, 1990, p. 248 [★530★]
$555, average Maine retired worker Social Security benefit, 1990, p. 248 [★530★]
$555, average Rhode Island disabled worker Social Security benefit, 1990, p. 249 [★531★]
$555, federal aid per capita in Kentucky, 1990, p. 252 [★536★]
$556, average Minnesota widow(er) Social Security benefit, 1990, p. 250 [★532★]
$556, average North Dakota disabled worker Social Security benefit, 1990, p. 249 [★531★]
$556, average Social Security benefit for disabled workers, 1989, p. 246 [★527★]
$556, average Tennessee disabled worker Social Security benefit, 1990, p. 249 [★531★]
$557, average South Dakota retired worker Social Security benefit, 1990, p. 248 [★530★]
$557, Colorado hospital care expenditure per capita, 1982, p. 307 [★652★]
$557, federal aid per capita in Tennessee, 1990, p. 252 [★536★]
$558, average Idaho widow(er) Social Security benefit, 1990, p. 250 [★532★]
$559, annual expenditures per capita for police protection and correction in Alaska, 1988, p. 136 [★310★]
$559, average Colorado widow(er) Social Security benefit, 1990, p. 250 [★532★]
$559, average Georgia disabled worker Social Security benefit, 1990, p. 249 [★531★]
$559, average Montana widow(er) Social Security benefit, 1990, p. 250 [★532★]
$559, average South Carolina disabled worker Social Security benefit, 1990, p. 249 [★531★]
$560, average Alabama disabled worker Social Security benefit, 1990, p. 249 [★531★]
$560, average Georgia retired worker Social Security benefit, 1990, p. 248 [★530★]
$560, average Louisiana retired worker Social Security benefit, 1990, p. 248 [★530★]
$560, federal aid per capita in Utah 1990, p. 269 [★573★]
$561, average North Carolina retired worker Social Security benefit, 1990, p. 248 [★530★]
$561, average South Carolina retired worker Social Security benefit, 1990, p. 248 [★530★]
$561, average Tennessee retired worker Social Security benefit, 1990, p. 248 [★530★]
$561, average Vermont widow(er) Social Security benefit, 1990, p. 250 [★532★]
$562, federal aid per capita in West Virginia, 1990, p. 252 [★536★]
$563, federal aid per capita in Maine 1990, p. 269 [★573★]
$563, Kentucky, hospital care cost per day, 1990, p. 305 [★649★]

Numerical Locator

Numbers following p. are page references. Numbers in [ ] are table references.

Money - continued

Dollars

$564, Illinois hospital care expenditure per capita, 1982, p. 307 [★652★]
$565, federal aid per capita in Idaho, 1990, p. 252 [★536★]
$565, West Virginia, hospital care cost per day, 1990, p. 305 [★649★]
$566, San Antonio, TX government revenue per capita, 1990, p. 273 [★579★]
$567, annual household expenditures for telephone service, 1989, p. 212 [★457★]
$567, average Iowa widow(er) Social Security benefit, 1990, p. 250 [★532★]
$567, average North Dakota retired worker Social Security benefit, 1990, p. 248 [★530★]
$567, average Social Security benefit for retired workers, 1989, p. 246 [★527★]
$567, average Social Security benefits for retired worker and wife 1980, p. 247 [★529★]
$567, average Virginia retired worker Social Security benefit, 1990, p. 248 [★530★]
$568, average Maryland widow(er) Social Security benefit, 1990, p. 250 [★532★]
$568, average New Mexico retired worker Social Security benefit, 1990, p. 248 [★530★]
$568, average Wyoming widow(er) Social Security benefit, 1990, p. 250 [★532★]
$568, federal aid per capita in Louisiana 1990, p. 269 [★573★]
$568, Florida hospital care expenditure per capita, 1982, p. 307 [★652★]
$569, average Nebraska disabled worker Social Security benefit, 1990, p. 249 [★531★]
$569, Florida hospital care expenditure per capita, 1982, p. 307 [★652★]
$570, average Kansas disabled worker Social Security benefit, 1990, p. 249 [★531★]
$570, average Nebraska widow(er) Social Security benefit, 1990, p. 250 [★532★]
$570, average Oklahoma disabled worker Social Security benefit, 1990, p. 249 [★531★]
$572, average Minnesota disabled worker Social Security benefit, 1990, p. 249 [★531★]
$572, average New Mexico disabled worker Social Security benefit, 1990, p. 249 [★531★]
$572, average Virginia disabled worker Social Security benefit, 1990, p. 249 [★531★]
$574, average Florida widow(er) Social Security benefit, 1990, p. 250 [★532★]
$574, Maine, hospital care cost per day, 1990, p. 305 [★649★]
$575, average Oklahoma retired worker Social Security benefit, 1990, p. 248 [★530★]
$575, average Social Security benefits for disabled workers 1980, p. 247 [★529★]
$575, average Utah disabled worker Social Security benefit, 1990, p. 249 [★531★]
$576, average Missouri disabled worker Social Security benefit, 1990, p. 249 [★531★]
$576, federal aid per capita in Oregon 1990, p. 269 [★573★]
$577, average Massachusetts disabled worker Social Security benefit, 1990, p. 249 [★531★]

Money - continued

Dollars

$578, average New Hampshire widow(er) Social Security benefit, 1990, p. 250 [★532★]
$578, average Rhode Island widow(er) Social Security benefit, 1990, p. 250 [★532★]
$578, average Vermont disabled worker Social Security benefit, 1990, p. 249 [★531★]
$578, Connecticut hospital care expenditure per capita, 1982, p. 307 [★652★]
$578, Delaware hospital care expenditure per capita, 1982, p. 307 [★652★]
$579, average Iowa disabled worker Social Security benefit, 1990, p. 249 [★531★]
$579, average Kansas widow(er) Social Security benefit, 1990, p. 250 [★532★]
$579, average Nevada widow(er) Social Security benefit, 1990, p. 250 [★532★]
$579, average Texas disabled worker Social Security benefit, 1990, p. 249 [★531★]
$579, federal aid per capita in Mississippi 1990, p. 269 [★573★]
$580, average Wyoming disabled worker Social Security benefit, 1990, p. 249 [★531★]
$581, average Arizona widow(er) Social Security benefit, 1990, p. 250 [★532★]
$581, average California widow(er) Social Security benefit, 1990, p. 250 [★532★]
$581, average Oregon widow(er) Social Security benefit, 1990, p. 250 [★532★]
$581, average Utah widow(er) Social Security benefit, 1990, p. 250 [★532★]
$582, average Colorado disabled worker Social Security benefit, 1990, p. 249 [★531★]
$582, average Hawaii disabled worker Social Security benefit, 1990, p. 249 [★531★]
$582, average Social Security benefits for retired workers 1983, p. 247 [★529★]
$583, average New Hampshire disabled worker Social Security benefit, 1990, p. 249 [★531★]
$583, average Texas retired worker Social Security benefit, 1990, p. 248 [★530★]
$584, average Alaska disabled worker Social Security benefit, 1990, p. 249 [★531★]
$584, average Kentucky disabled worker Social Security benefit, 1990, p. 249 [★531★]
$584, average Wisconsin widow(er) Social Security benefit, 1990, p. 250 [★532★]
$585, average AFDC payment by family in 1970, p. 251 [★535★]
$585, average Delaware widow(er) Social Security benefit, 1990, p. 250 [★532★]
$585, average Ohio widow(er) Social Security benefit, 1990, p. 250 [★532★]
$585, high school prom costs for "her," 1993, p. 241 [★516★]
$586, average Idaho retired worker Social Security benefit, 1990, p. 248 [★530★]
$586, average Massachusetts widow(er) Social Security benefit, 1990, p. 250 [★532★]
$586, personal funds expenditure per capita on health care, 1976, p. 284 [★598★]

Numbers following p. are page references. Numbers in [ ] are table references.

Money - continued

Dollars

$587, average Minnesota retired worker Social Security benefit, 1990, p. 248 [★530★]
$587, average Pennsylvania widow(er) Social Security benefit, 1990, p. 250 [★532★]
$588, Alabama, hospital care cost per day, 1990, p. 305 [★649★]
$588, average Colorado retired worker Social Security benefit, 1990, p. 248 [★530★]
$588, average Montana retired worker Social Security benefit, 1990, p. 248 [★530★]
$588, property tax per capita in New York, NY, p. 262 [★557★]
$589, average Indiana widow(er) Social Security benefit, 1990, p. 250 [★532★]
$589, average Missouri retired worker Social Security benefit, 1990, p. 248 [★530★]
$589, average Vermont retired worker Social Security benefit, 1990, p. 248 [★530★]
$590, AFDC benefits by family in Hawaii, 1990, p. 251 [★534★]
$590, average Connecticut disabled worker Social Security benefit, 1990, p. 249 [★531★]
$590, South Carolina, hospital care cost per day, 1990, p. 305 [★649★]
$591, average California disabled worker Social Security benefit, 1990, p. 249 [★531★]
$591, average Washington widow(er) Social Security benefit, 1990, p. 250 [★532★]
$591, public school expenditure per capita, 1985, p. 175 [★377★]
$593, AFDC benefits by family in Connecticut, 1990, p. 251 [★534★]
$593, average Hawaii retired worker Social Security benefit, 1990, p. 248 [★530★]
$593, average Idaho disabled worker Social Security benefit, 1990, p. 249 [★531★]
$593, Kansas hospital care expenditure per capita, 1982, p. 307 [★652★]
$594, average Wisconsin disabled worker Social Security benefit, 1990, p. 249 [★531★]
$595, average Louisiana disabled worker Social Security benefit, 1990, p. 249 [★531★]
$595, average Nebraska retired worker Social Security benefit, 1990, p. 248 [★530★]
$595, average West Virginia retired worker Social Security benefit, 1990, p. 248 [★530★]
$595, North Carolina, hospital care cost per day, 1990, p. 305 [★649★]
$597, average value per acre of farmland, 1989, p. 38 [★88★]
$598, Arizona government debt per capita, 1990, p. 272 [★577★]
$598, average Florida disabled worker Social Security benefit, 1990, p. 249 [★531★]
$598, average New York widow(er) Social Security benefit, 1990, p. 250 [★532★]
$598, Vermont, hospital care cost per day, 1990, p. 305 [★649★]
$599, Alabama hospital care expenditure per capita, 1982, p. 307 [★652★]
$600, annual personal federal income tax exemption, 1950, p. 281 [★593★]
$600, average automobile insurance claim, p. 237 [★506★]
$600, average Michigan widow(er) Social Security benefit, 1990, p. 250 [★532★]

Money - continued

Dollars

$600, average Oregon disabled worker Social Security benefit, 1990, p. 249 [★531★]
$600, cost of hand-held cellular telephones, 1988, p. 212 [★458★]
$600, federal aid per capita in Connecticut, 1990, p. 252 [★536★]
$600, Kuwaiti exiles' duty-free purchases, 1990, p. 442 [★959★]
$600, Lifeline (emergency medical service) flight, p. 317 [★680★]
$601, average Illinois widow(er) Social Security benefit, 1990, p. 250 [★532★]
$601, average Maryland retired worker Social Security benefit, 1990, p. 248 [★530★]
$601, average Rhode Island retired worker Social Security benefit, 1990, p. 248 [★530★]
$601, federal aid per capita in Oregon, 1990, p. 252 [★536★]
$602, average Alaska retired worker Social Security benefit, 1990, p. 248 [★530★]
$602, average Florida retired worker Social Security benefit, 1990, p. 248 [★530★]
$602, average Montana disabled worker Social Security benefit, 1990, p. 249 [★531★]
$602, average Social Security benefits for disabled workers, 1983, p. 247 [★529★]
$602, average Washington disabled worker Social Security benefit, 1990, p. 249 [★531★]
$603, average Delaware disabled worker Social Security benefit, 1990, p. 249 [★531★]
$604, average Nevada retired worker Social Security benefit, 1990, p. 248 [★530★]
$604, average spent per recreational fisherman, 1985, p. 441 [★955★]
$604, average Wyoming retired worker Social Security benefit, 1990, p. 248 [★530★]
$605, average Iowa retired worker Social Security benefit, 1990, p. 248 [★530★]
$605, average Massachusetts retired worker Social Security benefit, 1990, p. 248 [★530★]
$605, average New Hampshire retired worker Social Security benefit, 1990, p. 248 [★530★]
$605, hospital care cost per stay, 1970, p. 305 [★648★]
$606, Arizona hospital care expenditure per capita, 1982, p. 307 [★652★]
$606, average Indiana disabled worker Social Security benefit, 1990, p. 249 [★531★]
$606, average Maryland disabled worker Social Security benefit, 1990, p. 249 [★531★]
$608, average Pennsylvania disabled worker Social Security benefit, 1990, p. 249 [★531★]
$610, average Arizona retired worker Social Security benefit, 1990, p. 248 [★530★]
$610, average New Jersey disabled worker Social Security benefit, 1990, p. 249 [★531★]
$610, city government debt per capita, 1980, p. 277 [★584★]
$611, average Utah retired worker Social Security benefit, 1990, p. 248 [★530★]
$611, property tax per capita in Boston, MA, p. 262 [★557★]
$612, average Illinois disabled worker Social Security benefit, 1990, p. 249 [★531★]
$612, average New Jersey widow(er) Social Security benefit, 1990, p. 250 [★532★]

Money - continued

Dollars

$613, average New York disabled worker Social Security benefit, 1990, p. 249 [★531★]
$613, average Ohio disabled worker Social Security benefit, 1990, p. 249 [★531★]
$613, New Jersey, hospital care cost per day, 1990, p. 305 [★649★]
$614, average Oregon retired worker Social Security benefit, 1990, p. 248 [★530★]
$615, average California retired worker Social Security benefit, 1990, p. 248 [★530★]
$616, average Nevada disabled worker Social Security benefit, 1990, p. 249 [★531★]
$617, average Kansas retired worker Social Security benefit, 1990, p. 248 [★530★]
$618, average Ohio retired worker Social Security benefit, 1990, p. 248 [★530★]
$618, average Wisconsin retired worker Social Security benefit, 1990, p. 248 [★530★]
$618, public school expenditure per capita in Tennessee, 1991, p. 176 [★378★]
$619, average Arizona disabled worker Social Security benefit, 1990, p. 249 [★531★]
$620, federal aid per capita in Mississippi, 1990, p. 252 [★536★]
$621, average Connecticut widow(er) Social Security benefit, 1990, p. 250 [★532★]
$621, average Pennsylvania retired worker Social Security benefit, 1990, p. 248 [★530★]
$621, federal aid per capita in Maine, 1990, p. 252 [★536★]
$623, California hospital care expenditure per capita, 1982, p. 307 [★652★]
$624, average Washington retired worker Social Security benefit, 1990, p. 248 [★530★]
$624, United States hospital care expenditure per capita, 1982, p. 307 [★652★]
$625, monthly mortgage payment, 1993, p. 223 [★483★]
$626, California hospital care expenditure per capita, 1982, p. 307 [★652★]
$627, average Delaware retired worker Social Security benefit, 1990, p. 248 [★530★]
$627, average West Virginia disabled worker Social Security benefit, 1990, p. 249 [★531★]
$628, average Indiana retired worker Social Security benefit, 1990, p. 248 [★530★]
$628, California hospital care expenditure per capita, 1982, p. 307 [★652★]
$629, tax per capita 1960, p. 263 [★559★]
$630, federal aid per capita in Louisiana, 1990, p. 252 [★536★]
$630, Georgia hospital care expenditure per capita, 1982, p. 307 [★652★]
$630, Georgia, hospital care cost per day, 1990, p. 305 [★649★]
$631, hospital care cost per day, 1989, p. 306 [★650★]
$632, Oklahoma, hospital care cost per day, 1990, p. 305 [★649★]
$633, federal aid per capita in New Mexico, 1990, p. 252 [★536★]
$633, loss per street/highway robbery, 1990, p. 118 [★282★]
$633, Tennessee, hospital care cost per day, 1990, p. 305 [★649★]
$634.72, average cost of Switzerland periodical subscription, 1992, p. 166 [★364★]
$635, Virginia, hospital care cost per day, 1990, p. 305 [★649★]

Money - continued

Dollars

$638, Hawaii, hospital care cost per day, 1990, p. 305 [★649★]
$639, expenditure per student in Turkey, 1987-1988, p. 175 [★376★]
$640, AFDC benefits by family in California, 1990, p. 251 [★534★]
$641, average Illinois retired worker Social Security benefit, 1990, p. 248 [★530★]
$641, federal aid per capita in Massachusetts, 1990, p. 252 [★536★]
$641, New York, hospital care cost per day, 1990, p. 305 [★649★]
$643, average Michigan retired worker Social Security benefit, 1990, p. 248 [★530★]
$644, average Michigan disabled worker Social Security benefit, 1990, p. 249 [★531★]
$645, average New York retired worker Social Security benefit, 1990, p. 248 [★530★]
$647, federal aid per capita in South Dakota, 1990, p. 269 [★573★]
$653, District of Columbia tuition revenues per student for higher education, 1989, p. 148 [★330★]
$653, elderly family out-of-pocket health care costs for private insurance, 1991, p. 287 [★606★]
$659, average New Jersey retired worker Social Security benefit, 1990, p. 248 [★530★]
$661, average Connecticut retired worker Social Security benefit, 1990, p. 248 [★530★]
$662, Pennsylvania, hospital care cost per day, 1990, p. 305 [★649★]
$663, Rhode Island, hospital care cost per day, 1990, p. 305 [★649★]
$665, federal aid per capita in Rhode Island 1990, p. 269 [★573★]
$667, hospital care cost per stay, 1971, p. 306 [★651★]
$667, Indiana, hospital care cost per day, 1990, p. 305 [★649★]
$668, public school expenditure per capita in Arkansas, 1991, p. 176 [★378★]
$669, average Social Security benefits for retired worker and wife 1970, p. 247 [★529★]
$670, federal aid per capita in Vermont, 1990, p. 252 [★536★]
$671, New Hampshire, hospital care cost per day, 1990, p. 305 [★649★]
$672, city government revenue per capita, 1980, p. 276 [★582★]
$674.25, average cost of Ireland periodical subscription, 1992, p. 166 [★364★]
$675, Arkansas hospital care expenditure per capita, 1982, p. 307 [★652★]
$675, Iowa government debt per capita, 1990, p. 272 [★577★]
$678, Maryland, hospital care cost per day, 1990, p. 305 [★649★]
$679, Delaware hospital care expenditure per capita, 1982, p. 307 [★652★]
$679, Indiana hospital care expenditure per capita, 1982, p. 307 [★652★]
$679, Missouri, hospital care cost per day, 1990, p. 305 [★649★]
$681, annual household energy costs of five or more unit bldg., 1987, p. 204 [★437★]
$683, Idaho tuition revenues per student for higher education, 1989, p. 148 [★330★]
$684, city government expenditure per capita, 1980, p. 276 [★583★]
$685, federal aid per capita in Montana, 1990, p. 269 [★573★]

Money - continued

Dollars

$687, hospital care cost per day, 1990, p. 305 [★648★]
$687, United States, hospital care cost per day, 1990, p. 305 [★649★]
$689, monthly nursing home charge per resident, 1977, p. 322 [★689★]
$690, military contracts per capita in Maine, 1990, p. 260 [★550★]
$690, monthly personal nursing home payment per resident, 1977, p. 326 [★694★]
$691, farmer's gross average income per acre of peanuts, 1991, p. 387 [★829★]
$693, federal aid per capita in North Dakota, 1990, p. 269 [★573★]
$694, public school expenditure per capita in Alabama, 1991, p. 176 [★378★]
$699, federal aid per capita in Vermont, 1990, p. 269 [★573★]
$700, average Japanese employer's health care premium expenditures per employee, p. 288 [★609★]
$700, Illinois hospital care expenditure per capita, 1982, p. 307 [★652★]
$701, Louisiana, hospital care cost per day, 1990, p. 305 [★649★]
$706, lifetime energy costs of average dishwasher, p. 201 [★431★]
$716, Michigan, hospital care cost per day, 1990, p. 305 [★649★]
$716, public school expenditure per capita in South Dakota, 1991, p. 176 [★378★]
$717, federal aid per capita in New York, 1990, p. 269 [★573★]
$717, Hawaii tuition revenues per student for higher education, 1989, p. 148 [★330★]
$717, Illinois, hospital care cost per day, 1990, p. 305 [★649★]
$718, state tax per capita in South Dakota, 1990, p. 268 [★572★]
$719, public school expenditure per capita in North Dakota, 1991, p. 176 [★378★]
$720, AFDC benefits by family in Alaska, 1990, p. 251 [★534★]
$720, monthly Medicaid nursing home payment per resident, 1977, p. 324 [★691★]
$720, Ohio, hospital care cost per day, 1990, p. 305 [★649★]
$725, Colorado, hospital care cost per day, 1990, p. 305 [★649★]
$725, public school expenditure per capita in Idaho, 1991, p. 176 [★378★]
$725, public school expenditure per capita in Kentucky, 1991, p. 176 [★378★]
$726, cost of adenoidectomy, p. 292 [★619★]
$727, property tax per capita in Washington, DC, p. 262 [★557★]
$730, San Antonio, TX government expenditure per capita, 1990, p. 274 [★580★]
$730, Spain expenditure per capita on health care, 1990, p. 282 [★594★]
$734, federal aid per capita in South Dakota, 1990, p. 252 [★536★]
$734, New Mexico, hospital care cost per day, 1990, p. 305 [★649★]
$735, Colorado government debt per capita, 1990, p. 272 [★577★]
$735, loss per miscellaneous robbery, 1990, p. 118 [★282★]
$737, federal aid per capita in North Dakota, 1990, p. 252 [★536★]
$738, military contracts per capita in California, 1990, p. 260 [★550★]
$740, federal aid per capita in Montana, 1990, p. 252 [★536★]
$742, public school expenditure per capita in Missouri, 1991, p. 176 [★378★]
$743, Arkansas government debt per capita, 1990, p. 272 [★577★]

Money - continued

Dollars

$743, average Social Security benefits for retired worker and wife 1983, p. 247 [★529★]
$747, Indiana government debt per capita, 1990, p. 272 [★577★]
$750, annual costs per capita of substance abuse, p. 293 [★621★]
$750, annual expenditures on cigarettes per smoker, p. 242 [★519★]
$750, average Japanese per-day honeymoon travel expenses abroad, 1988, p. 442 [★960★]
$750, monthly home health care charge, p. 317 [★678★]
$752, Texas, hospital care cost per day, 1990, p. 305 [★649★]
$757, public school expenditure per capita in Nebraska, 1991, p. 176 [★378★]
$758, public school expenditure per capita in Louisiana, 1991, p. 176 [★378★]
$763, Houston, TX government revenue per capita, 1990, p. 273 [★579★]
$769, Florida government debt per capita, 1990, p. 272 [★577★]
$769, Florida, hospital care cost per day, 1990, p. 305 [★649★]
$770, federal aid per capita in Rhode Island, 1990, p. 252 [★536★]
$771, Delaware, hospital care cost per day, 1990, p. 305 [★649★]
$772, public school expenditure per capita, 1988-1989, p. 178 [★381★]
$773, public school expenditure per student, 1970, p. 177 [★379★]
$776, public school expenditure per capita in Oklahoma, 1991, p. 176 [★378★]
$783, loss per robbery, 1990, p. 118 [★282★]
$784, public school expenditure per capita in Hawaii, 1991, p. 176 [★378★]
$787, annual health care cost per South Dakota prison inmate, 1989, p. 121 [★287★]
$788, Massachusetts, hospital care cost per day, 1990, p. 305 [★649★]
$788, public school expenditure per capita in Illinois, 1991, p. 176 [★378★]
$792, annual health care cost per Alabama prison inmate, 1989, p. 121 [★287★]
$795, Dallas, TX government expenditure per capita, 1990, p. 274 [★580★]
$797, Montana tuition revenues per student for higher education, 1989, p. 148 [★330★]
$800, cost of hospital care for seat-belted victims of automobile crashes, p. 311 [★664★]
$800, employee health care expenditure per automobile, Ford Motor Co., 1993, p. 286 [★602★]
$800, Oregon, hospital care cost per day, 1990, p. 305 [★649★]
$801, Wyoming tuition revenues per student for higher education, 1989, p. 148 [★330★]
$802, military contracts per capita in Alaska, 1990, p. 260 [★550★]
$810, Arkansas hospital care expenditure per capita, 1982, p. 307 [★652★]
$811, San Diego, CA government expenditure per capita, 1990, p. 274 [★580★]
$813, public school expenditure per capita in North Carolina, 1991, p. 176 [★378★]
$817, Houston, TX government expenditure per capita, 1990, p. 274 [★580★]
$817, Washington, hospital care cost per day, 1990, p. 305 [★649★]

Numbers following p. are page references. Numbers in [ ] are table references.

Money - continued

Dollars

$819, San Jose, CA government revenue per capita, 1990, p. 273 [★579★]
$822, Dallas, TX government revenue per capita, 1990, p. 273 [★579★]
$825, Connecticut, hospital care cost per day, 1990, p. 305 [★649★]
$828, public school expenditure per capita in Utah, 1991, p. 176 [★378★]
$830, average spent per recreational hunter, 1985, p. 441 [★955★]
$831, annual health care cost per Louisiana prison inmate, 1989, p. 121 [★287★]
$831, North Carolina tuition revenues per student for higher education, 1989, p. 148 [★330★]
$832, Utah, hospital care cost per day, 1990, p. 305 [★649★]
$835, Columbus, OH government revenue per capita, 1990, p. 273 [★579★]
$835, public school expenditure per capita in South Carolina, 1991, p. 176 [★378★]
$841, public school expenditure per capita in Colorado, 1991, p. 176 [★378★]
$845, public school expenditure per capita, 1990, p. 175 [★377★]
$846, public school expenditure per capita in Iowa, 1991, p. 176 [★378★]
$850, cost of hospital care per day, rural hospitals, p. 310 [★662★]
$852, city government taxes per capita in Baltimore, MD 1990, p. 279 [★587★]
$853, Columbus, OH government expenditure per capita, 1990, p. 274 [★580★]
$854, Nevada, hospital care cost per day, 1990, p. 305 [★649★]
$855, expenditure per student in Spain, 1987-1988, p. 175 [★376★]
$857, Connecticut expenditures per capita on health care, 1982, p. 283 [★595★]
$859, public school expenditure per capita in Georgia, 1991, p. 176 [★378★]
$860, Milwaukee, WI government debt per capita, 1990, p. 275 [★581★]
$860, Minnesota government debt per capita, 1990, p. 272 [★577★]
$861, public school expenditure per capita in Ohio, 1991, p. 176 [★378★]
$862, public school expenditure per capita in Florida, 1991, p. 176 [★378★]
$863, monthly welfare nursing home payment per resident, 1985, p. 324 [★692★]
$863, Nebraska government debt per capita, 1990, p. 272 [★577★]
$866, public school expenditure per capita in Massachusetts, 1991, p. 176 [★378★]
$866, state tax per capita in Texas 1990, p. 268 [★572★]
$867, Arizona, hospital care cost per day, 1990, p. 305 [★649★]
$868, Idaho expenditures per capita on health care, 1982, p. 283 [★595★]
$869.25, average cost of Netherlands periodical subscription, 1992, p. 166 [★364★]
$870, state tax per capita in Tennessee, 1990, p. 268 [★572★]
$871, median monthly fee in continuing-care retirement communities, p. 222 [★480★]
$873, Kansas expenditures per capita on health care, 1982, p. 283 [★595★]

Money - continued

Dollars

$876, federal aid per capita in New York, 1990, p. 252 [★536★]
$879, Alabama government revenue per capita, 1990, p. 271 [★576★]
$879, average Social Security benefits for retired worker and wife 1980, p. 247 [★529★]
$881, public school expenditure per capita in New Hampshire, 1991, p. 176 [★378★]
$883, public school expenditure per capita in West Virginia, 1991, p. 176 [★378★]
$889, public school expenditure per capita in Pennsylvania, 1991, p. 176 [★378★]
$891, annual household energy costs in the Western United States, 1987, p. 203 [★436★]
$891, public school expenditure per capita in Delaware, 1991, p. 176 [★378★]
$892, public school expenditure per capita in Arizona, 1991, p. 176 [★378★]
$892, public school expenditure per capita in Rhode Island, 1991, p. 176 [★378★]
$896, Georgia expenditures per capita on health care, 1982, p. 283 [★595★]
$897, Delaware expenditures per capita on health care, 1982, p. 283 [★595★]
$899, average charitable gift, 1991, p. 195 [★417★]
$900, average refund (federal income tax), p. 267 [★570★]
$900, reduction in patient charges when physicians use computers to order services, p. 312 [★667★]
$903, military contracts per capita in Maryland, 1990, p. 260 [★550★]
$904, Indiana expenditures per capita on health care, 1982, p. 283 [★595★]
$905, annual two-to four-unit bldg. household energy costs, 1987, p. 204 [★437★]
$905, public school expenditure per capita in Texas, 1991, p. 176 [★378★]
$906, public school expenditure per capita in Kansas, 1991, p. 176 [★378★]
$907, annual health care cost per Missouri prison inmate, 1989, p. 121 [★287★]
$907, military contracts per capita in Arizona, 1990, p. 260 [★550★]
$907, public school expenditure per capita in U.S., 1991, p. 176 [★378★]
$909, annual health care cost per Oklahoma prison inmate, 1989, p. 121 [★287★]
$909, hospital care Medicare payments per enrollee, 1980, p. 308 [★655★]
$909, United Kingdom expenditure per capita on health care, 1990, p. 282 [★594★]
$915, public school expenditure per capita in New Mexico, 1991, p. 176 [★378★]
$915.75, average cost of chemistry periodical subscription, 1992, p. 165 [★363★]
$918, public school expenditure per capita in California, 1991, p. 176 [★378★]
$920, average labor cost per Honda vehicle, 1992, p. 30 [★69★]
$920, Milwaukee, WI government expenditure per capita, 1990, p. 274 [★580★]

Numbers following p. are page references. Numbers in [ ] are table references.

Money - continued

Dollars

$920, Pennsylvania government debt per capita, 1990, p. 272 [★577★]
$921, San Jose, CA government expenditure per capita, 1990, p. 274 [★580★]
$925, public school expenditure per capita in Indiana, 1991, p. 176 [★378★]
$925, public school expenditure per capita in Oregon, 1991, p. 176 [★378★]
$926, public school expenditure per capita in Michigan, 1991, p. 176 [★378★]
$927, federal income tax per capita in Delaware, 1989, p. 267 [★571★]
$927.27, average cost of physics periodical subscription, 1992, p. 165 [★363★]
$928, public school expenditure per capita in Wisconsin, 1991, p. 176 [★378★]
$931, Kansas expenditures per capita on health care, 1982, p. 283 [★595★]
$931, state tax per capita in Colorado, 1990, p. 268 [★572★]
$931, state tax per capita in Mississippi, 1990, p. 268 [★572★]
$932, Milwaukee, WI government revenue per capita, 1990, p. 273 [★579★]
$939, California, hospital care cost per day, 1990, p. 305 [★649★]
$941, public school expenditure per capita in Virginia, 1991, p. 176 [★378★]
$944, public school expenditure per capita in Maryland, 1991, p. 176 [★378★]
$945, state tax per capita in Alabama, 1990, p. 268 [★572★]
$948, annual mobile home household energy costs, 1987, p. 204 [★437★]
$954, city government expenditure per capita, 1985, p. 276 [★583★]
$955, New Mexico tuition revenues per student for higher education, 1989, p. 148 [★330★]
$957, Alabama expenditures per capita on health care, 1982, p. 283 [★595★]
$958, annual expenditures per capita for police protection and correction in District of Columbia, 1988, p. 136 [★310★]
$958, state tax per capita in Nebraska, 1990, p. 268 [★572★]
$960, Texas tuition revenues per student for higher education, 1989, p. 148 [★330★]
$961, city government debt per capita, 1985, p. 277 [★584★]
$961, state tax per capita in Arkansas, 1990, p. 268 [★572★]
$962, Chicago, IL government expenditure per capita, 1990, p. 274 [★580★]
$963, Denmark expenditure per capita on health care, 1990, p. 282 [★594★]
$963, Los Angeles, CA government expenditure per capita, 1990, p. 274 [★580★]
$964, city government taxes per capita in Philadelphia, PA 1990, p. 279 [★587★]
$965, state tax per capita in Missouri, 1990, p. 268 [★572★]
$966, average family Social Security benefit for retired worker and wife, p. 246 [★526★]
$968, state tax per capita in Louisiana, 1990, p. 268 [★572★]
$970, California government debt per capita, 1990, p. 272 [★577★]
$970, Idaho government debt per capita, 1990, p. 272 [★577★]

Money - continued

Dollars

$972, average family Social Security benefit for children, p. 246 [★526★]
$978, Hawaii expenditures per capita on health care, 1982, p. 283 [★595★]
$980, cost of hospital care per day, medicare enrollees, 1989, p. 308 [★654★]
$980, expenditure per student in Portugal, 1987-1988, p. 175 [★376★]
$980, state tax per capita in Oregon, 1990, p. 268 [★572★]
$981, average Social Security benefits for retired worker and wife 1983, p. 247 [★529★]
$983, lifetime energy costs of average dryer, p. 201 [★431★]
$983, public school expenditure per capita in Montana, 1991, p. 176 [★378★]
$983, Virginia government debt per capita, 1990, p. 272 [★577★]
$984, annual household energy costs, buildings built 1980-1984, 1987, p. 204 [★438★]
$985, Alabama government debt per capita, 1990, p. 272 [★577★]
$986, Idaho expenditures per capita on health care, 1982, p. 283 [★595★]
$987, Michigan government debt per capita, 1990, p. 272 [★577★]
$990, public school expenditure per capita in Minnesota, 1991, p. 176 [★378★]
$991, military contracts per capita in Colorado, 1990, p. 260 [★550★]
$992, public school expenditure per capita in Vermont, 1991, p. 176 [★378★]
$994, Arkansas expenditures per capita on health care, 1982, p. 283 [★595★]
$995, District of Columbia, hospital care cost per day, 1990, p. 305 [★649★]
$996, city government taxes per capita in Boston, MA 1990, p. 279 [★587★]
$1,000, average cost of testing a school's radon level, p. 418 [★897★]
$1,000, cost of cellular telephones, 1988, p. 212 [★458★]
$1,004, public school expenditure per capita in Nevada, 1991, p. 176 [★378★]
$1,005, average annual Catholic elementary school tuition, 1988, p. 179 [★382★]
$1,007, city government revenue per capita, 1985, p. 276 [★582★]
$1,010, public school expenditure per capita in District of Columbia, 1991, p. 176 [★378★]
$1,010, state government expenditure in 1980, p. 270 [★575★]
$1,016, San Diego, CA government revenue per capita, 1990, p. 273 [★579★]
$1,019, Chicago, IL government revenue per capita, 1990, p. 273 [★579★]
$1,019, cost of hospital care per day, 1990, p. 309 [★657★]
$1,020 to 1,170, annual cost of owning a dog, p. 210 [★454★]
$1,022, average tuition costs for two-year public college, 1991-1992, p. 145 [★326★]
$1,026, Missouri government debt per capita, 1990, p. 272 [★577★]
$1,026, state tax per capita in Utah, 1990, p. 268 [★572★]
$1,027, state tax per capita in Florida, 1990, p. 268 [★572★]
$1,029, average condominium timesharer's timesharing vacation expenditures, p. 105 [★257★]

Numbers following p. are page references. Numbers in [ ] are table references.

Money - continued

Dollars

$1,030, state government debt in 1986, p. 270 [★575★]
$1,033, Alabama expenditures per capita on health care, 1982, p. 283 [★595★]
$1,033, Ohio government debt per capita, 1990, p. 272 [★577★]
$1,034, state government revenue in 1980, p. 270 [★575★]
$1,035, annual health care cost per West Virginia prison inmate, 1989, p. 121 [★287★]
$1,036, Florida expenditures per capita on health care, 1982, p. 283 [★595★]
$1,036, Phoenix, AZ government revenue per capita, 1990, p. 273 [★579★]
$1,039, Utah government debt per capita, 1990, p. 272 [★577★]
$1,041, federal income tax per capita in Arkansas, 1989, p. 267 [★571★]
$1,042, annual household energy costs, buildings built 1940-1949, 1987, p. 204 [★438★]
$1,044, Illinois tuition revenues per student for higher education, 1989, p. 148 [★330★]
$1,048, Georgia expenditures per capita on health care, 1982, p. 283 [★595★]
$1,049, annual household energy costs, buildings built 1985 or later, 1987, p. 204 [★438★]
$1,049, robbery loss per residence, 1990, p. 118 [★282★]
$1,050, Los Angeles, CA government revenue per capita, 1990, p. 273 [★579★]
$1,054, Idaho expenditures per capita on health care, 1982, p. 283 [★595★]
$1,054, state tax per capita in Ohio, 1990, p. 268 [★572★]
$1,057, Indiana expenditures per capita on health care, 1982, p. 283 [★595★]
$1,059, state tax per capita in North Dakota, 1990, p. 268 [★572★]
$1,060, annual household energy costs, buildings built 1960-1969, 1987, p. 204 [★438★]
$1,060, Jacksonville, FL government revenue per capita, 1990, p. 273 [★579★]
$1,061, Florida tuition revenues per student for higher education, 1989, p. 148 [★330★]
$1,062, annual household energy costs, buildings built 1970-1974, 1987, p. 204 [★438★]
$1,062, public school expenditure per capita in Maine, 1991, p. 176 [★378★]
$1,066, state tax per capita in Virginia, 1990, p. 268 [★572★]
$1,070, Alaska, hospital care cost per day, 1990, p. 305 [★649★]
$1,073, state tax per capita in Montana, 1990, p. 268 [★572★]
$1,075, federal income tax per capita in Indiana, 1989, p. 267 [★571★]
$1,075, Washington tuition revenues per student for higher education, 1989, p. 148 [★330★]
$1,077, state tax per capita in Kansas, 1990, p. 268 [★572★]
$1,080, annual household energy costs, 1987, p. 203 [★435★]
$1,081, annual household energy costs in southern United States, 1987, p. 203 [★436★]
$1,086, Arizona expenditures per capita on health care, 1982, p. 283 [★595★]
$1,091, Arizona expenditures per capita on health care, 1982, p. 283 [★595★]
$1,092, state tax per capita in Georgia, 1990, p. 268 [★572★]

Money - continued

Dollars

$1,094, Jacksonville, FL government expenditure per capita, 1990, p. 274 [★580★]
$1,094, state government debt in 1987, p. 270 [★575★]
$1,095, public school expenditure per capita in Washington, 1991, p. 176 [★378★]
$1,099, Nevada tuition revenues per student for higher education, 1989, p. 148 [★330★]
$1,100, American goods bought by Hong Kong residents annually, p. 97 [★230★]
$1,100, state tax per capita in Indiana, 1990, p. 268 [★572★]
$1,101, Indiana expenditures per capita on health care, 1982, p. 283 [★595★]
$1,105, annual household energy costs, buildings built 1939 or earlier, 1987, p. 204 [★438★]
$1,105, state tax per capita in Oklahoma, 1990, p. 268 [★572★]
$1,106, Alaska expenditures per capita on health care, 1982, p. 283 [★595★]
$1,110, federal aid per capita in Wyoming, 1990, p. 269 [★573★]
$1,110, Florida expenditures per capita on health care, 1982, p. 283 [★595★]
$1,112, Arizona expenditures per capita on health care, 1982, p. 283 [★595★]
$1,112, Memphis, TN government expenditure per capita, 1990, p. 274 [★580★]
$1,112, state tax per capita in Pennsylvania, 1990, p. 268 [★572★]
$1,113, Italy expenditure per capita on health care, 1990, p. 282 [★594★]
$1,113, Japan expenditure per capita on health care, 1990, p. 282 [★594★]
$1,113, Oklahoma tuition revenues per student for higher education, 1989, p. 148 [★330★]
$1,115, Illinois expenditures per capita on health care, 1982, p. 283 [★595★]
$1,117, South Carolina government debt per capita, 1990, p. 272 [★577★]
$1,120, average family Social Security benefit for widowed mother or father, 2 children, p. 246 [★526★]
$1,121, annual household energy costs, buildings built 1950-1959, 1987, p. 204 [★438★]
$1,124, annual Midwest household energy costs, 1987, p. 203 [★436★]
$1,127, monthly mortgage payment, 1990, p. 225 [★487★]
$1,127, state tax per capita in Illinois, 1990, p. 268 [★572★]
$1,128, state tax per capita in South Carolina, 1990, p. 268 [★572★]
$1,129, state government debt in 1988, p. 270 [★575★]
$1,130, state tax per capita in Idaho, 1990, p. 268 [★572★]
$1,131, annual household energy costs, buildings built 1975-1979, 1987, p. 204 [★438★]
$1,131, federal aid per capita in Alaska, 1990, p. 269 [★573★]
$1,131, federal income tax per capita in Georgia, 1989, p. 267 [★571★]
$1,135, annual single-family attached household energy costs, 1987, p. 204 [★437★]
$1,141, Memphis, TN government debt per capita, 1990, p. 275 [★581★]
$1,143, per capita athletic expenditures on women in 1987, NCAA Div. II, p. 146 [★328★]

Money - continued

Dollars

$1,144, District of Columbia expenditures per capita on health care, 1982, p. 283 [★595★]
$1,145, federal income tax per capita in Alaska 1989, p. 267 [★571★]
$1,150, difference in automobile purchase price, white and black men, p. 193 [★413★]
$1,151, Australia expenditure per capita on health care, 1990, p. 282 [★594★]
$1,151, public school expenditure per capita in Connecticut, 1991, p. 176 [★378★]
$1,153, Delaware expenditures per capita on health care, 1982, p. 283 [★595★]
$1,154, annual health care cost per Colorado prison inmate, 1989, p. 121 [★287★]
$1,154, Delaware expenditures per capita on health care, 1982, p. 283 [★595★]
$1,156, state tax per capita in Kentucky, 1990, p. 268 [★572★]
$1,157, Nebraska tuition revenues per student for higher education, 1989, p. 148 [★330★]
$1,159, federal income tax per capita in Indiana, 1989, p. 267 [★571★]
$1,162, Memphis, TN government revenue per capita, 1990, p. 273 [★579★]
$1,165, Arkansas expenditures per capita on health care, 1982, p. 283 [★595★]
$1,165, Illinois expenditures per capita on health care, 1982, p. 283 [★595★]
$1,167, monthly Medicare nursing home payment per resident, 1977, p. 323 [★690★]
$1,168, Washington government debt per capita, 1990, p. 272 [★577★]
$1,171, West Virginia tuition revenues per student for higher education, 1989, p. 148 [★330★]
$1,174, annual health care cost per Utah prison inmate, 1989, p. 121 [★287★]
$1,176, Iowa expenditures per capita on health care, 1982, p. 283 [★595★]
$1,178, military contracts per capita in Missouri, 1990, p. 260 [★550★]
$1,178, public school expenditure per capita in Wyoming, 1991, p. 176 [★378★]
$1,181, Oklahoma government debt per capita, 1990, p. 272 [★577★]
$1,182, state tax per capita in Vermont, 1990, p. 268 [★572★]
$1,184, federal income tax per capita in Idaho, 1989, p. 267 [★571★]
$1,185, city government taxes per capita in San Francisco, CA, 1990, p. 279 [★587★]
$1,186, public school expenditure per capita in New York, 1991, p. 176 [★378★]
$1,186, state tax per capita in North Carolina, 1990, p. 268 [★572★]
$1,187, Alaska expenditures per capita on health care, 1982, p. 283 [★595★]
$1,191, Indianapolis, IN government revenue per capita, 1990, p. 273 [★579★]
$1,193, state tax per capita in Iowa, 1990, p. 268 [★572★]
$1,194, elderly family out-of-pocket health care costs for nursing home, 1991, p. 287 [★606★]

Money - continued

Dollars

$1,194, state government debt in 1989, p. 270 [★575★]
$1,194, state tax per capita in Arizona, 1990, p. 268 [★572★]
$1,201, federal income tax per capita in Delaware, 1989, p. 267 [★571★]
$1,208, New Mexico government debt per capita, 1990, p. 272 [★577★]
$1,209, Colorado expenditures per capita on health care, 1982, p. 283 [★595★]
$1,210, annual health care cost per Kentucky prison inmate, 1989, p. 121 [★287★]
$1,210, Utah tuition revenues per student for higher education, 1989, p. 148 [★330★]
$1,211, federal income tax per capita in Alabama, 1989, p. 267 [★571★]
$1,211, state tax per capita in United States, 1990, p. 268 [★572★]
$1,216, federal income tax per capita in Florida, 1989, p. 267 [★571★]
$1,216, Georgia expenditures per capita on health care, 1982, p. 283 [★595★]
$1,219, Iowa expenditures per capita on health care, 1982, p. 283 [★595★]
$1,220, New York tuition revenues per student for higher education, 1989, p. 148 [★330★]
$1,220, state tax per capita in Michigan, 1990, p. 268 [★572★]
$1,220, United States expenditures per capita on health care, 1982, p. 283 [★595★]
$1,223, public school expenditure per capita in New Jersey, 1991, p. 176 [★378★]
$1,226, annual health care cost per Maryland prison inmate, 1989, p. 121 [★287★]
$1,226, annual single-family detached household energy costs, 1987, p. 204 [★437★]
$1,228, Florida expenditures per capita on health care, 1982, p. 283 [★595★]
$1,228, Hawaii expenditures per capita on health care, 1982, p. 283 [★595★]
$1,229, Connecticut expenditures per capita on health care, 1982, p. 283 [★595★]
$1,229, Georgia tuition revenues per student for higher education, 1989, p. 148 [★330★]
$1,229, state tax per capita in Rhode Island, 1990, p. 268 [★572★]
$1,232, Arkansas expenditures per capita on health care, 1982, p. 283 [★595★]
$1,237, Indianapolis, IN government expenditure per capita, 1990, p. 274 [★580★]
$1,243, state tax per capita in West Virginia, 1990, p. 268 [★572★]
$1,244, federal income tax per capita in Arizona, 1989, p. 267 [★571★]
$1,246, federal income tax per capita in Alabama, 1989, p. 267 [★571★]
$1,246, Phoenix, AZ government expenditure per capita, 1990, p. 274 [★580★]
$1,247, Alaska expenditures per capita on health care, 1982, p. 283 [★595★]
$1,251, expenditure per student in Ireland, 1987-1988, p. 175 [★376★]
$1,251, Wisconsin government debt per capita, 1990, p. 272 [★577★]

Numbers following p. are page references. Numbers in [ ] are table references.

Money - continued

Dollars

$1,252, federal income tax per capita in Alabama 1989, p. 267 [★571★]
$1,253, federal aid per capita in Wyoming, 1990, p. 252 [★536★]
$1,256, federal income tax per capita in Connecticut, 1989, p. 267 [★571★]
$1,261, Kansas tuition revenues per student for higher education, 1989, p. 148 [★330★]
$1,264, annual health care cost per Wyoming prison inmate, 1989, p. 121 [★287★]
$1,264, lifetime energy costs of average washing machine, p. 201 [★431★]
$1,266, military contracts per capita in Virginia, 1990, p. 260 [★550★]
$1,271, Kansas expenditures per capita on health care, 1982, p. 283 [★595★]
$1,271, state tax per capita in Maine, 1990, p. 268 [★572★]
$1,273, California expenditures per capita on health care, 1982, p. 283 [★595★]
$1,276, annual Northeast household energy costs, 1987, p. 203 [★436★]
$1,281, Colorado expenditures per capita on health care, 1982, p. 283 [★595★]
$1,281, Norway expenditure per capita on health care, 1990, p. 282 [★594★]
$1,283, state government debt in 1990, p. 270 [★575★]
$1,285, District of Columbia expenditures per capita on health care, 1982, p. 283 [★595★]
$1,286, public school expenditure per student, 1975, p. 177 [★379★]
$1,287, Germany expenditure per capita on health care, 1990, p. 282 [★594★]
$1,293, military contracts per capita in Connecticut, 1990, p. 260 [★550★]
$1,299, city government expenditure per capita, 1990, p. 276 [★583★]
$1,300, average monthly military retirement income, p. 348 [★751★]
$1,303, federal aid per capita in Alaska, 1990, p. 252 [★536★]
$1,308, Illinois expenditures per capita on health care, 1982, p. 283 [★595★]
$1,308, Nevada government debt per capita, 1990, p. 272 [★577★]
$1,311, Indianapolis, IN government debt per capita, 1990, p. 275 [★581★]
$1,314, San Diego, CA government debt per capita, 1990, p. 275 [★581★]
$1,317, state tax per capita in Nevada, 1990, p. 268 [★572★]
$1,318, military contracts per capita in Massachusetts, 1990, p. 260 [★550★]
$1,323, city government revenue per capita, 1990, p. 276 [★582★]
$1,325, Alabama expenditures per capita on health care, 1982, p. 283 [★595★]
$1,329, state tax per capita in New Mexico, 1990, p. 268 [★572★]
$1,335, Illinois government debt per capita, 1990, p. 272 [★577★]
$1,339, Minnesota tuition revenues per student for higher education, 1989, p. 148 [★330★]
$1,340, Oregon tuition revenues per student for higher education, 1989, p. 148 [★330★]

Money - continued

Dollars

$1,340, state tax per capita in Wisconsin, 1990, p. 268 [★572★]
$1,341, loss per commercial house robbery, 1990, p. 118 [★282★]
$1,347, state tax per capita in Wyoming, 1990, p. 268 [★572★]
$1,348, Connecticut expenditures per capita on health care, 1982, p. 283 [★595★]
$1,349, state tax per capita in Maryland, 1990, p. 268 [★572★]
$1,349, state tax per capita in New Jersey, 1990, p. 268 [★572★]
$1,350, cost of hospital care per day, urban hospitals, p. 310 [★662★]
$1,351, Colorado expenditures per capita on health care, 1982, p. 283 [★595★]
$1,352, cost of two-year supply of diapers, disposable, p. 200 [★430★]
$1,361, Kentucky tuition revenues per student for higher education, 1989, p. 148 [★330★]
$1,365, North Dakota government debt per capita, 1990, p. 272 [★577★]
$1,366, annual health care cost per Ohio prison inmate, 1989, p. 121 [★287★]
$1,378, West Virginia government debt per capita, 1990, p. 272 [★577★]
$1,379, France expenditure per capita on health care, 1990, p. 282 [★594★]
$1,380, Hawaii expenditures per capita on health care, 1982, p. 283 [★595★]
$1,382, per capita athletic expenditures on women in 1987, all AIAW Colleges, p. 146 [★328★]
$1,387, annual health care cost per South Carolina prison inmate, 1989, p. 121 [★287★]
$1,389, Arkansas tuition revenues per student for higher education, 1989, p. 148 [★330★]
$1,389, Switzerland expenditure per capita on health care, 1990, p. 282 [★594★]
$1,390, Maryland government debt per capita, 1990, p. 272 [★577★]
$1,391, Texas government expenditure per capita, 1990, p. 273 [★578★]
$1,392, city government debt per capita, 1990, p. 277 [★584★]
$1,397, Tennessee tuition revenues per student for higher education, 1989, p. 148 [★330★]
$1,399, federal income tax per capita in Iowa, 1989, p. 267 [★571★]
$1,400, balance on all-purpose credit cards, p. 214 [★462★]
$1,400, cost of hand-held cellular telephones, 1988, p. 212 [★458★]
$1,400, Detroit, MI government debt per capita, 1990, p. 275 [★581★]
$1,408, Detroit, MI government expenditure per capita, 1990, p. 274 [★580★]
$1,410, New Hampshire government revenue per capita, 1990, p. 271 [★576★]
$1,412, San Jose, CA government debt per capita, 1990, p. 275 [★581★]
$1,417, Iowa expenditures per capita on health care, 1982, p. 283 [★595★]
$1,421, Sweden expenditure per capita on health care, 1990, p. 282 [★594★]
$1,426, Arizona tuition revenues per student for higher education, 1989, p. 148 [★330★]

Numbers following p. are page references. Numbers in [ ] are table references.

Money - continued

Dollars

$1,429, annual health care cost per Pennsylvania prison inmate, 1989, p. 121 [★287★]
$1,437, Kentucky government debt per capita, 1990, p. 272 [★577★]
$1,443, Chicago, IL government debt per capita, 1990, p. 275 [★581★]
$1,445, North Dakota tuition revenues per student for higher education, 1989, p. 148 [★330★]
$1,449, state government expenditure in 1985, p. 270 [★575★]
$1,450, monthly personal nursing home payment per resident, 1985, p. 326 [★694★]
$1,451, California expenditures per capita on health care, 1982, p. 283 [★595★]
$1,456, monthly nursing home charge per resident, 1985, p. 322 [★688★]
$1,458, state tax per capita in California, 1990, p. 268 [★572★]
$1,459, federal income tax per capita in Georgia, 1989, p. 267 [★571★]
$1,461, federal income tax per capita in District of Columbia, 1989, p. 267 [★571★]
$1,463, Mississippi tuition revenues per student for higher education, 1989, p. 148 [★330★]
$1,465, Detroit, MI government revenue per capita, 1990, p. 273 [★579★]
$1,468, Alaska tuition revenues per student for higher education, 1989, p. 148 [★330★]
$1,478, Boston, MA government debt per capita, 1990, p. 275 [★581★]
$1,490, federal income tax per capita in Arizona, 1989, p. 267 [★571★]
$1,492, Maine tuition revenues per student for higher education, 1989, p. 148 [★330★]
$1,495, Alabama tuition revenues per student for higher education, 1989, p. 148 [★330★]
$1,500, annual health care cost per Virginia prison inmate, 1989, p. 121 [★287★]
$1,500, cremation funeral cost, 1989, p. 240 [★514★]
$1,501, Massachusetts tuition revenues per student for higher education, 1989, p. 148 [★330★]
$1,504, Medicaid nursing home payment per resident, 1985, p. 324 [★691★]
$1,505, Missouri government expenditure per capita, 1990, p. 273 [★578★]
$1,505, Missouri tuition revenues per student for higher education, 1989, p. 148 [★330★]
$1,508, California expenditures per capita on health care, 1982, p. 283 [★595★]
$1,512, New Hampshire government expenditure per capita, 1990, p. 273 [★578★]
$1,522, federal income tax per capita in Kansas, 1989, p. 267 [★571★]
$1,524, annual health care cost per Delaware prison inmate, 1989, p. 121 [★287★]
$1,525, state tax per capita in Washington, 1990, p. 268 [★572★]
$1,528, Texas government revenue per capita, 1990, p. 271 [★576★]
$1,530, average cost of toys per child in Tokyo, p. 438 [★949★]

Money - continued

Dollars

$1,534, federal income tax per capita in Arkansas, 1989, p. 267 [★571★]
$1,536, state government revenue in 1985, p. 270 [★575★]
$1,553, New Jersey tuition revenues per student for higher education, 1989, p. 148 [★330★]
$1,555, federal income tax per capita in Kansas, 1989, p. 267 [★571★]
$1,557, state tax per capita in Massachusetts, 1990, p. 268 [★572★]
$1,558, annual health care cost per Vermont prison inmate, 1989, p. 121 [★287★]
$1,558, state tax per capita in Minnesota, 1990, p. 268 [★572★]
$1,560, annual health care cost per Idaho prison inmate, 1989, p. 121 [★287★]
$1,563, federal income tax per capita in Indiana, 1989, p. 267 [★571★]
$1,565, state government expenditure in 1986, p. 270 [★575★]
$1,566, Missouri government revenue per capita, 1990, p. 271 [★576★]
$1,566, Philadelphia, PA government expenditure per capita, 1990, p. 274 [★580★]
$1,570, annual health care cost per Illinois prison inmate, 1989, p. 121 [★287★]
$1,571, federal income tax per capita in Iowa, 1989, p. 267 [★571★]
$1,571, Philadelphia, PA government revenue per capita, 1990, p. 273 [★579★]
$1,572, federal income tax per capita in Georgia 1989, p. 267 [★571★]
$1,572, South Carolina tuition revenues per student for higher education, 1989, p. 148 [★330★]
$1,575, federal income tax per capita in District of Columbia, 1989, p. 267 [★571★]
$1,580, employers' share of household annual health care expenditure, p. 285 [★601★]
$1,589, Florida government expenditure per capita, 1990, p. 273 [★578★]
$1,590, Dallas, TX government debt per capita, 1990, p. 275 [★581★]
$1,590, state tax per capita in New York, 1990, p. 268 [★572★]
$1,594, Florida government revenue per capita, 1990, p. 271 [★576★]
$1,595, annual health care cost per Arkansas prison inmate, 1989, p. 121 [★287★]
$1,595, Connecticut tuition revenues per student for higher education, 1989, p. 148 [★330★]
$1,600, bridal dress rental cost, Japan, p. 240 [★512★]
$1,602, state tax per capita in Connecticut, 1990, p. 268 [★572★]
$1,616, Tennessee government expenditure per capita, 1990, p. 273 [★578★]
$1,618, annual health care cost per Iowa prison inmate, 1989, p. 121 [★287★]
$1,619, average annual religious (non-Catholic) elementary school tuition, 1988, p. 179 [★382★]
$1,626, federal income tax per capita in Alaska, 1989, p. 267 [★571★]
$1,626, public school expenditure per capita in Alaska, 1991, p. 176 [★378★]

Money - continued

Dollars

$1,627, state and local generated revenue per capita, 1980, p. 278 [★586★]
$1,629, average labor cost per Ford vehicle, 1992, p. 30 [★69★]
$1,629, federal income tax per capita in Florida, 1989, p. 267 [★571★]
$1,636, state government revenue in 1986, p. 270 [★575★]
$1,638, Tennessee government revenue per capita, 1990, p. 271 [★576★]
$1,640, annual health care cost per Kansas prison inmate, 1989, p. 121 [★287★]
$1,648, annual health care cost per Georgia prison inmate, 1989, p. 121 [★287★]
$1,651, federal income tax per capita in Hawaii, 1989, p. 267 [★571★]
$1,652.20, average social security taxes, 1990, p. 264 [★560★]
$1,653, federal income tax per capita in Kansas, 1989, p. 267 [★571★]
$1,657, Maryland tuition revenues per student for higher education, 1989, p. 148 [★330★]
$1,660, Louisiana tuition revenues per student for higher education, 1989, p. 148 [★330★]
$1,664, state government expenditure in 1987, p. 270 [★575★]
$1,665, annual health care cost per Montana prison inmate, 1989, p. 121 [★287★]
$1,670, expenditure per student in New Zealand, 1987-1988, p. 175 [★376★]
$1,672, Arkansas government expenditure per capita, 1990, p. 273 [★578★]
$1,673, Baltimore, MD government debt per capita, 1990, p. 275 [★581★]
$1,680, individual health care costs per retiree, 1991, p. 288 [★607★]
$1,686, per capita athletic expenditures on women in 1987, All NCAA, p. 146 [★328★]
$1,695, annual health care cost per Wisconsin prison inmate, 1989, p. 121 [★287★]
$1,696, state tax per capita in Delaware, 1990, p. 268 [★572★]
$1,708, Colorado government expenditure per capita, 1990, p. 273 [★578★]
$1,708, Mississippi government expenditure per capita, 1990, p. 273 [★578★]
$1,711, annual health care cost per Rhode Island prison inmate, 1989, p. 121 [★287★]
$1,725, Arkansas government revenue per capita, 1990, p. 271 [★576★]
$1,726, public school expenditure per capita in Mississippi, 1991, p. 176 [★378★]
$1,727, Georgia government revenue per capita, 1990, p. 271 [★576★]
$1,728, state government revenue, 1987, p. 270 [★575★]
$1,731, Maine government debt per capita, 1990, p. 272 [★577★]
$1,740, federal income tax per capita in Colorado, 1989, p. 267 [★571★]
$1,745, Virginia tuition revenues per student for higher education, 1989, p. 148 [★330★]
$1,747, Kansas government expenditure per capita, 1990, p. 273 [★578★]

Money - continued

Dollars

$1,747, Montana government debt per capita, 1990, p. 272 [★577★]
$1,754, Illinois government expenditure per capita, 1990, p. 273 [★578★]
$1,756, Colorado government revenue per capita, 1990, p. 271 [★576★]
$1,756, federal income tax per capita in Connecticut, 1989, p. 267 [★571★]
$1,758, Mississippi government revenue per capita, 1990, p. 271 [★576★]
$1,759, Georgia government expenditure per capita, 1990, p. 273 [★578★]
$1,763, state government expenditure in 1988, p. 270 [★575★]
$1,764, annual health care cost per Nevada prison inmate, 1989, p. 121 [★287★]
$1,770, Kansas government revenue per capita, 1990, p. 271 [★576★]
$1,771, Wisconsin tuition revenues per student for higher education, 1989, p. 148 [★330★]
$1,780, federal income tax per capita in California, 1989, p. 267 [★571★]
$1,784, Nebraska government expenditure per capita, 1990, p. 273 [★578★]
$1,784, Oklahoma government expenditure per capita, 1990, p. 273 [★578★]
$1,787, Pennsylvania government expenditure per capita, 1990, p. 273 [★578★]
$1,795, annual health care cost per Nebraska prison inmate, 1989, p. 121 [★287★]
$1,795, Canada expenditure per capita on health care, 1990, p. 282 [★594★]
$1,802, Indiana government expenditure per capita, 1990, p. 273 [★578★]
$1,810, Nebraska government revenue per capita, 1990, p. 271 [★576★]
$1,811, annual expenditures for meals away from home, 1990, p. 208 [★447★]
$1,812, federal income tax per capita in Colorado, 1989, p. 267 [★571★]
$1,815, state government revenue in 1988, p. 270 [★575★]
$1,818, Idaho government expenditure per capita, 1990, p. 273 [★578★]
$1,821, South Dakota tuition revenues per student for higher education, 1989, p. 148 [★330★]
$1,823, Illinois government revenue per capita, 1990, p. 271 [★576★]
$1,824, lifetime energy costs of average refrigerator, p. 201 [★431★]
$1,831, Alabama government expenditure per capita, 1990, p. 273 [★578★]
$1,840, GNP per capita in Slovakia, p. 359 [★769★]
$1,841, Ohio government revenue per capita, 1990, p. 271 [★576★]
$1,841, South Dakota government expenditure per capita, 1990, p. 273 [★578★]
$1,847, Pennsylvania government revenue per capita, 1990, p. 271 [★576★]
$1,848, hospital care Medicare payments per enrollee, 1989, p. 308 [★655★]

Numbers following p. are page references. Numbers in [ ] are table references.

Money - continued

Dollars

$1,862, North Carolina government revenue per capita, 1990, p. 271 [★576★]
$1,868, annual health care cost per Oregon prison inmate, 1989, p. 121 [★287★]
$1,870, annual health care cost per Maine prison inmate, 1989, p. 121 [★287★]
$1,872, average labor cost per Chrysler vehicle, 1992, p. 30 [★69★]
$1,873, Los Angeles, CA government debt per capita, 1990, p. 275 [★581★]
$1,880, Virginia government revenue per capita, 1990, p. 271 [★576★]
$1,882, Columbus, OH government debt per capita, 1990, p. 275 [★581★]
$1,882, South Dakota government revenue per capita, 1990, p. 271 [★576★]
$1,883, per capita athletic expenditures on men in 1987, NCAA Div. II, p. 146 [★328★]
$1,885, hysterectomy cost, Atlanta, p. 329 [★702★]
$1,889, federal income tax per capita in Colorado, 1989, p. 267 [★571★]
$1,889, Nevada government revenue per capita, 1990, p. 271 [★576★]
$1,889, Ohio government expenditure per capita, 1990, p. 273 [★578★]
$1,894, North Carolina government expenditure per capita, 1990, p. 273 [★578★]
$1,895, state government expenditure in 1989, p. 270 [★575★]
$1,900, Oklahoma government revenue per capita, 1990, p. 271 [★576★]
$1,905, Arizona government revenue per capita, 1990, p. 271 [★576★]
$1,910, expenditure per student in Italy, 1987-1988, p. 175 [★376★]
$1,911, Iowa tuition revenues per student for higher education, 1989, p. 148 [★330★]
$1,912, Rhode Island tuition revenues per student for higher education, 1989, p. 148 [★330★]
$1,913, annual health care cost per Arizona prison inmate, 1989, p. 121 [★287★]
$1,920, Virginia government expenditure per capita, 1990, p. 273 [★578★]
$1,925, annual support payments to relatives, 1988, p. 238 [★509★]
$1,927, Kentucky government expenditure per capita, 1990, p. 273 [★578★]
$1,930, Indiana government revenue per capita, 1990, p. 271 [★576★]
$1,941, annual health care cost per New Hampshire prison inmate, 1989, p. 121 [★287★]
$1,943, South Carolina government expenditure per capita, 1990, p. 273 [★578★]
$1,948, state government revenue in 1989, p. 270 [★575★]
$1,953, annual health care cost per California prison inmate, 1989, p. 121 [★287★]
$1,957, Oregon government expenditure per capita, 1990, p. 273 [★578★]
$1,958, federal income tax per capita in Hawaii, 1989, p. 267 [★571★]

Money - continued

Dollars

$1,962, annual health care cost per Tennessee prison inmate, 1989, p. 121 [★287★]
$1,964, federal income tax per capita in Illinois, 1989, p. 267 [★571★]
$1,968, Nevada government expenditure per capita, 1990, p. 273 [★578★]
$1,969, West Virginia government expenditure per capita, 1990, p. 273 [★578★]
$1,973, annual health care cost per North Carolina prison inmate, 1989, p. 121 [★287★]
$1,976, federal income tax per capita in Florida, 1989, p. 267 [★571★]
$1,978, federal income tax per capita in Idaho, 1989, p. 267 [★571★]
$1,980, Idaho government revenue per capita, 1990, p. 271 [★576★]
$1,987, South Carolina government revenue per capita, 1990, p. 271 [★576★]
$1,989, Baltimore, MD government expenditure per capita, 1990, p. 274 [★580★]
$1,994, Kentucky government revenue per capita, 1990, p. 271 [★576★]
$2,000, average federally subsidized student loan, p. 144 [★323★]
$2,000, monthly nursing home charge per resident, p. 317 [★678★]
$2,009, federal income tax per capita in Delaware, 1989, p. 267 [★571★]
$2,010, Indiana tuition revenues per student for higher education, 1989, p. 148 [★330★]
$2,014, Utah government expenditure per capita, 1990, p. 273 [★578★]
$2,016, annual health care cost per New Jersey prison inmate, 1989, p. 121 [★287★]
$2,020, Louisiana government expenditure per capita, 1990, p. 273 [★578★]
$2,021, District of Columbia hospital care expenditure per capita, 1982, p. 307 [★652★]
$2,027, Colorado tuition revenues per student for higher education, 1989, p. 148 [★330★]
$2,045, average annual Catholic secondary school tuition, 1988, p. 179 [★382★]
$2,047, state government expenditure in 1990, p. 270 [★575★]
$2,049, Utah government revenue per capita, 1990, p. 271 [★576★]
$2,050, annual personal federal income tax exemption, 1989, p. 281 [★593★]
$2,051, Oregon government revenue per capita, 1990, p. 271 [★576★]
$2,052, average annual religious (non-Catholic) school tuition, 1988, p. 179 [★382★]
$2,056, Arizona government expenditure per capita, 1990, p. 273 [★578★]
$2,057, Maryland government expenditure per capita, 1990, p. 273 [★578★]
$2,063, city government taxes per capita in New York City, NY 1990, p. 279 [★587★]
$2,066, Montana government expenditure per capita, 1990, p. 273 [★578★]
$2,067, Wyoming government debt per capita, 1990, p. 272 [★577★]

Numerical Locator

Numbers following p. are page references. Numbers in [ ] are table references.

Money - continued

Dollars

$2,073, federal income tax per capita in Illinois, 1989, p. 267 [★571★]
$2,079, Ohio tuition revenues per student for higher education, 1989, p. 148 [★330★]
$2,086, state government revenue in 1990, p. 270 [★575★]
$2,087, federal income tax per capita in California, 1989, p. 267 [★571★]
$2,096, expenditure per student in Japan, 1987-1988, p. 175 [★376★]
$2,099, Iowa government revenue per capita, 1990, p. 271 [★576★]
$2,104, Michigan government expenditure per capita, 1990, p. 273 [★578★]
$2,107, state tax per capita in Hawaii, 1990, p. 268 [★572★]
$2,108, annual health care cost per Connecticut prison inmate, 1989, p. 121 [★287★]
$2,115, Louisiana government revenue per capita, 1990, p. 271 [★576★]
$2,120, Michigan government revenue per capita, 1990, p. 271 [★576★]
$2,126, Houston, TX government debt per capita, 1990, p. 275 [★581★]
$2,127, expenditure per student in Australia, 1987-1988, p. 175 [★376★]
$2,127, West Virginia government revenue per capita, 1990, p. 271 [★576★]
$2,137, average tuition costs for four-year public college, 1991-1992, p. 145 [★325★]
$2,137, Iowa government expenditure per capita, 1990, p. 273 [★578★]
$2,141, monthly Medicare nursing home payment per resident, 1985, p. 323 [★690★]
$2,146, Wisconsin government expenditure per capita, 1990, p. 273 [★578★]
$2,156, per capita athletic expenditures on women in 1987, All AIAW and NCAA Div. I, p. 146 [★328★]
$2,157, annual health care cost per Minnesota prison inmate, 1989, p. 121 [★287★]
$2,185, Maryland government revenue per capita, 1990, p. 271 [★576★]
$2,203, Baltimore, MD government revenue per capita, 1990, p. 273 [★579★]
$2,218, average Japanese per-person travel expenses abroad, 1988, p. 442 [★960★]
$2,226, federal income tax per capita in Idaho, 1989, p. 267 [★571★]
$2,228, Phoenix, AZ government debt per capita, 1990, p. 275 [★581★]
$2,230, public school expenditure per student, 1980, p. 177 [★379★]
$2,234, Maine government expenditure per capita, 1990, p. 273 [★578★]
$2,236, Vermont government debt per capita, 1990, p. 272 [★577★]
$2,239, employers' share of family health care costs, 1991, p. 285 [★599★]
$2,244, Wisconsin government revenue per capita, 1990, p. 271 [★576★]

Money - continued

Dollars

$2,246, federal income tax per capita in Iowa, 1989, p. 267 [★571★]
$2,249, annual health care cost per New York prison inmate, 1989, p. 121 [★287★]
$2,250, GNP per capita in the Czech Republic, p. 359 [★769★]
$2,254, potential cost per life saved annually by pap smear testing, p. 329 [★701★]
$2,259, expenditure per student in Netherlands, 1987-1988, p. 175 [★376★]
$2,262, annual health care cost per Texas prison inmate, 1989, p. 121 [★287★]
$2,284, federal income tax per capita in Arkansas, 1989, p. 267 [★571★]
$2,289, Montana government revenue per capita, 1990, p. 271 [★576★]
$2,297, median annual pensions (private firms) for women, 1990, p. 348 [★750★]
$2,300, expenditure per student in Germany, West (former), 1987-1988, p. 175 [★376★]
$2,300, household's share of annual health care expenditure, p. 285 [★601★]
$2,305, Philadelphia, PA government debt per capita, 1990, p. 275 [★581★]
$2,306, federal income tax per capita in Hawaii, 1989, p. 267 [★571★]
$2,308, Oregon government debt per capita, 1990, p. 272 [★577★]
$2,319, Maine government revenue per capita, 1990, p. 271 [★576★]
$2,327, California government revenue per capita, 1990, p. 271 [★576★]
$2,334, New Jersey government expenditure per capita, 1990, p. 273 [★578★]
$2,337, Vermont state and local appropriations per student for higher education, 1989, p. 147 [★329★]
$2,340, Washington government expenditure per capita, 1990, p. 273 [★578★]
$2,359, California government expenditure per capita, 1990, p. 273 [★578★]
$2,362, Washington government revenue per capita, 1990, p. 271 [★576★]
$2,364, annual urban expenditures for food eaten at home, p. 208 [★448★]
$2,364, New Jersey government revenue per capita, 1990, p. 271 [★576★]
$2,379, annual health care cost per Massachusetts prison inmate, 1989, p. 121 [★287★]
$2,379, Minnesota government expenditure per capita, 1990, p. 273 [★578★]
$2,388, average labor cost per General Motors vehicle, 1992, p. 30 [★69★]
$2,391, federal income tax per capita in California, 1989, p. 267 [★571★]
$2,393, expenditure per student in France, 1987-1988, p. 175 [★376★]
$2,400, workers compensation for on-the-job motor vehicle-related injury, p. 43 [★101★]
$2,417, Michigan tuition revenues per student for higher education, 1989, p. 148 [★330★]

Numbers following p. are page references. Numbers in [ ] are table references.

Money - continued

Dollars

$2,436, individual health care costs per retiree, 2001, p. 288 [★607★]
$2,446, New Jersey government debt per capita, 1990, p. 272 [★577★]
$2,470, federal income tax per capita in Alaska, 1989, p. 267 [★571★]
$2,472, average October 1990 earnings of state government employees, p. 408 [★874★]
$2,483, North Dakota government expenditure per capita, 1990, p. 273 [★578★]
$2,485, annual expenditures for food eaten at home, 1990, p. 208 [★447★]
$2,493, cost of hysterectomy, p. 292 [★619★]
$2,500, average annual cost per student of Catholic secondary school, p. 179 [★385★]
$2,500, cost of hospital care for non-seat-belted victims of automobile crashes, p. 311 [★664★]
$2,507, federal income tax per capita in District of Columbia, 1989, p. 267 [★571★]
$2,512, Minnesota government revenue per capita, 1990, p. 271 [★576★]
$2,545, Vermont government revenue per capita, 1990, p. 271 [★576★]
$2,548, annual rural expenditures for food eaten at home, p. 208 [★448★]
$2,549, North Dakota government revenue per capita, 1990, p. 271 [★576★]
$2,550, fee for assisting with normal delivery of a baby, 1986, p. 330 [★703★]
$2,566, United States expenditure per capita on health care, 1990, p. 282 [★594★]
$2,568, New Mexico government expenditure per capita, 1990, p. 273 [★578★]
$2,568, Rhode Island government revenue per capita, 1990, p. 271 [★576★]
$2,568, South Dakota government debt per capita, 1990, p. 272 [★577★]
$2,570, state and local generated revenue per capita, 1985, p. 278 [★586★]
$2,577, San Francisco, CA government expenditure per capita, 1990, p. 274 [★580★]
$2,587, New York government debt per capita, 1990, p. 272 [★577★]
$2,593, Boston, MA government expenditure per capita, 1990, p. 274 [★580★]
$2,603, Vermont government expenditure per capita, 1990, p. 273 [★578★]
$2,610, average library spending for videos, 1992, p. 159 [★352★]
$2,622, Massachusetts government revenue per capita, 1990, p. 271 [★576★]
$2,627, Connecticut government revenue per capita, 1990, p. 271 [★576★]
$2,634, Louisiana state and local appropriations per student for higher education, 1989, p. 147 [★329★]
$2,636, annual health care cost per Michigan prison inmate, 1989, p. 121 [★287★]
$2,636, cost of baby's birth by caesarean, p. 292 [★619★]

Money - continued

Dollars

$2,664, annual health care cost per Washington prison inmate, 1989, p. 121 [★287★]
$2,685, federal income tax per capita in Illinois 1989, p. 267 [★571★]
$2,687, expenditure per student in United Kingdom, 1987-1988, p. 175 [★376★]
$2,688, New Mexico government revenue per capita, 1990, p. 271 [★576★]
$2,702, Connecticut government expenditure per capita, 1990, p. 273 [★578★]
$2,706, annual health care cost per Florida prison inmate, 1989, p. 121 [★287★]
$2,731, Pennsylvania tuition revenues per student for higher education, 1989, p. 148 [★330★]
$2,741, Rhode Island government expenditure per capita, 1990, p. 273 [★578★]
$2,763, New York government expenditure per capita, 1990, p. 273 [★578★]
$2,775, Boston, MA government revenue per capita, 1990, p. 273 [★579★]
$2,782, New Hampshire state and local appropriations per student for higher education, 1989, p. 147 [★329★]
$2,793, New Hampshire government debt per capita, 1990, p. 272 [★577★]
$2,800, annual cost of child care, 1988, p. 198 [★424★]
$2,803, North Dakota state and local appropriations per student for higher education, 1989, p. 147 [★329★]
$2,811, state tax per capita in Alaska, 1990, p. 268 [★572★]
$2,831, federal aid per capita in District of Columbia, 1990, p. 252 [★536★]
$2,832, Massachusetts government expenditure per capita, 1990, p. 273 [★578★]
$2,838, District of Columbia expenditures per capita on health care, 1982, p. 283 [★595★]
$2,853, Colorado state and local appropriations per student for higher education, 1989, p. 147 [★329★]
$2,867, expenditure per student in Belgium, 1987-1988, p. 175 [★376★]
$2,900, annual health care cost per New Mexico prison inmate, 1989, p. 121 [★287★]
$2,910, Texas state and local appropriations per student for higher education, 1989, p. 147 [★329★]
$2,915, New York government revenue per capita, 1990, p. 271 [★576★]
$2,918, West Virginia state and local appropriations per student for higher education, 1989, p. 147 [★329★]
$2,993, public school expenditure per student in Utah, 1991, p. 178 [★380★]
$2,994, Delaware government expenditure per capita, 1990, p. 273 [★578★]
$2,995, annual child support payments, 1990, p. 238 [★508★]
$3,010, teacher's average salary, 1949-1950, p. 364 [★779★]
$3,013, per capita athletic expenditures on men in 1987, All AIAW Colleges, p. 146 [★328★]
$3,015, state and local government expenditures per capita, 1990, p. 258 [★544★]
$3,026, Louisiana government debt per capita, 1990, p. 272 [★577★]

Numbers following p. are page references. Numbers in [ ] are table references.

Money - continued

Dollars

$3,043, federal funds per capita in North Carolina, 1990, p. 253 [★537★]
$3,051, federal funds per capita in Indiana, 1990, p. 253 [★537★]
$3,052, federal funds per capita in Wisconsin, 1990, p. 253 [★537★]
$3,065, Hawaii government debt per capita, 1990, p. 272 [★577★]
$3,091, average annual secular private elementary school tuition, 1988, p. 179 [★382★]
$3,111, Massachusetts government debt per capita, 1990, p. 272 [★577★]
$3,122, South Dakota state and local appropriations per student for higher education, 1989, p. 147 [★329★]
$3,139, federal income tax per capita in Connecticut, 1989, p. 267 [★571★]
$3,142, federal funds per capita in Michigan, 1990, p. 253 [★537★]
$3,148, federal funds per capita in Vermont, 1990, p. 253 [★537★]
$3,159, Delaware government revenue per capita, 1990, p. 271 [★576★]
$3,191, Nebraska state and local appropriations per student for higher education, 1989, p. 147 [★329★]
$3,200, public school expenditure per student in Idaho, 1991, p. 178 [★380★]
$3,201, Hawaii government expenditure per capita, 1990, p. 273 [★578★]
$3,202, San Francisco, CA government revenue per capita, 1990, p. 273 [★579★]
$3,209, federal funds per capita in New Hampshire, 1990, p. 253 [★537★]
$3,210, federal funds per capita in Illinois, 1990, p. 253 [★537★]
$3,226, expenditure per student in Austria, 1987-1988, p. 175 [★376★]
$3,226, federal funds per capita in Delaware, 1990, p. 253 [★537★]
$3,237, lifetime energy costs of average water heater, p. 201 [★431★]
$3,244, loss per bank robbery, 1990, p. 118 [★282★]
$3,261, individual health care costs per retiree, 2011, p. 288 [★607★]
$3,264, expenditure per student in Finland, 1987-1988, p. 175 [★376★]
$3,265, federal funds per capita in Georgia, 1990, p. 253 [★537★]
$3,269, San Francisco, CA government debt per capita, 1990, p. 275 [★581★]
$3,270, Wyoming government expenditure per capita, 1990, p. 273 [★578★]
$3,322, public school expenditure per student in Mississippi, 1991, p. 178 [★380★]
$3,334, public school expenditure per student in Arkansas, 1991, p. 178 [★380★]
$3,343, Connecticut government debt per capita, 1990, p. 272 [★577★]
$3,343, Mississippi state and local appropriations per student for higher education, 1989, p. 147 [★329★]
$3,358, federal tax liability in Mississippi, 1990, p. 270 [★574★]
$3,364, government social welfare programs per capita, p. 258 [★543★]
$3,378, federal tax liability in Arkansas, 1990, p. 270 [★574★]
$3,381, annual health care cost per Alaska prison inmate, 1989, p. 121 [★287★]

Money - continued

Dollars

$3,382, average annual Catholic school tuition, 1988, p. 179 [★382★]
$3,387, average federal income tax, 1980, p. 265 [★565★]
$3,392, Ohio state and local appropriations per student for higher education, 1989, p. 147 [★329★]
$3,410, federal tax liability in North Dakota, 1990, p. 270 [★574★]
$3,428, federal funds per capita in Texas, 1990, p. 253 [★537★]
$3,432, New Hampshire tuition revenues per student for higher education, 1989, p. 148 [★330★]
$3,445, federal funds per capita in Minnesota, 1990, p. 253 [★537★]
$3,446, federal tax liability in South Dakota, 1990, p. 270 [★574★]
$3,448, federal funds per capita in Nebraska, 1990, p. 253 [★537★]
$3,448, federal funds per capita in Nevada, 1990, p. 253 [★537★]
$3,452, average American employer's health care premium expenditures per employee, p. 288 [★609★]
$3,457, federal funds per capita in Oregon, 1990, p. 253 [★537★]
$3,467, Hawaii government revenue per capita, 1990, p. 271 [★576★]
$3,483, public school expenditure per student, 1985, p. 177 [★379★]
$3,491, Montana state and local appropriations per student for higher education, 1989, p. 147 [★329★]
$3,496, federal funds per capita in Ohio, 1990, p. 253 [★537★]
$3,500, health care expenditure per employee, companies with fewer than 500 employees, 1992, p. 286 [★603★]
$3,509, federal funds per capita in Arkansas, 1990, p. 253 [★537★]
$3,525, federal tax liability in Montana, 1990, p. 270 [★574★]
$3,537, New York City, NY government debt per capita, 1990, p. 275 [★581★]
$3,542, new car cost, 1975, p. 189 [★404★]
$3,559, federal tax liability in Utah, 1990, p. 270 [★574★]
$3,577, Missouri state and local appropriations per student for higher education, 1989, p. 147 [★329★]
$3,579, health care expenditure per employee, companies with 2,500-4,999 employees, 1992, p. 286 [★603★]
$3,580, annual cost of owning and operating a subcompact automobile, 1991, p. 191 [★407★]
$3,582, federal funds per capita in Louisiana, 1990, p. 253 [★537★]
$3,587, federal funds per capita in Iowa, 1990, p. 253 [★537★]
$3,592, average annual Religious (non-Catholic) secondary school tuition, 1988, p. 179 [★382★]
$3,599, health care expenditure per employee, companies with 1,000-2,499 employees, 1992, p. 286 [★603★]
$3,600, state and local generated revenue per capita, 1990, p. 278 [★586★]
$3,602, federal tax liability in Idaho 1990, p. 270 [★574★]
$3,603, expenditure per student in United States, 1987-1988, p. 175 [★376★]
$3,605, Rhode Island government debt per capita, 1990, p. 272 [★577★]
$3,610, expenditure per student in Denmark, 1987-1988, p. 175 [★376★]
$3,619, federal tax liability in West Virginia, 1990, p. 270 [★574★]
$3,628, health care expenditure per employee, companies with 500-999 employees, 1992, p. 286 [★603★]
$3,633, federal tax liability in New Mexico, 1990, p. 270 [★574★]

Numbers following p. are page references. Numbers in [ ] are table references.

Money - continued

Dollars

$3,643, Kentucky state and local appropriations per student for higher education, 1989, p. 147 [★329★]
$3,648, public school expenditure per student in Alabama, 1991, p. 178 [★380★]
$3,649, Wyoming government revenue per capita, 1990, p. 271 [★576★]
$3,650, Illinois state and local appropriations per student for higher education, 1989, p. 147 [★329★]
$3,659, federal tax liability in South Carolina, 1990, p. 270 [★574★]
$3,664, federal funds per capita in New Jersey, 1990, p. 253 [★537★]
$3,670, federal funds per capita in Kentucky, 1990, p. 253 [★537★]
$3,685, federal funds per capita in West Virginia, 1990, p. 253 [★537★]
$3,685, public school expenditure per student in North Dakota, 1991, p. 178 [★380★]
$3,698, Oklahoma state and local appropriations per student for higher education, 1989, p. 147 [★329★]
$3,701, federal funds per capita in Tennessee, 1990, p. 253 [★537★]
$3,707, public school expenditure per student in Tennessee, 1991, p. 178 [★380★]
$3,714, health care expenditure per employee, companies with 5,000-9,999 employees, 1992, p. 286 [★603★]
$3,722, Kansas state and local appropriations per student for higher education, 1989, p. 147 [★329★]
$3,728, federal tax liability in Maine, 1990, p. 270 [★574★]
$3,730, public school expenditure per student in South Dakota, 1991, p. 178 [★380★]
$3,742, public school expenditure per student in Oklahoma, 1991, p. 178 [★380★]
$3,744, city government taxes per capita in Washington, DC, 1990, p. 279 [★587★]
$3,753, federal funds per capita in Oklahoma, 1990, p. 253 [★537★]
$3,755, per capita athletic expenditures on men in 1987, All NCAA, p. 146 [★328★]
$3,760, federal tax liability in, Iowa 1990, p. 270 [★574★]
$3,767, Arkansas state and local appropriations per student for higher education, 1989, p. 147 [★329★]
$3,775, health care expenditure per employee, companies with 40,000-employees, 1992, p. 286 [★603★]
$3,779, federal funds per capita in Utah, 1990, p. 253 [★537★]
$3,790, South Carolina state and local appropriations per student for higher education, 1989, p. 147 [★329★]
$3,796, Oregon state and local appropriations per student for higher education, 1989, p. 147 [★329★]
$3,800, funeral-burial service package, 1989, p. 240 [★514★]
$3,823, federal funds per capita in Pennsylvania, 1990, p. 253 [★537★]
$3,847, health care expenditure per employee, companies with 10,000-19,000 employees, 1992, p. 286 [★603★]
$3,850, federal funds per capita in Kansas, 1990, p. 253 [★537★]
$3,855, federal tax liability in Kentucky, 1990, p. 270 [★574★]
$3,856, federal tax liability in Nebraska, 1990, p. 270 [★574★]
$3,862, farmer's gross average income per acre of tobacco, 1991, p. 387 [★829★]

Money - continued

Dollars

$3,862, federal funds per capita in Idaho, 1990, p. 253 [★537★]
$3,865, Alabama state and local appropriations per student for higher education, 1989, p. 147 [★329★]
$3,880, federal tax liability in Oklahoma, 1990, p. 270 [★574★]
$3,885, Virginia state and local appropriations per student for higher education, 1989, p. 147 [★329★]
$3,890, Delaware tuition revenues per student for higher education, 1989, p. 148 [★330★]
$3,891, federal funds per capita in California, 1990, p. 253 [★537★]
$3,894, Indiana state and local appropriations per student for higher education, 1989, p. 147 [★329★]
$3,912, federal funds per capita in Mississippi, 1990, p. 253 [★537★]
$3,918, federal funds per capita in New York, 1990, p. 253 [★537★]
$3,919, federal funds per capita in South Carolina, 1990, p. 253 [★537★]
$3,932, average federal income tax, 1985, p. 265 [★565★]
$3,941, average annual secular private school tuition, 1988, p. 179 [★382★]
$3,944, Arizona state and local appropriations per student for higher education, 1989, p. 147 [★329★]
$3,946, federal tax liability in Alabama, 1990, p. 270 [★574★]
$3,970, federal funds per capita in Florida, 1990, p. 253 [★537★]
$3,980, Michigan state and local appropriations per student for higher education, 1989, p. 147 [★329★]
$3,994, federal funds per capita, 1990, p. 253 [★537★]
$3,996, Pennsylvania state and local appropriations per student for higher education, 1989, p. 147 [★329★]
$3,998, federal tax liability in Wisconsin, 1990, p. 270 [★574★]
$4,000-6,000, average annual educational expenditure per student, p. 180 [★388★]
$4,003, health care expenditure per employee, companies with 20,000-39,000 employees, 1992, p. 286 [★603★]
$4,007, federal tax liability in Louisiana, 1990, p. 270 [★574★]
$4,011, federal funds per capita in Maine, 1990, p. 253 [★537★]
$4,011, Maryland state and local appropriations per student for higher education, 1989, p. 147 [★329★]
$4,012, public school expenditure per student in Louisiana, 1991, p. 178 [★380★]
$4,028, Wisconsin state and local appropriations per student for higher education, 1989, p. 147 [★329★]
$4,037, federal tax liability in North Carolina, 1990, p. 270 [★574★]
$4,045, federal tax liability in Vermont, 1990, p. 270 [★574★]
$4,066, individual health care costs per retiree, 2021, p. 288 [★607★]
$4,079, Utah state and local appropriations per student for higher education, 1989, p. 147 [★329★]
$4,089, federal funds per capita in Wyoming, 1990, p. 253 [★537★]
$4,100, Nevada state and local appropriations per student for higher education, 1989, p. 147 [★329★]
$4,112, federal funds per capita in Arizona, 1990, p. 253 [★537★]
$4,114, federal funds per capita in South Dakota, 1990, p. 253 [★537★]
$4,140, federal funds per capita in Washington, 1990, p. 253 [★537★]
$4,140, federal tax liability in Oregon, 1990, p. 270 [★574★]
$4,155, Iowa state and local appropriations per student for higher education, 1989, p. 147 [★329★]

Numbers following p. are page references. Numbers in [ ] are table references.

Money - continued

Dollars

$4,165, cost of hysterectomy, New York City, p. 329 [*702*]
$4,182, federal tax liability in Ohio, 1990, p. 270 [*574*]
$4,186, federal funds per capita in Montana, 1990, p. 253 [*537*]
$4,194, federal tax liability in Tennessee, 1990, p. 270 [*574*]
$4,207, expenditure per student in Canada, 1987-1988, p. 175 [*376*]
$4,210, Tennessee state and local appropriations per student for higher education, 1989, p. 147 [*329*]
$4,218, federal tax liability in Arizona, 1990, p. 270 [*574*]
$4,225, federal tax liability in Indiana, 1990, p. 270 [*574*]
$4,231, Minnesota state and local appropriations per student for higher education, 1989, p. 147 [*329*]
$4,231, public school expenditure per student in Arizona, 1991, p. 178 [*380*]
$4,235, Georgia state and local appropriations per student for higher education, 1989, p. 147 [*329*]
$4,238, public school expenditure per student in Texas, 1991, p. 178 [*380*]
$4,256, New York City, NY government expenditure per capita, 1990, p. 274 [*580*]
$4,260, per capita GNP in Portugal, p. 358 [*768*]
$4,270, cost of caesarean section, 1986, p. 330 [*703*]
$4,272, federal funds per capita in Alabama, 1990, p. 253 [*537*]
$4,278, San Antonio, TX government debt per capita, 1990, p. 275 [*581*]
$4,295, expenditure per student in Sweden, 1987-1988, p. 175 [*376*]
$4,296, family's share of family health care costs, 1991, p. 285 [*599*]
$4,300, average income tax deduction for state and local taxes, p. 266 [*566*]
$4,303, federal funds per capita in Rhode Island, 1990, p. 253 [*537*]
$4,306, federal tax liability in Missouri, 1990, p. 270 [*574*]
$4,307, Washington state and local appropriations per student for higher education, 1989, p. 147 [*329*]
$4,327, public school expenditure per student in South Carolina, 1991, p. 178 [*380*]
$4,330, annual cost of raising children (birth to 2 years), household income below 29,900, p. 199 [*427*]
$4,335, average annual veterans' benefit payment, 1990, p. 257 [*542*]
$4,360, New York City, NY government revenue per capita, 1990, p. 273 [*579*]
$4,361, average underpayment penalty (federal income tax), 1991, p. 266 [*569*]
$4,364, annual cost of owning and operating a mid-size automobile, 1991, p. 191 [*407*]
$4,368, expenditure per student in Norway, 1987-1988, p. 175 [*376*]
$4,381, federal tax liability in Minnesota 1990, p. 270 [*574*]
$4,381, public school expenditure per student in Nebraska, 1991, p. 178 [*380*]
$4,390, public school expenditure per student in Kentucky, 1991, p. 178 [*380*]
$4,402, Rhode Island state and local appropriations per student for higher education, 1989, p. 147 [*329*]

Money - continued

Dollars

$4,405, federal tax liability in Georgia, 1990, p. 270 [*574*]
$4,405, New Mexico state and local appropriations per student for higher education, 1989, p. 147 [*329*]
$4,415, public school expenditure per student in Missouri, 1991, p. 178 [*380*]
$4,428, federal funds per capita in Colorado, 1990, p. 253 [*537*]
$4,431, Idaho state and local appropriations per student for higher education, 1989, p. 147 [*329*]
$4,446, public school expenditure per student in New Mexico, 1991, p. 178 [*380*]
$4,471, Delaware government debt per capita, 1990, p. 272 [*577*]
$4,471, federal tax liability in Wyoming, 1990, p. 270 [*574*]
$4,483, federal tax liability in Colorado, 1990, p. 270 [*574*]
$4,484, federal funds per capita in Connecticut, 1990, p. 253 [*537*]
$4,492, federal tax liability in Kansas, 1990, p. 270 [*574*]
$4,500, annual U.S. spending per child, p. 281 [*592*]
$4,515, expenditure per student in Switzerland, 1987-1988, p. 175 [*376*]
$4,541, cost of normal delivery, 1991, p. 330 [*703*]
$4,544, federal tax liability in Rhode Island, 1990, p. 270 [*574*]
$4,555, federal funds per capita in North Dakota, 1990, p. 253 [*537*]
$4,555, federal tax liability in Alaska, 1990, p. 270 [*574*]
$4,560, lifetime energy costs of average air conditioning, Central, p. 201 [*431*]
$4,564, public school expenditure per student in Nevada, 1991, p. 178 [*380*]
$4,572, hospital care cost per stay, 1989, p. 306 [*651*]
$4,591, federal tax liability in Pennsylvania, 1990, p. 270 [*574*]
$4,626, annual cost of operating an automobile in Sioux Falls, SD, p. 192 [*411*]
$4,649, Delaware state and local appropriations per student for higher education, 1989, p. 147 [*329*]
$4,658, federal tax liability in Hawaii, 1990, p. 270 [*574*]
$4,705, average federal income tax, 1988, p. 265 [*565*]
$4,741, federal funds per capita in Missouri, 1990, p. 253 [*537*]
$4,800, average federal income tax 1989, p. 265 [*565*]
$4,802, public school expenditure per student in North Carolina, 1991, p. 178 [*380*]
$4,809, public school expenditure per student in Colorado, 1991, p. 178 [*380*]
$4,817, federal tax liability in Michigan, 1990, p. 270 [*574*]
$4,820, federal tax liability in United States, 1990, p. 270 [*574*]
$4,826, individual health care costs per retiree, 2031, p. 288 [*607*]
$4,826, public school expenditure per student in California, 1991, p. 178 [*380*]
$4,839, public school expenditure per student in Iowa, 1991, p. 178 [*380*]
$4,840, federal tax liability in Delaware, 1990, p. 270 [*574*]
$4,860, public school expenditure per student in Georgia, 1991, p. 178 [*380*]
$4,896, federal tax liability in Virginia, 1990, p. 270 [*574*]
$4,907, California state and local appropriations per student for higher education, 1989, p. 147 [*329*]

Numbers following p. are page references. Numbers in [ ] are table references.

Money - continued

Dollars

$4,907, federal tax liability in Texas, 1990, p. 270 [★574★]
$4,913, average student commuter expenses at two-year public college, 1991-1992, p. 145 [★326★]
$4,913, federal tax liability in Washington, 1990, p. 270 [★574★]
$4,927, federal funds per capita in Hawaii, 1990, p. 253 [★537★]
$4,944, tax per capita in United States, p. 262 [★556★]
$4,947, hospital care cost per stay, 1990, p. 305 [★648★]
$4,949, federal funds per capita in Massachusetts, 1990, p. 253 [★537★]
$4,975, public school expenditure per student, 1990, p. 177 [★379★]
$4,984, Florida state and local appropriations per student for higher education, 1989, p. 147 [★329★]
$5,000-7,700, average cost to delead a house, p. 418 [★898★]
$5,008, public school expenditure per student in Hawaii, 1991, p. 178 [★380★]
$5,009, public school expenditure per student in Kansas, 1991, p. 178 [★380★]
$5,032, loss per motor vehicle theft, 1990, p. 119 [★284★]
$5,045, public school expenditure per student in Washington, 1991, p. 178 [★380★]
$5,046, public school expenditure per student in West Virginia, 1991, p. 178 [★380★]
$5,051, public school expenditure per student in Indiana, 1991, p. 178 [★380★]
$5,053, federal tax liability in New Hampshire, 1990, p. 270 [★574★]
$5,062, public school expenditure per student in Illinois, 1991, p. 178 [★380★]
$5,082, North Carolina state and local appropriations per student for higher education, 1989, p. 147 [★329★]
$5,086, federal government expenditures per capita, 1990, p. 258 [★544★]
$5,099, median annual pensions (private firms) for men, 1990, p. 348 [★750★]
$5,121, federal tax liability in Florida, 1990, p. 270 [★574★]
$5,154, public school expenditure per student in Florida, 1991, p. 178 [★380★]
$5,167, Maine state and local appropriations per student for higher education, 1989, p. 147 [★329★]
$5,184, public school expenditure per student in Montana, 1991, p. 178 [★380★]
$5,239, federal tax liability in Nevada, 1990, p. 270 [★574★]
$5,255, public school expenditure per student in Wyoming, 1991, p. 178 [★380★]
$5,257, per capita athletic expenditures on men in 1987, All AIAW and NCAA Div. I, p. 146 [★328★]
$5,257, public school expenditure per student in Michigan, 1991, p. 178 [★380★]
$5,260, public school expenditure per student in Minnesota, 1991, p. 178 [★380★]
$5,261, public school expenditure per student in U.S., 1991, p. 178 [★380★]
$5,273, Massachusetts state and local appropriations per student for higher education, 1989, p. 147 [★329★]
$5,290, average tuition costs for two-year private college, 1991-1992, p. 145 [★326★]

Money - continued

Dollars

$5,291, public school expenditure per student in Oregon, 1991, p. 178 [★380★]
$5,302, tax per capita in Austria, p. 262 [★556★]
$5,309, federal tax liability in Maryland, 1990, p. 270 [★574★]
$5,340, per capita GNP in Greece, p. 358 [★768★]
$5,360, public school expenditure per student in Virginia, 1991, p. 178 [★380★]
$5,374, federal tax liability in Massachusetts, 1990, p. 270 [★574★]
$5,406, federal tax liability in Illinois 1990, p. 270 [★574★]
$5,483, tax per capita in Netherlands, p. 262 [★556★]
$5,491, Vermont tuition revenues per student for higher education, 1989, p. 148 [★330★]
$5,499, tax per capita in Finland, p. 262 [★556★]
$5,504, public school expenditure per student in New Hampshire, 1991, p. 178 [★380★]
$5,526, federal tax liability in California, 1990, p. 270 [★574★]
$5,548, Washington, DC government debt per capita, 1990, p. 275 [★581★]
$5,559, annual cost of owning and operating a full-size automobile, 1991, p. 191 [★407★]
$5,639, public school expenditure per student in Ohio, 1991, p. 178 [★380★]
$5,641, federal tax liability in District of Columbia, 1990, p. 270 [★574★]
$5,648, New Jersey state and local appropriations per student for higher education, 1989, p. 147 [★329★]
$5,671, federal funds per capita in Maryland, 1990, p. 253 [★537★]
$5,703, federal funds per capita in New Mexico, 1990, p. 253 [★537★]
$5,730, federal tax liability in New York, 1990, p. 270 [★574★]
$5,740, public school expenditure per student in Vermont, 1991, p. 178 [★380★]
$5,802, tax per capita in France, p. 262 [★556★]
$5,867, federal funds per capita in Alaska, 1990, p. 253 [★537★]
$5,874, federal funds per capita in Virginia, 1990, p. 253 [★537★]
$5,894, public school expenditure per student in Maine, 1991, p. 178 [★380★]
$5,934, New York state and local appropriations per student for higher education, 1989, p. 147 [★329★]
$5,946, public school expenditure per student in Wisconsin, 1991, p. 178 [★380★]
$5,947, median annual pensions (government) for women, 1990, p. 348 [★750★]
$6,016, public school expenditure per student in Delaware, 1991, p. 178 [★380★]
$6,028, Wyoming state and local appropriations per student for higher education, 1989, p. 147 [★329★]
$6,036, average commuter student expenses at four-year public college, 1991-1992, p. 145 [★325★]
$6,100, added annual paycheck value of trade union membership, 1990, p. 413 [★884★]
$6,162, federal tax liability in New Jersey, 1990, p. 270 [★574★]
$6,184, public school expenditure per student in Maryland, 1991, p. 178 [★380★]
$6,345, Connecticut state and local appropriations per student for higher education, 1989, p. 147 [★329★]
$6,351, public school expenditure per student in Massachusetts, 1991, p. 178 [★380★]

Numbers following p. are page references. Numbers in [ ] are table references.

Money - continued

Dollars

$6,391, average annual secular private secondary school tuition, 1988, p. 179 [★382★]
$6,534, public school expenditure per student in Pennsylvania, 1991, p. 178 [★380★]
$6,535, family health care costs, 1991, p. 285 [★599★]
$6,540, tax per capita in Germany, p. 262 [★556★]
$6,600, average annual value of married men's housework, p. 401 [★858★]
$6,612, Washington, DC government revenue per capita, 1990, p. 273 [★579★]
$6,707, tax per capita in Switzerland, p. 262 [★556★]
$6,713, Washington, DC government expenditure per capita, 1990, p. 274 [★580★]
$6,756, federal tax liability in Connecticut, 1990, p. 270 [★574★]
$6,768, Jacksonville, FL government debt per capita, 1990, p. 275 [★581★]
$6,989, public school expenditure per student in Rhode Island, 1991, p. 178 [★380★]
$7,064, sales per capita (all industries), 1989, p. 58 [★136★]
$7,090, average interest deduction (federal income tax), 1991, p. 266 [★566★]
$7,103, saved per employee suggestion, 1990, p. 50 [★119★]
★$7,106, average per-employee spending on R^D, 1992, p. 53 [★126★]
$7,177, average household welfare payments to immigrants, late 1950s, p. 258 [★545★]
$7,200, average cost per student of private school, p. 179 [★385★]
$7,229, Hawaii state and local appropriations per student for higher education, 1989, p. 147 [★329★]
$7,280, cost of open-heart surgery, p. 292 [★619★]
$7,529, annual cost of operating an automobile in Los Angeles, p. 192 [★411★]
$7,584, average residential student expenses at four-year public college, 1991-1992, p. 145 [★325★]
$7,592, tax per capita 1990, p. 263 [★559★]
$7,695, District of Columbia state and local appropriations per student for higher education, 1989, p. 147 [★329★]
$7,790, Alaska government expenditure per capita, 1990, p. 273 [★578★]
$7,826, cost of caesarean section, 1991, p. 330 [★703★]
$7,887, public school expenditure per student in Alaska, 1991, p. 178 [★380★]
$7,907, average household welfare payments, late 1950s, p. 258 [★545★]
$7,914, public school expenditure per student in Connecticut, 1991, p. 178 [★380★]
$7,928, median income of nonmarried elderly person, 1988, p. 411 [★881★]
$8,000, annual health care expenditure per household, p. 285 [★601★]
$8,000, average college loan debt, 1985, p. 144 [★324★]
$8,000-12,000, average prostate operation, p. 293 [★622★]
$8,151, tax per capita in Denmark, p. 262 [★556★]
$8,210, public school expenditure per student in District of Columbia, 1991, p. 178 [★380★]
$8,346, tax per capita in Norway, p. 262 [★556★]
$8,385, tax per capita in Sweden, p. 262 [★556★]

Money - continued

Dollars

$8,500, public school expenditure per student in New York, 1991, p. 178 [★380★]
$8,770, annual cost of raising children (birth to 2 years), household income above 48,300, p. 199 [★427★]
$8,779, annual per capita money income in San Antonio, TX, 1987, p. 353 [★760★]
$8,804, Alaska government revenue per capita, 1990, p. 271 [★576★]
$8,910, household automobile insurance payments over past ten years, p. 237 [★506★]
$9,006, average commuter student expenses at two-year private college, 1991-1992, p. 145 [★326★]
$9,017, Black per capita income, 1990, p. 398 [★853★]
$9,159, public school expenditure per student in New Jersey, 1991, p. 178 [★380★]
$9,662, annual per capita money income in Detroit, MI, 1987, p. 353 [★760★]
$9,879, Alaska state and local appropriations per student for higher education, 1989, p. 147 [★329★]
$9,989, annual per capita money income in Baltimore, MD, 1987, p. 353 [★760★]
$10,000, average annual value of married women's housework, p. 401 [★858★]
$10,000, average increase in CEO's weekly compensation, p. 406 [★868★]
$10,000-14,000, average annual educational expenditure per handicapped student, p. 180 [★388★]
$10,002, annual per capita money income in Philadelphia, PA, 1987, p. 353 [★760★]
$10,017, average tuition costs for four-year private college, 1991-1992, p. 145 [★325★]
$10,065, Alaska government debt per capita, 1990, p. 272 [★577★]
$10,251, cleaning woman's average annual salary, 1989, p. 385 [★825★]
$10,347, annual per capita money income in Memphis, TN, 1987, p. 353 [★760★]
$10,379, average salary for 25-year-old high school dropout, 1988, p. 360 [★773★]
$10,593, annual per capita money income in Milwaukee, WI, 1987, p. 353 [★760★]
$10,806, annual per capita money income in Chicago, IL, 1987, p. 353 [★760★]
$10,914, average residential student expenses at two-year private college, 1991-1992, p. 145 [★326★]
$10,942, hospital revenue from cardiovascular admission, 1989, p. 290 [★613★]
$11,300, annual U.S. spending per elderly person, p. 281 [★592★]
$11,400, per capita disposable income, 1991, p. 404 [★864★]
$11,500, average winnings per winner per show on *Jeopardy!*, p. 431 [★931★]
$11,500, preschool teacher's average annual salary, p. 368 [★787★]
$11,514, annual per capita money income in Jacksonville, FL, 1987, p. 353 [★760★]
$11,650 to 13,350, lifetime cost of owning a dog, p. 210 [★454★]
$11,684, median annual pensions (government) for men, 1990, p. 348 [★750★]

Numbers following p. are page references. Numbers in [ ] are table references.

Money - continued

Dollars

$11,777, average salary for 25-year-old G.E.D. holder, 1988, p. 360 [*773*]
$11,923, average annual per capita money income, 1987, p. 351 [*757*]
$11,980, loss to arson, 1990, p. 120 [*285*]
$12,007, annual per capita money income in Houston, TX, 1987, p. 353 [*760*]
$12,111, annual per capita money income in Indianapolis, IN, 1987, p. 353 [*760*]
$12,375, annual per capita money income in Phoenix, AZ, 1987, p. 353 [*760*]
$12,411, average salary for 28-year-old high school dropout or G.E.D. holder, 1988, p. 360 [*773*]
$12,734, annual per capita money income in Honolulu, HI, 1987, p. 353 [*760*]
$12,926, annual per capita money income in New York, NY, 1987, p. 353 [*760*]
$12,978, annual per capita money income in San Diego, CA, 1987, p. 353 [*760*]
$12,984, annual per capita money income in Boston, MA, 1987, p. 353 [*760*]
$13,000, average annual cost of five-day boarding school, p. 179 [*384*]
$13,000, national debt per capita, p. 259 [*547*]
$13,092, median family income, single mothers, 1993, p. 411 [*880*]
$13,370, wedding and reception cost, 1988, p. 239 [*510*]
$13,489, annual per capita money income in Dallas, TX, 1987, p. 353 [*760*]
$13,552, average household welfare payments to immigrants, 1975-1980, p. 258 [*545*]
$13,592, annual per capita money income in Los Angeles, CA, 1987, p. 353 [*760*]
$13,700, average annual cost per student of boarding school, p. 179 [*385*]
$13,711, annual per capita money income in San Jose, CA, 1987, p. 353 [*760*]
$13,983, average commuter student expenses at four-year private college, 1991-1992, p. 145 [*325*]
$14,000, average college loan debt, 1990, p. 144 [*324*]
$14,570, per capita, GNP in Great Britain, p. 358 [*768*]
$14,718, retail sales per household in West Virginia, 1990, p. 55 [*128*]
$14,778, annual per capita money income in Washington, DC, 1987, p. 353 [*760*]
$15,000, hospitalization for tuberculosis patient, 1990, p. 291 [*616*]
$15,000, weekly funds needed for Senatorial reelection campaign, p. 40 [*92*]
$15,131, retail sales per household in Mississippi, 1990, p. 55 [*128*]
$15,137, annual per capita money income in San Francisco, CA, 1987, p. 353 [*760*]
$15,150 per capita in Italy, p. 358 [*768*]
$15,213, average salary for 25-year-old high school graduate, 1988, p. 360 [*773*]
$15,265, White per capita income, 1990, p. 398 [*853*]

Money - continued

Dollars

$15,360, retail sales per household in District of Columbia, 1990, p. 55 [*128*]
$16,000, average annual cost of seven-day boarding school, p. 179 [*384*]
$16,010, per capita GNP in Netherlands, p. 358 [*768*]
$16,012, new car cost, 1990, p. 189 [*404*]
$16,292, average residential student expenses at four-year private college, 1991-1992, p. 145 [*325*]
$16,314, retail sales per household in Wyoming, 1990, p. 55 [*128*]
$16,591, retail sales per household in Idaho, 1990, p. 55 [*128*]
$16,780, retail sales per household in Oklahoma, 1990, p. 55 [*128*]
$16,932, median female householder family income, 1990, p. 399 [*855*]
$17,000, average annual British school tuition and boarding expenses, p. 179 [*383*]
$17,044, retail sales per household in New Mexico, 1990, p. 55 [*128*]
$17,113, retail sales per household in Nebraska, 1990, p. 55 [*128*]
$17,205, retail sales per household in Arkansas, 1990, p. 55 [*128*]
$17,283, retail sales per household in Kentucky, 1990, p. 55 [*128*]
$17,360, per capita GNP in Austria, p. 358 [*768*]
$17,384, retail sales per household in Tennessee, 1990, p. 55 [*128*]
$17,420, sheriff's deputy's average starting salary, 1990, p. 382 [*817*]
$17,435, retail sales per household in Alabama, 1990, p. 55 [*128*]
$17,478, retail sales per household in Montana, 1990, p. 55 [*128*]
$17,540, retail sales per household in Kansas, 1990, p. 55 [*128*]
$17,724, retail sales per household in Iowa, 1990, p. 55 [*128*]
$17,830, per capita GNP in France, p. 358 [*768*]
$17,880, retail sales per household in Ohio, 1990, p. 55 [*128*]
$17,921, retail sales per household in South Dakota, 1990, p. 55 [*128*]
$17,995, retail sales per household in North Carolina, 1990, p. 55 [*128*]
$18,006, retail sales per household in Cleveland-Akron-Lorain, OH, 1990, p. 54 [*127*]
$18,083, median Black household income, 1989, p. 394 [*847*]
$18,153, retail sales per household in Indiana, 1990, p. 55 [*128*]
$18,281, retail sales per household in Missouri, 1990, p. 55 [*128*]
$18,366, per capita GNP in Germany, p. 358 [*768*]
$18,413, retail sales per household in Pennsylvania, 1990, p. 55 [*128*]
$18,630, retail sales per household in North Dakota, 1990, p. 55 [*128*]
$18,660, school custodian's median annual salary, 1992, p. 368 [*786*]
$18,667, retail sales per household in New York, 1990, p. 55 [*128*]

Money - continued

Dollars

$18,676, median Black household income, 1992, p. 396 [★850★]
$18,685, retail sales per household in South Carolina, 1990, p. 55 [★128★]
$18,703, retail sales per household in Arizona, 1990, p. 55 [★128★]
$18,864, retail sales per household in Colorado, 1990, p. 55 [★128★]
$18,910, police officer's average starting salary, 1990, p. 381 [★816★]
$19,092, average annual per capita income, 1991, p. 350 [★756★]
$19,230, annual expenses of one-parent families with children under 18, 1990., p. 218 [★470★]
$19,260, retail sales per household in Louisiana, 1990, p. 55 [★128★]
$19,276, retail sales per household in Rhode Island, 1990, p. 55 [★128★]
$19,344, wedding and reception cost, 1990, p. 239 [★511★]
$19,370, retail sales per household in Washington, 1990, p. 55 [★128★]
$19,417, retail sales per household in Georgia, 1990, p. 55 [★128★]
$19,488, retail sales per household in United States, 1990, p. 55 [★128★]
$19,501, retail sales per household in Utah, 1990, p. 55 [★128★]
$19,612, median per capita income of women, 1989, p. 402 [★859★]
$19,631, retail sales per household in Texas, 1990, p. 55 [★128★]
$19,691, retail sales per household in Wisconsin, 1990, p. 55 [★128★]
$19,765, retail sales per household in Michigan, 1990, p. 55 [★128★]
$19,800, price of single section mobile home, p. 223 [★482★]
$19,840, retail sales per household in Illinois, 1990, p. 55 [★128★]
$19,917, median family income, 1979, p. 351 [★758★]
$19,938, retail sales per household in Nevada, 1990, p. 55 [★128★]
$19,940, retail sales per household in Houston-Galveston-Brazoria, TX, 1990, p. 54 [★127★]
$20,000, annual cost of maintaining prison inmate, p. 120 [★286★]
$20,000, funeral service cost, Japan, p. 240 [★513★]
$20,025, average Native American household income, p. 394 [★845★]
$20,028, retail sales per household in NY-NJ-CT, 1990, p. 54 [★127★]
$20,030, starting salary for Retailing graduate, 1992, p. 361 [★774★]
$20,050, retail sales per household in Minnesota, 1990, p. 55 [★128★]
$20,056, retail sales per household in Florida, 1990, p. 55 [★128★]
$20,159, retail sales per household in Oregon, 1990, p. 55 [★128★]
$20,195, average salary for 25-year-old high school graduate, 1988, p. 360 [★773★]
$20,459, retail sales per household in Virginia, 1990, p. 55 [★128★]
$20,609, retail sales per household in PA-NJ-DE-MD, 1990, p. 54 [★127★]
$20,701, retail sales per household in Detroit-Ann Arbor, MI, 1990, p. 54 [★127★]
$20,790, retail sales per household in Maryland, 1990, p. 55 [★128★]
$21,045, retail sales per household in Chicago-Gary-Lake County, IL-IN-WI, 1990, p. 54 [★127★]

Money - continued

Dollars

$21,100, per capita GNP in United States, p. 358 [★768★]
$21,194, starting salary for Liberal Arts graduate, 1992, p. 361 [★774★]
$21,223, retail sales per household in Vermont, 1990, p. 55 [★128★]
$21,245, retail sales per household in California, 1990, p. 55 [★128★]
$21,340, retail sales per household in Seattle-Tacoma, WA, 1990, p. 54 [★127★]
$21,423, median Black family income, 1990, p. 399 [★855★]
$21,534, retail sales per household in Minneapolis-St. Paul, MN-WI, 1990, p. 54 [★127★]
$21,710, per capita GNP in Sweden, p. 358 [★768★]
$21,793, retail sales per household in Dallas-Fort Worth, TX, 1990, p. 54 [★127★]
$21,870, sheriff's department sergeant's average starting salary, 1990, p. 382 [★817★]
$21,918, retail sales per household in Miami-Fort Lauderdale, FL, 1990, p. 54 [★127★]
$21,921, median Hispanic household income, 1989, p. 394 [★847★]
$21,946, retail sales per household in San Francisco-Oakland-San Jose, CA, 1990, p. 54 [★127★]
$22,000, average amount spent in the U.S. by foreign visitors every minute, p. 444 [★964★]
$22,149, retail sales per household in Maine, 1990, p. 55 [★128★]
$22,194, starting salary for Advertising graduate, 1992, p. 361 [★774★]
$22,207, retail sales per household in Washington, DC-MD-VA, 1990, p. 54 [★127★]
$22,233, retail sales per household in Atlanta, GA, 1990, p. 54 [★127★]
$22,285, retail sales per household in Los Angeles-Anaheim-Riverside, CA, 1990, p. 54 [★127★]
$22,435, retail sales per household in Connecticut, 1990, p. 55 [★128★]
$22,488, retail sales per household in Massachusetts, 1990, p. 55 [★128★]
$22,579, retail sales per household in New Jersey, 1990, p. 55 [★128★]
$22,936, retail sales per household in Brockton, MA, 1990, p. 54 [★127★]
$23,401, starting salary for Education graduate, 1992, p. 361 [★774★]
$23,431, median Hispanic family income, 1990, p. 399 [★855★]
$23,558, retail sales per household in Burlington, VT, 1990, p. 54 [★127★]
$23,602, average annual pay, 1990, p. 352 [★759★]
$23,730, per capita GNP in Japan, p. 358 [★768★]
$23,897, retail sales per household in Orlando, FL, 1990, p. 54 [★127★]
$24,000, cost per worker's compensation claim, 1992, p. 43 [★102★]
$24,000, school maintenance staff's median annual salary, 1992, p. 368 [★786★]
$24,000, Shinto wedding cost, Japan, p. 240 [★512★]
$24,145, retail sales per household in Delaware, 1990, p. 55 [★128★]

Numbers following p. are page references. Numbers in [ ] are table references.

Money - continued

Dollars

$24,411, average annual blue-collar worker's pay, 1992, p. 406 [★869★]
$24,533, retail sales per household in Alaska, 1990, p. 55 [★128★]
$24,808, retail sales per household in Bangor, ME, 1990, p. 54 [★127★]
$24,934, retail sales per household in Rochester, MN, 1990, p. 54 [★127★]
$24,982, retail sales per household in Atlantic City, NJ, 1990, p. 54 [★127★]
$25,139, retail sales per household in Grand Forks, ND, 1990, p. 54 [★127★]
$25,211, median family income, single fathers, 1993, p. 411 [★880★]
$25,382, starting salary for Financial administration graduate, 1992, p. 361 [★774★]
$25,420, police sergeant's average starting salary, 1990, p. 381 [★816★]
$25,919, annual household expenses in the Midwest, 1990, p. 216 [★467★]
$26,151, retail sales per household in Danbury, CT, 1990, p. 54 [★127★]
$26,777, retail sales per household in Rapid City, SD, 1990, p. 54 [★127★]
$27,011, annual household expenses in the South, 1990, p. 216 [★467★]
$27,800, sales price of new mobile homes, 1990, p. 232 [★497★]
$28,000, temple lot for cremated remains, Japan, p. 240 [★513★]
$28,369, annual household expenditures, 1990, p. 215 [★465★]
$28,369, annual household expenditures, 1990, p. 215 [★466★]
$28,373, retail sales per household in New Hampshire, 1990, p. 55 [★128★]
$28,502, retail sales per household in Manchester-Nashua, NH, 1990, p. 54 [★127★]
$28,511, median income of men, 1989, p. 402 [★859★]
$28,592, federal funds per capita in District of Columbia, 1990, p. 253 [★537★]
$28,776, starting salary for geology graduate, 1992, p. 361 [★774★]
$28,906, median household income, 1989, p. 394 [★847★]
$29,032, retail sales per household in NH, 1990, p. 54 [★127★]
$29,159, starting salary for nursing graduate, 1992, p. 361 [★774★]
$29,489, annual Northeast household expenses, 1990, p. 216 [★467★]
$29,600, annual per-resident operating costs of juvenile facilities, 1988, p. 122 [★289★]
$29,663, retail sales per household in Terre Haute, IN, 1990, p. 54 [★127★]
$29,943, median annual household income, 1990, p. 355 [★763★]
$29,963, retail sales per household in Anchorage, AK, 1990, p. 54 [★127★]
$30,000, compensation for average truck driver, p. 393 [★844★]
$30,000, coronary bypass surgery operation, p. 290 [★613★]
$30,126, median household income, 1992, p. 411 [★879★]
$30,240, chief of police's average starting salary, 1990, p. 381 [★816★]
$30,270, per capita GNP in Switzerland, p. 358 [★768★]
$30,406, median White household income, 1989, p. 394 [★847★]
$31,000, annual nursing home charges, 1990, p. 223 [★481★]

Money - continued

Dollars

$31,121, retail sales per household in Hawaii, 1990, p. 55 [★128★]
$31,231, median White household income, 1992, p. 396 [★850★]
$31,300, costs per resident of private juvenile facilities, 1989, p. 122 [★290★]
$31,340, retail sales per household in Honolulu, HI, 1990, p. 54 [★127★]
$31,509, annual expenses of husband-and-wife households, 1990, p. 217 [★469★]
$32,000, suburban Blacks' average annual family income, 1990, p. 394 [★846★]
$32,445, annual household expenses in the Western United States, 1990, p. 216 [★467★]
$32,650, retail sales per household in Portland, ME, 1990, p. 54 [★127★]
$33,000, median household income of readers of women's magazines, p. 107 [★262★]
$33,100, annual after-tax expenses of two-earner families with two dependent children, 1990, p. 218 [★471★]
$33,200, annual household income of average motorcyclist, p. 105 [★254★]
$33,530, sheriff's average starting salary, 1990, p. 382 [★817★]
$34,098, teacher's average salary, 1992, p. 406 [★869★]
$34,413, teacher's average salary, 1991-1992, p. 365 [★780★]
$34,434, female librarian's median salary, 1992, p. 373 [★797★]
$34,886, man's mean money earnings, 1992, p. 360 [★772★]
$35,000, annual household income of average sewing enthusiast, p. 104 [★253★]
$35,000, casino gambler's median household income, p. 106 [★258★]
$35,000, male librarian's median salary, 1992, p. 373 [★797★]
$35,000, retirement savings of average baby boomer, p. 412 [★883★]
$35,081, gardener's median household income, p. 104 [★251★]
$35,353, median family income, 1990, p. 399 [★855★]
$36,000, price of double section mobile home, p. 223 [★482★]
$36,072, median household income of lesbians, p. 107 [★261★]
$36,640, retail sales per household in St. Cloud, MN, 1990, p. 54 [★127★]
$36,915, median White family income, 1990, p. 399 [★855★]
$37,100, productivity per worker in Great Britain, p. 49 [★118★]
$38,200, productivity per worker in Japan, p. 49 [★118★]
$39,648, city chief of police median salary, 1991, p. 382 [★818★]
$39,895, median married-couple family income, 1990, p. 399 [★855★]
$40,000, business travelers' median income, p. 106 [★259★]
$40,000, cost of hiring new employee, p. 41 [★95★]
$40,000, knee-replacement surgery, 1992, p. 292 [★618★]
$41,260, median family income, couples with children, 1993, p. 411 [★880★]
$42,300, annual local police department operating expenses per employee, 1990, p. 136 [★311★]
$42,374, debt for students completing medical school, 1989, p. 237 [★505★]
$43,100, annual local sheriffs' department operating expenses per employee, 1990, p. 138 [★314★]
$44,200, productivity per worker in Germany, p. 49 [★118★]
$46,689, median household income of gay men, p. 107 [★261★]

Numbers following p. are page references. Numbers in [ ] are table references.

Money - continued

Dollars

$47,000, productivity per worker in France, p. 49 [★118★]
$49,600, productivity per worker in United States, p. 49 [★118★]
$50,000-70,000, annual income of average franchisee, p. 105 [★255★]
$51,000, condominium timesharers' median household income, p. 105 [★257★]
$52,000, breast cancer treatment, p. 292 [★620★]
$52,000, national debt per family, p. 259 [★547★]
$52,900, female professor's average salary, 1992, p. 372 [★794★]
$55,000, price of house, 1983, p. 223 [★483★]
$56,000, compensation for average unionized railroad employee, p. 393 [★844★]
$58,240, average annual engineer's pay, 1992, p. 406 [★869★]
$58,637, median entrance fee to continuing-care retirement communities, p. 222 [★480★]
$59,240, male professor's average salary, 1992, p. 372 [★794★]
$69,000, annual health care cost per elderly prison inmate, p. 121 [★288★]
$74,000, median sales price of existing houses in the Midwest, 1990, p. 231 [★496★]
$79,100, median United States home value, 1990, p. 229 [★493★]
$79,874, school superintendent's average annual salary, 1991, p. 366 [★781★]
$80,677, college (2-year) president's average annual salary, 1991, p. 369 [★788★]
$85,900, median sales price of existing houses in the South, 1990, p. 231 [★496★]
$86,206, college athletic director's average annual income, p. 369 [★789★]
$91,570, heart transplant operation, 1988, p. 290 [★614★]
$92,000, woman executive's average total compensation, 1982, p. 407 [★873★]
$95,500, median sales price of existing houses, 1990, p. 231 [★496★]
$95,900, general/family practice physician's mean annual income, 1989, p. 375 [★802★]
$96,703, college (4-year) president's average annual salary, 1991, p. 369 [★788★]
$99,000, median purchase price of new houses in the South, 1990, p. 230 [★495★]
$100,000, AIDS patient care and treatment, p. 289 [★612★]
$100,000, price of house, 1993, p. 223 [★483★]
$102,000, AIDS patient care and treatment, p. 292 [★620★]
$104,700, pediatrics physician's mean annual income, 1989, p. 375 [★802★]
$105,200, price of house for first-time buyer, 1989, p. 226 [★488★]
$107,900, median purchase price of new houses in the Midwest, 1990, p. 230 [★495★]
$110,500, workers compensation for on-the-job motor vehicle related fatality, p. 43 [★101★]
$111,000, family practice physician's mean annual salary, 1991, p. 376 [★803★]
$119,579, baseball player's series share, 1991, p. 377 [★807★]
$126,000, life insurance per household, 1991, p. 238 [★507★]
$128,000, psychiatry physician's mean annual salary, 1991, p. 376 [★803★]
$128,670, cost of raising a child from birth to 18, p. 199 [★428★]
$139,600, median sales price of existing houses in the West, 1990, p. 231 [★496★]
$139,732, HMO or hospital physician's average compensation, 1992, p. 376 [★804★]
$141,200, median sales price of existing houses in the Northeast, 1990, p. 231 [★496★]
$144,700, price of house for repeat buyer, 1989, p. 226 [★488★]
$145,795, liver transplant operation, 1988, p. 290 [★615★]
$146,500, internal medicine physician's mean annual income, 1989, p. 375 [★802★]
$147,500, median purchase price of new houses in the West, 1990, p. 230 [★495★]
$149,800, price of site-built home, p. 223 [★482★]
$150,000, arthritis patient care and treatment, p. 289 [★611★]
$150,000, internal medicine physician's mean annual salary, 1991, p. 376 [★803★]
$154,800, average spending in House of Representatives on office expenses, 1991, p. 259 [★546★]
$155,000, median salary of foundation CEO, 1991, p. 406 [★870★]
$155,800, physician's mean annual income, 1989, p. 375 [★802★]
$158,000, average estate tax per return, 1992, p. 280 [★590★]
$159,000, median purchase price of new houses in the Northeast, 1990, p. 230 [★495★]
$170,000, physician's mean annual salary, 1991, p. 376 [★803★]
$176,900, average House of Representatives allowance for office expenses, 1991, p. 259 [★546★]
$185,000, kidney dialysis course of treatment, p. 292 [★620★]
$185,600, self-employed physician's net income, 1992, p. 376 [★804★]
$187,000, woman executive's average total compensation, 1992, p. 407 [★873★]
$192,200, sales per employee (all industries), 1992, p. 58 [★135★]
$194,300, obstetrics/gynecology physician's mean annual income, 1989, p. 375 [★802★]
$198,000, pathology physician's mean annual salary, 1991, p. 376 [★803★]
$200,000, average annual compensation of legal executives, p. 407 [★871★]
$200,000, average House of Representatives allowance for mailing expenses, 1991, p. 259 [★546★]
$200,000, average salary of hospital CEOs, 1991, p. 407 [★872★]
$206,543, average weekly supermarket sales, 1990, p. 55 [★129★]
$220,500, surgeon's mean annual income, 1989, p. 375 [★802★]
$221,000, anesthesiologist's mean annual salary, 1991, p. 376 [★803★]
$222,000, gynecologist's mean annual salary, 1991, p. 376 [★803★]
$230,000, radiologist's mean annual salary, 1991, p. 376 [★803★]
$234,000, surgeon's mean annual salary, 1991, p. 376 [★803★]
$235,000, hospital executive's average annual income, p. 376 [★805★]
$245,300, median Hawaii home value, 1990, p. 229 [★493★]
$250,000, average award for childhood vaccination fatalities, p. 143 [★322★]
$254,000, hockey player's average salary, 1990, p. 377 [★806★]
$350,000, football player's average salary, 1990, p. 377 [★806★]
$477,000, average bonus for baseball first round draft picks, 1992, p. 377 [★808★]

Numbers following p. are page references. Numbers in [ ] are table references.

Money - continued

Dollars

$525,000, average annual compensation of top CEOs in Japan, p. 405 [★865★]
$589,000, baseball player's average salary, 1990, p. 377 [★806★]
$800,000, average annual compensation of top CEOs in France, p. 405 [★865★]
$800,000, average annual compensation of top CEOs in Germany, p. 405 [★865★]
$817,000, basketball player's average salary, 1990, p. 377 [★806★]
$878,000, CEO's average salary and bonuses, p. 105 [★256★]
$891,909, baseball player's average salary, 1991, p. 377 [★807★]
$1,100,000, average annual compensation of top CEOs in Great Britain, p. 405 [★865★]
$1,100,000, average award for childhood vaccination neurological complications, p. 143 [★322★]
$1,200,000, tax money spent at Library of Congress on an average day, 1991, p. 160 [★353★]
$1,500,000, average publishing and media libel award, 1980s, p. 143 [★321★]
$1,675,000, annual local police department operating expenses, 1990, p. 138 [★313★]
$1,710,000, average lawsuit payment against corporate directors, 1989, p. 143 [★320★]
$2,500,000, million average annual compensation of CEOs of blue-chip companies, p. 406 [★867★]
$3,200,000, average annual compensation of top CEOs in United States, p. 405 [★865★]
$3,360,000, average lawsuit payment against corporate directors, 1992, p. 143 [★320★]
$3,842,247, average annual CEO's pay, 1992, p. 406 [★869★]
$4,000,000, annual cost to U.S. Postal Service of "dead letters", p. 43 [★103★]
$4,000,000, funds needed for Senatorial reelection campaign, p. 40 [★92★]
$9,000,000, public hospital costs not met by insurance or government subsidies, 1988, p. 309 [★658★]
$9,000,000, average publishing and media libel award, 1990s, p. 143 [★321★]
$10.0 million, average movie production costs, 1980, p. 46 [★111★]
$10.3 million, average net worth of America's wealthiest families, 1989, p. 98 [★234★]
$12 million, average movie advertising and promotion costs, 1992, p. 46 [★112★]
$26.8 million, average movie production costs, 1990, p. 46 [★111★]
$40 million, public hospital costs not met by private or public insurance, 1988, p. 309 [★658★]
$50 million, cost of preparation for a hurricane in the United States, p. 499 [★1080★]
$100 million, annual American property loss to lightning, p. 503 [★1095★]
$175,000,000, average daily cost of U.S. air combat in Persian Gulf War, p. 261 [★553★]
$200 million, annual business losses due to poor handwriting, p. 43 [★103★]
$230 million, development cost of new drug, p. 315 [★674★]
$500,000,000, average cost per day of U.S. combined air/ground combat in Persian Gulf War, p. 261 [★553★]

Money - continued

Dollars

$800 million, average cost of traffic congestion in Seattle-Everett, 1988, p. 446 [★971★]
$960 million, average cost of traffic congestion in Dallas, 1988, p. 446 [★971★]
$1,040 million, average cost of traffic congestion in Miami, 1988, p. 446 [★971★]
$1,280 million, average cost of traffic congestion in Boston, 1988, p. 446 [★971★]
$1,470 million, average cost of traffic congestion in Houston, 1988, p. 446 [★971★]
$1,510 million, average cost of traffic congestion in Detroit, 1988, p. 446 [★971★]
$1,550 million, average cost of traffic congestion in Philadelphia, 1988, p. 446 [★971★]
$1,730 million, average cost of traffic congestion in Washington, 1988, p. 446 [★971★]
$1,880 million, average cost of traffic congestion in Chicago, 1988, p. 446 [★971★]
$2 billion, average annual spending on classified help-wanted ads, p. 27 [★63★]
$2,340 million, average cost of traffic congestion in San Francisco-Oakland, 1988, p. 446 [★971★]
$3-4 billion, average annual American expenditure on benign prostate surgery, p. 293 [★622★]
$4 billion, value of employee time spent recovering "lost" data on PCs, p. 42 [★100★]
$6 billion, federally subsidized student loans, per year, p. 144 [★323★]
$6 billion, Los Angeles average annual cost of traffic congestion, p. 44 [★104★]
$6,040 million, average cost of traffic congestion in New York, 1988, p. 446 [★971★]
$6,880 million, average cost of traffic congestion in Los Angeles, 1988, p. 446 [★971★]
$13.7 billion, Americans' average yearly cost for alternative medicine treatments, 1990, p. 330 [★705★]
$16.4 billion, value of employee time lost annually to motor vehicle accidents, p. 43 [★101★]
$177 billion, annual costs of substance abuse, p. 293 [★621★]

Percentages

0.03, carbon dioxide in air breathed in, p. 341 [★734★]
0.35, potassium, sulphur, sodium, chlorine, magnesium in chemical composition of human body, p. 335 [★717★]
1, phosphorus in chemical composition of human body, p. 335 [★717★]
1-5, new HIV infections in developing countries due to tainted blood transfusions, p. 288 [★608★]
1.5, calcium in chemical composition of human body, p. 335 [★717★]
4.1, average family automobile usage devoted to civic affairs, p. 450 [★981★]
4.6, unemployment rate under Democratic administrations since WWII, p. 61 [★143★]
4.8, full-time workers absent per week, p. 42 [★99★]
5.1, average annual return on stocks, bonds, and real estate under Reagan's 1st term, p. 35 [★82★]

Percentages - continued

5.6, carbon dioxide in air breathed out, p. 341 [★734★]
5.7, households with VCR, 1982, p. 430 [★928★]
5.9, average annual return on stocks, 1970s, p. 35 [★83★]
6.2, unemployment rate under Republican administrations since WWII, p. 61 [★143★]
6.5, average annual return on stocks, bonds, and real estate under Bush, p. 35 [★82★]
6.6, average annual return on stocks, bonds, and real estate under Kennedy/Johnson, p. 35 [★82★]
6.9, average growth rate after recession, 1949-1992, p. 34 [★80★]
7, daily ideal saturated fat intake, p. 97 [★232★]
7.3, average annual return on stocks, bonds, and real estate under Truman, p. 35 [★82★]
7.5, mortgage interest rate, 1993, p. 223 [★483★]
7.8, average annual return on stocks, 1960s, p. 35 [★83★]
7.83, average interest rate of home equity loan, 1992, p. 237 [★504★]
8, cost of health care billing in Canadian hospital, p. 311 [★665★]
8.12, average interest rate of 30-year fixed mortgage, 1992, p. 237 [★504★]
8.2, average annual return on stocks, bonds, and real estate under Eisenhower's 1st, p. 35 [★82★]
9.2, average annual return on stocks, 1940s, p. 35 [★83★]
9.39, average interest rate of automobile loan, 1992, p. 237 [★504★]
9.5, hydrogen in chemical composition of human body, p. 335 [★717★]
10, daily recommended saturated fat intake, p. 97 [★232★]
10-20, average weight loss with commercial diet programs, p. 339 [★728★]
10.1, average annual return on stocks, 1990s, p. 35 [★83★]
10.4, population receiving food stamp aid, p. 261 [★552★]
11, brown-bag lunches, p. 84 [★200★]
11, canned food of total food consumed, p. 77 [★179★]
11.1, average annual return on stocks, bonds, and real estate under Regan's 2nd term, p. 35 [★82★]
12, daily saturated fat intake, p. 97 [★232★]
12.5, average new car loan interest rate, 1990, p. 194 [★415★]
13, mortgage interest rate, 1983, p. 223 [★483★]
13, reduction in patient charges when physicians use computers to order services, p. 312 [★667★]
13.2, Blacks who are affluent, p. 98 [★235★]
14, oxygen in air breathed out, p. 341 [★734★]
14.8, average new car loan interest rate, 1980, p. 194 [★415★]
15, average commercial time in one hour of television, p. 437 [★945★]
16, average used car loan interest rate, 1990, p. 195 [★416★]
16.1, Hispanics who are affluent, p. 98 [★235★]
17.5, average annual return on stocks, 1980s, p. 35 [★83★]
18, cost of health care billing, p. 311 [★665★]
18.45, average interest rate on credit cards, 1992, p. 237 [★504★]
18.5, carbon in chemical composition of human body, p. 335 [★717★]
19.1, average used car loan interest rate, 1980, p. 195 [★416★]
19.2, average annual return on stocks, 1920s, p. 35 [★83★]
19.4, average annual return on stocks, 1950s, p. 35 [★83★]
19.4, disposable income devoted to consumer installment credit, p. 214 [★464★]

Percentages - continued

20-50, of landfill volume is made up of yard waste, p. 421 [★904★]
21, oxygen in air breathed in, p. 341 [★734★]
23, humidity in Phoenix, AZ, p. 497 [★1076★]
23, share of income to housing cost, 1993, p. 223 [★483★]
24, average lifetime tax rate for parents of Baby Boomers, p. 280 [★589★]
24, health care providers' administrative and billing costs, p. 287 [★605★]
25, abortions performed on adolescents, p. 99 [★237★]
25, daily ideal total fat intake, p. 97 [★232★]
25, portion of a movie's revenue from theater showings, 1992, p. 430 [★928★]
26, employers' health care costs (share of earnings), p. 286 [★604★]
27, femur in the height of average person, p. 333 [★712★]
27, humidity in El Paso, TX, p. 497 [★1076★]
29, humidity in Albuquerque, NM, p. 497 [★1076★]
30, average family automobile usage devoted to social purposes, p. 450 [★981★]
30, daily recommended total fat intake, p. 97 [★232★]
30, difference between Atlantic Ocean wave heights 1960s to 1980s, p. 503 [★1096★]
30, drug-related emergency room visits, 1988, p. 310 [★660★]
30, muscle in average woman's body weight, p. 342 [★738★]
30, portion of average glass soft drink bottle made of recycled glass, p. 428 [★923★]
30, possible sunshine in Juneau, AK, p. 511 [★1119★]
30, share of income to housing cost, 1983, p. 223 [★483★]
30.4, average family automobile usage devoted to family business, p. 450 [★981★]
31, average lifetime tax rate for Baby Boomers, p. 280 [★589★]
32, humidity in Reno, NV, p. 497 [★1076★]
34, spent on meals away from home, urban vs. rural, p. 208 [★448★]
34.3, average family automobile usage devoted to earning a living, p. 450 [★981★]
35, Asian Americans who are affluent, p. 98 [★235★]
35, water content of bread, p. 95 [★225★]
36, daily total fat intake, p. 97 [★232★]
38, saved by boys out of weekly allowance, 1991, p. 347 [★748★]
40, humidity in Denver, CO, p. 497 [★1076★]
40, muscle in average man's body weight, p. 342 [★738★]
40, landfill volume made up of paper waste, p. 423 [★909★]
40, portion of a movie's revenue from video rentals, 1992, p. 430 [★928★]
40, possible sunshine in Charleston, WV, p. 511 [★1119★]
40, working mothers who pay for child care, 1988, p. 198 [★424★]
41, family Christmas cards sent to non-relatives, p. 101 [★243★]
41, increase in number of pages faxed per day, 1992-1993, p. 31 [★74★]
43, humidity in Boise, ID, p. 497 [★1076★]
43, humidity in Salt Lake City, UT, p. 497 [★1076★]
43, saved by girls out of weekly allowance, 1991, p. 347 [★748★]
44, humidity in Cheyenne, WY, p. 497 [★1076★]
45, humidity in Great Fall, MT, p. 497 [★1076★]
46, humidity in Sacramento, CA, p. 497 [★1076★]
46, possible sunshine in Pittsburgh, PA, p. 511 [★1119★]
46, possible sunshine in Seattle-Tacoma, WA, p. 511 [★1119★]

Numbers following p. are page references. Numbers in [ ] are table references.

Percentages - continued

47, possible sunshine in Sault Ste. Marie, MI, p. 511 [★1119★]
48, possible sunshine in Portland, OR, p. 511 [★1119★]
49, possible sunshine in Buffalo, NY, p. 511 [★1119★]
49, possible sunshine in Burlington, VT, p. 511 [★1119★]
49, possible sunshine in Cleveland, OH, p. 511 [★1119★]
49, possible sunshine in Columbus, OH, p. 511 [★1119★]
50, hand and foot bones to total in human body, p. 333 [★712★]
50-70, water content of meat, p. 95 [★225★]
58, brown-bag lunches that contained a sandwich, 1992, p. 84 [★200★]
61.7, average hotel room occupancy, 1992, p. 441 [★957★]
62, humidity in Cleveland, OH, p. 497 [★1077★]
62, humidity in Indianapolis, IN, p. 497 [★1077★]
62, humidity in San Diego, CA, p. 497 [★1077★]
62, humidity in Seattle-Tacoma, WA, p. 497 [★1077★]
63, humidity in Buffalo, NY, p. 497 [★1077★]
63, humidity in Duluth, MN, p. 497 [★1077★]
63, humidity in New Oreleans, LA, p. 497 [★1077★]
64, humidity in Los Angeles, CA, p. 497 [★1077★]
64, humidity in Milwaukee, WI, p. 497 [★1077★]
65, oxygen in chemical composition of human body, p. 335 [★717★]
67, humidity in Sault Ste. Marie, MI, p. 497 [★1077★]
68, possible sunshine in Oklahoma City, OK, p. 511 [★1118★]
68, possible sunshine in San Diego, CA, p. 511 [★1118★]
68.5, female manager's wage compared to a man's, 1992, p. 403 [★862★]
69, possible sunshine in Honolulu, HI, p. 511 [★1118★]
70, possible sunshine in Denver, CO, p. 511 [★1118★]
71, annual personal income reflected in national debt, p. 259 [★547★]
71, average lifetime tax rate for grandchildren of Baby Boomers, p. 280 [★589★]
72, brown-bag lunches that contained a sandwich, 1984, p. 84 [★200★]
72.2, female physician's wage compared to a man's, 1992, p. 403 [★862★]
73, humidity in Juneau, AK, p. 497 [★1077★]
73, possible sunshine in Los Angeles, CA, p. 511 [★1118★]
73, possible sunshine in Miami, FL, p. 511 [★1118★]
74, energy saved by recycling steel cans, instead of making from virgin materials., p. 428 [★924★]
75, homes built before 1980 may contain lead-based paint, p. 418 [★898★]
75, households with VCR, 1992, p. 430 [★928★]
76, possible sunshine in Albuquerque, NM, p. 511 [★1118★]
78, female lawyer's wage compared to a man's, 1992, p. 403 [★862★]
78, possible sunshine in Sacramento, CA, p. 511 [★1118★]
79, possible sunshine in Reno, NV, p. 511 [★1118★]
80, prison inmates who are high school dropouts, p. 120 [★286★]
81, humidity in Cincinnati, OH, p. 497 [★1077★]
83, possible sunshine in El Paso, TX, p. 511 [★1118★]
84.1, female programmer's wage compared to a man's, 1992, p. 403 [★862★]
85, female engineer's wage compared to a man's, 1992, p. 403 [★862★]
86, possible sunshine in Phoenix, AZ, p. 511 [★1118★]
87, water content of pineapple, p. 95 [★225★]
95, water content of a ripe tomato, p. 95 [★225★]
99, hospital bills containing overcharges, p. 310 [★663★]
99, military retirees who are men, p. 348 [★751★]
268, increase in cost of given drug, 1980 to 1991, p. 316 [★676★]
365, increase in cost of given drug, 1980 to 1995, p. 316 [★676★]
604, increase in cost of given drug, 1980 to 2000, p. 316 [★676★]

Pressure

pounds per square foot

4,500, force of an avalanche, p. 493 [★1063★]

Speed

Feet per hour

15-20, spread of treetop fire, p. 495 [★1070★]
120-180, spread of grass fire, p. 495 [★1070★]

Inch per month

0.5, hair growth, p. 336 [★720★]

Miles per hour

7, speed of a raindrop, p. 504 [★1099★]
8, average speed of horse-drawn carriage, 1900, p. 444 [★966★]
8, average traffic speed in central London, 1988, p. 444 [★966★]
9.9, average traffic speed in New York City, 1988, p. 444 [★966★]
10-15, early hurricane speed, p. 501 [★1086★]
10-15, speed of damp snow avalanche, p. 493 [★1064★]
10.5, average traffic speed in Paris, 1988, p. 444 [★966★]
11.1, wind speed in Duluth, MN, p. 519 [★1136★]
11.1, wind speed in Sioux Falls, SD, p. 519 [★1136★]
11.4, wind speed in Honolulu, HI, p. 519 [★1136★]
11.6, wind speed in Milwaukee, WI, p. 519 [★1136★]
12, forward movement of hurricane system, p. 499 [★1082★]
12, wind speed in Buffalo, NY, p. 519 [★1136★]
12.3, wind speed in Wichita, KS, p. 519 [★1136★]
12.4, wind speed in Oklahoma City, OK, p. 519 [★1136★]
12.5, wind speed in Boston, MA, p. 519 [★1136★]
12.8, wind speed in Great Falls, MT, p. 519 [★1136★]
13, wind speed in Cheyenne, WY, p. 519 [★1136★]
17, average traffic speed on Los Angeles freeways, 1988, p. 445 [★967★]
20-30, mature hurricane speed, p. 501 [★1086★]
20-30, speed of dry loose-snow avalanche, p. 493 [★1064★]
22, speed of wet loose-snow avalanche, p. 493 [★1064★]
30-50, wave speeds in hurricanes, p. 499 [★1082★]
35, average wind speed at Mount Washington, NH, p. 519 [★1137★]
40, tornado speed, p. 516 [★1130★]
60, cyclone speed, p. 519 [★1135★]
60, top hurricane speed, p. 501 [★1086★]
73, wind speed of weakest tornadoes, p. 516 [★1131★]
74, hurricane middle wind speed, p. 500 [★1083★]
100, hurricane wind speed, p. 499 [★1082★]
100, speed of powder snow avalanche, p. 493 [★1064★]
120-150, hurricane inner wind speed, p. 500 [★1083★]
231, highest surface wind speed recorded, p. 519 [★1137★]
261, wind speed of strongest tornadoes, p. 517 [★1132★]
350-400, hurricane outer wind speed, p. 500 [★1083★]

Numbers following p. are page references. Numbers in [ ] are table references.

Speed - continued

Millimeter per day

0.5, daily beard growth, p. 336 [★722★]

Temperature

Degrees Fahrenheit

14, average January temperatures in Montreal, Canada, p. 515 [★1127★]
16, average January temperatures in Moscow, p. 515 [★1127★]
23, average January temperatures in Peking, China, p. 515 [★1127★]
27, average January temperatures in Stockholm, p. 515 [★1127★]
30, average January temperatures in Vienna, p. 515 [★1127★]
32, average January temperatures in Geneva, p. 515 [★1127★]
32, average January temperatures in New York, p. 515 [★1127★]
34, average January temperatures in Prague, p. 515 [★1127★]
36, average January temperatures in Amsterdam, p. 515 [★1127★]
37, average January temperatures in Paris, p. 515 [★1127★]
39, average April temperature in Moscow, p. 515 [★1127★]
39, average April temperature in Stockholm, p. 515 [★1127★]
40.6, annual average temperature in Juneau, p. 514 [★1125★]
41, average January temperatures in Dublin, p. 515 [★1127★]
41, average January temperatures in London, p. 515 [★1127★]
41, average January temperatures in Madrid, p. 515 [★1127★]
41.6, annual average temperature in Bismarck, p. 514 [★1125★]
43, average April temperature in Montreal, Canada, p. 515 [★1127★]
44.6, annual average temperature in Burlington, p. 514 [★1125★]
44.8, annual average temperature in Great Falls, p. 514 [★1125★]
44.9, annual average temperature in St. Paul, p. 514 [★1125★]
45.1, annual average temperature in Concord, p. 514 [★1125★]
45.4, annual average temperature in Portland, p. 514 [★1125★]
45.5, annual average temperature in Sioux Falls, p. 514 [★1125★]
45.6, annual average temperature in Cheyenne, p. 514 [★1125★]
46, average April temperature in Amsterdam, p. 515 [★1127★]
46, average April temperature in Berlin, p. 515 [★1127★]
46, average April temperature in Dublin, p. 515 [★1127★]
46, average April temperature in Prague, p. 515 [★1127★]
46, average January temperatures in Rome, p. 515 [★1127★]
46.1, annual average temperature in Milwaukee, p. 514 [★1125★]
47.4, annual average temperature in Albany, p. 514 [★1125★]
48, average April temperature in Geneva, p. 515 [★1127★]
48, average April temperature in Vienna, p. 515 [★1127★]
48, average January temperatures in Athens, p. 515 [★1127★]
48, average January temperatures in Berlin, p. 515 [★1127★]
48, average November temperature in London, p. 515 [★1127★]
48.6, annual average temperature in Detroit, p. 514 [★1125★]
49, annual average temperature in Chicago, p. 514 [★1125★]
49.6, annual average temperature in Cleveland, p. 514 [★1125★]
49.9, annual average temperature in Des Moines, p. 514 [★1125★]
49.9, annual average temperature in Hartford, p. 514 [★1125★]
50, average April temperature in London, p. 515 [★1127★]
50, average April temperature in Paris, p. 515 [★1127★]
50.3, annual average temperature in Denver, p. 514 [★1125★]
50.4, annual average temperature in Providence, p. 514 [★1125★]
50.6, annual average temperature in Omaha, p. 514 [★1125★]
50.72 degrees Fahrenheit, lowest average annual U.S. temperature, 1895-1990, p. 516 [★1128★]
50.8, annual average temperature in Reno, p. 514 [★1125★]

Temperature - continued

Degrees Fahrenheit

50.9, annual average temperature in Boise, p. 514 [★1125★]
51.3, annual average temperature in Boston, p. 514 [★1125★]
52, annual average temperature in Salt Lake City, p. 514 [★1125★]
52, annual average temperature in Seattle-Tacoma, p. 514 [★1125★]
52, average April temperature in New York, p. 515 [★1127★]
52, average September temperature in Moscow, p. 515 [★1127★]
52.3, annual average temperature in Indianapolis, p. 514 [★1125★]
52.5 degrees Fahrenheit, average annual U.S. temperature, 1895-1990, p. 516 [★1128★]
53, annual average temperature in Atlantic City, p. 514 [★1125★]
53.6, annual average temperature in Portland, p. 514 [★1125★]
54.2, annual average temperature in Wilmington, p. 514 [★1125★]
54.3, annual average temperature in Philadelphia, p. 514 [★1125★]
54.67 degrees Fahrenheit, highest average annual U.S. temperature, 1895-1990, p. 516 [★1128★]
54.7, annual average temperature in New York, p. 514 [★1125★]
55, annual average temperature in Charleston, p. 514 [★1125★]
55, average April temperature in Madrid, p. 515 [★1127★]
55.1, annual average temperature in Baltimore, p. 514 [★1125★]
56.1, annual average temperature in Louisville, p. 514 [★1125★]
56.1, annual average temperature in St. Louis, p. 514 [★1125★]
56.2, annual average temperature in Albuquerque, p. 514 [★1125★]
56.2, annual average temperature in Wichita, p. 514 [★1125★]
57, average April temperature in Rome, p. 515 [★1127★]
57.7, annual average temperature in Richmond, p. 514 [★1125★]
58, annual average temperature in Washington, p. 514 [★1125★]
59, average April temperature in Athens, p. 515 [★1127★]
59, average April temperature in Peking, China, p. 515 [★1127★]
59, average August temperature in Dublin, p. 515 [★1127★]
59, average July temperature in Dublin, p. 515 [★1127★]
59.30, average annual temperature in 1977, 1986, p. 514 [★1126★]
59.31, average annual temperature in 1973, 1986, p. 514 [★1126★]
59.3, annual average temperature in Raleigh, p. 514 [★1125★]
59.31, average annual temperature in 1973, p. 514 [★1126★]
59.30, average annual temperature, 1986, p. 514 [★1126★]
59.45, average annual temperature in 1989, p. 514 [★1126★]
59.51, average annual temperature in 1980, 1983, p. 514 [★1126★]
59.56, average annual temperature in 1987, p. 514 [★1126★]
59.64, average annual temperature in 1981, 1988, p. 514 [★1126★]
59.81, average annual temperature in 1990, p. 514 [★1126★]
60, annual average temperature in Oklahoma City, p. 514 [★1125★]
61.3, annual average temperature in Atlanta, p. 514 [★1125★]
61.8, annual average temperature in Little Rock, p. 514 [★1125★]
62.3, annual average temperature in Memphis, p. 514 [★1125★]
63, annual average temperature in Los Angeles, p. 514 [★1125★]
63, average July temperature in Amsterdam, p. 515 [★1127★]
63.1, annual average temperature in Columbia, p. 514 [★1125★]
64, average July temperature in Berlin, p. 515 [★1127★]
64, average July temperature in Geneva, p. 515 [★1127★]
64, average July temperature in London, p. 515 [★1127★]
64, average July temperature in Moscow, p. 515 [★1127★]
64, average July temperature in Stockholm, p. 515 [★1127★]
64.2, annual average temperature in Jackson, p. 514 [★1125★]
65.4, annual average temperature in Ft. Worth, p. 514 [★1125★]

Numbers following p. are page references. Numbers in [ ] are table references.

Temperature - continued

Degrees Fahrenheit

66, average July temperature in Paris, p. 515 [★1127★]
66, average July temperature in Prague, p. 515 [★1127★]
66, average July temperature in Vienna, p. 515 [★1127★]
67.5, annual average temperature in Mobile, p. 514 [★1125★]
68, average August temperature in Montreal, Canada, p. 515 [★1127★]
68.1, annual average temperature in New Orleans, p. 514 [★1125★]
70, average July temperature in Montreal, Canada, p. 515 [★1127★]
72, average July temperature in Douala, Cameroon, p. 515 [★1127★]
72.6, annual average temperature in Phoenix, p. 514 [★1125★]
73, average January temperatures in Jeddah, Saudi Arabia, p. 515 [★1127★]
75, average April temperature in Douala, Cameroon, p. 515 [★1127★]
75, average January temperatures in Bombay, India, p. 515 [★1127★]
75, average January temperatures in Douala, Cameroon, p. 515 [★1127★]
75, average July temperature in Madrid, p. 515 [★1127★]
75, average July temperature in Rome, p. 515 [★1127★]
75.9, annual average temperature in Miami, p. 514 [★1125★]
77, average July temperature in Darwin, Australia, p. 515 [★1127★]
77, average July temperature in New York, p. 515 [★1127★]
77, average June temperature in Athens, p. 515 [★1127★]
77.2, annual average temperature in Honolulu, p. 514 [★1125★]
81, average April temperature in Bombay, India, p. 515 [★1127★]
81, average July temperature in Athens, p. 515 [★1127★]
82, average January temperatures in Darwin, Australia, p. 515 [★1127★]
82, average July temperature in Bombay, India, p. 515 [★1127★]
84, average April temperature in Darwin, Australia, p. 515 [★1127★]
84, average April temperature in Jeddah, Saudi Arabia, p. 515 [★1127★]
86, heat increase produced in 12 gallons of water by a fire from 1 pound of wood, p. 495 [★1069★]
88, average July temperature in Peking, China, p. 515 [★1127★]
91, average July temperature in Jeddah, Saudi Arabia, p. 515 [★1127★]
98.1-98.6, body temperature of healthy adults, p. 344 [★743★]

Time

Age

19, twice-married mothers at first marriage, p. 458 [★998★]
27.3, twice-married mothers at divorce, p. 458 [★998★]
30.9, twice-married mothers at remarriage, p. 458 [★998★]

Days

0.5, relaxation time around Christmas, p. 10 [★17★]
2, stomach cell life span, p. 334 [★716★]
2-3, spermatozoa life span, p. 334 [★716★]
3-4, colon cell life span, p. 334 [★716★]
3.7, average length of business trip, 1990, p. 59 [★139★]
3.8, business trip length, 1989, p. 23 [★49★]
4.2, business trip length, 1990, p. 23 [★49★]
4.3, sick (work-loss) days per man, p. 20 [★42★]
4.4, average days per pleasure trip, 1990, p. 443 [★961★]

Time - continued

Days

4.4, business trip length, 1991, p. 23 [★49★]
5.5, sick (bed-rest) days per married man, p. 21 [★43★]
5.5, sick (work-loss) days per woman, p. 20 [★42★]
5.8, length of hospital stay for children, 1990, p. 302 [★644★]
6, vacation days per year in Mexico, p. 4 [★7★]
6.2, length of hospital stay for women, 1988, p. 303 [★645★]
6.4, annual hospitalization time for white children, p. 301 [★643★]
6.5, length of hospital stay for women, 1990, p. 302 [★644★]
6.7, length of hospital stay, 1990, p. 302 [★644★]
6.8, length of hospital stay for men, 1990, p. 302 [★644★]
6.8, length of hospital stay for whites, 1990, p. 302 [★644★]
7, sick (bed-rest) days per married woman, p. 21 [★43★]
7.1, length of hospital stay for men, 1988, p. 303 [★645★]
7.6, length of hospital stay for women, 1970, p. 303 [★645★]
7.8, length of hospital stay for blacks, 1990, p. 302 [★644★]
8, length of honeymoon, 1987 and 1992, p. 11 [★21★]
8, sick (bed-rest) days per single man, p. 21 [★43★]
8.2, annual hospitalization time for whites, p. 301 [★643★]
8.2, length of hospital stay, 1985, p. 307 [★653★]
8.7, length of hospital stay for men, 1970, p. 303 [★645★]
8.8, annual hospitalization time for black children, p. 301 [★643★]
8.9, length of hospital stay for elderly, p. 302 [★644★]
8.9, length of hospital stay, 1989, p. 307 [★653★]
9, lifespans of Atlantic Ocean hurricanes, p. 499 [★1081★]
9.5, annual hospitalization time for blacks, p. 301 [★643★]
10, average lifespan of a firefly, p. 341 [★735★]
10, average lifespan of a taste bud, p. 341 [★735★]
10, platelet life span, p. 334 [★716★]
10, vacation days per year in Japan, p. 4 [★7★]
10, vacation days per year in United States, p. 4 [★7★]
10.6, length of hospital stay, 1980, p. 307 [★653★]
11-12, sick (bed-rest) days per single woman, p. 21 [★43★]
11.7, annual hospitalization time for white elderly, p. 301 [★643★]
12, lifespans of Atlantic Ocean hurricanes, p. 499 [★1081★]
13.2, Hispanics' average disability days per person, 1989, p. 25 [★56★]
14, annual hospitalization time for black elderly, p. 301 [★643★]
15, whites' average disability days per person, 1989, p. 25 [★56★]
16.1, days off per full-time worker, 1989, p. 23 [★48★]
17.1, blacks' average disability days per person, 1989, p. 25 [★56★]
18, vacation days per year in Germany, p. 4 [★7★]
19-30, average fly life expectancy, p. 488 [★1054★]
19-34, skin cell life span, p. 334 [★716★]
19.8, days off per full-time worker, 1981, p. 23 [★48★]
21, sick (work-loss) days due to colds, p. 23 [★50★]
22, vacation days per year in Britain, p. 4 [★7★]
23, hospital admission for heart transplant, 1988, p. 290 [★614★]
23, sick (work-loss) days due to fractures, p. 23 [★50★]
25, vacation days per year in France, p. 4 [★7★]
26, condominium timesharers' annual leisure travel, p. 105 [★257★]
30, sick (work-loss) days due to sprains, p. 23 [★50★]
30, vacation days per year in Sweden, p. 4 [★7★]
33, hospital admission for liver transplants, 1988, p. 290 [★615★]
40, average cockroach life expectancy, p. 488 [★1054★]
52, fox gestation period, p. 520 [★1139★]

Numbers following p. are page references. Numbers in [ ] are table references.

Time - continued

Days

61, dog gestation period, p. 520 [★1139★]
63, cat gestation period, p. 520 [★1139★]
76, sick (work-loss) days due to flu, p. 23 [★50★]
87, days lost to workers compensation injuries, per 100 employees, 1992, p. 43 [★102★]
93 days, worked to pay taxes 1950, p. 264 [★561★]
99 days, worked to pay taxes 1955, p. 264 [★561★]
105 days, worked to pay taxes 1965, p. 264 [★561★]
107 days, worked to pay taxes 1960, p. 264 [★561★]
118 days, worked to pay taxes 1970, p. 264 [★561★]
118 days, worked to pay taxes 1975, p. 264 [★561★]
120, red blood cell life span, p. 334 [★716★]
121 days, worked to pay taxes 1980, p. 264 [★561★]
121 days, worked to pay taxes 1985, p. 264 [★561★]
122, beaver gestation period, p. 520 [★1139★]
125 days, worked to pay taxes 1990, p. 264 [★561★]
128 days, worked to pay taxes 1991, p. 264 [★561★]
151, goat gestation period, p. 520 [★1139★]
151, with at least 0.01 inch of precipitation in Charleston, WV, p. 507 [★1108★]
154, with at least 0.01 inch of precipitation in Cleveland, OH, p. 507 [★1108★]
156, with at least 0.01 inch of precipitation in Seattle-Tacoma, WA, p. 507 [★1108★]
169, with at least 0.01 inch of precipitation in Buffalo, NY, p. 507 [★1108★]
175, average time needed to hike the full Appalachian Trail, p. 440 [★952★]
178, average length of school year, p. 9 [★16★]
201, deer gestation period, p. 520 [★1139★]
219, bear (black) gestation period, p. 520 [★1139★]
220, with at least 0.01 inch of precipitation in Juneau, AK, p. 507 [★1108★]
257, gorilla gestation period, p. 520 [★1139★]
278, buffalo gestation period, p. 520 [★1139★]
284, cow gestation period, p. 520 [★1139★]
406, Bactrian camel gestation period, p. 520 [★1139★]
425, giraffe gestation period, p. 520 [★1139★]
645, elephant gestation period, p. 520 [★1139★]
157 million, sick (work-loss) days nationally due to migraine headaches, p. 298 [★639★]

Hours

0.9, children spend reading per week, p. 8 [★14★]
1 and 46 min., in each working day to pay federal taxes, p. 264 [★562★]
1.6, teen-agers spend reading per week, p. 8 [★14★]
1.8, children spend studying per week, p. 8 [★14★]
1.9, daily time parents spend with children, p. 12 [★23★]
2.7, children spend performing household chores per week, p. 8 [★14★]
3.21, average daily time spent listening to the radio, p. 433 [★936★]
3.48, average daily time spent watching television, p. 433 [★936★]
3-4, adolescent's average weekly exposure to television advertising, p. 27 [★60★]
3.8, teen-agers spend studying per week, p. 8 [★14★]
4-5, hurricane duration, p. 500 [★1084★]

Time - continued

Hours

4.2, volunteer time per week, p. 19 [★38★]
4.8, teen-agers spend performing household chores per week, p. 8 [★14★]
5.1, weekly unproductive time per PC worker, p. 42 [★100★]
5.3, worker-hours per ton of steel in United States, p. 4 [★5★]
5.4, U.S. average time to produce and ship one metric ton of steel, 1991, p. 50 [★120★]
5.4, worker-hours per ton of steel in Japan, p. 4 [★5★]
5.6, Japanese average time to produce and ship one metric ton of steel, 1991, p. 50 [★120★]
5.6, wait for public hospital admission from emergency room, 1991, p. 310 [★661★]
5.6, worker-hours per ton of steel in Britain, p. 4 [★5★]
5.6, worker-hours per ton of steel in Germany, p. 4 [★5★]
6, shopping for Christmas gifts, p. 10 [★17★]
6, worked at home in addition to job, p. 22 [★47★]
6.4, worker-hours per ton of steel in South Korea, p. 4 [★5★]
7.1, average daily television viewing time per household, 1989, p. 436 [★941★]
7.5, baking and cooking for Christmas, p. 10 [★17★]
8, spent on crafts or hobbies per week, p. 11 [★20★]
8, weapons training for average security guard, p. 393 [★843★]
9, income tax computation time (1040 with Schedule A), 1991, p. 16 [★33★]
10, arguing related to Christmas, p. 10 [★17★]
10, daily time parents spend working and commuting, p. 12 [★23★]
10, hurricane duration, p. 500 [★1084★]
10, white blood cell life span, p. 334 [★716★]
10.01, Japanese average time to produce and ship one metric ton of steel, 1982, p. 50 [★120★]
10.59, U.S. average time to produce and ship one metric ton of steel, 1982, p. 50 [★120★]
13, worked at home per self-employed man, p. 22 [★47★]
13.5, daily confinement time for jail inmates, 1983 and 1988, p. 125 [★293★]
21, worked at home per self-employed woman, p. 22 [★47★]
23:17, average television viewing time of children, 6-11, p. 434 [★939★]
25:43, average television viewing time of children, 2-5, p. 434 [★939★]
27.4, income tax computation time, 1990, p. 16 [★32★]
28:53, average television viewing time per week of women, 55 and older, p. 434 [★939★]
33, worked at home per woman, p. 22 [★47★]
37:32, average television viewing time of men, 55 and older, p. 434 [★939★]
39, worked at home per man, p. 22 [★47★]
43, worked per week per man in Western Europe, p. 4 [★6★]
47.5, worked per week per woman in North America and Australia, p. 4 [★6★]
48, worked per week per man in Asia (except Japan), p. 4 [★6★]
48, worked per week per woman in Western Europe, p. 4 [★6★]
49, average weekly hours worked by women at "Big Three" auto manufacturers, p. 45 [★107★]
49, worked per week per man in North America and Australia, p. 4 [★6★]

Numbers following p. are page references. Numbers in [ ] are table references.

Time - continued

Hours

53, worked per week per man in Africa, p. 4 [★6★]
54, worked per week per man in Japan, p. 4 [★6★]
54, worked per week per man in Latin America, p. 4 [★6★]
56, worked per week per woman in Japan, p. 4 [★6★]
60, worked per week per woman in Latin America, p. 4 [★6★]
62, worked per week per woman in Asia (except Japan), p. 4 [★6★]
67, worked per week per woman in Africa, p. 4 [★6★]
138, increased work hours per full-time worker, 1989, p. 23 [★48★]
543, training time required for local police departments, 1990, p. 141 [★318★]
750, lifespan of 75-watt light bulbs, p. 491 [★1059★]
1,000, lifespan of low wattage light bulbs, p. 491 [★1059★]
1,539, worked yearly in Sweden per person, p. 3 [★4★]
1,572, worked yearly in Belgium per person, p. 3 [★4★]
1,592, worked yearly in Netherlands per person, p. 3 [★4★]
1,595, worked yearly in Denmark per person, p. 3 [★4★]
1,603, worked yearly in Germany per person, p. 3 [★4★]
1,610, worked yearly in France per person, p. 3 [★4★]
1,614, worked yearly in Norway per person, p. 3 [★4★]
1,856, worked yearly in Great Britain per person, p. 3 [★4★]
1,858, worked yearly in Italy per person, p. 3 [★4★]
1,887, worked yearly in Canada per person, p. 3 [★4★]
1,951, worked yearly in United States per person, p. 3 [★4★]
2,155, worked yearly in Japan per person, p. 3 [★4★]
3,000, lifespan of 75-watt long-life light bulbs, p. 491 [★1059★]
3,000, lifespan of Halogen light bulbs, p. 491 [★1059★]
10,000, lifespan of Fluorescent (compact) light bulbs, p. 491 [★1059★]
22,000, average television viewing time of adolescent by high school graduation, p. 437 [★946★]

Minutes

1, Human breath-holding capability, p. 522 [★1143★]
1.5, Polar bear breath-holding capability, p. 522 [★1143★]
2.5, Pearl diver (human) breath-holding capability, p. 522 [★1143★]
5, kidneys filter all blood in body, p. 333 [★711★]
5, Sea otter breath-holding capability, p. 522 [★1143★]
6, increase in physicians' time on patient paperwork with computer use, p. 312 [★667★]
6, spent daily on child care by employed men in Belgium, p. 1 [★1★]
7, spent by physicians per patient on paperwork, p. 311 [★666★]
7, spent daily on child care by employed men in Germany, Federal Republic,1,, p. 1 [★1★]
8, spent daily on child care by employed men in France, p. 1 [★1★]
8, spent daily on child care by employed men in United States,1,, p. 1 [★1★]
8, waiting in grocery check-out line, p. 13 [★26★]
9 minutes 17 seconds, average commercial time in 1 hour of FM radio, p. 26 [★59★]
9, average commercial time in one hour of television, p. 437 [★945★]
9, playing with children on Christmas, per parent, p. 10 [★17★]
10 minutes 6 seconds, average commercial time in 1 hour of AM radio, p. 26 [★59★]
10, for cloud to expand to thundercloud, p. 509 [★1112★]
10, platypus breath-holding capability, p. 522 [★1143★]
12, muskrat breath-holding capability, p. 522 [★1143★]

Time - continued

Minutes

13, spent daily on child care by employed men in Bulgaria, p. 1 [★1★]
13, spent daily on child care by employed men in Yugoslavia,1,, p. 1 [★1★]
14, grooming time (men), 1988, p. 12 [★22★]
14, spent daily on child care by employed women in Belgium, p. 1 [★1★]
15 to 28, seal breath-holding capability, p. 522 [★1143★]
15, average daily time executives spend on telephone hold, p. 24 [★54★]
15, hippopotamus breath-holding capability, p. 522 [★1143★]
15, porpoise breath-holding capability, p. 522 [★1143★]
15, spent daily on child care by employed men in German Democratic Republic, p. 1 [★1★]
16, sea cow breath-holding capability, p. 522 [★1143★]
17, average visit to children's section of college/university bookstores, p. 161 [★356★]
17, spent daily on child care by employed men in Hungary, p. 1 [★1★]
17, spent daily on child care by employed women in United States,1,, p. 1 [★1★]
17.1, average visit to children's section of general/specialty bookstores, p. 161 [★356★]
18, spent daily on child care by employed men in Czechoslovakia, p. 1 [★1★]
19.66, average one-way commute, p. 60 [★140★]
20, average daily time spent reading magazines, p. 433 [★936★]
20, beaver breath-holding capability, p. 522 [★1143★]
20, spent daily on child care by employed men in Poland, p. 1 [★1★]
20-30, hurricane eye duration over a given spot, p. 500 [★1084★]
21, average life extension by sedentary person walking or running a mile, p. 331 [★707★]
21, spent daily on child care by employed women in Bulgaria, p. 1 [★1★]
21.5, average visit to children's section of national/regional chain bookstores, p. 161 [★356★]
22, grocery-shopping time per trip, 1992, p. 13 [★25★]
23.2, commuting time, p. 20 [★41★]
23.6, average visit to children's bookstores, p. 161 [★356★]
24, spent daily on child care by employed women in France, p. 1 [★1★]
25, spent daily on child care by employed women in Yugoslavia,1,, p. 1 [★1★]
25, thunderstorm duration, p. 509 [★1113★]
26, spent daily on child care by employed women in Germany, Federal Republic,1,, p. 1 [★1★]
26, spent daily on child care by employed women in Hungary, p. 1 [★1★]
26.1, commuting time in Houston, p. 20 [★41★]
26.4, commuting time in Los Angeles, p. 20 [★41★]
27, spent daily on child care by employed women in Poland, p. 1 [★1★]
28, grocery-shopping time per trip, 1982, p. 13 [★25★]
28.1, commuting time in Chicago, p. 20 [★41★]
29.5, commuting time in Washington, D.C., p. 20 [★41★]
30, spent daily on child care by employed men in USSR (former), p. 1 [★1★]

Numbers following p. are page references. Numbers in [ ] are table references.

Time - continued

Minutes

30, spent daily on child care by employed women in Czechoslovakia, p. 1 [★1★]
30, spent daily on child care by employed women in USSR (former), p. . [★1★]
30-60, orb-weaver spider web completion, p. 524 [★1147★]
30.6, commuting time in New York City, p. 20 [★41★]
33, spent daily on child care by employed women in German Democratic Republic, p. 1 [★1★]
34, average daily time spent reading newspapers, p. 433 [★936★]
34, spent daily on housework by employed men in Belgium, p. 1 [★1★]
44, grooming time (men), 1990, p. 12 [★22★]
46, spent daily on housework by employed men in United States,1,, p. 1 [★1★]
48, spent daily on housework by employed men in Germany, Federal Republic,1,, p. 1 [★1★]
50, average daily time office workers spend looking for lost items, p. 24 [★53★]
53, grooming time (men under 25), 1990, p. 12 [★22★]
58, spent daily on housework by employed men in France, p. 1 [★1★]
59, in each working day to pay non-federal taxes, p. 264 [★562★]
60, Greenland whale breath-holding capability, p. 522 [★1143★]
60, spent daily on housework by employed men in Poland, p. 1 [★1★]
66, spent daily on housework by employed men in Bulgaria, p. 1 [★1★]
67, spent daily on housework by employed men in USSR (former), p. 1 [★1★]
76, spent daily on housework by employed men in Yugoslavia,1,, p. 1 [★1★]
78, spent daily on housework by employed men in Czechoslovakia, p. 1 [★1★]
78, spent daily on housework by employed men in Hungary, p. 1 [★1★]
80, spent daily on housework by employed men in German Democratic Republic, p. 1 [★1★]
90, sperm whale breath-holding capability, p. 522 [★1143★]
120, battlenose whale breath-holding capability, p. 522 [★1143★]
149, spent daily on housework by employed women in Bulgaria, p. 1 [★1★]
162, spent daily on housework by employed women in United States,1,, p. 1 [★1★]
163, spent daily on housework by employed women in Belgium, p. 1 [★1★]
173, spent daily on housework by employed women in France, p. 1 [★1★]
197, spent daily on housework by employed women in USSR (former), p. 1 [★1★]
200, spent daily on housework by employed women in Poland, p. 1 [★1★]
216, spent daily on housework by employed women in Germany, Federal Republic,1,, p. 1 [★1★]
217, spent daily on housework by employed women in Hungary, p. 1 [★1★]
220, spent daily on housework by employed women in German Democratic Republic, p. 1 [★1★]

Time - continued

Minutes

228, average weekly mathematics instruction, p. 18 [★36★]
231, spent daily on housework by employed women in Yugoslavia,1,, p. 1 [★1★]
233, average weekly science instruction, p. 19 [★37★]
255, spent daily on housework by employed women in Czechoslovakia, p. 1 [★1★]
9.5 billion, telephone calling on average business day, p. 31 [★72★]

Months

1, outer skin replaced, p. 343 [★740★]
2, average mosquito life expectancy, p. 488 [★1054★]
4.7, average house-hunting time for repeat-buyer, 1992, p. 15 [★30★]
5, added life expectancy for controlling blood pressure (woman), p. 486 [★1048★]
5, added life expectancy for controlling weight (woman), p. 486 [★1048★]
5, average course of psychotherapy treatment, p. 320 [★685★]
5-6, period of unemployment, p. 20 [★40★]
5.4, average house-hunting time for first-time buyer, 1992, p. 15 [★30★]
6, average worker bee life expectancy, p. 488 [★1054★]
7.3, survival time of melanoma patients who receive standard treatment, p. 296 [★630★]
8, added life expectancy for not smoking (woman), p. 486 [★1048★]
8, average prison sentence length for public-order offenses, 1989, p. 135 [★309★]
9, average course of psychiatric treatment, p. 320 [★685★]
10, added life expectancy for cholesterol under 200 (woman), p. 486 [★1048★]
10, added life expectancy for not smoking (man), p. 485 [★1047★]
16, time served in prison for women released 1986, p. 134 [★308★]
18-24, survival time of AIDS patients, p. 297 [★635★]
19, average prison sentence length for drug offenses, 1989, p. 135 [★309★]
21, average prison sentence length for property offenses, 1989, p. 135 [★309★]
22.3, length of prison sentence for property offenses, 1990, p. 128 [★298★]
23, survival time of melanoma patients who receive new vaccine treatment, p. 296 [★630★]
35, average prison sentence length for violent offenses, 1989, p. 135 [★309★]
36.6, length of prison sentence for female offenders, 1988, p. 130 [★300★]
37.3, length of probation for all offenses, 1990, p. 129 [★299★]
38, length of prison sentence (16- to 18-year-olds), 1988, p. 131 [★302★]
40.4, length of probation for property offenses, 1990, p. 129 [★299★]
40.6, length of probation for drug offenses, 1990, p. 129 [★299★]
43, length of probation for violent offenses, 1990, p. 129 [★299★]
44.2, length of prison sentence (19- to 20-year-olds), 1988, p. 131 [★302★]
47.8, length of prison sentence (no previous convictions), 1988, p. 133 [★306★]

Numbers following p. are page references. Numbers in [ ] are table references.

Time - continued

Months

50.8, length of prison sentence (no known history of drug abuse), 1988, p. 133 [★307★]
52, average length of economic expansion after a recession, p. 34 [★80★]
56.8, length of prison sentence for Hispanics, 1988, p. 131 [★301★]
57.4, length of prison sentence for all offenses, 1990, p. 128 [★298★]
57.6, length of prison sentence for offenders over 40, 1988, p. 131 [★302★]
58.6, length of prison sentence for male offenders, 1988, p. 130 [★300★]
58.7, length of prison sentence (prior felony conviction), 1988, p. 133 [★306★]
58.8, length of prison sentence (history of drug abuse), 1988, p. 133 [★307★]
63.2, length of prison sentence for blacks, 1988, p. 131 [★301★]
66, maximum length of prison sentence for women admitted 1986, p. 134 [★308★]
73.7, length of prison sentence (prior misdemeanor conviction), 1988, p. 133 [★306★]
81.2, length of prison sentence for drug offenses, 1990, p. 128 [★298★]
89.8, length of prison sentences for violent offenses, 1990, p. 128 [★298★]

Seconds

22, average stock car racing pit stop, 1990, p. 454 [★990★]
30, spent by Canadian physicians per patient on paperwork, p. 311 [★666★]

Weeks

1, cyclone lifespan, p. 519 [★1135★]
2, average annual time executives spend on telephone hold, p. 24 [★54★]
2.08, average temporary employee assignment in France, 1989, p. 59 [★137★]
2.7, average family leave time men estimate they would take for birth or adoption, p. 24 [★52★]
6, average annual time office workers spend looking for lost items, p. 24 [★53★]
6.9, median duration of unemployment, 1991, p. 61 [★142★]
8.4, average family leave time women estimate they would take for spouse/parent critical illness, p. 24 [★52★]
8.5, average family leave time men estimate they would take for spouse/parent critical illness, p. 24 [★52★]
8.5, average family leave time women estimates they would take for birth or adoption, p. 24 [★52★]
9-14, average cricket life expectancy, p. 488 [★1054★]
11.9, average duration of unemployment, 1980, p. 60 [★141★]
13.8, average duration of unemployment, 1991, p. 60 [★141★]
13.8, mean duration of unemployment, 1991, p. 61 [★142★]
18.7, time to earn money to pay for new car, 1970, p. 190 [★405★]
24.5, time to earn money to pay for new car, 1990, p. 190 [★405★]

Years

1, added life expectancy for controlling blood pressure (man), p. 485 [★1047★]
1, added life expectancy for controlling cholesterol (man), p. 485 [★1047★]
1, added life expectancy for reducing weight (man), p. 485 [★1047★]

Numbers following p. are page references. Numbers in [ ] are table references.

Time - continued

Years

1, lymphocyte life span, p. 334 [★716★]
1.3, outside of labor force due to caregiving (men), p. 7 [★13★]
2, frequency of flooding on a given river, p. 496 [★1075★]
2-3, survival time of Lou Gehrig's disease patients, p. 297 [★633★]
2-6, life span of hair follicle, p. 336 [★720★]
2.3, saving to buy house in Seattle, p. 11 [★19★]
2.4, saving to buy house in Cleveland, p. 11 [★19★]
2.9, saving to buy house in Detroit, p. 11 [★19★]
3, automobile loan period, 1975, p. 192 [★410★]
3-10, lifespan of Composite fillings, p. 312 [★668★]
3.1, added life expectancy for avoiding heart disease (man), p. 485 [★1047★]
3.3, added life expectancy for avoiding heart disease (woman), p. 486 [★1048★]
3.7, saving to buy house in Boston, p. 11 [★19★]
4.2, saving to buy house in New York, p. 11 [★19★]
4.5, automobile loan period, 1992, p. 192 [★410★]
4.8, average age of automobiles in Japan, p. 451 [★983★]
4.8, saving to buy house in Los Angeles, p. 11 [★19★]
6, average queen bee life expectancy, p. 488 [★1054★]
6.5, median duration of marriage, 1975, p. 476 [★1028★]
6.7, median duration of marriage, 1970, p. 476 [★1028★]
6.8, median duration of marriage, 1980, p. 476 [★1028★]
7.1, median duration of marriage, 1988, p. 476 [★1028★]
7.6, average age of automobiles, p. 451 [★983★]
7.8, average lifespan of an automobile, 1990, p. 490 [★1057★]
9.1, median school years completed by Hispanics, 1970, p. 186 [★399★]
9.8, median school years completed by blacks, 1970, p. 186 [★399★]
10, frequency of major flood on a given river, p. 496 [★1075★]
10, lifespan of a dishwasher, p. 491 [★1058★]
10, lifespan of ceramic fillings, p. 312 [★668★]
10-20, lifespan of amalgam fillings, p. 312 [★668★]
11, life expectancy for 7-year-old black woman in 1989, p. 482 [★1041★]
11, lifespan of a air-conditioner, p. 491 [★1058★]
11, lifespan of a gas water heater, p. 491 [★1058★]
11, lifespan of a microwave, p. 491 [★1058★]
11.5, outside of labor force due to caregiving (women), p. 7 [★13★]
12, development time for new drug, p. 315 [★674★]
12, lifespan of dogs, p. 486 [★1050★]
12, median school years completed by Hispanics, 1991, p. 186 [★399★]
12.1, median school years completed by whites, 1970, p. 186 [★399★]
12.4, median school years completed by blacks, 1991, p. 186 [★399★]
12.67, age of passenger airplanes in use, p. 489 [★1056★]
12.8, median school years completed by whites, 1991, p. 186 [★399★]
13, lifespan of a automatic washer, p. 491 [★1058★]
13, lifespan of a gas dryer, p. 491 [★1058★]
13-15, life expectancy for cats, p. 486 [★1050★]
13.9, average number of years a gardener has gardened, p. 104 [★251★]
14, lifespan of an electric dryer, p. 491 [★1058★]

Time - continued

Years

14, lifespan of an electric water heater, p. 491 [*1058*]
14.9, age of juveniles held in private custody, 1989, p. 128 [*296*]
16, age of juveniles held in public custody, 1989, p. 128 [*296*]
16, lifespan of a electric furnace, p. 491 [*1058*]
16, lifespan of a freezer, p. 491 [*1058*]
16.3, median age in Nigeria 1990, p. 456 [*993*]
16.7, median age, 1820, p. 455 [*992*]
17, child-care obligation per woman, p. 7 [*13*]
17, life expectancy for 6-year-old black woman in 1989, p. 482 [*1041*]
17, lifespan of a electric stove, p. 491 [*1058*]
17, lifespan of a refrigerator, p. 491 [*1058*]
17.9, women at first marriage in 1981, India, p. 473 [*1024*]
18, elder-care obligation per woman, p. 7 [*13*]
18.1, median age in Nigeria 2010, p. 456 [*993*]
19, lifespan of a gas furnace, p. 491 [*1058*]
19, lifespan of a gas stove, p. 491 [*1058*]
20 or more, lifespan of gold fillings, p. 312 [*668*]
20, average worker termite life expectancy, p. 488 [*1054*]
20, lifespan of a oil furnace, p. 491 [*1058*]
20, median age in Mexico 1990, p. 456 [*993*]
20, women at first marriage in 1981, Hungary, p. 473 [*1024*]
20.3, women at first marriage, 1950, p. 474 [*1025*]
20.4, divorced mothers at first marriage, p. 458 [*999*]
20.6, women at first marriage, 1970, p. 475 [*1026*]
20.9, once-married mothers at first marriage, p. 458 [*997*]
21.1, women at first marriage, 1975, p. 474 [*1025*]
21.2, women at first marriage, 1920, p. 474 [*1025*]
21.3, women at first marriage, 1930, p. 474 [*1025*]
21.5, median age in Provo-Orem, UT, p. 457 [*996*]
21.5, women at first marriage, 1940, p. 474 [*1025*]
21.6, women at first marriage, 1910, p. 474 [*1025*]
21.8, women at first marriage in 1980, Israel, p. 473 [*1024*]
21.9, women at first marriage, 1900, p. 474 [*1025*]
22, age of average Peace Corps volunteer, 1960s, p. 103 [*250*]
22, remaining life expectancy for a 55-year-old man on retirement, p. 486 [*1049*]
22, women at first marriage in 1981, Canada, p. 473 [*1024*]
22, women at first marriage, 1890, p. 474 [*1025*]
22, women at first marriage, 1980, p. 474 [*1025*]
22.2, women at first marriage in 1982, China, p. 473 [*1024*]
22.5, men at first marriage, 1970, p. 475 [*1027*]
22.7, men at first marriage, 1975, p. 475 [*1027*]
22.8, men at first marriage, 1950, p. 474 [*1025*]
22.8, women at first marriage, 1984, p. 475 [*1026*]
22.9, median age in Brazil 1990, p. 456 [*993*]
22.9, once-married mothers at birth of first child, p. 458 [*997*]
23, men at first marriage, 1977, p. 475 [*1027*]
23, women at first marriage in 1981, Australia, p. 473 [*1024*]
23, women at first marriage in 1982, United States, p. 473 [*1024*]
23, women at first marriage, 1985, p. 475 [*1026*]
23.2, median age of Hispanics, 1980, p. 455 [*991*]
23.2, men at first marriage, 1970, p. 474 [*1025*]
23.3, men at first marriage in 1981, India, p. 473 [*1024*]
23.3, women at first marriage, 1985, p. 474 [*1025*]
23.9, women at first marriage, 1990, p. 474 [*1025*]

Time - continued

Years

24, men at first marrage in 1981, Hungary, p. 473 [*1024*]
24, women at first marriage in 1982, Germany, Federal Republic, p. 473 [*1024*]
24.3, median age in Jacksonville, NC, p. 457 [*996*]
24.3, men at first marriage, 1930, p. 474 [*1025*]
24.3, men at first marriage, 1940, p. 474 [*1025*]
24.5, men at first marriage in 1982, China, p. 473 [*1024*]
24.6, men at first marriage, 1920, p. 474 [*1025*]
24.8, median age in Laredo, TX, p. 457 [*996*]
24.8, men at first marriage, 1985, p. 475 [*1027*]
24.9, median age of blacks, 1980, p. 455 [*991*]
25, men at first marriage in 1981, Australia, p. 473 [*1024*]
25, men at first marriage in 1981, Canada, p. 473 [*1024*]
25, men at first marriage in 1982, United States, p. 473 [*1024*]
25, women at first marriage in 1980, Japan, p. 473 [*1024*]
25-30, bone cell life span, p. 334 [*716*]
25-45, age of average sewing enthusiast, p. 104 [*253*]
25.1, median age in Bryan-College Station, TX, p. 457 [*996*]
25.1, men at first marriage, 1910, p. 474 [*1025*]
25.3, median age in McAllen-Edinburg-Mission, TX, p. 457 [*996*]
25.3, men at first marriage in 1980, Israel, p. 473 [*1024*]
25.4, median age in China 1990, p. 456 [*993*]
25.5, median age in Salt Lake City-Ogden, UT, p. 457 [*996*]
25.7, median age in Brownsville-Harlingen, TX, p. 457 [*996*]
25.7, median age in Korea 1990, p. 456 [*993*]
25.9, men at first marriage, 1900, p. 474 [*1025*]
26.1, men at first marriage, 1890, p. 474 [*1025*]
26.1, men at first marriage, 1990, p. 474 [*1025*]
26.2, men at first marriage in 1982, France, p. 473 [*1024*]
26.4, median age in State College, PA, p. 457 [*996*]
26.5, median age in Killeen-Temple, TX, p. 457 [*996*]
26.5, median age in Mexico 2010, p. 456 [*993*]
26.6, median age in Champaign-Urbana, IL, p. 457 [*996*]
27.5, age of average motorcyclist, 1980s, p. 105 [*254*]
28, median age, 1970, p. 455 [*992*]
28, men at first marriage in 1980, Japan, p. 473 [*1024*]
28, men at first marriage in 1982, Germany, Federal Republic, p. 473 [*1024*]
28.1, median age of blacks, 1990, p. 455 [*991*]
29, average life expectancy of Neanderthal man, p. 484 [*1045*]
29.2, median age in Brazil 2010, p. 456 [*993*]
29.2, once-married mothers at birth of last child, p. 458 [*997*]
29.3, median age of Hispanics, 1990, p. 455 [*991*]
29.6, age of first-time house buyer, 1989, p. 459 [*1000*]
29.8, women at divorce, 1970, p. 476 [*1029*]
30, average life expectancy of Fifth-century man (England), p. 484 [*1045*]
30, median age, 1950, p. 455 [*992*]
30, median age, 1980, p. 455 [*992*]
30.1, divorced women at remarriage, 1970, p. 475 [*1026*]
30.3, women at divorce, 1980, p. 476 [*1029*]
30.6, women at divorce, 1981, p. 476 [*1029*]
31, divorced women at remarriage, 1980, p. 475 [*1026*]
32, age of average Peace Corps volunteer, 1990, p. 103 [*250*]
32, average life expectancy of Cro-Magnon man, p. 484 [*1045*]
32, average waiting period for telephone installation, Russia, p. 102 [*247*]

Numbers following p. are page references. Numbers in [ ] are table references.

Time - continued

Years

32.2, men at divorce, 1975, p. 476 [★1029★]
32.5, age of average motorcyclist, p. 105 [★254★]
32.7, men at divorce, 1980, p. 476 [★1029★]
32.8, median age, 1990, p. 455 [★991★]
32.9, median age in U.S. 1990, p. 456 [★993★]
32.9, men at divorce, 1970, p. 476 [★1029★]
33, median age, 1990, p. 455 [★992★]
33.6, divorced men at remarriage, 1975, p. 475 [★1027★]
33.9, median age in China 2010, p. 456 [★993★]
34, divorced men at remarriage, 1980, p. 475 [★1027★]
34.2, divorced mothers at divorce, p. 458 [★999★]
34.4, median age in Korea 2010, p. 456 [★993★]
34.4, men at divorce, 1985, p. 476 [★1029★]
34.5, divorced men at remarriage, 1970, p. 475 [★1027★]
35, median age of average lesbian, p. 107 [★261★]
35.7, median age in Great Britain 1990, p. 456 [★993★]
36, average life expectancy of Greek and Roman man, p. 484 [★1045★]
36, average life expectancy of Man in the Copper Age, p. 484 [★1045★]
36.1, divorced men at remarriage, 1985, p. 475 [★1027★]
36.2, median age in Italy 1990, p. 456 [★993★]
37, median age of average gay man, p. 107 [★261★]
37.2, median age in Japan 1990, p. 456 [★993★]
37.4, median age in U.S. 2010, p. 456 [★993★]
38, average life expectancy of Fourteenth-century man (England), p. 484 [★1045★]
38, average life expectancy of Man in the Bronze Age, p. 484 [★1045★]
38.7, median age in Denver, CO, p. 457 [★995★]
39, age of average business traveler, p. 106 [★259★]
39, median age, 2010, p. 455 [★992★]
39.4, age of repeat house buyer, 1989, p. 459 [★1000★]
39.8, life expectancy in Chad, 1991, p. 481 [★1039★]
39.9, median age in New Orleans, LA, p. 457 [★995★]
40, age of average franchisee, p. 105 [★255★]
40, median age in Great Britain 2010, p. 456 [★993★]
40-60, age of average casino visitor, p. 106 [★258★]
40.5, median age in Atlanta, GA, p. 457 [★995★]
40.9, median age in Dallas, TX, p. 457 [★995★]
41.1, median age in Los Angeles, CA, p. 457 [★995★]
41.8, median age, 2030, p. 455 [★992★]
42.1, median age in Philadelphia, PA, p. 457 [★995★]
42.2, median age in Japan 2010, p. 456 [★993★]
42.2, median age in New York, NY, p. 457 [★995★]
42.4, median age in Italy 2010, p. 456 [★993★]
42.8, life expectancy in Guinea, 1991, p. 481 [★1039★]
43, median age of readers of women's magazines, p. 107 [★262★]
43.1, median age in Detroit, MI, p. 457 [★995★]
43.5, life expectancy in Afghanistan, 1991, p. 481 [★1039★]
43.5, projected life expectancy in Chad, 2000, p. 481 [★1039★]
44.3, life expectancy in Angola, 1991, p. 481 [★1039★]
45, average life expectancy of Eighteenth century man (Europe), p. 484 [★1045★]
45.7, median age in Chicago, IL, p. 457 [★995★]
46.1, life expectancy in Mali, 1991, p. 481 [★1039★]
47, age of average condominium timesharer, p. 105 [★257★]

Time - continued

Years

47, average age of gardener, p. 104 [★251★]
47, projected life expectancy in Guinea, 2000, p. 481 [★1039★]
47.1, life expectancy in Central African Republic, 1991, p. 481 [★1039★]
47.4, life expectancy in Mozambique, 1991, p. 481 [★1039★]
47.4, life expectancy worldwide, 1950-1955, p. 482 [★1040★]
47.9, projected life expectancy in Afghanistan, 2000, p. 481 [★1039★]
48.9, life expectancy in Nigeria, 1991, p. 481 [★1039★]
48.9, projected life expectancy in Angola, 2000, p. 481 [★1039★]
49.2, life expectancy in Malawi, 1991, p. 481 [★1039★]
49.3, life expectancy in Cambodia, 1991, p. 481 [★1039★]
50, average queen termite life expectancy, p. 488 [★1054★]
50.2, life expectancy in Laos, 1991, p. 481 [★1039★]
50.3, projected life expectancy in Mali, 2000, p. 481 [★1039★]
50.5, life expectancy in Benin, 1991, p. 481 [★1039★]
50.6, life expectancy in Nepal, 1991, p. 481 [★1039★]
51, average life expectancy of Seventeenth century man (England), p. 484 [★1045★]
51, life expectancy in Cameroon, 1991, p. 481 [★1039★]
51, life expectancy in Niger, 1991, p. 481 [★1039★]
51.1, median age in Sarasota, FL, p. 457 [★995★]
51.2, widowed women at remarriage, 1970, p. 475 [★1026★]
51.3, life expectancy in Ethiopia, 1991, p. 481 [★1039★]
51.5, projected life expectancy in Central African Republic, 2000, p. 481 [★1039★]
51.5, projected life expectancy in Mozambique, 2000, p. 481 [★1039★]
52.2, life expectancy in Burkina, 1991, p. 481 [★1039★]
52.4, life expectancy in Burundi, 1991, p. 481 [★1039★]
52.4, widowed women at remarriage, 1975, p. 475 [★1026★]
52.6, life expectancy in Madagascar, 1991, p. 481 [★1039★]
52.7, projected life expectancy in Nigeria, 2000, p. 481 [★1039★]
52.9, projected life expectancy in Malawi, 2000, p. 481 [★1039★]
53, life expectancy in Bangladesh, 1991, p. 481 [★1039★]
53.1, widowed women at remarriage, 1977, p. 475 [★1026★]
53.6, life expectancy in Haiti, 1991, p. 481 [★1039★]
53.6, widowed women at remarriage, 1980, p. 475 [★1026★]
54.2, life expectancy in Congo, 1991, p. 481 [★1039★]
54.3, life expectancy in Cote d'Ivoire, 1991, p. 481 [★1039★]
54.6, life expectancy in Ghana, 1991, p. 481 [★1039★]
54.6, projected life expectancy in Cambodia, 2000, p. 481 [★1039★]
54.6, widowed women at remarriage, 1985, p. 475 [★1026★]
54.7, projected life expectancy in Benin, 2000, p. 481 [★1039★]
54.7, projected life expectancy in Laos, 2000, p. 481 [★1039★]
54.9, life expectancy in Burma, 1991, p. 481 [★1039★]
55.2, projected life expectancy in Cameroon, 2000, p. 481 [★1039★]
55.2, projected life expectancy in Nepal, 2000, p. 481 [★1039★]
55.4, projected life expectancy in Ethiopia, 2000, p. 481 [★1039★]
55.5, projected life expectancy in Niger, 2000, p. 481 [★1039★]
56, age of average CEO, p. 105 [★256★]
56.1, projected life expectancy in Haiti, 2000, p. 481 [★1039★]
56.4, life expectancy in Liberia, 1991, p. 481 [★1039★]
56.4, projected life expectancy in Burkina, 2000, p. 481 [★1039★]
56.4, projected life expectancy in Burundi, 2000, p. 481 [★1039★]

Time - continued

Years

56.6, life expectancy in Pakistan, 1991, p. 481 [★1039★]
56.7, projected life expectancy in Madagascar, 2000, p. 481 [★1039★]
57.2, life expectancy in India, 1991, p. 481 [★1039★]
57.5, projected life expectancy in Bangladesh, 2000, p. 481 [★1039★]
57.5, projected life expectancy in Ghana, 2000, p. 481 [★1039★]
58, projected life expectancy in Burma, 2000, p. 481 [★1039★]
58.3, projected life expectancy in Congo, 2000, p. 481 [★1039★]
58.3, projected life expectancy in Cote d'Ivoire, 2000, p. 481 [★1039★]
58.7, widowed men at remarriage, 1970, p. 475 [★1027★]
59.4, widowed men at remarriage, 1975, p. 475 [★1027★]
59.6, projected life expectancy in Pakistan, 2000, p. 481 [★1039★]
60.4, projected life expectancy in Liberia, 2000, p. 481 [★1039★]
60.7, life expectancy of blacks 1950,2,, p. 483 [★1043★]
60.8, life expectancy in Egypt, 1991, p. 481 [★1039★]
61, life expectancy in Indonesia, 1991, p. 481 [★1039★]
61.2, widowed men at remarriage, 1980, p. 475 [★1027★]
61.4, projected life expectancy in India, 2000, p. 481 [★1039★]
61.5, life expectancy in Bolivia, 1991, p. 481 [★1039★]
61.5, life expectancy in Kenya, 1991, p. 481 [★1039★]
62-63, average retirement age, 1992, p. 24 [★51★]
62.5, life expectancy in Nicaragua, 1991, p. 481 [★1039★]
62.7, widowed men at remarriage, 1985, p. 475 [★1027★]
63.2, life expectancy in Guatemala, 1991, p. 481 [★1039★]
64.1, life expectancy of blacks, 1970, p. 483 [★1042★]
64.5, life expectancy in Iran, 1991, p. 481 [★1039★]
64.5, projected life expectancy worldwide, 1995-2000, p. 482 [★1040★]
64.6, life expectancy in Morocco, 1991, p. 481 [★1039★]
64.6, projected life expectancy in Kenya, 2000, p. 481 [★1039★]
64.8, life expectancy for black man in 1989, p. 482 [★1041★]
65, projected life expectancy in Egypt, 2000, p. 481 [★1039★]
65.1, life expectancy in Mongolia, 1991, p. 481 [★1039★]
65.2, life expectancy in Brazil, 1991, p. 481 [★1039★]
65.5, life expectancy in El Salvador, 1991, p. 481 [★1039★]
65.5, projected life expectancy in Bolivia, 2000, p. 481 [★1039★]
66, average life expectancy of left-handers, p. 485 [★1046★]
66, life expectancy in Honduras, 1991, p. 481 [★1039★]
66.2, life expectancy in Ecuador, 1991, p. 481 [★1039★]
66.6, projected life expectancy in Indonesia, 2000, p. 481 [★1039★]
66.7, life expectancy in Algeria, 1991, p. 481 [★1039★]
66.8, life expectancy of blacks 1975, p. 483 [★1043★]
66.8, projected life expectancy in Nicaragua, 2000, p. 481 [★1039★]
66.9, projected life expectancy in Guatemala, 2000, p. 481 [★1039★]
67, life expectancy in Iraq, 1991, p. 481 [★1039★]
67.2, life expectancy in Dominican Republic, 1991, p. 481 [★1039★]
67.5, projected life expectancy in Brazil, 2000, p. 481 [★1039★]
68.1, life expectancy in Libya, 1991, p. 481 [★1039★]
68.1, life expectancy in Malaysia, 1991, p. 481 [★1039★]
68.1, life expectancy of blacks 1980, p. 483 [★1043★]
68.4, life expectancy in Lebanon, 1991, p. 481 [★1039★]
68.4, projected life expectancy in Mongolia, 2000, p. 481 [★1039★]

Time - continued

Years

69, life expectancy in North Korea, 1991, p. 481 [★1039★]
69, projected life expectancy in Ecuador, 2000, p. 481 [★1039★]
69.1, projected life expectancy in Iran, 2000, p. 481 [★1039★]
69.2, life expectancy of blacks, 1989, p. 483 [★1042★]
69.6, projected life expectancy in Algeria, 2000, p. 481 [★1039★]
69.6, projected life expectancy in El Salvador, 2000, p. 481 [★1039★]
69.8, projected life expectancy in Morocco, 2000, p. 481 [★1039★]
70, life expectancy in China: Mainland, 1991, p. 481 [★1039★]
70, projected life expectancy in Honduras, 2000, p. 481 [★1039★]
70.2, projected life expectancy in Dominican Republic, 2000, p. 481 [★1039★]
70.8, life expectancy, 1970, p. 483 [★1042★]
70.9, life expectancy in Argentina, 1991, p. 481 [★1039★]
70.9, projected life expectancy in Malaysia, 2000, p. 481 [★1039★]
71, life expectancy in Colombia, 1991, p. 481 [★1039★]
71, projected life expectancy in Iraq, 2000, p. 481 [★1039★]
71.2, life expectancy in Jordan, 1991, p. 481 [★1039★]
71.4, projected life expectancy in Lebanon, 2000, p. 481 [★1039★]
71.5, projected life expectancy in North Korea, 2000, p. 481 [★1039★]
71.6, life expectancy in Hungary, 1991, p. 481 [★1039★]
72.2, life expectancy in Mexico, 1991, p. 481 [★1039★]
72.3, projected life expectancy in Argentina, 2000, p. 481 [★1039★]
72.4, projected life expectancy in China: Mainland, 2000, p. 481 [★1039★]
72.4, projected life expectancy of blacks, 1995, p. 483 [★1042★]
72.5, projected life expectancy in Libya, 2000, p. 481 [★1039★]
72.6, life expectancy 1975, p. 483 [★1043★]
72.7, life expectancy in Bulgaria, 1991, p. 481 [★1039★]
72.9, life expectancy in Czechoslovakia, 1991, p. 481 [★1039★]
73.2, projected life expectancy in Jordan, 2000, p. 481 [★1039★]
73.4, life expectancy in Chile, 1991, p. 481 [★1039★]
73.5, projected life expectancy of Blacks, 2000, p. 483 [★1042★]
73.6, life expectancy in Jamaica, 1991, p. 481 [★1039★]
73.6, life expectancy in Kuwait, 1991, p. 481 [★1039★]
73.5, life expectancy for A-year-old black woman in 1989, p. 482 [★1041★]
73.7, life expectancy, 1980, p. 483 [★1042★]
74.1, projected life expectancy in Colombia, 2000, p. 481 [★1039★]
74.6, life expectancy in Taiwan, 1991, p. 481 [★1039★]
74.6, projected life expectancy of blacks, 2005, p. 483 [★1042★]
74.8, life expectancy 1986, p. 483 [★1043★]
75, average life expectancy of right-handers, p. 485 [★1046★]
75, projected life expectancy of Blacks, 2010, p. 483 [★1042★]
75.1, life expectancy in Albania, 1991, p. 481 [★1039★]
75.1, projected life expectancy in Chile, 2000, p. 481 [★1039★]
75.3, life expectancy, 1989, p. 483 [★1042★]
75.4, projected life expectancy in Mexico, 2000, p. 481 [★1039★]
75.5, life expectancy in Ireland, 1991, p. 481 [★1039★]
75.5, life expectancy in New Zealand, 1991, p. 481 [★1039★]
75.5, projected life expectancy in Hungary, 2000, p. 481 [★1039★]
75.6, life expectancy in Cuba, 1991, p. 481 [★1039★]
75.7, life expectancy in United States, 1991, p. 481 [★1039★]
75.8, life expectancy in Finland, 1991, p. 481 [★1039★]
75.8, life expectancy in Germany, 1991, p. 481 [★1039★]
75.9, life expectancy in Denmark, 1991, p. 481 [★1039★]

Numbers following p. are page references. Numbers in [ ] are table references.

Time - continued

Years

75.9, projected life expectancy in Jamaica, 2000, p. 481 [★1039★]
76, projected life expectancy in Bulgaria, 2000, p. 481 [★1039★]
76.3, projected life expectancy in Czechoslovakia, 2000, p. 481 [★1039★]
76.3, projected life expectancy in Taiwan, 2000, p. 481 [★1039★]
76.3, projected life expectancy, 1995, p. 483 [★1042★]
76.4, projected life expectancy in Cuba, 2000, p. 481 [★1039★]
76.7, projected life expectancy in Kuwait, 2000, p. 481 [★1039★]
76.8, life expectancy in Costa Rica, 1991, p. 481 [★1039★]
77, life expectancy in Australia, 1991, p. 481 [★1039★]
77, life expectancy in Israel, 1991, p. 481 [★1039★]
77, projected life expectancy in United States, 2000, p. 481 [★1039★]
77, projected life expectancy, 2000, p. 483 [★1042★]
77.1, life expectancy in Belgium, 1991, p. 481 [★1039★]
77.1, life expectancy in Norway, 1991, p. 481 [★1039★]
77.3, life expectancy in Austria, 1991, p. 481 [★1039★]
77.5, life expectancy in Canada, 1991, p. 481 [★1039★]
77.5, projected life expectancy in Albania, 2000, p. 481 [★1039★]
77.6, projected life expectancy, 2005, p. 483 [★1042★]
77.7, life expectancy in Greece, 1991, p. 481 [★1039★]
77.8, life expectancy in France, 1991, p. 481 [★1039★]
77.8, life expectancy in Netherlands, 1991, p. 481 [★1039★]
77.9, projected life expectancy in Germany, 2000, p. 481 [★1039★]
77.9, projected life expectancy in Ireland, 2000, p. 481 [★1039★]
77.9, projected life expectancy, 2010, p. 483 [★1042★]
78, projected life expectancy in Denmark, 2000, p. 481 [★1039★]
78, projected life expectancy in New Zealand, 2000, p. 481 [★1039★]
78.1, life expectancy in Italy, 1991, p. 481 [★1039★]
78.1, projected life expectancy in Finland, 2000, p. 481 [★1039★]
78.5, projected life expectancy in Israel, 2000, p. 481 [★1039★]
78.7, projected life expectancy in Australia, 2000, p. 481 [★1039★]
78.8, projected life expectancy in Belgium, 2000, p. 481 [★1039★]
78.8, projected life expectancy in Norway, 2000, p. 481 [★1039★]
78.9, projected life expectancy in Austria, 2000, p. 481 [★1039★]
78.9, projected life expectancy in Costa Rica, 2000, p. 481 [★1039★]
79.2, life expectancy in Japan, 1991, p. 481 [★1039★]
79.2, projected life expectancy in Canada, 2000, p. 481 [★1039★]
79.3, projected life expectancy in France, 2000, p. 481 [★1039★]
79.3, projected life expectancy in Greece, 2000, p. 481 [★1039★]
79.3, projected life expectancy in Netherlands, 2000, p. 481 [★1039★]
79.6, projected life expectancy in Italy, 2000, p. 481 [★1039★]
80.8, projected life expectancy in Japan, 2000, p. 481 [★1039★]
100, estimated waiting period for telephone installation, Albania, p. 102 [★247★]
33,3,, life expectancy of blacks, 1900, p. 483 [★1043★]
1,000, average redwood tree life expectancy, p. 489 [★1055★]

Volume

Cubic feet

61.1, timber consumed per capita, 1970, p. 96 [★227★]
76.2, timber consumed per capita, 1985, p. 96 [★227★]
13 million, air breathed in lifetime by average human, p. 340 [★731★]

Volume - continued

Cubic feet

250 million, annual molded (loose) plastic peanuts used for packaging in the U.S., p. 415 [★890★]

Cubic feet per second

426,000, daily flow of water in Potomac River flood in 1936, p. 496 [★1073★]

Cubic yards

3.3, landfill space saved by recycling 1 ton of wastepaper, p. 428 [★922★]

Cup

0.25, milk consumed per person in South Korea daily, 1989, p. 83 [★196★]

Gallons

0.3, hot water used per minute for bathroom sink, p. 93 [★220★]
1.1, wine consumed per capita in Ireland, 1989, p. 82 [★193★]
1.2, sports drink consumed per capita, 1991, p. 80 [★189★]
1.5, hot water used per minute for dishwasher, p. 93 [★220★]
1.6, hot water used per minute for kitchen sink, p. 93 [★220★]
1.6, wine consumed per capita in Finland, 1989, p. 82 [★193★]
1.7, capacity of human lungs, p. 341 [★733★]
1.7, wine consumed per capita in Norway, 1989, p. 82 [★193★]
1.7, wine consumed per capita in USSR (former), 1989, p. 82 [★193★]
1.8, wine consumed per capita in Iceland, 1989, p. 82 [★193★]
2, water used daily for brushing teeth per capita, p. 91 [★215★]
2, wine consumed per capita in Poland, 1989, p. 82 [★193★]
2.1, wine consumed per capita in United States, 1989, p. 82 [★193★]
2.2, distilled spirits consumed per capita, 1990, p. 80 [★187★]
2.2, wine consumed per capita, 1970, p. 80 [★187★]
2.3, wine consumed per capita in Canada, 1989, p. 82 [★193★]
2.4, wine consumed per capita in South Africa, 1989, p. 82 [★193★]
2.5, hot water used per minute for shower, p. 93 [★220★]
2.9, wine consumed per capita, 1990, p. 80 [★187★]
3, distilled spirits consumed per capita, 1970, p. 80 [★187★]
3-4, water used for single toilet flush, p. 94 [★221★]
3.2, wine consumed per capita in East Germany (former), 1989, p. 82 [★193★]
3.3, hot water used per minute for laundry, p. 93 [★220★]
3.4, wine consumed per capita in Sweden, 1989, p. 82 [★193★]
3.4, wine consumed per capita in United Kingdom, 1989, p. 82 [★193★]
3.6, citrus juice consumed per capita, 1970, p. 80 [★187★]
3.6, hot water used per minute for bathtub, p. 93 [★220★]
3.6, wine consumed per capita in Cyprus, 1989, p. 82 [★193★]
3.6, wine consumed per capita in Czechoslovakia (former), 1989, p. 82 [★193★]
3.8, wine consumed per capita in New Zealand, 1989, p. 82 [★193★]
4, citrus juice consumed per capita, 1990, p. 80 [★187★]
4.2, wine consumed per capita in Netherlands, 1989, p. 82 [★193★]
5, water used daily for utility sink by a family of four, p. 93 [★218★]
5, water used per minute for shower, p. 94 [★221★]
5, water used per toilet flush, p. 213 [★461★]
5.1, wine consumed per capita in Australia, 1989, p. 82 [★193★]
5.1, wine consumed per capita in Denmark, 1989, p. 82 [★193★]
5.3, wine consumed per capita in Hungary, 1989, p. 82 [★193★]

Volume - continued

Gallons

5.6, wine consumed per capita in Yugoslavia (former), 1989, p. 82 [★193★]
5.8, wine consumed per capita in Bulgaria, 1989, p. 82 [★193★]
6.8, tea consumed per capita, 1970, p. 80 [★187★]
6.9, tea consumed per capita, 1990, p. 80 [★187★]
6.9, wine consumed per capita in Belgium, 1989, p. 82 [★193★]
6.9, wine consumed per capita in West Germany (former), 1989, p. 82 [★193★]
7.4, wine consumed per capita in Romania, 1989, p. 82 [★193★]
7.4, wine consumed per capita in Uruguay, 1989, p. 82 [★193★]
7.9, wine consumed per capita in Greece, 1989, p. 82 [★193★]
8, water used daily for bathroom sink by a family of four, p. 93 [★218★]
8, water used for cooking, p. 94 [★221★]
9.2, wine consumed per capita in Chile, 1989, p. 82 [★193★]
9.3, wine consumed per capita in Austria, 1989, p. 82 [★193★]
10, water used daily for dishwasher per capita, p. 91 [★215★]
10, water used per minute to water lawn, p. 94 [★221★]
10, wine consumed per capita in Spain, 1989, p. 82 [★193★]
10-15, water used daily for shave per capita, p. 91 [★215★]
13.1, wine consumed per capita in Argentina, 1989, p. 82 [★193★]
13.2, wine consumed per capita in Switzerland, 1989, p. 82 [★193★]
15, water used daily for dishwashing by a family of four, p. 93 [★218★]
16.2, wine consumed per capita in Luxembourg, 1989, p. 82 [★193★]
17.4, wine consumed per capita in Italy, 1989, p. 82 [★193★]
20, water used daily for washing dishes per capita, p. 91 [★215★]
20-30, water used for washing clothes, p. 94 [★221★]
20.8, soft drinks consumed per capita, 1970, p. 80 [★187★]
20.8, wine consumed per capita in Portugal, 1989, p. 82 [★193★]
21.2, wine consumed per capita in France, 1989, p. 82 [★193★]
25-30, water used daily for shower per capita, p. 91 [★215★]
25.7, milk consumed per capita, 1990, p. 80 [★187★]
26.7, coffee consumed per capita, 1990, p. 80 [★187★]
30-40, water used for bath, p. 94 [★221★]
30.6, beer consumed per capita, 1970, p. 80 [★187★]
31.2, milk consumed per capita, 1970, p. 80 [★187★]
32, water used to produce 1 pound steel, p. 94 [★223★]
33.4, coffee consumed per capita, 1970, p. 80 [★187★]
34.4, beer consumed per capita, 1990, p. 80 [★187★]
35, water used daily for laundry by a family of four, p. 93 [★218★]
40, water used to produce 1 egg, p. 94 [★222★]
42.5, soft drinks consumed per capita, 1990, p. 80 [★187★]
49, water used for 18-lb laundry load, p. 213 [★461★]
50, gasoline used to produce meat and poultry consumed per person per year, p. 90 [★213★]
50, water used for shower, p. 213 [★461★]
80, water used daily for bathing by a family of four, p. 93 [★218★]
80, water used to produce 1 ear of corn, p. 94 [★222★]
100, water used daily for flushing toilet by a family of four, p. 93 [★218★]
152, water used daily per capita Rhode Island, in 1985, p. 92 [★217★]
160, water used to produce 1 loaf of bread, p. 94 [★222★]
168, water used daily per capita, p. 91 [★215★]

Volume - continued

Gallons

222, water used daily per capita Delaware, in 1985, p. 92 [★217★]
235, water used daily per capita Vermont, in 1985, p. 92 [★217★]
280, water used to produce 1 Sunday paper, p. 94 [★223★]
300, water used to produce 1 pound synthetic rubber, p. 94 [★223★]
307, water used daily per capita New Jersey, in 1985, p. 92 [★217★]
321, water used daily per capita Maryland, in 1985, p. 92 [★217★]
339, water used daily per capita, 1960, p. 91 [★216★]
375, water used daily per capita Connecticut, in 1985, p. 92 [★217★]
380, water used daily per capita, 1985, p. 91 [★216★]
386, water used daily per capita Oklahoma, in 1985, p. 92 [★217★]
403, water used daily per capita, 1965, p. 91 [★216★]
427, water used daily per capita, 1970, p. 91 [★216★]
440, water used daily per capita, 1980, p. 91 [★216★]
451, water used daily per capita, 1975, p. 91 [★216★]
505, fuel consumed per car, 1990, p. 66 [★154★]
508, water used daily per capita, New York, in 1985, p. 92 [★217★]
525, fuel consumed per car, 1985, p. 66 [★154★]
554, water used daily per capita Florida, in 1985, p. 92 [★217★]
556, water used daily per capita District of Columbia, in 1985, p. 92 [★217★]
591, fuel consumed per car, 1980, p. 66 [★154★]
676, water used daily per capita Minnesota, in 1985, p. 92 [★217★]
688, water used daily per capita New Hampshire, in 1985, p. 92 [★217★]
716, fuel consumed per car, 1975, p. 66 [★154★]
727, water used daily per capita Alaska, in 1985, p. 92 [★217★]
733, water used daily per capita Maine, in 1985, p. 92 [★217★]
760, fuel consumed per car, 1970, p. 66 [★154★]
853, water used daily per capita Virginia, in 1985, p. 92 [★217★]
885, water used daily per capita Mississippi, in 1985, p. 92 [★217★]
899, water used daily per capita Georgia, in 1985, p. 92 [★217★]
956, water used daily per capita South Dakota, in 1985, p. 92 [★217★]
960, water used daily per capita Iowa, in 1985, p. 92 [★217★]
1,000, water used to produce 1 pound aluminum, p. 94 [★223★]
1,070, water used daily per capita Massachusetts, in 1985, p. 92 [★217★]
1,100, water used daily per capita Hawaii, in 1985, p. 92 [★217★]
1,130, water used daily per capita Kentucky, in 1985, p. 92 [★217★]
1,180, water used daily per capita Ohio, in 1985, p. 92 [★217★]
1,191, fuel consumed per truck, 1982, p. 66 [★154★]
1,210, water used daily per capita Missouri, in 1985, p. 92 [★217★]
1,210, water used daily per capita Pennsylvania, in 1985, p. 92 [★217★]
1,217, fuel consumed per truck, 1975, p. 66 [★154★]
1,230, water used daily per capita Texas, in 1985, p. 92 [★217★]
1,243, fuel consumed per truck, 1980, p. 66 [★154★]
1,250, water used daily per capita Illinois, in 1985, p. 92 [★217★]
1,257, fuel consumed per truck, 1970, p. 66 [★154★]
1,260, water used daily per capita North Carolina, in 1985, p. 92 [★217★]
1,270, water used daily per capita Michigan, in 1985, p. 92 [★217★]
1,300, water contained in 1 inch of dry snow, p. 509 [★1111★]
1,305, fuel consumed per truck, 1990, p. 66 [★154★]

Numbers following p. are page references. Numbers in [ ] are table references.

Volume - continued

Gallons

1,400, water used daily per capita United States, in 1985, p. 92 [*217*]
1,400, water used daily per capita Wisconsin, in 1985, p. 92 [*217*]
1,407, fuel consumed per bus, 1985, p. 66 [*154*]
1,420, water used daily per capita California, in 1985, p. 92 [*217*]
1,436, fuel consumed per bus, 1990, p. 66 [*154*]
1,470, water used daily per capita Indiana, in 1985, p. 92 [*217*]
1,600, water used daily per capita Washington, in 1985, p. 92 [*217*]
1,690, water used daily per capita North Dakota, in 1985, p. 92 [*217*]
1,770, water used daily per capita Tennessee, in 1985, p. 92 [*217*]
1,926, fuel consumed per bus, 1980, p. 66 [*154*]
1,960, water used daily per capita Arizona, in 1985, p. 92 [*217*]
2,040, water used daily per capita South Carolina, in 1985, p. 92 [*217*]
2,140, water used daily per capita Alabama, in 1985, p. 92 [*217*]
2,172, fuel consumed per bus, 1970, p. 66 [*154*]
2,210, water used daily per capita Louisiana, in 1985, p. 92 [*217*]
2,279, fuel consumed per bus, 1975, p. 66 [*154*]
2,310, water used daily per capita Kansas, in 1985, p. 92 [*217*]
2,320, water used daily per capita New Mexico, in 1985, p. 92 [*217*]
2,450, water used daily per capita Oregon, in 1985, p. 92 [*217*]
2,500, water used daily per capita Arkansas, in 1985, p. 92 [*217*]
2,500, water used to produce 1 pound of beef, p. 94 [*222*]
2,540, water used daily per capita Utah, in 1985, p. 92 [*217*]
2,810, water used daily per capita West Virginia, in 1985, p. 92 [*217*]
3,860, water used daily per capita Nevada, in 1985, p. 92 [*217*]
4,190, water used daily per capita Colorado, in 1985, p. 92 [*217*]
4,488, water used monthly per household in Santa Barbara (dry), p. 95 [*224*]
5,000, air breathed daily by average human, p. 340 [*731*]
5,300, water contained in 1 inch of wet snow, p. 509 [*1111*]
6,250, water used daily per capita Nebraska, in 1985, p. 92 [*217*]
7,000, water saved by recycling 1 ton of wastepaper, p. 428 [*922*]
7,480, water used monthly per household in Chicago (lake-fed), p. 95 [*224*]
7,650, water used monthly per household in New York City, p. 95 [*224*]
8,600, water used monthly per household in Tucson (dusty), p. 95 [*224*]
8,750, water used monthly per household in Miami (aquifer-fed), p. 95 [*224*]
10,000, saliva produced in average lifetime, p. 342 [*736*]
10,500, water used daily per capita Montana, in 1985, p. 92 [*217*]
12,200, water used daily per capita Wyoming, in 1985, p. 92 [*217*]
22,200, water used daily per capita Idaho, in 1985, p. 92 [*217*]

Volume - continued

Gallons

100,000, water used to produce 1 automobile, p. 94 [*223*]
107,000, water used annually per residence, p. 91 [*215*]
2 million, gasoline needed to melt an average iceberg, p. 502 [*1092*]
13 million, blood circulated by heart in average lifetime, p. 337 [*724*]
200-400 million, annual waste oil generated by consumer oil-changes, p. 423 [*910*]
4.8 billion, water used daily for toilet flushing, p. 93 [*219*]
1,100 billion water absorbed annually by soil in U.S., p. 506 [*1105*]
4,400 billion precipitation daily in continental U.S., p. 506 [*1104*]
40,000 billion, atmospheric water over U.S. daily, p. 507 [*1107*]

Pints

2-3, average daily urine output, p. 333 [*711*]
2.25, blood pumped through kidneys in a minute, p. 333 [*711*]
7-10, blood in an average-sized adult, p. 333 [*710*]

Quarts

1, daily human water intake from food items, p. 95 [*225*]
1.5, daily human water intake from liquids, p. 95 [*225*]
2.5, daily human water requirements, p. 95 [*225*]
3.5, blood in an average-sized woman, p. 333 [*710*]
5.5, blood in an average-sized man, p. 333 [*710*]

Teaspoon

0.25, milk consumed per person in South Korea daily, 1969, p. 83 [*196*]

Weight

Grams

20, daily recommended fiber intake, women, p. 97 [*231*]
55, daily recommended sugar intake, women, p. 97 [*231*]
250, daily recommended complex carbohydrate intake, women, p. 97 [*231*]

Kilograms

24, coal equivalent of energy consumed per capita in Ethiopia, 1989, p. 85 [*202*]
28, electricity energy used per capita, 1960, p. 87 [*205*]
36, coal equivalent of energy consumed per capita in Tanzania, 1989, p. 85 [*202*]
62, coal equivalent of energy consumed per capita in Burma, 1989, p. 85 [*202*]
64, coal equivalent of energy consumed per capita in Sudan, 1989, p. 85 [*202*]
67, coal equivalent of energy consumed per capita in Zaire, 1989, p. 85 [*202*]
69, coal equivalent of energy consumed per capita in Bangladesh, 1989, p. 85 [*202*]
95, electricity energy used per capita, 1988, p. 87 [*205*]
132, coal equivalent of energy consumed per capita in Vietnam, 1989, p. 85 [*202*]
197, natural gas energy used per capita, 1960, p. 87 [*205*]
198, coal equivalent of energy consumed per capita in Zambia, 1989, p. 85 [*202*]
207, coal equivalent of energy consumed per capita in Nigeria, 1989, p. 85 [*202*]
265, coal equivalent of energy consumed per capita in Pakistan, 1989, p. 85 [*202*]

Numbers following p. are page references. Numbers in [ ] are table references.

Weight - continued

Kilograms

295, coal equivalent of energy consumed per capita in Philippines, 1989, p. 85 [★202★]
307, coal equivalent of energy consumed per capita in India, 1989, p. 85 [★202★]
311, coal equivalent of energy consumed per capita in Indonesia, 1989, p. 85 [★202★]
366, coal equivalent of energy consumed per capita in Moracco, 1989, p. 85 [★202★]
433, liquid fuels energy used per capita, 1960, p. 87 [★205★]
449, natural gas energy used per capita, 1988, p. 87 [★205★]
505, coal equivalent of energy consumed per capita in Peru, 1989, p. 85 [★202★]
637, coal equivalent of energy consumed per capita in Thailand, 1989, p. 85 [★202★]
644, solid fuels energy used per capita, 1960, p. 87 [★205★]
645, solid fuels energy used per capita, 1988, p. 87 [★205★]
662, coal equivalent of energy consumed per capita in Ecuador, 1989, p. 85 [★202★]
739, coal equivalent of energy consumed per capita in Egypt, 1989, p. 85 [★202★]
761, coal equivalent of energy consumed per capita in Tunisia, 1989, p. 85 [★202★]
773, coal equivalent of energy consumed per capita in Colombia, 1989, p. 85 [★202★]
791, coal equivalent of energy consumed per capita in Brazil, 1989, p. 85 [★202★]
794, liquid fuels energy used per capita, 1988, p. 87 [★205★]
819, coal equivalent of energy consumed per capita in China:Mainland, 1989, p. 85 [★202★]
936, coal equivalent of energy consumed per capita in Algeria, 1989, p. 85 [★202★]
978, coal equivalent of energy consumed per capita in Syria, 1989, p. 85 [★202★]
1,027, coal equivalent of energy consumed per capita in Turkey, 1989, p. 85 [★202★]
1,059, coal equivalent of energy consumed per capita in Iraq, 1989, p. 85 [★202★]
1,206, coal equivalent of energy consumed per capita in Chile, 1989, p. 85 [★202★]
1,278, coal equivalent of energy consumed per capita in Malaysia, 1989, p. 85 [★202★]
1,532, coal equivalent of energy consumed per capita in Cuba, 1989, p. 85 [★202★]
1,650, coal equivalent of energy consumed per capita in Iran, 1989, p. 85 [★202★]
1,736, coal equivalent of energy consumed per capita in Mexico, 1989, p. 85 [★202★]
1,811, coal equivalent of energy consumed per capita in Portugal, 1989, p. 85 [★202★]
1,899, coal equivalent of energy consumed per capita in Argentina, 1989, p. 85 [★202★]
1,975, coal equivalent of energy consumed per capita in World, total, 1989, p. 85 [★202★]
1,996, coal equivalent of energy consumed per capita in Hong Kong, 1989, p. 85 [★202★]
2,195, coal equivalent of energy consumed per capita in South Korea, 1989, p. 85 [★202★]

Weight - continued

Kilograms

2,485, coal equivalent of energy consumed per capita in Spain, 1989, p. 85 [★202★]
2,644, coal equivalent of energy consumed per capita in South Africa, 1989, p. 85 [★202★]
2,656, coal equivalent of energy consumed per capita in Yugoslavia (former), 1989, p. 85 [★202★]
2,814, coal equivalent of energy consumed per capita in North Korea, 1989, p. 85 [★202★]
3,012, coal equivalent of energy consumed per capita in Venezuela, 1989, p. 85 [★202★]
3,040, coal equivalent of energy consumed per capita in Israel, 1989, p. 85 [★202★]
3,074, coal equivalent of energy consumed per capita in Taiwan, 1989, p. 85 [★202★]
3,113, coal equivalent of energy consumed per capita in Greece, 1989, p. 85 [★202★]
3,632, coal equivalent of energy consumed per capita in Ireland, 1989, p. 85 [★202★]
3,662, coal equivalent of energy consumed per capita in Switzerland, 1989, p. 85 [★202★]
3,697, coal equivalent of energy consumed per capita in Hungary, 1989, p. 85 [★202★]
3,813, coal equivalent of energy consumed per capita in Italy, 1989, p. 85 [★202★]
3,915, coal equivalent of energy consumed per capita in France, 1989, p. 85 [★202★]
3,995, coal equivalent of energy consumed per capita in Japan, 1989, p. 85 [★202★]
4,014, coal equivalent of energy consumed per capita in Austria, 1989, p. 85 [★202★]
4,286, coal equivalent of energy consumed per capita in Libya, 1989, p. 85 [★202★]
4,413, coal equivalent of energy consumed per capita in Denmark, 1989, p. 85 [★202★]
4,486, coal equivalent of energy consumed per capita in Romania, 1989, p. 85 [★202★]
4,531, coal equivalent of energy consumed per capita in Poland, 1989, p. 85 [★202★]
5,043, coal equivalent of energy consumed per capita in United Kingdom, 1989, p. 85 [★202★]
5,052, coal equivalent of energy consumed per capita in Bulgaria, 1989, p. 85 [★202★]
5,053, coal equivalent of energy consumed per capita in New Zealand, 1989, p. 85 [★202★]
5,070, coal equivalent of energy consumed per capita in Sweden, 1989, p. 85 [★202★]
5,391, coal equivalent of energy consumed per capita in West Germany (former), 1989, p. 85 [★202★]
5,787, coal equivalent of energy consumed per capita in Belgium, 1989, p. 85 [★202★]
5,839, coal equivalent of energy consumed per capita in Finland, 1989, p. 85 [★202★]
6,003, coal equivalent of energy consumed per capita in Czechoslovakia (former), 1989, p. 85 [★202★]
6,160, coal equivalent of energy consumed per capita in Trinidad and Tobago, 1989, p. 85 [★202★]
6,362, coal equivalent of energy consumed per capita in Saudi Arabia, 1989, p. 85 [★202★]

Numbers following p. are page references. Numbers in [ ] are table references.

Weight - continued

Kilograms

6,553, coal equivalent of energy consumed per capita in Soviet Union (former), 1989, p. 85 [★202★]
6,639, coal equivalent of energy consumed per capita in Netherlands, 1989, p. 85 [★202★]
7,181, coal equivalent of energy consumed per capita in Norway, 1989, p. 85 [★202★]
7,208, coal equivalent of energy consumed per capita in Australia, 1989, p. 85 [★202★]
7,631, coal equivalent of energy consumed per capita in East Germany (former), 1989, p. 85 [★202★]
8,825, coal equivalent of energy consumed per capita in Kuwait, 1989, p. 85 [★202★]
10,124, coal equivalent of energy consumed per capita in United States, 1989, p. 85 [★202★]
10,927, coal equivalent of energy consumed per capita in Canada, 1989, p. 85 [★202★]
15,601, coal equivalent of energy consumed per capita in Bahrain, 1989, p. 85 [★202★]
20,361, coal equivalent of energy consumed per capita in United Arab Emirates, 1989, p. 85 [★202★]

Miligrams

2400, daily recommended sodium intake, women, p. 97 [★231★]

Ounces

2.82, average human hair can support, p. 336 [★723★]

Pounds

0.07, breakfast cereal consumed per capita in South Korea, 1993, p. 69 [★160★]
0.3, cured fish consumed per capita, 1990, p. 76 [★174★]
0.59, breakfast cereal consumed per capita in Taiwan, 1993, p. 69 [★160★]
0.67, breakfast cereal consumed per capita in South Africa, 1993, p. 69 [★160★]
0.8, yogurt consumed per capita, 1970, p. 73 [★169★]
0.9, veal consumed per capita, 1990, p. 72 [★167★]
1, edible protein produced by 21.4 pounds of protein in a cow, p. 95 [★226★]
1, edible protein produced by 5.5 pounds of protein in a chicken, p. 95 [★226★]
1, edible protein produced by 8.3 pounds of protein in a pig, p. 95 [★226★]
1.1, lamb consumed per capita, 1990, p. 72 [★167★]
1.1, sour cream and dip consumed per capita, 1970, p. 73 [★169★]
1.22, pretzels consumed per capita, 1993, p. 71 [★164★]
1.4, broccoli consumed per capita, 1980, p. 76 [★177★]
1.64, pies consumed per capita, 1993, p. 70 [★162★]
1.71, doughnuts consumed per capita, 1993, p. 70 [★162★]
1.78, breakfast cereal consumed per capita in France, 1993, p. 69 [★160★]
1.79, breakfast cereal consumed per capita in Germany, 1993, p. 69 [★160★]
2, average monthly plastic container waste per person, p. 425 [★914★]
2, veal consumed per capita, 1970, p. 72 [★167★]
2.1, lamb consumed per capita, 1970, p. 72 [★167★]
2.5, weight of average human lungs, p. 340 [★732★]
2.5, daily waste generated per American, 1960, p. 420 [★901★]
2.5, sour cream and dip consumed per capita, 1990, p. 73 [★169★]

Weight - continued

Pounds

2.8, feed to produce 1 pound of chicken, p. 96 [★228★]
2.87, municipal solid waste landfilled per person per day, 1988, p. 427 [★918★]
3, weight of average human brain, p. 334 [★714★]
3, corn and soybeans to produce 1 pound of milk, p. 96 [★228★]
3-5, weight of average woman's handbag, p. 111 [★269★]
3.36, peanut butter consumed per capita, p. 77 [★181★]
3.4, broccoli consumed per capita, 1990, p. 76 [★177★]
3.4, cottage cheese consumed per capita, 1990, p. 73 [★169★]
3.43, daily waste generated per American, p. 420 [★901★]
3.56, bagels consumed per capita, 1993, p. 70 [★162★]
3.8, cream consumed per capita, 1970, p. 73 [★169★]
4.1, yogurt consumed per capita, 1990, p. 73 [★169★]
4.36, frozen bakery products per capita, 1993, p. 71 [★163★]
4.6, canned fish consumed per capita, 1981, p. 76 [★174★]
4.6, cream consumed per capita, 1990, p. 73 [★169★]
4.7, canned fish consumed per capita, 1983, p. 76 [★174★]
4.8, corn and soybeans to produce 1 pound of beef, p. 96 [★228★]
4.9, canned fish consumed per capita, 1988, p. 76 [★174★]
5, potato chips consumed per person, p. 79 [★186★]
5, canned fish consumed per capita, 1985, p. 76 [★174★]
5.1, canned fish consumed per capita, 1990, p. 76 [★174★]
5.2, canned fish consumed per capita, 1987, p. 76 [★174★]
5.2, cottage cheese consumed per capita, 1970, p. 73 [★169★]
5.5, protein required by a chicken to produce 1 pound of edible protein, p. 95 [★226★]
5.9, weight of skin of average adult, p. 343 [★739★]
6, average monthly steel can waste per person, p. 425 [★914★]
6-8, weight gain over 5 years after quitting smoking, p. 328 [★699★]
6.02, breakfast cereal consumed per capita in Canada, 1993, p. 69 [★160★]
6.4, turkey consumed per capita, 1970, p. 72 [★167★]
7 lb. 4 oz., birth weight babies born 1970, p. 472 [★1020★]
7 lb. 7 oz., birth weight babies born 1980-1989, p. 472 [★1020★]
7, breakfast cereal consumed per capita in Australia, 1993, p. 69 [★160★]
7, corn and soybeans to produce 1 pound of pork, p. 96 [★228★]
7.32, breakfast cereal consumed per capita in Netherlands, 1993, p. 69 [★160★]
7.34, breakfast cereal consumed per capita in Great Britain, 1993, p. 69 [★160★]
7.8, fresh and frozen fish consumed per capita, 1981, p. 76 [★174★]
8.23, crackers consumed per capita, 1993, p. 71 [★164★]
8.3, protein required by a pig to produce 1 pound of edible protein, p. 95 [★226★]
8.34, soft cakes consumed per capita, 1993, p. 70 [★162★]
8.4, fresh and frozen fish consumed per capita, 1983, p. 76 [★174★]
8.51, breakfast cereal consumed per capita in New Zealand, 1993, p. 69 [★160★]
9.8, fresh and frozen fish consumed per capita, 1985, p. 76 [★174★]
10, candy consumed per person annually, p. 79 [★186★]
10.1, fresh and frozen fish consumed per capita, 1990, p. 76 [★174★]

Numbers following p. are page references. Numbers in [ ] are table references.

Weight - continued

Pounds

11.4, cheese (not cottage) consumed per capita, 1970, p. 73 [*169*]
11.90, breakfast cereal consumed per capita in United States, 1993, p. 69 [*160*]
12.29, cookies and crackers consumed per capita in United States, 1993, p. 71 [*165*]
12.29, cookies consumed per capita, 1993, p. 71 [*164*]
12.7, fish consumed per capita, 1981, p. 76 [*174*]
13.2, annual consumption of apples in juice form, p. 81 [*190*]
13.4, fish consumed per capita, 1983, p. 76 [*174*]
13.46, hamburger and hotdog rolls consumed per capita, 1993, p. 70 [*162*]
14.4, turkey consumed per capita, 1990, p. 72 [*167*]
15, annual hazardous waste generated in the average home, p. 421 [*905*]
15.1, fish consumed per capita, 1985, p. 76 [*174*]
15.5, fish consumed per capita, 1990, p. 76 [*174*]
15.68, cookies and crackers consumed per capita in Germany, 1993, p. 71 [*165*]
15.7, ice cream consumed per capita, 1990, p. 73 [*169*]
16.2, fish consumed per capita, 1987, p. 76 [*174*]
16.51, cookies and crackers consumed per capita in Canada, 1993, p. 71 [*165*]
17.8, ice cream consumed per capita, 1970, p. 73 [*169*]
17.90, cookies and crackers consumed per capita in Austria, 1993, p. 71 [*165*]
18, fresh fruit consumed per capita, 1970-1974, p. 77 [*178*]
18.25, cookies and crackers consumed per capita in Australia, 1993, p. 71 [*165*]
19, carbon dioxide emitted for every gallon of gasoline burned, p. 415 [*888*]
19, pasta consumed per capita, p. 77 [*180*]
20 gallons, ice cream consumed per person, p. 79 [*186*]
20.7, candy consumed per capita, 1991, p. 78 [*182*]
21.4, protein required by a cow to produce 1 pound of edible protein, p. 95 [*226*]
23, average weight lost by hiking the full Appalachian Trail, p. 440 [*952*]
23.91, rolls (all) consumed per capita, 1993, p. 70 [*162*]
23.95, cookies and crackers consumed per capita in New Zealand, 1993, p. 71 [*165*]
24.7, cheese (not cottage) consumed per capita, 1990, p. 73 [*169*]
27.7, chicken consumed per capita, 1970, p. 72 [*167*]
29.34, cookies and crackers consumed per capita in Great Britain, 1993, p. 71 [*165*]
40, dead skin shed in average lifetime, p. 343 [*741*]
40, weekly waste generated per suburban family of three, p. 418 [*899*]
43, muscle in average woman, p. 342 [*738*]
43.6, cheese consumed in France per capita, 1983, p. 76 [*175*]
45.23, cookies and crackers consumed per capita in Belgium/Luxembourg, 1993, p. 71 [*165*]
46.3, pork consumed per capita, 1990, p. 72 [*167*]
48.2, pork consumed per capita, 1970, p. 72 [*167*]
49.2, poultry consumed per capita in Hungary, 1991, p. 75 [*172*]
49.3, chicken consumed per capita, 1990, p. 72 [*167*]

Weight - continued

Pounds

50, cookies and cakes consumed per person, p. 79 [*186*]
50.7, poultry consumed per capita in Spain, 1991, p. 75 [*172*]
51.19, breads consumed per capita, 1993, p. 70 [*162*]
51.6, poultry consumed per capita in Taiwan, 1991, p. 75 [*172*]
55, fat and oil consumed per person, p. 79 [*186*]
56.2, beef and veal consumed per capita in Switzerland, 1991, p. 75 [*173*]
56.2, poultry consumed per capita in Australia, 1991, p. 75 [*172*]
56.7, poultry consumed per capita in Saudi Arabia, 1991, p. 75 [*172*]
58.28, cookies and crackers consumed per capita in Netherlands, 1993, p. 71 [*165*]
58.9, beef and veal consumed per capita in Italy, 1991, p. 75 [*173*]
60, pasta consumed per capita (Italians), p. 77 [*180*]
62.7, oil and fat consumed per capita, 1990, p. 74 [*170*]
63 dozen, doughnuts consumed per person, p. 79 [*186*]
63.5, poultry consumed per capita in Canada, 1991, p. 75 [*172*]
64, beef consumed per capita, 1990, p. 72 [*167*]
65.5, pork consumed per capita in United States, 1991, p. 74 [*171*]
65.9, beef and veal consumed per capita in Soviet Union (former), 1991, p. 75 [*173*]
66.2, beef and veal consumed per capita in France, 1991, p. 75 [*173*]
69, muscle in average man, p. 342 [*738*]
70-150, Chinese potbellied pig, p. 523 [*1145*]
75.2, poultry consumed per capita in Singapore, 1991, p. 75 [*172*]
76.5, poultry consumed per capita in Hong Kong, 1991, p. 75 [*172*]
76.7, beef and veal consumed per capita in New Zealand, 1991, p. 75 [*173*]
79.6, beef consumed per capita, 1970, p. 72 [*167*]
80.3, beef and veal consumed per capita in Canada, 1991, p. 75 [*173*]
81.8, poultry consumed per capita in Israel, 1991, p. 75 [*172*]
84, beef and veal consumed per capita in Australia, 1991, p. 75 [*173*]
85, average annual glass waste per person, p. 425 [*914*]
92.4, pork consumed per capita in Netherlands, 1991, p. 74 [*171*]
94, fresh fruit consumed per capita, 1989, p. 77 [*178*]
94.8, poultry consumed per capita in United States, 1991, p. 75 [*172*]
97, beef and veal consumed per capita in United States, 1991, p. 75 [*173*]
97.7, pork consumed per capita in Bulgaria, 1991, p. 74 [*171*]
100 pounds, refined sugar consumed per person, p. 79 [*186*]
103.2, pork consumed per capita in Spain, 1991, p. 74 [*171*]
104.7, pork consumed per capita in Belgium-Luxembourg, 1991, p. 74 [*171*]
106, potatoes consumed per capita, 1960, p. 76 [*176*]
112.3, red meat, total consumed per capita, 1990, p. 72 [*167*]
116.4, pork consumed per capita in Austria, 1991, p. 74 [*171*]
116.4, pork consumed per capita in Poland, 1991, p. 74 [*171*]
117, weight of the average Miss America, 1980, p. 339 [*729*]

Numbers following p. are page references. Numbers in [ ] are table references.

Weight - continued

Pounds

121.5, pork consumed per capita in Germany, 1991, p. 74 [*171*]
122.8, beef and veal consumed per capita in Uruguay, 1991, p. 75 [*173*]
126, potatoes consumed per capita, 1989, p. 76 [*176*]
132, red meat, total consumed per capita, 1970, p. 72 [*167*]
132, weight of the average Miss America, 1954, p. 339 [*729*]
135, average woman, p. 337 [*725*]
137, red meat and poultry consumed per capita, 1955, p. 72 [*166*]
144.2, pork consumed per capita in Denmark, 1991, p. 74 [*171*]
147.5, pork consumed per capita in Hungary, 1991, p. 74 [*171*]
150, canned food consumed per capita, p. 77 [*179*]
153.9, beef and veal consumed per capita in Argentina, 1991, p. 75 [*173*]
162, average man, p. 337 [*725*]
178, red meat and poultry consumed per capita, 1992, p. 72 [*166*]
180, annual wastepaper generated by the average office worker, p. 417 [*896*]
190, average annual plastic waste per person, p. 425 [*914*]
221.5, milk (fluid) consumed per capita, 1990, p. 73 [*169*]
238, average daily waste generated by a McDonald's restaurant, p. 416 [*891*]
269.1, milk (fluid) consumed per capita, 1970, p. 73 [*169*]
550, annual wastepaper generated by an average newspaper subscription, p. 417 [*894*]
563.8, total (milk equivalent) consumed per capita, 1970, p. 73 [*169*]
570.6, total (milk equivalent) consumed per capita, 1990, p. 73 [*169*]
600, annual U.S. paper waste generation per capita, p. 423 [*909*]
1,200-1,500, farm pig, p. 523 [*1145*]
1,900, annual waste generated per American, p. 418 [*899*]
90,000, lifetime waste generated by average American, p. 420 [*901*]

Weight - continued

Pounds

8 million, wastepaper generated by an average *New York Times* Sunday edition, p. 417 [*894*]
98 million, annual paper used to print government documents, p. 32 [*76*]
8 billion, average annual hazardous substances shipped in U.S., p. 415 [*889*]
350 billion, average annual toxic chemicals produced in U.S., p. 415 [*889*]

Tons

3, annual grass clippings generated by mowing 1/2 acre lawn, p. 421 [*904*]
9.2, average annual carbon dioxide emitted per person by the transportation sector, p. 425 [*913*]
10, Minke whale weight, p. 523 [*1146*]
17, Bryde's whale weight, p. 523 [*1146*]
20, gray whale weight, p. 523 [*1146*]
33, humpback whale weight, p. 523 [*1146*]
50, finback whale weight, p. 523 [*1146*]
84, blue whale weight, p. 523 [*1146*]
2,000, TNT needed to break up an average iceberg, p. 502 [*1092*]
0.5 million, rain dropped by a thunderstorm, p. 509 [*1113*]
20 million particles falling to earth daily, p. 524 [*1148*]
148.1 million, annual waste generated by Americans, p. 420 [*901*]

Numbers following p. are page references. Numbers in [ ] are table references.